*Thyristor Phase-Controlled
Converters and Cycloconverters*

Thyristor Phase-Controlled Converters and Cycloconverters

OPERATION, CONTROL, AND PERFORMANCE

B. R. Pelly

Research and Development Laboratories
Westinghouse Electric Corporation
Pittsburgh, Pennsylvania

WILEY–INTERSCIENCE
a Division of John Wiley & Sons, Inc.
New York · London · Sydney · Toronto

Copyright © 1971, by John Wiley & Sons, Inc.

All rights reserved. Published simultaneously in Canada.

No part of this book may be reproduced by any means, nor transmitted, nor translated into a machine language without the written permission of the publisher.

Library of Congress Catalogue Card Number: 70-125276

ISBN 0 471 67790 6

Printed in the United States of America

10 9 8 7 6 5 4 3 2 1

Preface

Since its introduction some ten years ago, the silicon-controlled rectifier, or thyristor, has become firmly established as the active power control element of static electrical power conversion equipments of many types, ranging in power ratings from a few hundred watts to several megawatts.

The subjects of this book are two particular, closely related, types of static power conversion circuit which today employ thyristors—the phase-controlled converter, which converts alternating current to direct current (or vice versa), and the cycloconverter, which converts alternating current at one frequency to alternating current at another frequency.

Neither of these basic types of equipment is new; indeed each has been employed, using older types of grid-controlled rectifier, for many years prior to the introduction of the thyristor. Nonetheless, the emergence of the thyristor, in conjunction with the rapid technological advances that have been made in the area of transistors, integrated circuits, and solid state devices of all types, has given birth to a much wider field of applications, for the phase-controlled converter as well as the cycloconverter. As a result, many innovations have been made, in terms of both circuitry and control concepts.

The main objective in this book is to present a detailed description and explanation of the fundamental principles of operation and control of the phase-controlled converter and the cycloconverter, and to derive analytical data that describe quantitatively their basic external performance characteristics. The underlying philosophy is to convey an understanding of basic principles, and to demonstrate the fundamental similarities of operation and performance that characterize this general class of power conversion circuitry.

The book should be useful to engineers and students working in the field of Power Electronics, as a source of reference and analytical design data, as well as to users of these types of equipment, and to all who require an understanding of basic principles.

To the best of my knowledge, until the publication of the present work, no "exact" quantitative data, that generally describe the external performance of the cycloconverter, have been available. I suspect that the main reason for this is the seeming complexity of the mathematics involved in such an analysis, together with an apparent absence of a simple general analytical method, which have acted as effective deterrents to the would-be cycloconverter analyst. Thus the analytical work related to the cycloconverter which so far has appeared in the literature is limited in scope, and generally is based on simplifying, and often none too valid, assumptions. Another simple reason, which should not be disregarded, for the previous lack of precise analytical data, has been the lack of sufficient demand for such information. Thus it has always been possible, through a combination of intelligent estimation and practical observation, to arrive at a satisfactory, though possibly less than optimum, cycloconverter system design. Today, however, all indications are that the cycloconverter will find use in applications involving higher and higher levels of power. Hence the need for a precise understanding of the basic nature of the operating characteristics of the cycloconverter, together with the need for "exact" fundamental analytical data, is becoming increasingly felt. I hope that the hitherto unpublished information that I have presented in my book will fulfill this need.

An "historical" note may be appropriate here. At the outset of this project, my intention was to write a book devoted almost entirely to the cycloconverter, based mainly on my own original theoretical work related to the analysis of its external performance characteristics. The immediate question that arose was how to present the cycloconverter in its proper perspective, without first discussing the fundamental principles of the phase-controlled converter and the dual-converter—of which the cycloconverter is so close a relative. A book that sets out to deal with the cycloconverter either has to assume a prior knowledge on the part of the reader of the phase-controlled converter and the dual-converter; or it can cover these topics in the barest detail required for a decent introduction of the cycloconverter; or it can treat the phase-controlled converter and the dual-converter with about the same amount of attention afforded to the cycloconverter. I have chosen the last approach.

My reason for choosing to do this is that I feel this gives the book a sense of completeness that otherwise would have eluded it. More important, however, I hope that those chapters that deal with the phase-controlled converter and dual-converter will, in their own right, prove to be of equally much interest and usefulness as those devoted to the cycloconverter. Thus, although the analytical information related to the phase-controlled converter presented in this book is not original, in the sense that much, if not all, of it has previously appeared in the literature, in some form or other, I believe, nonetheless, that the material as a whole constitutes a unified treatment and concentration of information not hitherto available all in one place.

Finally, I should anticipate a criticism which may arise in the minds of some readers—namely, that generally I have ignored the practical presence of a-c source impedance throughout most of this book, and it is only in the final chapter that this topic is examined. My reason for doing this is simply that my main purpose has been to convey an understanding of *basic* principles and operating characteristics. Since the practical imperfections arising from the presence of a-c source impedance are generally of a second-order nature, it is my opinion that this topic is best set aside, until after fundamental principles and operating characteristics have been established.

B. R. PELLY

Murrysville, Pennsylvania
February 1970

Acknowledgments

My thanks are due to the Westinghouse Electric Corporation, and in particular to the Aerospace Electrical, Industrial Systems, and General Control Divisions, for permission to publish this book.

I am indebted to Mr. A. H. B. Walker, Director, Communication, Control, and Power Electronics Research, for his most valuable encouragement and support in this project, without which this book could not have been brought to fruition.

It has been my good fortune to have worked with colleagues having a very high level of expertise in the field of Solid State Power Electronics, amongst whom there has always been a free exchange of ideas. Thus many of the concepts presented in this book are really the outcome of the combined work and ideas of several people.

In this vein, I would like to acknowledge the influence of Mr. J. Rosa, who has made many contributions towards the advancement of the static converter and cycloconverter arts. And, in particular, I wish to acknowledge the part played by Mr. L. Gyugyi, with whom I have been closely associated, and who has done much original pioneering work in connection with cycloconverters. It was he who first sowed the seeds of the idea for this book, and prompted me to write it. He has continually encouraged me in this project, and finally has performed a most valuable service in reading and criticizing the complete manuscript, as well as making several suggestions for improvements (all of which I have adopted).

My thanks are due to Misses A. Tomasic and J. Woffenden, who (especially the former) have borne the brunt of the typing of the manuscript, and have uncomplainingly and successfully carried out the often difficult task of deciphering the original.

Finally, and most of all, I thank my wife, Dinah, for making this book possible.

B. R. Pelly

Contents

NOMENCLATURE   xxi

ASSUMPTIONS   xxv

CHAPTER 1   THE THYRISTOR   1

        The Static Anode-Cathode Characteristics, 2
        Dynamic Switching Characteristics, 4
           Turn-On, 4
           Turn-Off, 6

CHAPTER 2   FUNCTIONS CHARACTERISTICS, AND APPLICATIONS OF THYRISTOR PHASE-CONTROLLED CONVERTERS AND CYCLOCONVERTERS   9

        The Thyristor Phase-Controlled Converter—Its Basic Functions and Characteristics, 9
        Applications of Thyristor Phase-Controlled Converters, 12
           General, 12
           Variable speed control of d-c machines, 13
        The Thyristor Phase-Controlled Cycloconverter—Its Basic Functions and Characteristics, 16
        Applications of Thyristor Cycloconverters, 18

xii   CONTENTS

Variable speed control of a-c machines, 18
Constant frequency power supplies, 25

CHAPTER 3   PHASE-CONTROLLED CONVERTERS—CIRCUITS AND
            OPERATING PRINCIPLES   27

Two-Quadrant Converters, 27
  The two-pulse midpoint converter, 27
  The two-pulse bridge converter, 32
  Definition of converter firing angle, 33
  Converters with three-phase input, 33
  The three-pulse midpoint converter, 34
  The simple six-pulse midpoint converter, 38
  The six-pulse midpoint converter with interphase
    reactor, 38
  The twelve-pulse midpoint converter with interphase
    reactors, 41
  The six-pulse bridge converter, 43
  The twelve-pulse bridge converter, 47
  Other "multipulse" converter connections, 47
Operation of the Two-Quadrant Converter with Discontinuous Current 49
One-Quadrant Converters, 54
  The two-pulse half-controlled bridge converter, 55
  The two-pulse midpoint converter with freewheel
    diode, 58
  The three-pulse half-controlled bridge converter, 59
  The three-pulse midpoint converter with freewheel
    diode, 61
  The six-pulse half-controlled converter, 62
  The six-pulse converter with one freewheel diode, 65
  The six-pulse converter with two freewheel diodes, 67

CHAPTER 4   THE EXTERNAL PERFORMANCE CHARACTERISTICS OF PHASE-
            CONTROLLED CONVERTERS   69

Definition of Converter Performance Parameters, 69
  The d-c voltage ratio, 69
  The input displacement angle, 70

CONTENTS xiii

The input displacement factor, 70

The input power factor, 70

The input current distortion factor, 70

Basic Relationships between a-c Input Parameters, 70

Relationship between mean a-c power and displacement factor, 71

Relationship between displacement, power, and distortion factors, 71

Method of Analysis of Converter Performance Characteristics, 72

The Performance Characteristics of Two-Quadrant Converters, 73

Analysis of the performance characteristics of the three-pulse converter, 73

Analysis of the performance characteristics of multi-pulse converters, containing combinations of the three-pulse group, 81

The universal relationships between firing angle, d-c voltage ratio, and input displacement factor for all two-quadrant converters, 92

The universal harmonic relationships for all two-quadrant converters, 94

The Performance Characteristics of One-Quadrant Converters, 95

Harmonic content of the d-c terminal voltage, and of the a-c input current, of various one-quadrant converters, 96

Relationships between firing angle, d-c voltage ratio, and input displacement factor for one-quadrant converters, 96

Reduction of Reactive Loading of the Supply by the Two-Quadrant Converter by Means of Consecutive Firing Angle Control, 109

CHAPTER 5 THE DUAL-CONVERTER 111

Basic Principle of the Dual-Converter, 111

The Circulating Current-Free Mode of Operation, 114
  The basic principle, 114
  A simple "open-loop" control scheme for a dual-converter operating with no circulating current, 116
  Control difficulties due to discontinuous load current, 118
  A closed-loop control scheme for a dual-converter operating with no circulating current, 121
The Circulating Current Mode of Operation, 126
  The basic principle, 126
  Details of operation, 128
  Operation under dynamic load changing conditions, 139
  The waveform of the d-c terminal voltage of the dual-converter operating in the circulating current mode, 142

CHAPTER 6 THE PHASE-CONTROLLED CYCLOCONVERTER 145

The Basic Principle of Operation of the Cycloconverter, 146
The Circulating Current-Free Mode of Operation, 151
The "Natural" Circulating Current Mode of Operation, 151
Discussion on the Harmonic Distortion of the Cycloconverter Output Voltage, 161
  General considerations, 161
  Necessary distortion terms, 161
  Unnecessary distortion terms, 165
  "Practical" distortion terms, 165
Discussion on the Loading Effect of the Cycloconverter on the Input System, 166
  General considerations, 166
  Estimation of the approximate in-phase and quadrature composition of the input current, 173

CONTENTS   XV

CHAPTER 7   CYCLOCONVERTER CONTROL PROBLEMS AND TECHNIQUES ASSOCIATED WITH DISCONTINUOUS CURRENT   181

Basic Control Difficulties Associated with Discontinuous Current, 182
Approaches to the Voltage Distortion Problem, 186
    Closed-loop control of the output voltage, 186
    Operation with circulating current, 187
Approaches to the Bank Selection Problem, 198
    "First current-zero" bank selection, 199
    "Fundamental current-zero" bank selection, 201
    Voltage-sensing bank selection, 203

CHAPTER 8   CYCLOCONVERTER CIRCUITS   207

"Symmetrical" Cycloconverter Circuits, 208
    "Symmetrical" three-pulse midpoint circuit, 208
    "Symmetrical" six-pulse midpoint circuit, 210
    "Symmetrical" twelve-pulse midpoint circuit, 210
    "Symmetrical" six-pulse bridge circuit with isolated loads, 210
    "Symmetrical" six-pulse bridge circuit with non-isolated loads, 210
    "Symmetrical" twelve-pulse bridge circuit, 215
Open-Delta Cycloconverter Circuits, 215
    Open-delta three-pulse midpoint circuit, 219
    Open-delta six-pulse bridge circuit, 219
"Ring-Connected" Cycloconverter Circuits, 220
    "Ring-connected" three-pulse midpoint circuit, 227
    "Ring-connected" six-pulse bridge circuit, 228

CHAPTER 9   THEORETICAL PRINCIPLES OF FIRING PULSE TIMING CONTROL   229

The "Cosine Wave Crossing" Pulse Timing Method, 229
    The basic principle, 229

xvi CONTENTS

Self-regulating property of the "cosine wave crossing" control method, 231

The "cosine wave crossing" control method applied to a cycloconverter, 234

Limitations of Open-Loop Pulse Timing for a Cycloconverter, 238

The Use of Feedback for Suppression of "Objectionable" Distortion, 240

Other Pulse Timing Control Principles, 242

"Integral control," 242

Phase-locked oscillator, 245

CHAPTER 10 FIRING PULSE GENERATOR CIRCUIT PRINCIPLES   248

Assumptions and Scope of Discussion, 248

Some Functional Schemes, 249

Schemes using cosine wave crossing control, 249

Scheme using "integral control," 252

Scheme using phase-locked oscillator, 253

End-Stop Control, 259

Simple method using reference voltage clamp, 259

Methods using time-dependent information, 260

Complementary Firing Pulse Generators for Dual-Converters, 271

The Pulse Isolating Output Stage, 272

Simple output stage, 273

Blocking oscillator output stage, 274

Output stages using square-wave carrier oscillators, 275

CHAPTER 11 HARMONIC ANALYSIS OF THE OUTPUT VOLTAGE OF THE CYCLOCONVERTER   278

Scope of the Analysis, 278

The Basic Analytical Problem, and the Approach Used, 279

Selection of the Three-Pulse Waveform for Detailed Analysis, and the Reasons for This, 281

Derivation of the General Expression for the Three-Pulse Waveform, for an Arbitrary Firing Angle Control Method, 281
   The circulating current mode, 285
   The circulating current-free mode, 287
Selection of the Cosine Wave Crossing Control Method for Detailed Analysis, and the Reasons for This, 289
Mathematical Definition of the Modulating Function for the Cosine Wave Crossing Control Method, 290
Derivation of the Harmonic Series for the Three-Pulse Voltage, 292
   The circulating current mode, 292
   The circulating current-free mode, 295
Harmonic Series for the Six-Pulse Voltage Waveform, 302
   The circulating current mode, 302
   The circulating current-free mode, 302
Harmonic Series for the Twelve-Pulse Voltage Waveform, 303
   The circulating current mode, 303
   The circulating current-free mode, 304
Harmonic Series for Voltage Waveforms With Other Pulse Numbers, 305
The Wanted Component of the Output Voltage, 305
The Harmonic Frequencies, 306
   General comments, 306
   Frequency spectrum for the circulating current-free mode, 307
   Frequency spectrum for the circulating current mode, 311
The Amplitudes of the Harmonic Components, 313
   General comments, 313
   Harmonic amplitudes for the circulating current-free mode, 314
   Harmonic amplitudes for the circulating current mode, 323
Assessment of the Limits of Performance of Circuits of Different Pulse Numbers, as Dictated by the Distortion of the Output Voltage, 323

Unnecessary Distortion Generated by Other Control Methods, 327

CHAPTER 12 ANALYSIS OF THE INPUT CURRENT OF THE CYCLOCONVERTER 328

Assumptions, 329
Analytical Approach, 329
Notation, 330
Derivation of the Harmonic Series for Three-Pulse Current Waveforms, 330
  Current at converter input terminals, one-phase output, 330
  Transformer primary currents, one-phase output, 337
  Rules for determining harmonic series for input current with three-phase output, 338
  Converter input current, three-phase output, 339
Harmonic Series for Six-Pulse Current Waveforms, 341
  One-phase output, 341
  Three-phase output, 343
Harmonic Series for Twelve-Pulse Current Waveforms, 345
  One-phase output, 345
  Three-phase output, 348
The Fundamental Component of Input Current, 349
  General comments, 349
  The in-phase component, 350
  The lagging quadrature component, 351
  The total fundamental component, 355
  The universal input/output displacement factor characteristics, 358
The Harmonic Frequencies, 359
  General comments, 359
  The characteristic cycloconverter harmonic frequencies, 360
  Circuit-dependent harmonic frequencies, 362
  Harmonic frequency charts, 363

CONTENTS xix

The Universal Relationships between the Output Voltage and Input Current Harmonics of the Cycloconverter, 365

The Amplitudes of the Harmonic Components, 366

General comments, 366

Quantitative data, 368

The Total RMS Value of the Input Current Waveform, 368

The Distortion Factor of the Input Current Waveform, 375

General Conclusions on the Distortion of the Input Current Waveform, 375

Cycloconverter with one-phase output, 375

Cycloconverter with balanced three-phase output, 377

Input Current Unbalance at Discrete Frequency Ratios, 382

Reduction of the Quadrature Input Current by Means of Special Control Techniques, 385

CHAPTER 13 THE EFFECT OF INPUT SOURCE IMPEDANCE 389

Assumptions and Scope of Discussion, 389

The Commutation Process, 390

Effect of Source Inductance on the External Performance Characteristics of the Phase-Controlled Converter, 395

Some general observations, 395

The loss of direct voltage, 398

Harmonic distortion of the d-c terminal voltage, 406

Relationships between per-unit equivalent output resistance and per-unit input reactance, 409

Effect of Source Inductance on the External Performance Characteristics of the Cycloconverter, 409

General comments, 409

Cycloconverter supplying one-phase output, 410

Cycloconverter supplying three-phase output, 418

BIBLIOGRAPHY 423

INDEX 427

Nomenclature

The following is a list of the principal symbols used throughout this book.

$\hat{V}_N$ peak value of phase-to-neutral voltage at converter input

V_N rms value of phase-to-neutral voltage at converter input

V_d mean value of voltage at d-c terminals of phase-controlled converter, at any firing angle, with or without commutation overlap

$V_{do\alpha}$ mean value of voltage at d-c terminals of phase-controlled converter, at firing angle α, with no commutation overlap

$V_{d_{\max}}$ maximum possible mean value of voltage at d-c terminals of phase-controlled converter, obtained at $\alpha = 0°$, with no commutation overlap

v_d instantaneous value of voltage at d-c terminals of phase-controlled converter

$\hat{V}_{W_{\max}}$ peak value of the maximum possible wanted sinusoidal component of output voltage of the cycloconverter, obtained with "full" firing angle modulation, and no commutation overlap (i.e. $\hat{V}_{W_{\max}} = V_{d_{\max}}$)

$V_{W_{\max}}$ rms value of the maximum possible wanted sinusoidal component of output voltage of the cycloconverter, obtained with "full" firing angle modulation, and no commutation overlap

v_W instantaneous value of the wanted sinusoidal component of the output voltage of the cycloconverter

v_o instantaneous value of the raw output voltage of the cycloconverter

r	for phase-controlled converter: ratio of d-c terminal voltage of converter at firing angle α, to maximum possible mean d-c terminal voltage, obtained at $\alpha = 0°$, with no commutation overlap
r	for cycloconverter: ratio of amplitude of wanted sinusoidal component of output voltage, to the maximum possible wanted component of output voltage, obtained with "full" firing angle modulation, with no commutation overlap (i.e. $\hat{V}_{W_{\max}}$)
I_{RMS}	rms value of the converter input line current
$\hat{I}_1$	peak value of the fundamental component of the converter input line current
I_1	rms value of the fundamental component of the converter input line current
$\hat{I}_P$	peak value of the fundamental "in-phase" or "power" component of converter input line current
I_P	rms value of the fundamental "in-phase" or "power" component of converter input line current
$\hat{I}_Q$	peak value of the fundamental "quadrature" component of converter input line current
I_Q	rms value of the fundamental "quadrature" component of converter input line current
$i_{A(B,C)}$	instantaneous value of current in converter input line $A(B,C)$
I_d	direct current at output of phase-controlled converter
i_d	instantaneous value of current at d-c terminals of phase-controlled converter
$\hat{I}_o$	peak value of output current of cycloconverter
I_o	rms value of output current of cycloconverter
i_o	instantaneous value of output current of cycloconverter
P_i	average power at input of converter
p_i	instantaneous power at input of converter
P_d	average power at output of phase-controlled converter
p_d	instantaneous power at output of phase-controlled converter
P_o	average power at output of cycloconverter
p_o	instantaneous power at output of cycloconverter
ϕ_i	displacement angle of fundamental component of converter input current
ϕ_o	output load displacement angle of cycloconverter
λ	power factor at input of converter

NOMENCLATURE

μ	distortion factor of converter input line current
f_i	frequency of supply at input of converter
ω_i	$2\pi f_i$
θ_i	$2\pi f_i t$
f_o	wanted output frequency of cycloconverter
ω_o	$2\pi f_o$
θ_o	$2\pi f_o t$
f_H	harmonic frequency
α	converter firing angle
u	commutation overlap angle
δ	thyristor "extinction" or "recovery" angle (inverter operation)
P	pulse number of converter
R_e	equivalent resistance at output of phase-controlled converter due to input source inductance
$R_{e_{\text{pu}}}$	per-unit value of R_e
R_o	equivalent resistance at output of cycloconverter due to input source inductance
X_o	equivalent reactance at output of cycloconverter due to input source inductance
X	reactance of the input source feeding the phase-controlled converter or cycloconverter
X_{pu}	per-unit value of X.

Assumptions

The following simplifying assumptions are made throughout this book, unless stated otherwise.

1. Thyristors and diodes are regarded as being "ideal" circuit elements. Thus forward voltage drop while in conduction and leakage current while blocking are considered to be negligible. By the same token, "ideal" device switching characteristics are assumed; that is, both turn-on and turn-off are considered to be instantaneous. Furthermore, because of simplifying circuit assumptions, it is often tacitly implied that the waveshapes of current and voltage handled by the rectifier devices have theoretically infinite rate of change. In practice—as mentioned in Chapter 1—such extreme waveshapes are not tolerable (and usually it is necessary to provide auxiliary circuitry for "rounding off" the steep edges of the waveforms).

2. "Stray" circuit resistance, inductance, and capacitance are assumed to be negligible.

3. It is generally assumed, unless stated otherwise, that the current flowing at the output terminals of the converter is continuous, and perfectly smooth. Because of the inevitable presence of ripple voltage components, this assumption necessarily implies that an "ideal" filter circuit is connected at the output terminals.

4. Whenever considering the operation of a converter power circuit, it is assumed that the control equipment is capable of providing a set of synchronized gate firing pulses, with perfect "tracking" accuracy, the phase delay of which is controllable at will.

5. Transformers and inductors are assumed to be "ideal." Thus magnetizing current, core losses, winding resistance, and leakage reactance are neglected, unless stated otherwise.

6. A-c input voltages are assumed to be sinusoidal, and perfectly balanced, both in amplitude and in phase.

7. The details of circuit operation, the waveform diagrams, and the quantitative performance data are generally presented on the basis of the assumption of zero impedance of the a-c source, to which the converter input terminals are connected. In Chapter 13, however, the practical effects of supply reactance on the converter circuit operation, and the modifying effect of this on the external performance characteristics, are examined.

*Thyristor Phase-Controlled
Converters and Cycloconverters*

Chapter One

The Thyristor

The term thyristor belongs to a family of semiconductor switching devices, all of which are characterized by a bistable switching action depending upon *p-n-p-n* regenerative feedback. The silicon-controlled rectifier (SCR) is the most widely used and important member of the thyristor family. It is the only member to be considered in this chapter, and generally throughout this book. For this reason, although not strictly correct, the terms thyristor and silicon-controlled rectifier in this book are assumed to be synonymous.

The SCR is a static solid state switch, the external electrical behavior of which is similar to that of the older thyratron and grid-controlled mercury arc rectifier. However, it has many practical advantages. The most significant of these are its very small size and weight, relative to its power-handling capacity; its capability for providing extremely reliable, maintenance-free operation, with virtually unlimited lifetime; and its physical robustness, which makes it virtually immune to the effects of mechanical vibration and shock.

The SCR is available in a variety of voltage and current ratings. At the time of writing, SCR's capable of blocking voltages (in both forward and reverse directions), from $50V$ (or lower), up to $2000V$, and even higher, are available; current ratings range from $5A$ or so, up to around $800A$ rms. Thus the SCR naturally finds use in a wide field of applications, ranging from those requiring a few hundreds of watts, to those requiring several thousands of kilowatts. Over the years, as manufacturing techniques have been perfected, the cost of the SCR has steadily decreased. Simultaneously, continuous developments in device technology have resulted in the production of SCR's with ever increasing power-handling capability. These trends are likely to continue for some years hence. Thus it is certain that the SCR, together with the many types of power conversion equipment, including the phase-controlled converter and the cycloconverter, of which it forms so vital a part, have become firmly established items of electrical apparatus.

Since the SCR is the vital element of the modern phase-controlled converter

2 THE THYRISTOR

and cycloconverter, it is logical, at the outset, that the device itself should receive some attention.

Much has been written about the SCR; the basic theory of operation of the device, the methods of fabrication and manufacture, the external electrical characteristics, the relationship of these to its internal physical properties, the effects of temperature upon operation, and so on. It is not the intention here to repeat this wealth of information. Rather, a very brief description of the external electrical characteristics of the SCR, as most directly related to its operation as a circuit element, will be given. This will suffice to provide a foundation for an understanding of the discussion, presented in subsequent chapters, which deals with the operating principles of thyristor converter circuits.

THE STATIC ANODE-CATHODE CHARACTERISTICS

The SCR is a 4-layer *p-n-p-n* switching device. It has three external terminals called anode, cathode, and gate. The anode and cathode are the main current carrying connections; the gate carries only low level control current, the circuit for this current being from gate to cathode. The graphical symbol for the SCR is shown in Fig. 1.1, and the static anode-cathode voltage/current characteristics are illustrated in Fig. 1.2.

In the reverse, cathode to anode, direction, the SCR exhibits a blocking characteristic similar to that of a silicon diode. Thus, with an applied voltage which is less than the reverse breakdown voltage, only a relatively small reverse leakage current flows. At a critical reverse voltage breakdown level, the current increases sharply; this results in a rapidly increasing power dissipation in the device, as a consequence of which there is a possibility of complete destruction. In practical applications, the reverse breakdown voltage of the SCR is arranged to be greater than the peak reverse voltage applied to it, by a suitable safety margin. Thus, so far as practical circuit operation is concerned (with proper design), the SCR always appears as a high impedance in the reverse direction, and "blocks" whatever reverse voltage the circuit applies to it.

In the forward, anode to cathode, direction, the voltage-current relationship depends upon the amount of gate current, if any, delivered to the device.

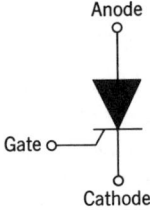

Figure 1.1. Graphical symbol for the SCR.

THE STATIC ANODE-CATHODE CHARACTERISTICS

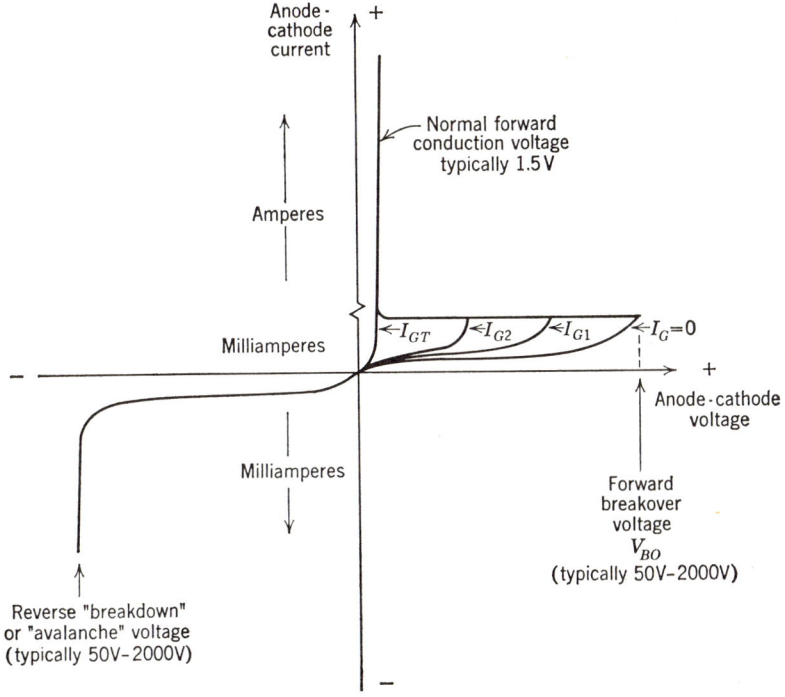

Figure 1.2. Diagrammatic representation of static anode-cathode characteristics of the SCR.

With no gate current ($I_G = 0$), application of forward voltage results in the flow of a relatively small forward leakage current, up to the level of the critical forward breakover voltage (V_{BO}). At this point, the SCR triggers into a low impedance condition; that is, the anode voltage falls from, typically, say several hundreds of volts (maximum values of V_{BO} at present are in the region of 2 kV) to 1.5 V or so. In this low impedance condition, the amplitude of the forward anode current is determined essentially only by the conditions existing in the external circuit. Thus, for most practical purposes, the thyristor when carrying forward anode current can be regarded as being a perfectly closed switch.

Increasing amounts of gate current (I_{G1} and I_{G2}), result in a reduction of the critical forward breakover voltage of the SCR, (as well as somewhat increasing the forward leakage current) until finally, with a sufficient amount of gate current (I_{GT}, or higher), the forward anode breakover voltage is reduced to substantially zero.

In practical applications, the "graded" forward breakover voltage/gate current characteristic of the SCR is rarely used. Almost invariably, (and certianly this is the case for the circuits considered in this book), the circuit operation is such that the applied forward anode voltage is less than V_{BO}, by a suitable safety margin. When blocking forward voltage, the gate current is held at zero (or possibly, a negative gate current may be delivered).

At the desired instant of initiation of conduction, the SCR is triggered into the low forward impedance condition, by means of delivering a pulse of firing current to the gate. The amplitude of this current pulse is at least equal to, but normally is arranged to be in excess of, the minimum gate current required for triggering (I_{GT}).

Typically, for a SCR rated at 500A rms, I_{GT} might be, say, 200 mA, with a corresponding applied gate voltage of say 4V. (Thus it is clear that the SCR is capable of exhibiting an extremely high "power gain.")

Once the SCR has been triggered into the conducting state, the gate loses control, and removal of the gate current has no effect upon the anode current—provided that this is greater than a relatively small minimum value, called the "latching" current. In order to return the SCR to its original forward voltage blocking condition, it is necessary, by dint of the operation of the external anode circuit, to reduce the forward anode current to a relatively low level, called the "holding" current. For most practical purposes this "holding" current can be regarded as being essentially zero.

DYNAMIC SWITCHING CHARACTERISTICS

The static characteristics give no indication as to the speed at which the SCR is capable of being switched from the "forward voltage blocking" to the "conducting" state, and vice versa. In point of fact, the transition from one state to the other does not take place instantaneously, but it occupies a finite period of time. In order to ensure reliable and correct SCR circuit operation, it is necessary to take proper account of the imperfections in the dynamic switching characteristics. It is pertinent, then, to make a brief review here of the external dynamic switching characteristics of the SCR.

Turn-On

When firing current is delivered to the gate of the SCR, it does not switch immediately from the forward voltage blocking to the final "fully on" condition. Indeed, for a short period of time, the SCR continues to block the anode voltage applied to it, in almost the same way as it would had the firing pulse not yet been delivered. Thereafter, the forward impedance starts to decrease, but not until a further time period has elapsed does the SCR become "fully turned on."

DYNAMIC SWITCHING CHARACTERISTICS

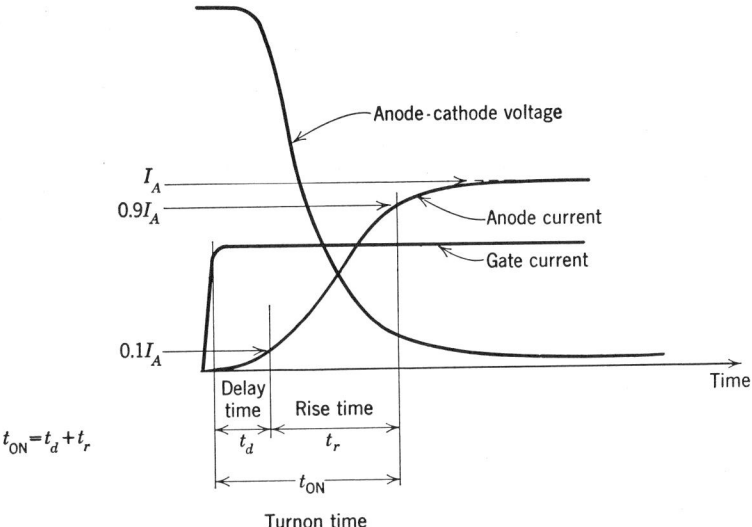

Figure 1.3. Waveforms defining SCR turnon time. Waveforms are for pure resistance anode load.

As illustrated in Fig. 1.3, the total turn-on time of the SCR is subdivided into two distinct periods, called the "delay time" and the "rise time." These time periods are defined in terms of the waveforms of the anode voltage and current obtained in a circuit in which the anode load consists of a pure resistance. The delay time is the time between the point at which the gate current reaches 90% of its final value, and the point at which the ensuing anode current reaches 10% of its final value. The rise time is the time taken for the anode current to rise from 10 to 90% of its final value.

Both the delay time and the rise time are related to the rise time and amplitude of the gate firing current. If it is required to keep these times to the practical minimum, then it is necessary to deliver a gate firing pulse with a fast rise time—ideally, in the order of 0.1 μsec, but certainly not in excess of say 1 μsec—and with an amplitude considerably greater than—say 3 to 5 times—the minimum gate current required to trigger the device. The width of the firing pulse should be at least 10 to 20 μsec. Delay and rise times obtained with such "hard" firing pulses might typically be in the order of 0.5 and 3.5 μsec respectively. With slow rising, relatively low amplitude, "soft" firing pulses, on the other hand, both these times may be increased severalfold.*

* But, at the time of writing, SCR's with improved switching characteristics, for which "soft firing" is claimed always to be sufficient, are beginning to appear on the market.

From a practical viewpoint, the delay time, per se, even with "soft" firing, is generally of little consequence in SCR circuits operating at normal power frequencies (i.e., up to, say, 400 Hz). The rise time, however, may be of significance, since during this time the SCR simultaneously supports appreciable forward voltage, and carries forward anode current. The instantaneous power dissipation therefore can be very high. This can lead to local internal "hot spots," and eventual burn out of the SCR. For this reason, it is necessary to ensure that the rate of rise of anode current at switch-on does not exceed a specified limiting value. Often, this is achieved by inserting a special "di/dt inductor" in the anode circuit of the SCR.

Turn-Off

The SCR cannot block forward anode voltage immediately after the anode current has been reduced to zero. Thus it is necessary to apply a reverse anode voltage for a finite period of time, before forward anode voltage can be reapplied. Premature reapplication of forward voltage results in the resumption of forward conduction.

The turn-off time of the SCR is defined by the waveforms shown in Fig. 1.4. The total turn-off time is subdivided into two regions, called the reverse recovery time, t_{rr}, and the gate recovery time, t_{gr}.

During the reverse recovery time, anode current actually flows in the *reverse* direction, whilst the SCR remains in the low impedance condition, and continues to develop a small positive voltage. At time t, the SCR begins to exhibit a reverse blocking impedance. Reverse anode voltage is developed across it, and the reverse "recovery" or "sweep out" current decreases towards zero.

For a given SCR, the reverse recovery time is a function of the forward current and the rate of decay of forward current. Its duration might typically be 1 or 2 μsec for relatively low current SCR's (say less than 100A rating), up to possibly 6 or 7 μsec for "high current" devices.

The reverse recovery effect has practical significance, because the relatively sudden interruption of the reverse anode current tends to create a high amplitude induced voltage transient in the associated anode circuit inductance. Thus it is necessary to provide appropriate resistance-capacitance "snubber" circuitry, to absorb the energy trapped in the anode circuit inductance at the instant of reverse voltage blocking of the SCR.

During the gate recovery time, reverse voltage must be maintained across the SCR; however, the amplitude of this reverse voltage is not too critical. Not until the end of the gate recovery time is the SCR capable of blocking reapplied forward anode voltage; but even then, the rate of rise of this forward anode voltage (dv/dt) must be less than a specified limit, in order to avoid

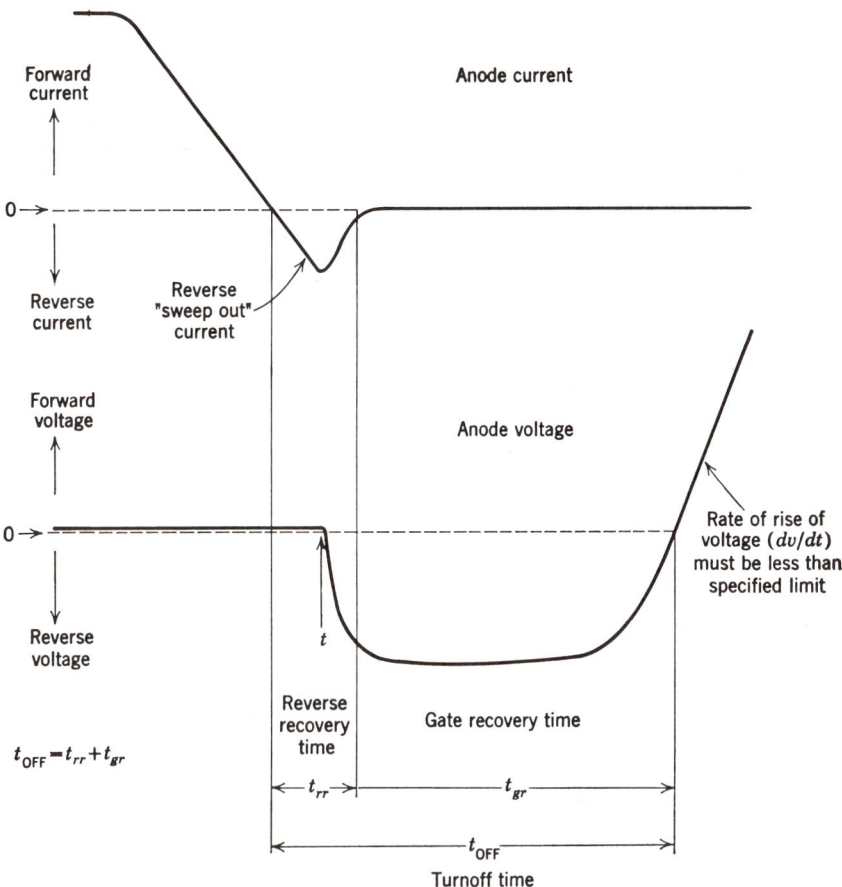

Figure 1.4. Waveforms defining SCR turnoff time.

spurious triggering of the SCR. Typically, this critical value of dv/dt might be say 100 V/μsec.

For a given SCR, to a lesser or greater extent, the gate recovery time is dependent upon several independent factors. The most important of these are the junction temperature of the device, and the rate of reapplication of forward voltage.

Typical SCR gate recovery times presently range from say, 10 μsec, for relatively low current, "fast switching" devices, up to possibly 200 μsec, for high current, "slow switching" SCR's.

In practical applications, it is necessary to ensure that the turn-off time afforded to the SCR by the circuit is greater than the critical device turn-off time, by a suitable safety margin. Thus, for a high current, "slow switching" SCR, it might be necessary for the circuit to allow say 300 μsec for turn-off. For a 60 Hz application, this time corresponds to about 6.5° of the voltage wave, and it is therefore comparatively small by comparison with the total cycle time. For higher frequency applications, however, the required circuit turn-off time can become an appreciable portion of the total cycle time; here, the selection and use of special "fast switching" SCR's may become an important consideration.

Chapter Two

Functions, Characteristics, and Applications of Thyristor Phase-Controlled Converters and Cycloconverters

The main objective of this book is to present a detailed description and explanation of the fundamental principles of operation and control of the phase-controlled converter, and of the cycloconverter, and to derive analytical data which describe quantitatively their basic external performance characteristics. At the outset, however, before proceeding to the more detailed technical discussion, it is pertinent to review briefly the basic functions, characteristics, and most typical present-day applications, of these types of power conversion equipment.

THE THYRISTOR PHASE-CONTROLLED CONVERTER—ITS BASIC FUNCTIONS AND CHARACTERISTICS

The basic function of the phase-controlled converter is to convert an alternating input voltage to a controllable direct output voltage. It consists, basically, of a rectifier circuit arrangement in which some, or all, of the rectifier devices have a gate-controlled forward voltage blocking capability. More specifically, if the discussion here is limited to thyristors, then the controlled rectifier devices of the phase-controlled converter would be thyristors.

The basic principle of operation of the phase-controlled converter is to control the point in time at which conduction in each thyristor is allowed to commence during each a-c cycle. By this means, it is possible to choose the time-segments of the a-c voltage waves which appear at the d-c terminals, and, thereby, to continuously control the mean value of the d-c terminal voltage. In this way, the essentially bistable switching characteristic of the thyristor itself is utilized to provide a linear control of the mean output quantity.

Turn-off of the thyristors is achieved by so called natural means; that is to say, the commutation of current from one thyristor to the next is made to occur at a point at which the incoming a-c voltage wave has a higher instantaneous potential than that of the outgoing wave. Thus, there is always a natural tendency for the current to be commutated from the outgoing to the incoming thyristor, without the aid of any additional commutating circuitry. This commutation process is often referred to as a "natural" or "line" commutation.

By appropriate circuit design, it is possible for the phase-controlled converter to provide either a so-called 1-, 2-, or 4-quadrant operating characteristic at its d-c terminals. These three types of operation, together with the associated circuit type, are illustrated diagrammatically by Fig. 2.1.

Consider, firstly, the 2-quadrant converter, illustrated at *b*. This consists basically of a conventional rectifier circuit arrangement, which contains thyristors in all positions of the circuit. Due to the unidirectional current carrying property of the thyristors, the current at the d-c terminals of the converter, of course, can flow only in one direction. However, it is possible, by means of suitable control of the phase position of the firing pulses applied to the thyristors, with respect to the a-c input voltage waves, for the mean voltage at the d-c terminals to be continuously controlled from maximum positive to maximum negative. Thus, power can be made to flow either from the a-c to the d-c side of the converter, or vice versa, and the 2-quadrant converter is capable of operation both as a rectifier and as an inverter.

Many applications actually require only a 1-quadrant operating characteristic—that is to say, only one polarity of voltage and current at the d-c terminals. In this case, although, of course, the desired range of operation could be obtained from the 2-quadrant converter, simply by using only half of its control range, usually there are both technical and economic advantages to be gained, by arranging the converter circuit so that intrinsically it is capable only of 1-quadrant operation. This involves the use of uncontrolled diodes in certain positions of the circuit, as illustrated at *a*.

Again, other applications require a full 4-quadrant operation—that is to say, the converter is required to operate with either polarity of voltage and current at the d-c terminals. One method for obtaining a 4-quadrant operation is to equip the two quadrant converter with a reversing switch at its d-c terminals, as illustrated at *c* in Fig. 2.1. This reversing switch provides the facility for both polarities of current in the load, although, of course, the current through the converter itself still is inherently unidirectional. Such

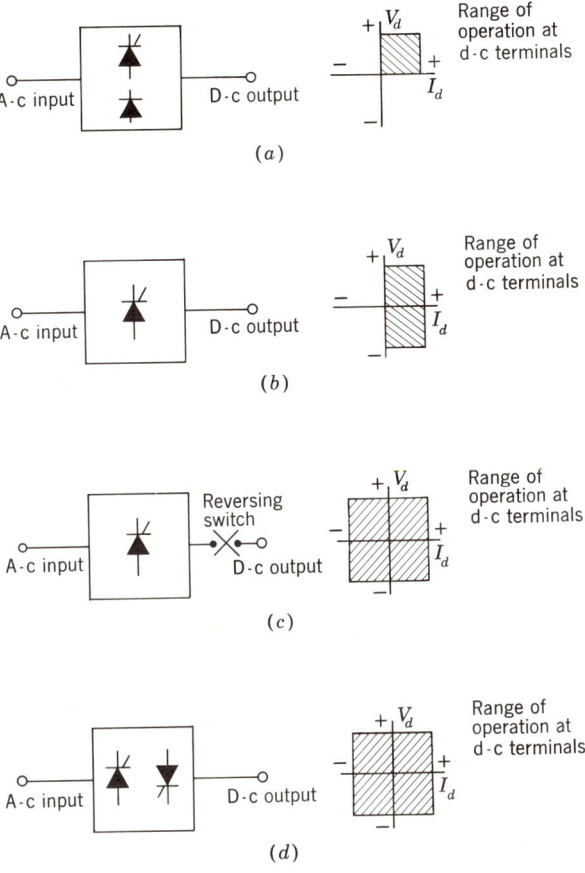

Figure 2.1. Simplified diagrammatic representation of the various types of thyristor phase-controlled converter. (a) One-quadrant converter; converter contains controlled and uncontrolled rectifiers in different circuit positions. (b) Two-quadrant converter; converter contains controlled rectifiers in all circuit positions. (c) Four-quadrant converter; consisting of a 2-quadrant converter, and a d-c reversing switch. (d) Four-quadrant converter—"dual-converter"; consisting of two 2-quadrant converters, connected "back-to-back."

an arrangement is satisfactory, provided that the time delay inevitably associated with the change-over of the mechanical reversing switch can be tolerated.

An alternative, more sophisticated, approach, illustrated at d, is to connect two similar 2-quadrant converters back-to-back with one another, thus providing the facility for bidirectional current flow through the load, without the need for a reversing switch. This type of circuit arrangement, referred to in this book as a "dual-converter," offers the facility for virtually instantaneous reversals of current at the d-c terminals.

In common with conventional uncontrolled rectifier circuits, the external operation of the phase-controlled converter is characterized by the production of harmonic voltage distortion components, in addition to the direct component, at the d-c side, and by the production of harmonic components of current, in addition to the fundamental component, at the a-c side. Moreover, due to the process of phase delay of the firing pulses, inherent in the basic control mechanism, the displacement angle between the voltage and the fundamental component of current at the a-c side is generally lagging, and hence the input displacement factor is generally less than unity. On the other hand, due to the relatively small losses associated with the solid state rectifier devices, the power conversion efficiency is generally high, and might typically be in the order of 96 to 98% (or higher).

Because of the relatively low trigger power requirements of the thyristor, the control and firing circuits can be designed to operate at a relatively low power level, and these, therefore, readily lend themselves to the use of transistors, integrated circuits, and other low power solid state devices. Thus the rapid advances in the area of low power solid state device technology which have taken place over recent years can be well utilized in the design of the control and firing pulse circuits for thyristor converters, and it is possible to realize extremely compact, robust, reliable, and flexible control modules. Moreover, due to the possibility for designing the firing pulse timing circuits so that they respond to milliwatt level analog control input signals, it is possible to design control systems which provide virtually any desired complex closed loop system control function. Thus, the thyristor phase-controlled converter, taken in conjunction with its control circuits, can be looked upon as being, in effect, a very versatile high power d-c amplifier, the performance of which can be designed to meet virtually any practical control requirements.

APPLICATIONS OF THYRISTOR PHASE-CONTROLLED CONVERTERS

General

The thyristor is available in a wide variety of power ratings; in addition, it is possible to connect individual thyristors, as well as complete thyristor equipments, either in series or in parallel with one another. Thus, thyristor phase-controlled converter systems are feasible with power ratings varying from less than a kilowatt, to tens of megawatts, and the corresponding range of applications embraces a very wide field.

Thus, at one end of the scale, the thyristor phase-controlled converter finds use in low power equipments such as speed controllers for fractional horsepower d-c motors, constant potential power supplies, battery chargers, and the like. At the other end of the scale, the thyristor phase-controlled converter has now become the unquestioned successor to the high power grid-controlled mercury arc converter, as well as to the rotating a-c to d-c converter—in particular the classical Ward-Leonard system—in applications such as high power variable speed reversing drives for industrial rolling mills, having ratings of many megawatts. Indeed, in competing with the grid-controlled mercury arc rectifier, it is true to say, at the time of writing, that there is but one remaining practical application, that of high voltage d-c transmission, in which the thyristor has not yet found preference. But, even here, due to the now proven feasibility of operating large numbers of thyristors connected in series with one another, there are strong indications that the thyristor will eventually replace the mercury arc converter.

Probably the most widespread present day application for the thyristor phase-controlled converter is in the speed control of d-c machines. This particular subject is therefore considered to be worthy of a slightly more detailed discussion. This is presented in the following section.

Variable Speed Control of d-c Machines

The d-c machine has, in the past, been generally preferred over other types, in most industrial applications requiring a variable speed drive. This is because the speed of the d-c machine can be controlled relatively easily, by control of either its armature voltage, or its field voltage, or both, depending upon the desired performance characteristics of the drive.

Hitherto, the required supply of variable d-c voltage has generally been obtained either from a grid-controlled mercury arc converter, or from a rotary Ward-Leonard converter. Today, the thyristor phase-controlled converter is economically competitive with, and in most respects, is technically superior to these older types of a-c to d-c converter. Its application to the speed control of d-c machines in a variety of industrial applications, such as rolling mill drives, printing press drives, mine winders, and so on, can be seen, therefore, to be a natural process of evolution.

Various alternative drive systems for d-c machines, using thyristor converters to control the applied armature voltage, are illustrated in simplified diagrammatic form in Fig. 2.2. It is assumed here that the field winding of the machine is separately excited, and that the steady state field current is nominally fixed. Thus, by continuously controlling the armature voltage, the speed of the machine can be continuously controlled, with a "constant torque" characteristic. Actually, in practice, the field current of the machine

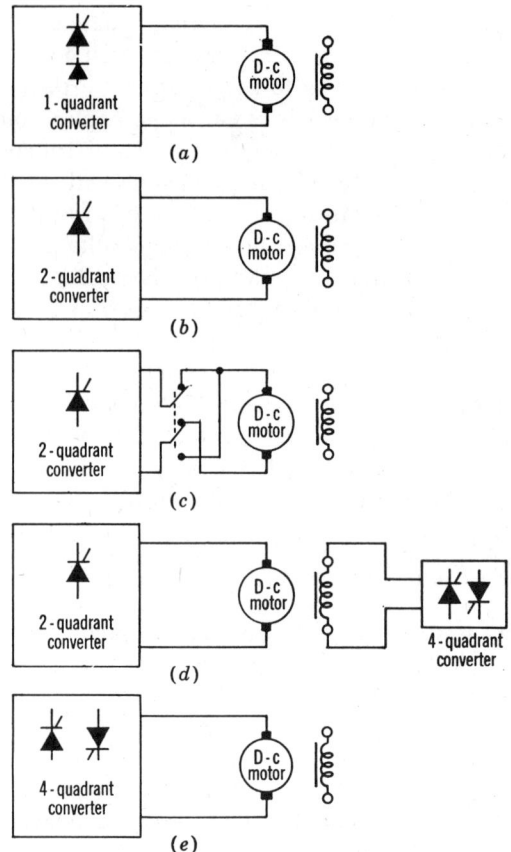

Figure 2.2. Simplified diagrammatic representation of various types of thyristor phase-controlled converter d-c drives. (a) One-quadrant control; speed control in forward direction only, no regenerative braking. (b) Two-quadrant control; speed control in the forward direction with "positive" power flow, and in the reverse direction with "negative" power flow. (c) Four-quadrant control–armature reversing switch; speed control in both the forward and reverse directions, with regenerative braking. (d) Four-quadrant control–reversing field control; speed control in both the forward and reverse directions, with regenerative braking. (e) Four-quadrant control–dual-converter armature control; speed control in both the forward and reverse directions, with regenerative braking.

may also be controlled, thus providing an additional degree of flexibility in the operating characteristics of the drive.

The simplest, and most economical, type of armature-controlled d-c drive is shown at *a* in Fig. 2.2. Here, a 1-quadrant converter is used to provide a unidirectional speed control of the machine. Due to the inability of the converter to operate in its inverting region, it is not possible for energy to be returned from the machine to the supply, and hence this simple drive system does not have the facility for regenerative braking.

At *b*, a 2-quadrant converter is connected to the armature of a d-c machine. This type of system finds use in special applications, such as mine winders, in which the polarity of the induced emf of the machine reverses, whenever regenerative braking is required, as a natural outcome of the particular duty cycle. Thus, in a mine winder, for example, when the load is being raised, the converter supplies power to the machine, the induced emf of which is "positive." When the load is being lowered, on the other hand, the direction of rotation, and hence also the induced emf, is reversed; hence the polarity of the machine emf is now "negative," and therefore it is in the correct direction for the converter to operate in its inverting region, thus returning regenerative power from the machine to the a-c system.

At *c*, a 2-quadrant converter is connected to the armature of a d-c machine through a reversing switch. Here, by appropriate control of the switch, it is possible to control the speed of the machine in both the forward and reverse directions, with the facility for regenerative braking. This type of system is satisfactory, provided that the time delay, typically in the order of 0.1 sec, required for the operation of the reversing switch, can be tolerated. During this "dead time" period, of course, the machine "coasts" freely, and is not under the control of the converter.

At *d*, a 2-quadrant converter is connected to the armature of a d-c machine, and a second, relatively low power, 4-quadrant dual-converter supplies current to its field winding. Here, in effect, the mechanical reversing switch in the armature circuit of Fig. 2.2*c* has been replaced by a static reversing switch in the field circuit. Thus, by suitable control of the dual-converter connected to the field winding, the field current can be made to flow in either direction, and hence the polarity of the induced armature emf can be reversed, relative to the 2-quadrant armature converter, without actually reversing the electrical connections in the armature circuit. Thus, a full 4-quadrant operation is possible, and the speed of the machine can be controlled in both the forward and reverse directions, with the facility for regenerative braking.

Since the field winding is inherently highly inductive, it is often necessary, in order to achieve a sufficiently rapid reversal of field current, to design the converter connected to the field winding to produce a maximum voltage at its d-c terminals which is several times in excess of the steady state field voltage. Thus, under transient conditions during which the field current is being changed from one polarity to the other, the voltage of the field converter is adjusted to the maximum level of which it is capable, thereby producing the maximum possible rate of change of current; and, under steady state conditions, the d-c terminal voltage of the field converter is reduced to the rated voltage of the field winding, which is considerably less than the maximum "forcing" voltage.

At *e*, a 4-quadrant dual-converter is connected to the armature of a d-c machine. This system provides a full reversing drive, with regenerative braking, without the need for manipulation of either the armature circuit connections, or the polarity of the field current. Due to the use of two separate 2-quadrant converters connected "back to back," each of which must generally be rated for the full power of the machine, this system is inevitably more expensive than the others. However, it is also capable of providing the most sophisticated system performance. Thus, by means of suitable control, it is possible to achieve very rapid reversals of speed and torque, with little or no "dead time." The behavior and control of this type of system is discussed in some detail in Chapter 5.

In each of the various types of drive of Fig. 2.2, it is possible to control the speed of the machine to virtually any desired accuracy, by means of appropriate closed-loop control circuits. If the required accuracy is not too great, it may be sufficient to use a closed-loop control of the applied armature voltage, possibly with an armature current compensation. If greater accuracy is required, this can be obtained by means of closed-loop control of the feedback signal obtained from a tacho-generator coupled directly to the drive motor. It is also possible, of course, to provide closed-loop control of virtually any desired output parameter, such as torque, acceleration, position, and so on. Often in this type of drive, an automatic closed loop armature current-limit control is incorporated. Such a control is typically arranged to override all other control inputs, and to produce an essentially constant current mode of operation, in the event that the armature current should attempt to exceed a preset limiting value. By this means, it is possible to provide a fast automatic protection against current overloads, thus ensuring fully-controlled operation of the system under all conditions.

THE THYRISTOR PHASE-CONTROLLED CYCLOCONVERTER—ITS BASIC FUNCTIONS AND CHARACTERISTICS

The basic function of the thyristor cycloconverter is to convert an alternating input voltage of one frequency to an alternating output voltage of another frequency. Its essential feature is that it contains only one stage

THE PHASE-CONTROLLED CYCLOCONVERTER

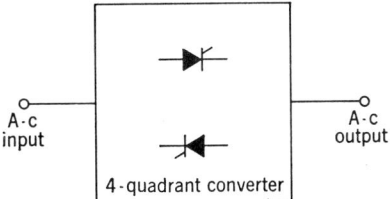

Figure 2.3 Simplified diagrammatic representation of the phase-controlled cycloconverter. Output frequency is continuously controllable from zero up to some practical limit which is normally less than the input frequency. Amplitude of the output voltage is continuously controllable from zero to maximum.

of power conversion, which consists, in its basic form, simply of an array of thyristor switches. The basic principle of operation is to fabricate a "rough" output voltage waveform, having the desired frequency, directly from the input voltage waves, simply by opening and closing the switches within the converter in an appropriate manner. This rough output waveform is then filtered, if required, and thus it is possible to produce a high quality final output voltage waveform from the cycloconverter. Due to the use of thyristors, with their relatively low losses, the cycloconverter is capable of high power-conversion efficiency.

The cycloconverter is illustrated in simplified diagrammatic form in Fig. 2.3. In essence, it is simply a dual-converter, which is controlled, through a time-varying phase modulation of its firing pulses, so that it produces an alternating, rather than a direct, output voltage. By appropriate control, it is possible to produce a continuous variation of both the amplitude and frequency of the output voltage wave. However, due to the necessity for always producing a natural commutation of current from one thyristor to the next, the maximum attainable useful output frequency is, for most practical purposes, less than the input frequency. Typically, the maximum "usable" output frequency may be somewhere between $\frac{1}{3}$ and $\frac{2}{3}$ of the input frequency, depending upon the particular converter circuit, and upon the distortion which can be tolerated at the output.

Just as the dual-converter provides a full 4-quadrant operation, so the cycloconverter is able to operate with loads of any displacement factor at its output terminals, and power can flow either from the input to the output side, or vice versa.

In common with the phase-controlled converter, the external operation of the cycloconverter is characterized by the production of harmonic voltage distortion components, in addition to the wanted sinusoidal component, at the output side, and by the production of quadrature and harmonic components of current, in addition to the in-phase component, at the input side. Moreover, the displacement angle between the fundamental component of the input current and the input voltage is invariably lagging, regardless of whether the power factor of the load at the output is lagging, leading, or unity.

APPLICATIONS OF THYRISTOR CYCLOCONVERTERS

There are presently two main applications for the thyristor cycloconverter. The first is in the area of variable frequency variable speed drives for a-c machines. Here, the input power supply to the cycloconverter has a fixed frequency, and the variable frequency output of the cycloconverter is connected to the machine to be driven. The second is in the area of constant frequency power supplies. Here, the function of the cycloconverter is to provide a closely regulated fixed frequency power output, from a variable frequency power source connected to its input.

Each of these applications of the cycloconverter is described in more detail in the following sections.

Variable Speed Control of a-c Machines

The d-c machine finds wide application in variable speed drives, for the reason that its speed can be efficiently controlled by adjusting its voltage. It does, however, have the practical disadvantage of employing a commutator and brushgear, which is relatively costly, and also requires periodic maintenance. Moreover, the use of a commutator makes operation difficult with voltages or currents beyond a certain practical limit; furthermore, the commutator actually precludes the use of a d-c motor in certain extreme environmental conditions.

The a-c machine, on the other hand, does not require a commutator. As a result, it is generally considerably less expensive than the d-c machine, as well as being much more robust mechanically and, therefore, much less in need of maintenance. However, the a-c motor is not generally regarded as being a variable speed machine, because its speed is a function of its applied frequency, which is normally fixed. Of course, various means for controlling the speed of a-c induction motors, connected to a fixed frequency supply, have been devised. These methods, although satisfactory within their own limitations, either are inefficient, or are not generally applicable to a drive which is required to provide a high performance over a wide range of load and speed, with, perhaps, the added requirement for reversing. An exception to this rule is the a-c commutator motor; but here, of course, because of the use of the commutator, the machine is similar in construction to the d-c motor, and does not therefore have the mechanical, or economical, advantages of the commutatorless a-c machine.

In point of fact, if a supply of variable frequency, variable voltage, power is connected to a commutatorless a-c machine, then it is possible, by means of appropriate control of the frequency and amplitude of the voltage, to provide an efficient variable speed drive, with a performance characteristic which is, for all practical purposes, equivalent to that of a high performance 4-quadrant armature voltage-controlled d-c drive.

As has been mentioned, the cycloconverter has the facility for continuous control of both its output frequency and voltage, independently of one another, and it is able to operate with loads of any power factor, including regenerative loads, at its output terminals. Moreover, with the cycloconverter, it is a simple matter to reverse the phase sequence of the output voltages, by means of suitable electrical manipulation of low level signals in the firing control circuits. The output operating characteristics of the cycloconverter are, therefore, ideally suited to the efficient, variable speed, reversing control of a-c machines. Since, moreover, in industrial applications, the primary source of power is almost invariably a fixed frequency a-c supply, it is possible to connect this supply directly to the input terminals of the cycloconverter, thus providing a variable speed a-c drive system with only a single stage of power conversion. Of course, the upper output frequency limit of the cycloconverter is usually less than the supply frequency—typically, for a 6-pulse cycloconverter, the maximum output frequency may be around $\frac{2}{3}$ of the supply frequency. This restriction does, then, place a practical limitation on the range of applications for this type of drive system.*

In principle, either a synchronous or an asynchronous machine can be used; but, in practice, the latter, in the form of the squirrel cage induction motor, with its extreme simplicity, and mechanical robustness, is often to be preferred.

The principles for efficient variable speed control of an induction motor are outlined briefly in the following section.

PRINCIPLES OF VARIABLE FREQUENCY SPEED CONTROL OF AN INDUCTION MOTOR. Under steady state operating conditions, with full rated voltage applied to the stator windings, the speed of an induction motor is nearly proportional to the frequency of the applied voltage, and is practically independent of the load on the machine, within the normal load range. By

---

* It should, of course, be mentioned that the cycloconverter is not the only type of solid state variable frequency power supply with characteristics suitable for a-c drives. In particular, systems using an a-c to d-c converter, followed by a variable frequency inverter, can be made to produce the desired operating characteristics. However, such systems are not the subject of consideration here.

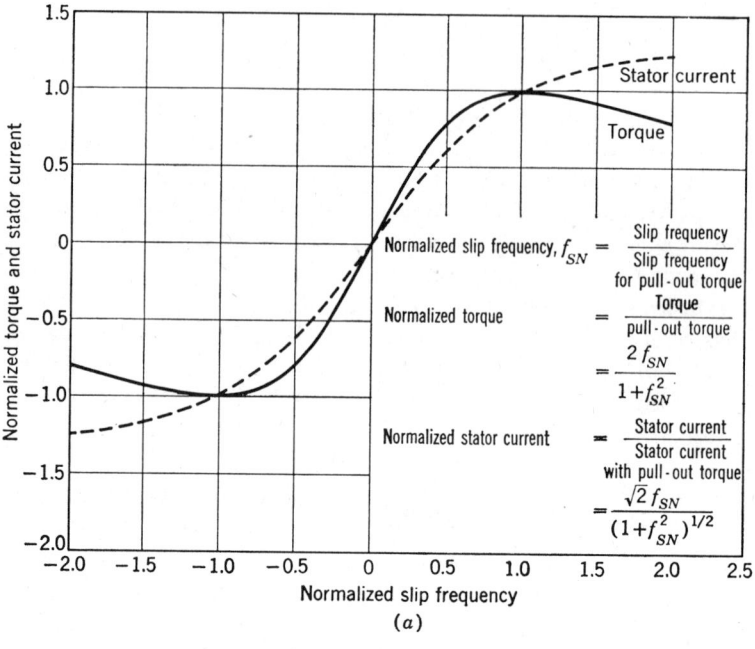

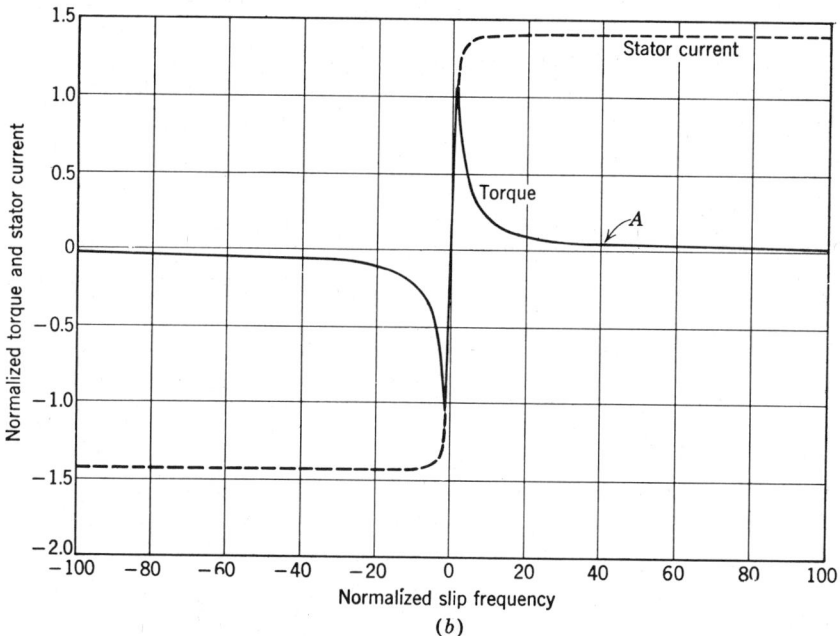

Figure 2.4. "Universal" characteristics for an induction motor with constant excitation flux.

APPLICATIONS OF CYCLOCONVERTERS 21

controlling the frequency and amplitude of the stator voltage, in such a relationship to one another as to maintain a constant flux in the machine, the speed can be controlled in a manner which is akin to that of the speed control of a separately excited d-c machine, with armature voltage control.

In the case of the d-c machine, the torque developed is proportional to the armature current, which is determined by the difference between the applied voltage and the induced emf of the machine—which is itself proportional to the machine speed. In the case of the induction motor, with the voltage to frequency ratio controlled so as to maintain a constant flux in the machine, the torque developed is determined only by the absolute difference or "slip" frequency between the applied stator frequency and the output "shaft frequency," irrespective of the speed.

The torque/slip frequency characteristic for an induction motor with constant excitation flux is shown in Fig. 2.4a and b. Also shown plotted alongside is the corresponding stator current/slip frequency characteristic. Both of these characteristics are "universal," and are generally applicable to any induction motor, independently of the applied frequency or speed. It should be noted that these characteristics show both positive and negative values of slip, torque, and current; that is to say, they are applicable to operation of the machine both in its "motoring" and in its "regenerating" mode.

Considering the "motoring" region of operation, it can be seen that as the slip frequency is increased, so the torque and current increase, up to a certain maximum torque. Thereafter, further increase in the slip frequency results in a reduced torque, but with increased current. The absolute slip frequency which results in the maximum torque of which the machine is capable—the "pullout torque"—depends upon the resistance of the rotor circuit, and is usually a few percent (or less) of the rated frequency. This frequency is referred to here as the "pullout" slip frequency.

The normal steady state operating range of an induction machine is over the portion of the torque/slip characteristic which has a positive slope. However, in the case where the supply frequency is fixed, it is clear that, at starting, the slip frequency must inevitably be much greater than that which results in maximum torque. A typical operating point, with a fixed frequency supply, with the rotor at standstill, is at A (Fig. 2.4b). At this point, the torque is considerably less than the pullout torque, and the current is relatively high. It can be seen, in order to increase the starting torque, that it is necessary for the starting point "A" to be moved to the left. This implies that the absolute value of the pullout slip frequency must be increased, which, in turn, implies that the resistance of the rotor circuit must also be increased. This is undesirable for a squirrel cage induction motor, since, necessarily, it is accompanied by a reduced full-load efficiency. Thus, it can be seen that the squirrel cage induction motor, when fed from a fixed frequency supply, has the undesirable, yet inevitable, characteristic that the available starting torque is less than the pullout torque, and the starting current is relatively high.

If, on the other hand, the frequency and voltage of the power source can be controlled at will, then there is no longer any need, under any circumstances, to operate the machine outside of that portion of the torque-slip characteristic having a positive slope, either in the "motoring" or "regenerating" region of operation. In other words, the applied frequency can now be controlled so as to always keep the absolute slip frequency less than (or equal to) the "pullout" slip frequency, and the machine can therefore be made to operate on the most favorable portion of its torque-slip characteristic, under all conditions. Thus, it is possible to realize the maximum driving and braking torques of which the machine is capable, at all speeds, including standstill. Moreover, due to this possibility, there is no longer any need for the resistance of the rotor circuit to have an artificially high value, simply to provide a given starting torque capability. Thus it is possible to design the rotor circuit to have the minimum possible resistance—which implies, also the minimum possible pullout slip frequency. These features are desirable, not only from the viewpoint of increased efficiency, but also, because of the low slip, the speed of the machine becomes almost exactly proportional to the applied frequency, over a wide range of operation. Thus, it is possible to control the speed, with a fairly high degree of accuracy, without resort to a closed loop speed control, by accurately controlling the applied frequency.

As has been mentioned, in order to maintain a constant flux in the machine, it is necessary to control the amplitude of the applied voltage in accordance with the frequency. For an "ideal" machine, in which the resistance and leakage inductance of the stator winding are assumed to be zero, and hence the "airgap" voltage is equal to the applied voltage, a constant ratio between the voltage and the frequency would be required. For a practical machine, in order to maintain a constant ratio between the "airgap" voltage and the frequency, it is necessary to adjust the applied voltage beyond the proportional value, in order to offset the voltage drop across the series impedance of the winding, due to the stator current. This voltage drop is actually a relatively small proportion of the full rated voltage of the machine, and hence the required amount of adjustment of the voltage with load is small by comparison with the full voltage. Thus under conditions when the applied stator voltage and frequency are relatively high, the effect of a small adjustment of the amplitude of the applied voltage with load may have little practical effect upon the operation. At low operating frequencies, however, under which conditions the applied voltage is relatively small, the voltage drop across the series resistance of the stator winding becomes a large proportion of the applied voltage, and the adjustment of the voltage with load, so as

APPLICATIONS OF CYCLOCONVERTERS

to maintain the full airgap flux, and hence also the maximum torque capability of the drive, now has a critical effect upon the performance of the system. Thus, in order to realize the maximum torque capability of the machine at low speeds, it is essential to adjust the voltage to frequency ratio beyond the proportional value. The cycloconverter is inherently capable of producing any desired voltage, independently of the frequency, and hence this requirement does not represent a limitation. Herein lies a significant advantage of the cycloconverter over the rotary frequency changer, which inherently does not have this capability.

A CYCLOCONVERTER VARIABLE FREQUENCY SPEED CONTROL SYSTEM FOR AN INDUCTION MOTOR. The basic elements of a variable frequency speed control scheme for an induction motor, using a cycloconverter, are illustrated in Fig. 2.5. The cycloconverter provides 3-phase variable frequency output power, with reversible phase sequence, from a fixed frequency 3-phase power input. Taken in conjunction with its pulse timing and firing circuits, the cycloconverter can be regarded as comprising a high power 3-phase a-c amplifier. Thus the output voltages of the cycloconverter are exactly proportional to the 3-phase analog sine wave reference voltages, obtained from the reference voltage generator, and connected as the "driving" input signals to the pulse timing and firing circuits.

In this simplified representation, it is assumed that a constant flux can be maintained in the machine by providing a linear ratio between the output voltage and the output frequency. As has been explained, in order to achieve this result in practice, it is necessary to adjust the applied voltage beyond the proportional value, especially at low frequency.

Under steady state operating conditions, the output signal from the "excess slip" amplifier is zero, and hence the only input signal to the 3-phase variable frequency sine wave reference voltage generator is the d-c analog reference voltage representing the desired output frequency, and the speed of the machine is appropriate to this frequency. A "positive" frequency reference at the input results in a "positive" phase sequence at the output of the cycloconverter, and hence a "positive" direction of rotation of the machine. A "negative" frequency reference, on the other hand, results in a "negative" phase sequence at the output, and hence a "negative" direction of rotation.

The slip measuring circuit provides a d-c analog signal proportional to the slip frequency. For the "positive" direction of rotation, a positive voltage signal proportional to the slip frequency is provided when the speed of the machine is less than the synchronous speed, that is, the machine is motoring. Conversely, when the machine acts as an induction generator, and delivers power to the cycloconverter, the speed is greater than the synchronous speed, and a negative voltage signal proportional to the "negative" slip

24 FUNCTIONS, CHARACTERISTICS, APPLICATIONS

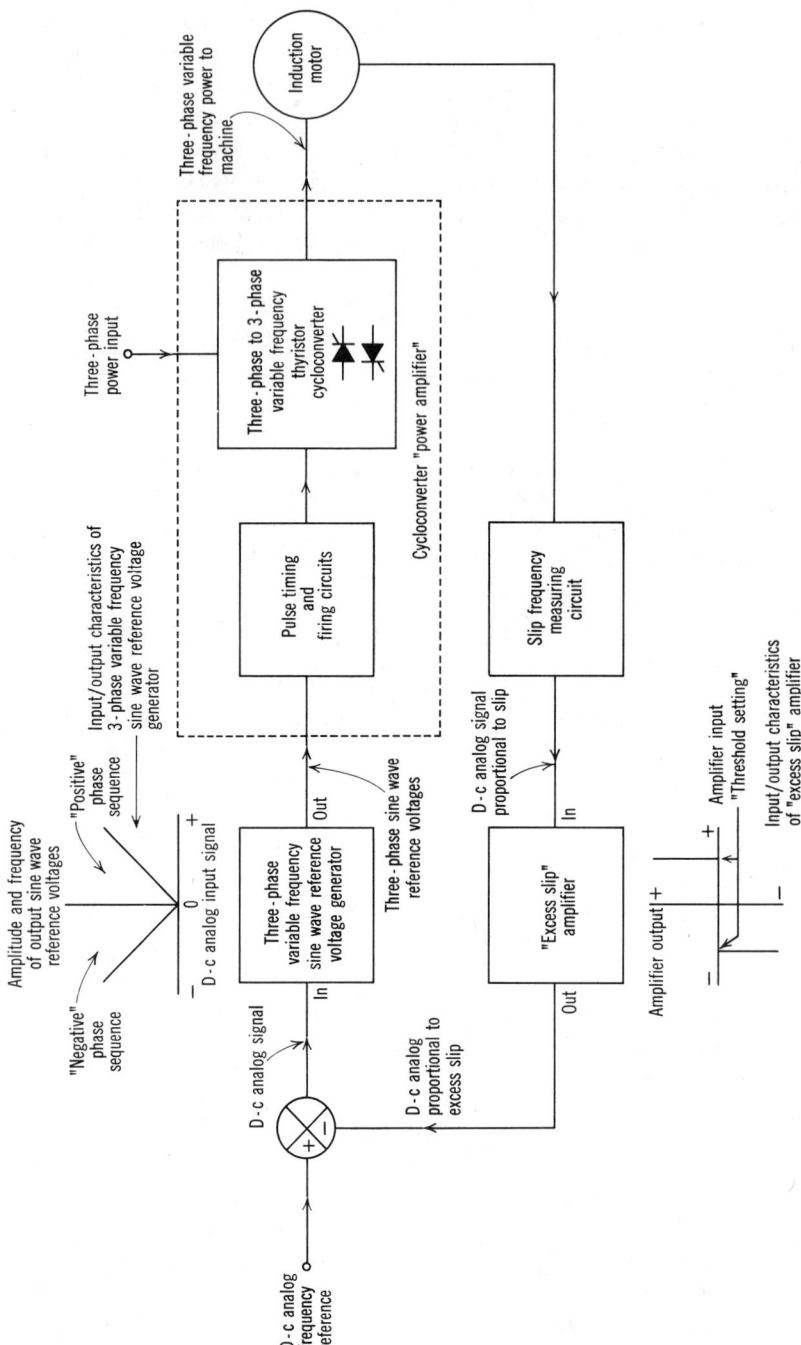

Figure 2.5. Schematic representation of a variable frequency speed control scheme for an induction motor, using a thyristor phase-controlled cycloconverter.

frequency is provided. For the "negative" direction of rotation, the polarities of the slip signal for the motoring and generating conditions are interchanged.

The "threshold" setting of the "excess slip" amplifier is less than, or, in the limiting case, equal to, the "pullout" slip frequency of the machine. Since, under normal steady state conditions, the actual slip frequency is less than the "pullout" slip frequency, the output voltage of the slip measuring circuit is therefore less than the threshold setting of the "excess slip" amplifier, and thus there is no output from this amplifier.

If, however, the slip frequency should attempt to exceed a level corresponding to the threshold setting of the "excess slip" amplifier—this might be caused, for example, by a sudden change in the demanded stator frequency—then this amplifier produces a sharply increasing output signal, which opposes the analog input voltage to the sine wave reference generator. The gain of the "excess slip" amplifier is sufficiently high that, under this condition, its output signal effectively overrides the frequency reference, and the system now operates with a virtually constant slip frequency, corresponding to the threshold setting of the "excess slip" amplifier. Thus, it can be seen that the actual slip frequency can never exceed the "pullout" slip frequency, and therefore the machine always operates on the most favorable "positive slope" portion of its torque-slip characteristic. It therefore produces the maximum torque of which it is capable, at all speeds, in both directions of rotation, in both the "motoring" and "regenerating" modes of operation. Furthermore, since the magnitude of the stator current is directly related to the slip frequency—as shown by the curve of Fig. 2.4—it is clear that this automatic slip limit control also constitutes, in effect, an automatic stator current limit. Thus it can be seen that the performance characteristics of this variable speed a-c drive are, for all practical purposes, equivalent to those of a high performance d-c drive.

Constant Frequency Power Supplies

Several applications require the production of an accurately regulated fixed frequency power output, from a variable frequency power source. In this type of application, the thyristor cycloconverter often constitutes an ideal means of frequency conversion.

One application in this category is aircraft power conversion. Here, the prime source of electrical power is an alternator, which receives its mechanical power input from the engine of the aircraft. Clearly, since the engine speed is not constant, it is not possible for the alternator to produce a constant frequency power output, if it is coupled directly to the engine. Hitherto, the generally accepted practice has been to insert a hydraulic constant speed

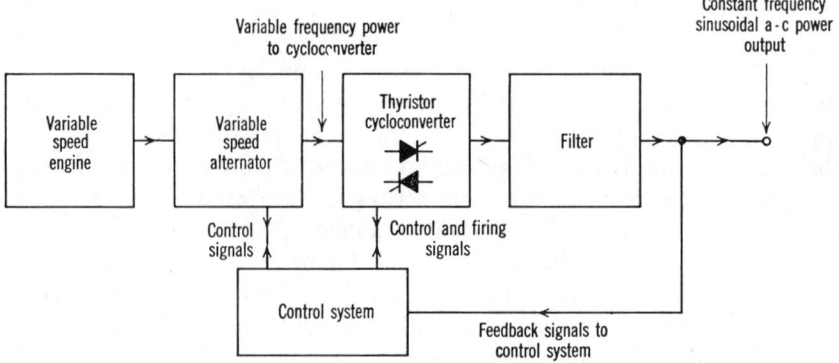

Figure 2.6. Schematic representation of a variable speed constant frequency (VSCF) system using a thyristor cycloconverter.

coupling device between the engine and the alternator, thereby enabling the alternator to be driven at a constant speed, and hence to deliver a constant frequency output power. Such a system has several disadvantages, not least of which is the frequent, and costly, maintenance required.

An alternative, more modern, system approach to aircraft power conversion, is illustrated in simplified schematic form in Fig. 2.6. Here, the alternator is coupled directly to the aircraft engine, so that it produces variable frequency power at its output, as dictated by the engine speed. This variable frequency power is converted to accurately regulated constant frequency power output, by means of a thyristor cycloconverter, in conjunction with a suitable electrical filter. This type of system is generally known as a "variable speed constant frequency" (VSCF) power converter.

Here, of course, such factors as weight and reliability are of prime importance. At the time of writing, VSCF cycloconverters, having individual output ratings of up to 60 kVA, have been developed. The main advantages of the VSCF system over the hydraulic constant speed drive are that it requires much less maintenance, and its electrical performance is generally superior. Thus, for example, the speed of response of the VSCF cycloconverter is virtually instantaneous; and it is possible to provide a very close control of the output frequency, as well as to provide a very close amplitude and phase balance between output phases, even with unbalanced loads. These features greatly facilitate the paralleling of individual converters, and generally enable the VSCF system to meet the stringent requirements of modern aircraft power systems.

Chapter Three

Phase-Controlled Converters—Circuits and Operating Principles

In Chapter 2, it was seen that the function of the thyristor phase-controlled converter is to produce a continuously controllable d-c output voltage from an a-c input, and that it consists, essentially, of a conventional rectifier circuit arrangement, containing thyristors.

In order to provide a 2-quadrant operation—that is to say, both polarities of voltage, with a given polarity of current at the d-c terminals—it is necessary to connect thyristors in all positions of the circuit. On the other hand, for a 1-quadrant operation—that is to say, only one polarity of voltage and current at the d-c terminals—it is possible, and generally desirable, to connect uncontrolled diodes in certain circuit positions.

In this chapter, the basic principles and operating features of both 1- and 2-quadrant phase-controlled converters are discussed. The most commonly used converter circuits, together with associated waveform diagrams, are presented.

TWO-QUADRANT CONVERTERS

The Two-Pulse Midpoint Converter

As will be seen, there are many possible circuit arrangements of the phase-controlled converter. As an introduction to the subject, a simple circuit using two thyristors, in conjunction with a single phase supply transformer, with a midpoint connection, will be considered.

The circuit is shown in Fig. 3.1, and typical idealized waveforms associated with this circuit, which assume a perfectly smooth direct current, are shown in Fig. 3.2.

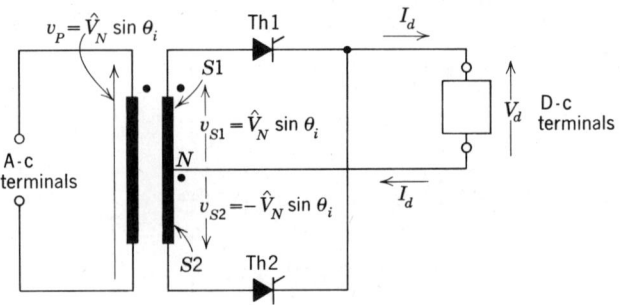

Figure 3.1. Two-pulse midpoint converter.

Consider firstly the waveforms shown at *a*. Here, each thyristor has a firing pulse applied to it at the instant its anode voltage becomes positive, and forward conduction commences at this point. In this case the thyristors do not at any time block forward voltage, and the operation of the circuit is, in appearance, the same as it would be if the thyristors were replaced by "uncontrolled" rectifiers. The voltage waveform at the d-c terminals is comprised of a steady d-c component, represented by the dashed line, onto which is superimposed an a-c ripple component, having a fundamental frequency equal to twice that of the a-c supply. The input line current has a square waveform of amplitude I_d, and the fundamental component of this waveform is in phase with the input voltage.

At *b*, the firing pulses are retarded by an angle of 45° from the "fully advanced" position shown at *a*. Consider the point in time marked t_1; thyristor 2 is carrying the current to the load via transformer secondary winding $S2$, and the voltage across the load is v_{S2}. At time t_2 the polarity of v_{S2} reverses, and the anode voltage of thyristor 1 now becomes positive. Since thyristor 1 does not yet have a firing pulse applied, however, it blocks this forward anode voltage, and the load current continues to flow through thyristor 2. The voltage at the d-c terminals is still equal to v_{S2}, and therefore, for the time being, it *reverses its polarity*. It is important to understand that the voltage at the d-c terminals is free to reverse, so long as the current continues to flow. At time t_3, a firing pulse is applied to thyristor 1. This thyristor is already blocking forward voltage, and therefore it now switches on; this has the effect of applying the total transformer secondary voltage, in the reverse direction, across thyristor 2, as a result of which this latter thyristor is commutated off. Thus, the flow of load current is now taken over by thyristor 1, via the transformer winding $S1$, and the d-c terminal voltage now becomes positive once more, being equal to v_{S1}. Thus, as is always the case in this type of converter circuit, the commutation of current from one thyristor to the next is brought about "naturally," without the need for any additional commutating means, by virtue of the polarity of the a-c input voltage, at the instant of commutation, being in the correct direction to commutate the current away from the outgoing thyristor.

Comparing the d-c terminal voltage waveform at *b* with that at *a*, it is seen that the effect of retarding the firing pulses has been to withhold a portion of positive voltage from the d-c terminals, and, in this case, a portion of negative voltage has been substituted; it is clear therefore that the mean component of the d-c terminal voltage has been reduced. It is also evident, from an inspection of this waveform, that the ripple voltage component appearing at the d-c terminals has been increased.

If it is assumed that the direct current I_d is the same at *a* and *b*, it is clear that at *b* the power in the d-c circuit, which is equal to the product of the mean d-c terminal voltage and the direct current I_d, has been reduced. Since the converter losses are assumed to be zero, the mean power at the a-c side of the circuit must also have been reduced. This can be seen to be the case, since although the a-c line current still has the same amplitude and waveshape, the phase of the fundamental component of this waveform now lags the voltage by an angle of 45°.

At *c*, the firing pulses are retarded by an angle of 90° with respect to the "fully advanced" position at *a*. In this case, the d-c terminal voltage consists entirely of a-c ripple components, and the mean component of this voltage is zero. The d-c power, and the mean a-c power, must therefore also be zero. This is confirmed by the fact that the fundamental component of the input current waveform now lags the voltage by an angle of exactly 90°.

At *d*, the firing pulses are retarded by an angle of 135°. The d-c terminal voltage waveform now contains a mean *negative* component, represented by the dashed line, and the fundamental component of the a-c line current waveform lags the voltage by an angle of 135°. Since the mean d-c terminal voltage is negative, the d-c power, and hence also the mean a-c power, must also be negative. In other words, power is now being delivered from the d-c side of the converter to the a-c side, and the converter is operating as a "line commutated" inverter. This is confirmed by the fact that the a-c line current is displaced from the a-c voltage by an angle greater than 90°, which indicates that a mean component of power is being delivered to the a-c side of the circuit.

In order to achieve this situation in practice, it is necessary for a source of voltage to be present in the d-c circuit, as shown in Fig. 3.3. It is this source of external voltage which drives the direct current into the converter against the "counter" voltage produced at its d-c terminals. In the ideal case, the amplitude of the voltage E would be exactly equal to the mean "counter" voltage at the d-c terminals of the converter; in practice, a small

30 PHASE-CONTROLLED CONVERTERS

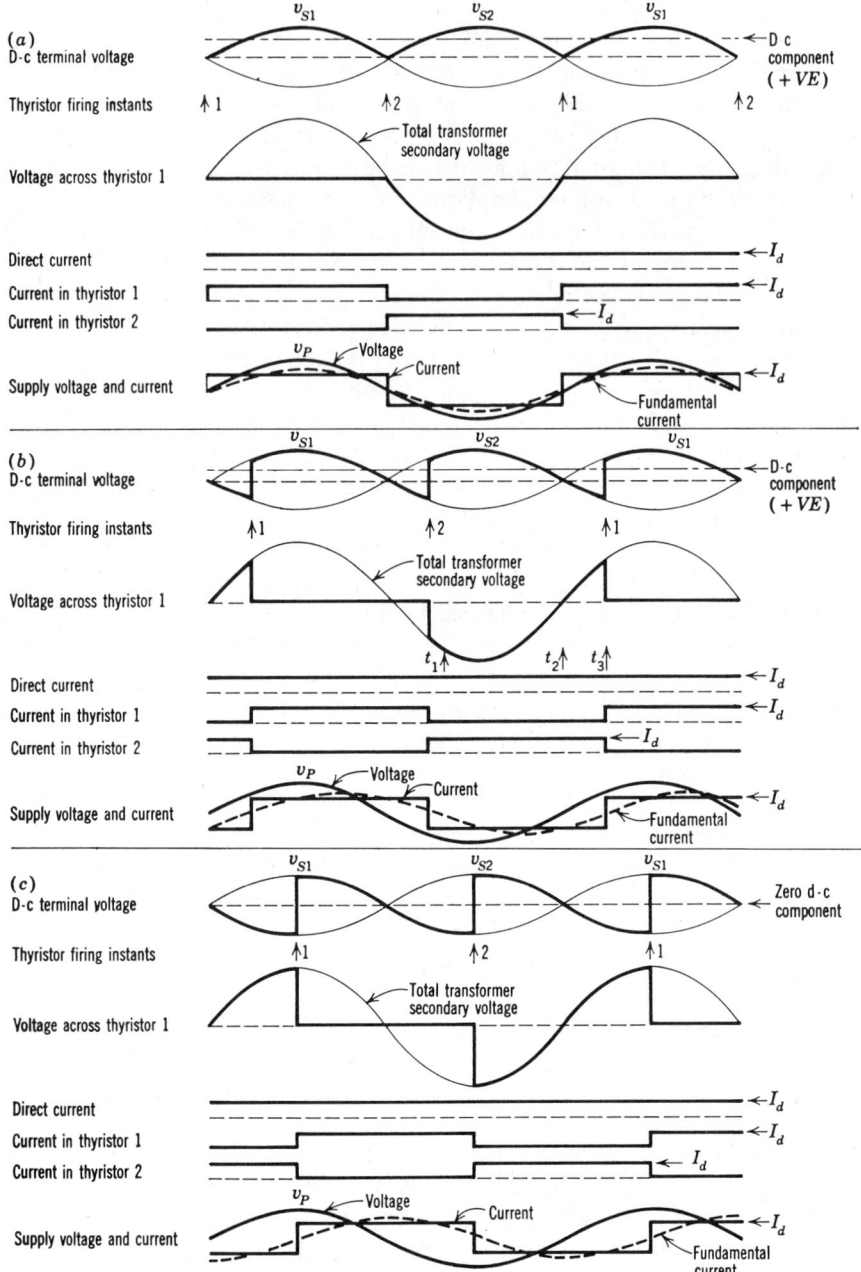

Figure 3.2. Theoretical waveforms associated with the 2-pulse midpoint converter. (a) $\alpha = 0°$; (b) $\alpha = 45°$; (c) $\alpha = 90°$; (d) $\alpha = 135°$; (e) $\alpha \approx 180°$.

TWO-QUADRANT CONVERTERS

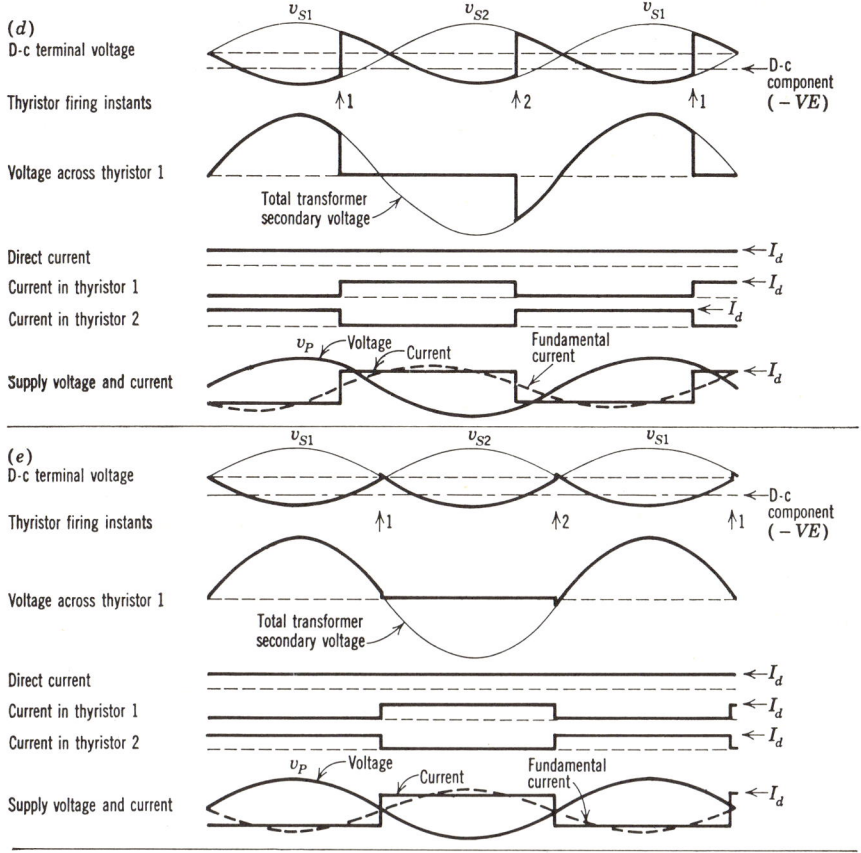

Figure 3.2. Continued.

difference exists between these voltages, which is just sufficient to drive the direct current I_d through the series resistance of the circuit.

Finally, at *e*, the firing pulses are retarded by an angle of almost 180° with respect to *a*. The voltage waveform at the d-c terminals is now almost entirely in the negative direction, and the mean component of this "counter" voltage is virtually equal and opposite to that at *a*. Assuming the direct current is the same in both cases, it is clear that the d-c power at *e* is almost equal, but in the opposite sense, to that at *a*. The mean a-c power therefore also must be almost equal and opposite to that at *a*, and this is confirmed by the fact that whilst the a-c line current has the same waveshape, its fundamental component now lags the voltage by an angle of nearly 180°.

The reason that the commutation instant in Fig. 3.2*e* is shown to be slightly in advance of 180° is a practical one; if the firing pulses were placed at

32   PHASE-CONTROLLED CONVERTERS

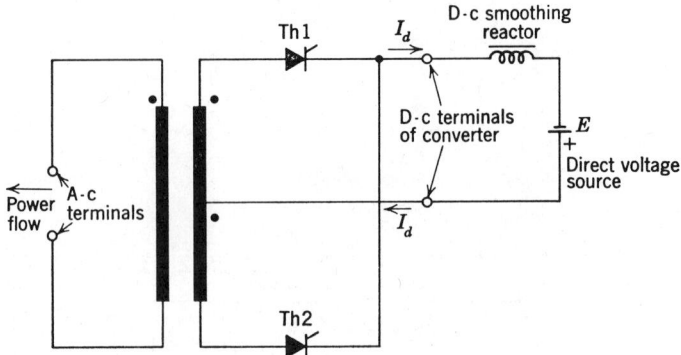

Figure 3.3. Two-pulse midpoint converter operating as a "line-commutated" inverter.

exactly 180°, then the voltage available for commutation would be exactly zero. Moreover, the thyristor anode voltage becomes positive immediately after this point, and therefore it would no longer be possible to achieve a natural commutation. In practice, therefore, the firing pulses are not permitted to come too close to this limiting point. The practical boundary of the firing angle control range depends upon several factors, but typically with a 60 Hz a-c supply, it might be 15 to 20° in advance of the 180° point.

At this stage the following summary can be made: By controlling the phase of the firing pulses applied to the gates of the thyristors through a theoretical range of 180°, the mean component of the d-c terminal voltage can be continuously controlled from maximum positive to maximum negative, assuming that a continuous flow of current is maintained at the d-c terminals. The current flowing at the d-c terminals is inherently unidirectional, but the power flow through the converter can be in either direction, by virtue of the fact that the mean d-c terminal voltage is reversible. Thus the d-c terminal voltage of the converter may be either a positive "driving" voltage, or a negative "counter" voltage. In this latter case, the "counter" voltage of the converter just balances an equal and opposite "driving" voltage connected in the external d-c circuit.

The Two-Pulse Bridge Converter

An alternative circuit arrangement of a 2-quadrant converter, operating from a single phase supply, is the full wave bridge circuit shown in Fig. 3.4. The operation of this circuit is in principle similar to that of the 2-pulse midpoint circuit of Fig. 3.1.

In the bridge circuit, diagonally opposite pairs of thyristors are made to conduct, and are commutated, simultaneously. Once again, it is possible to

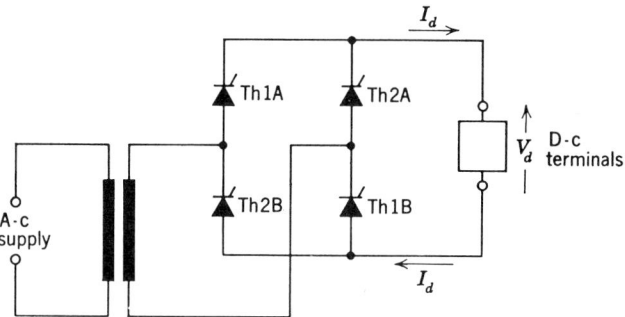

Figure 3.4. Two-pulse 2-quadrant bridge converter.

control the mean voltage at the d-c terminals, from maximum positive to maximum negative, by controlling the phase of the firing pulses through a theoretical range of 180°.

For a given peak thyristor anode voltage, the maximum mean voltage at the d-c terminals of the bridge circuit is twice that of the midpoint circuit; and whereas a transformer is an integral part of the latter circuit, the bridge does not require a transformer.

Definition of Converter Firing Angle

At this stage, it is convenient to define what is meant by the *converter firing angle;* this is the angle, measured with respect to a given reference point, at which the firing pulses are applied to the thyristor gates. The reference point is the point at which application of the gate pulses results in the maximum mean positive d-c terminal voltage of which the converter is capable. In other words, a firing angle of 0° corresponds to the condition when each thyristor in the circuit is fired at the instant its anode voltage first becomes positive in each cycle; under this condition, therefore, the converter operates in exactly the same manner as if it was an uncontrolled rectifier circuit. Thus the firing angles illustrated in Fig. 3.1*a* through *e* are respectively 0, 45, 90, 135, and 180° (very nearly). The symbol for the firing angle is α.

This definition is applicable to all phase-controlled converter and cycloconverter circuits.

Converters with Three-Phase Input

The converter operating from a single phase supply produces a relatively high proportion of a-c ripple voltage at its d-c terminals. This ripple is generally undesirable, and may necessitate the use of quite large filter components in the d-c circuit. Moreover, the use of a single phase supply is frequently unsuitable for applications involving power levels in excess of a few tens of kilowatts.

By using a 3-phase input supply, and by means of appropriate transformer and converter circuit connections, the "pulse number," that is, the number of discrete segments of the d-c terminal voltage waveform which are fabricated during each cycle of the a-c input voltage, can be increased to any desired extent. The higher is the pulse number of the circuit, the more "naturally perfect" is the a-c–d-c conversion process. This is because the relative amount of the a-c ripple voltage at the d-c terminals, as well as the relative distortion of the current at the a-c input terminals, both decrease with increasing pulse number.

The Three-Pulse Midpoint Converter

The simplest type of phase-controlled converter which operates from a three-phase supply is the 3-pulse midpoint circuit, shown in Fig. 3.5.

Although this simple "commutating group" by itself does not have wide application, nevertheless, as will be seen, it is the basic "building block" of almost all practical "multipulse" phase-controlled converter and cycloconverter circuits. The operation of this simple circuit therefore deserves some consideration.

Typical waveforms associated with the 3-pulse converter, assuming a

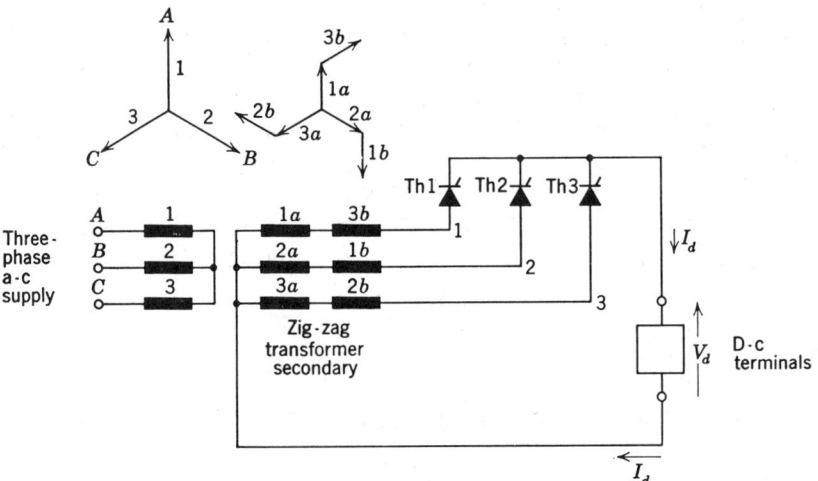

Figure 3.5. Three-pulse midpoint converter.

perfectly smooth direct current, are illustrated in Fig. 3.6a through e. Consider firstly the waveforms shown at a. Here the firing angle is 0°, and each thyristor is fired at the instant its anode voltage first becomes positive. Thus the a-c input voltage with the highest instantaneous potential is always the one which is connected to the d-c terminals, and the mean d-c terminal voltage has its maximum positive value. Each thyristor conducts in turn for a period of 120° of the input a-c supply, and the fundamental frequency of the ripple voltage at the d-c terminals is $3X$ the input frequency; that is, the pulse number of the circuit is 3. At this firing angle, the thyristors each block reverse voltage for a period of 240°; but they do not block forward voltage at any time. The blocking voltage developed across each thyristor is equal to one, or other, of the appropriate line to line voltages at the input terminals; thus, considering thyristor 1, this blocks the 1 to 2 line voltage whilst thyristor 2 is conducting, and the 1 to 3 line voltage whilst thyristor 3 is conducting. This, of course, is generally true, regardless of the firing angle.

The current waveform in each transformer secondary winding, which is the same as the associated thyristor current waveform, consists of a unidirectional or "half wave" rectangular "block," having a duration of 120°. This waveform contains a direct component having an amplitude of $I_d/3$. If a simple wye-connected transformer secondary is used to carry this current, then a d-c magnetization of the transformer core results. Whilst this sometimes can be tolerated, it is common practice to eliminate the d-c magnetization, by means of the secondary "zigzag" connection shown in Fig. 3.5.

Examining the transformer primary voltage and current waveforms in Fig. 3.6a, it is seen that the fundamental component of the current is in phase with the voltage; this is the same situation, of course, as with the single phase converter, with a firing angle of 0°.

At b, the firing angle is delayed by 45°, and each thyristor blocks forward voltage for a 45° period, prior to the instant at which it is fired. It is clear that the mean voltage at the d-c terminals has been reduced, and the fundamental component of the a-c input current waveform now lags the voltage by 45°.

At c, the firing angle is 90°, and each thyristor now blocks forward and reverse voltage for equal periods of time. The mean voltage at the d-c terminals is zero, and the fundamental component of the a-c input current waveform lags the voltage by 90°.

At d, the firing angle is 135°, and the blocking voltage waveform developed across each thyristor is now predominantly in the forward direction. The mean voltage at the d-c terminals is negative, and this implies that a source of voltage must be present in the d-c circuit, which is driving the current against the "counter" voltage developed at the d-c terminals of the converter.

Finally, at e, the firing angle is nearly 180°. The thyristor blocking voltage

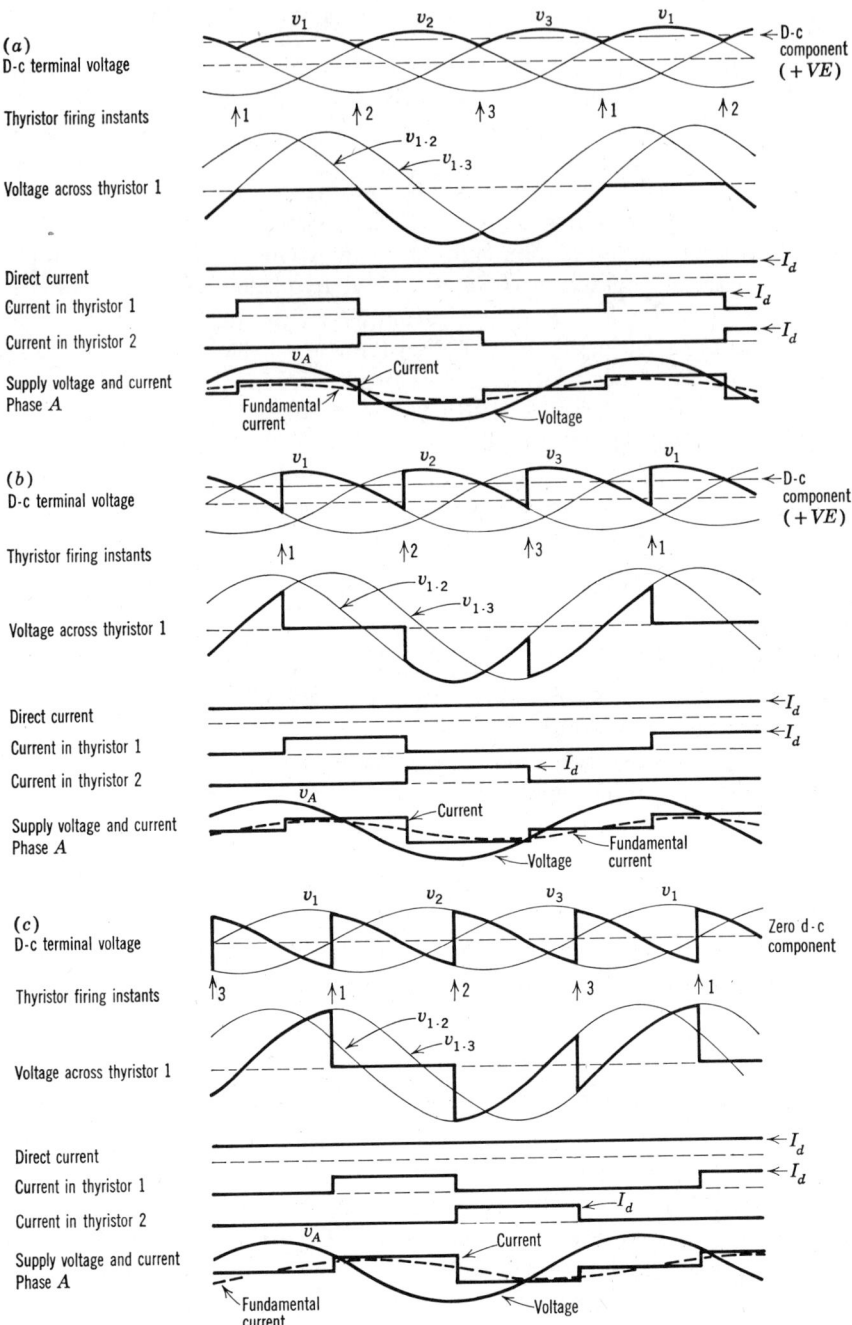

Figure 3.6. Theoretical waveforms associated with the 3-pulse midpoint converter. (a) $\alpha = 0°$; (b) $\alpha = 45°$; (c) $\alpha = 90°$; (d) $\alpha = 135°$; (e) $\alpha \approx 180°$.

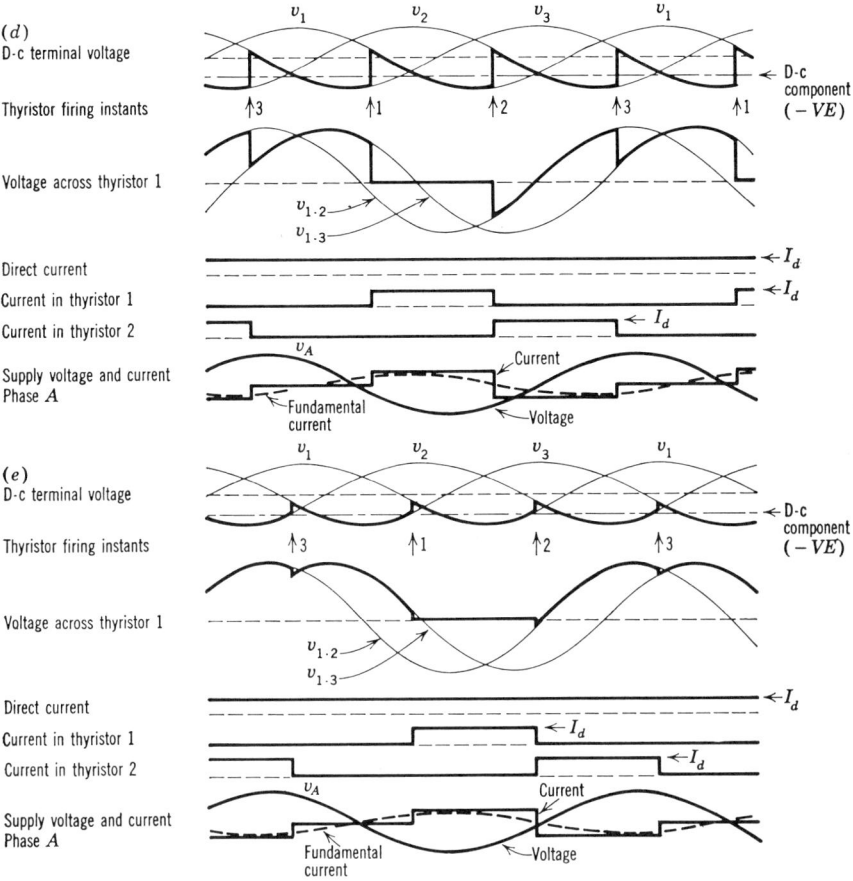

Figure 3.6. Continued.

is now almost entirely in the forward direction, and this voltage swings "negative" only for a short "turnoff" period immediately following the conduction interval. The mean voltage at the d-c terminals has its maximum negative value, which is almost equal and opposite to the maximum positive value obtained at $\alpha = 0°$. The converter now is operating at "full inversion," and the phase displacement between the fundamental component of the a-c input current waveform and the voltage is almost 180°.

From the foregoing description of the operation of the 3-pulse converter, it is evident that this has essentially the same basic operating characteristics as the 2-pulse converter. That is to say, it has the facility for continuous control of the mean d-c terminal voltage from maximum positive to maximum negative; and this is achieved by means of controlling the phase of the thyristor firing pulses through a theoretical range of 180°. Thus, the converter is capable of operation either as a controlled rectifier, or as a controlled inverter.

The Simple Six-Pulse Midpoint Converter

In order to construct a circuit with a higher pulse number than 3, perhaps the most logical extension of the 3-pulse converter would be the 6-pulse circuit shown in Fig. 3.7. Here, 6 alternating voltages, which are phase displaced from one another at 60° intervals, are connected to the anodes of 6 thyristors, the common cathode connection of which constitutes the d-c terminal of the converter.

Although a 6-pulse voltage waveform is obtained at the d-c terminals, this circuit has the practical disadvantage that each thyristor conducts for a period of only 60° during each cycle. This results in a relatively high ratio of rms to average current in the thyristors, as well as in the windings of the transformer. Consequently, the "utilization factor" of the circuit is relatively poor. For this reason, this simple 6-pulse circuit is not often used in practice.

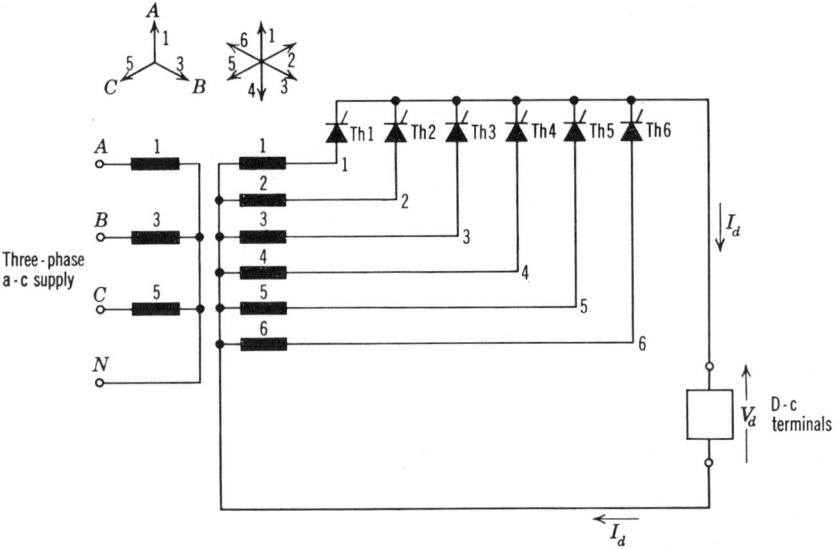

Figure 3.7. Simple 6-pulse midpoint converter.

The Six-Pulse Midpoint Converter with Interphase Reactor

A better utilization of thyristors and transformer windings can be achieved by interposing an interphase reactor between the d-c terminals of two 3-pulse

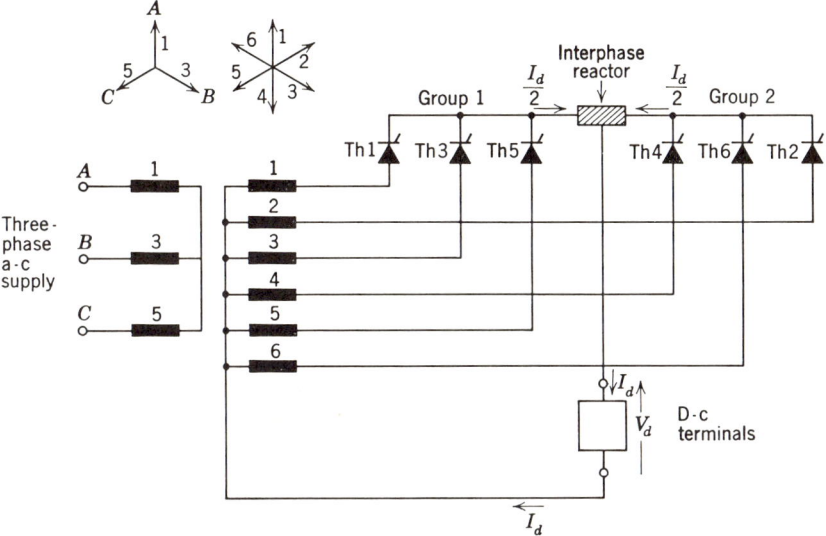

Figure 3.8. Six-pulse midpoint converter with interphase reactor.

circuits, having mutually displaced a-c input voltages, as shown in Fig. 3.8. Typical waveforms associated with this circuit are illustrated in Fig. 3.9.

Each 3-pulse commutating group now operates independently of the other, in the same way as if it were by itself, and therefore each thyristor conducts for a period of 120°. The mean d-c terminal voltages of each group are the same as one another, and since there is theoretically no mean voltage difference across the interphase reactor, the value of the mean d-c terminal voltage of the combination is equal to the mean d-c terminal voltage of either group. In the ideal case, the direct load current is shared equally between the two groups, and therefore there is no d-c magnetization of the core of the interphase reactor. In practice, some current unbalance may exist, but this can be kept relatively small.

Because of the phase displacement between the a-c ripple voltages at the d-c terminals of the individual groups, an a-c ripple voltage having a fundamental frequency of $3X$ the line frequency appears between these two points; it is the function of the interphase reactor to support this alternating voltage, and, in so doing, to maintain independent operation of the two groups.

The instantaneous voltage at the midpoint of the interphase reactor consists of the arithmetic mean of the instantaneous voltages of the individual groups. Due to the phase displacement between the individual group voltages, the resulting ripple voltage waveform at the midpoint of the interphase reactor has a fundamental frequency of $6X$ the input frequency.

40 PHASE-CONTROLLED CONVERTERS

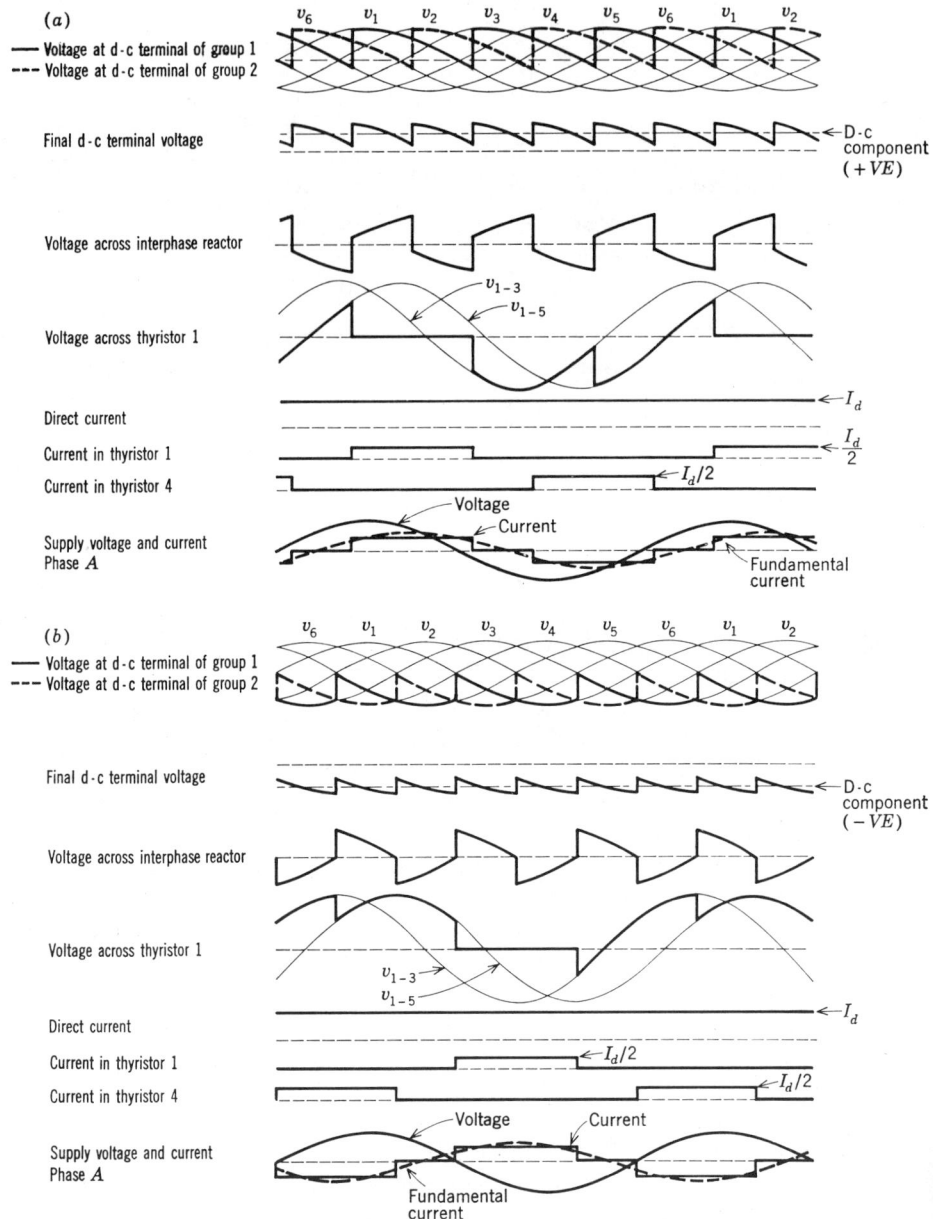

Figure 3.9. Theoretical waveforms associated with the 6-pulse midpoint converter with interphase reactor. (a) $\alpha = 45°$; (b) $\alpha = 150°$.

TWO-QUADRANT CONVERTERS 41

It should be noted that the mode of operation described is achieved only so long as the direct load current in the interphase reactor exceeds the peak exciting current of the reactor, due to the ripple voltage across it; in other words, so long as this exciting current is free to flow in both directions. The peak exciting current of the interphase reactor might typically be about 1 or 2% of the full load direct current, and therefore independent operation of the two groups is maintained over the major portion of the load range. At light load, the operation of the circuit reverts to a more complex "intermittent conduction" mode, and at no load, it becomes essentially that of the simple 6-pulse circuit of Fig. 3.7.

Since the firing angle control range for the 6-pulse converter with interphase reactor is 30° advanced with respect to the firing angle control range for the simple 6-pulse converter, this means, with a fixed phase position of the firing pulses, that the firing angle effectively advances by 30° between "light load" and no load. Thus, depending upon the particular operating point in the control range, a substantial increase in the d-c terminal voltage may be obtained, if the phase position of the firing pulses remains unaltered as the load is removed. Even if a 30° adjustment of firing angle is made, to compensate against this effect, the mean value of the d-c terminal voltage still rises by about 15% between "light load" and no load. This is because the voltage at the d-c terminals follows the actual input voltage waves at no load, whereas it follows the mean voltage waves resulting from two 60° displaced input waves at increased load, and the amplitudes of these latter waves are only 0.866 of the input waves. Thus, in order to maintain a constant d-c terminal voltage between "light load" and no load, a retardation of the firing angle by slightly more than 30° is required.

Although the individual input line current waveforms consist of unidirectional 120° rectangular "blocks," there is no d-c magnetization of the transformer core, because the d-c mmf's of the 2 windings on each phase of the transformer cancel one another.

The Twelve-Pulse Midpoint Converter with Interphase Reactors

It has just been seen that the d-c terminals of 2 individual 3-pulse groups, having mutually displaced a-c voltages, can be connected in parallel with one another through an interphase reactor, the object being to increase the overall circuit pulse number from 3 to 6, but at the same time maintaining a 120° thyristor conduction angle. This same principle, of connecting individual circuit groups in parallel through an interphase reactor, can be further applied to produce circuits with higher pulse numbers, but still with individual thyristor conduction angles of 120°. This, of course, is desirable from the

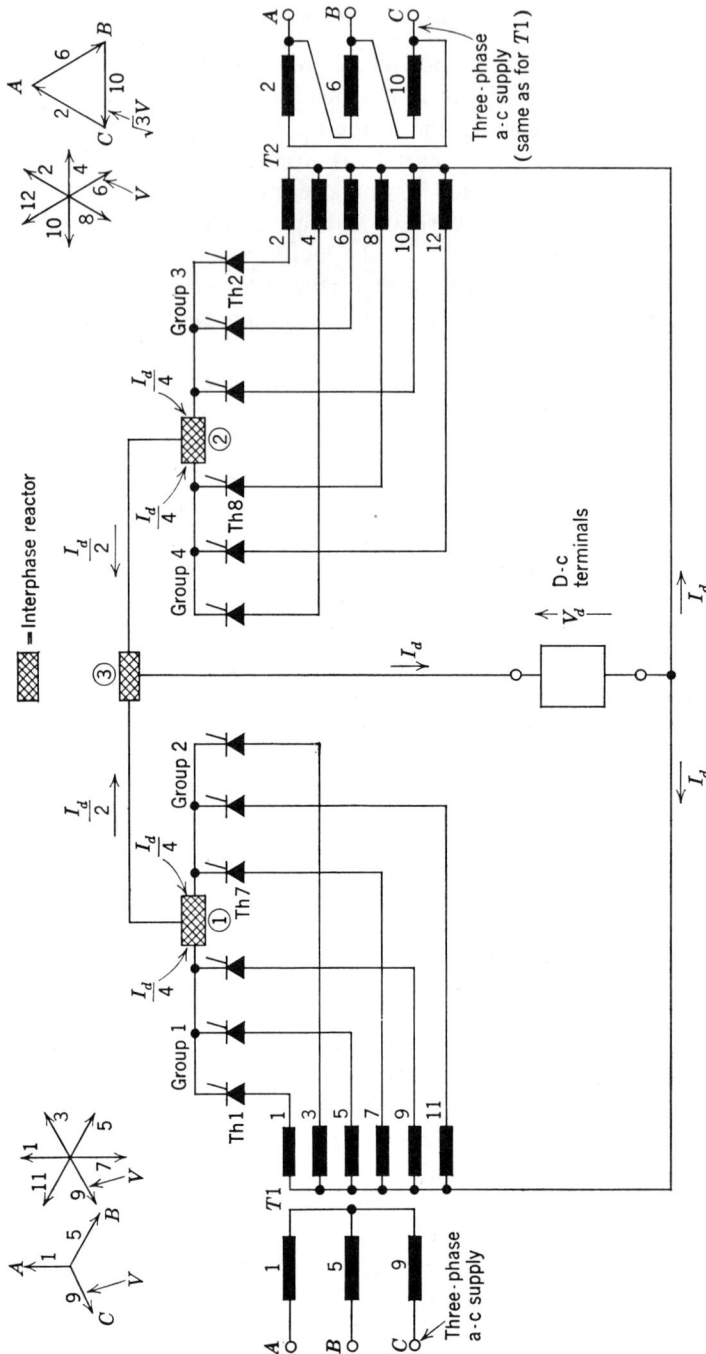

Figure 3.10. Twelve-pulse midpoint converter with interphase reactors.

TWO-QUADRANT CONVERTERS 43

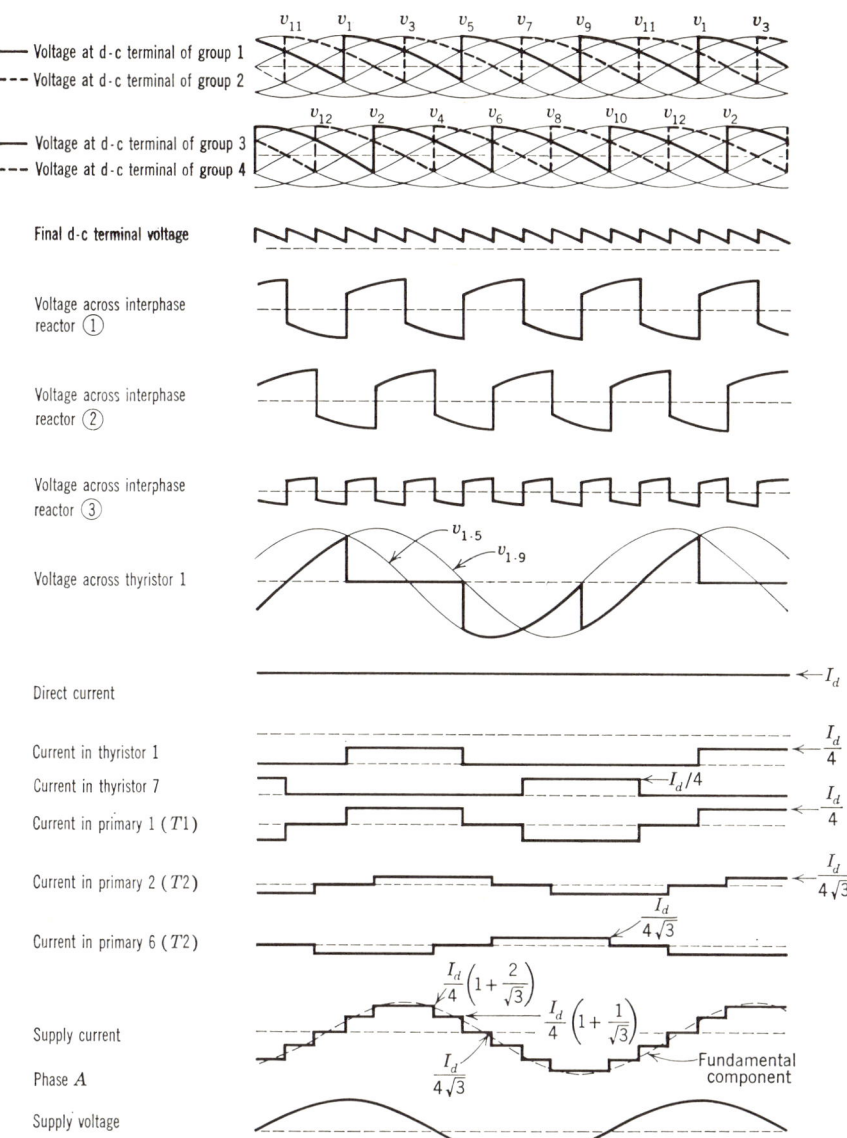

Figure 3.11. Theoretical waveforms associated with the 12-pulse midpoint converter with interphase reactors. $\alpha = 60°$.

viewpoint of maintaining a favorable form factor for the thyristor and a-c input current waveforms.

The next logical extension of the principle results in the 12-pulse midpoint circuit with interphase reactors, shown in Fig. 3.10. The circuit consists of two 6-pulse converters, the a-c voltages connected to the individual groups being symmetrically displaced with respect to one another. Each 6-pulse circuit has its own interphase reactor, and the d-c terminals of each of these 6-pulse groups are connected in parallel with one another, through a third interphase reactor, the function of which is to support the ripple voltages existing between the two 6-pulse groups, thus permitting these to operate independently of one another.

Typical waveforms associated with this circuit are shown in Fig. 3.11. The "final" d-c terminal voltage waveform is the arithmetic mean of the individual d-c terminal voltage waveforms of each of the 4 basic 3-pulse commutating groups, and it has a fundamental ripple frequency of $12X$ the input frequency. It should be noted, too, that the primary line current waveform now has a "staircase" appearance, the general outline of which approximates fairly closely to a sinusoid. Comparing this waveform, for example, with the primary alternating current waveform of the 6-pulse midpoint converter, shown in Fig. 3.9, it is clear that the effect of increasing the circuit pulse number has been to decrease the distortion of the a-c input current wave.

The Six-Pulse Bridge Converter

It has been seen that the d-c terminals of two 3-pulse groups can be connected in parallel with one another, to give a circuit with an overall 6-pulse operation. By the same token, it is also possible to connect the d-c terminals of two 3-pulse groups in series with one another, to give an overall 6-pulse operation, and in this case the circuit formed is known as a bridge connection.

In Fig. 3.12a, a 3-pulse group is shown which provides a positive d-c terminal voltage with respect to the neutral connection at $\alpha = 0°$. At b, a second 3-pulse group is shown, which, by virtue of the "reversed" connection of the thyristors, provides a negative voltage with respect to the neutral connection at $\alpha = 0°$. It is clear that if the direct currents of the "positive" and "negative" groups are equal to one another, then the net current flowing in the neutral connection is zero, and this connection could be removed, without changing the operation of the circuit. Conversely, if the neutral is omitted, then the two groups are forced to share a common d-c load, and clearly the direct currents of each must be the same. The resulting

TWO-QUADRANT CONVERTERS 45

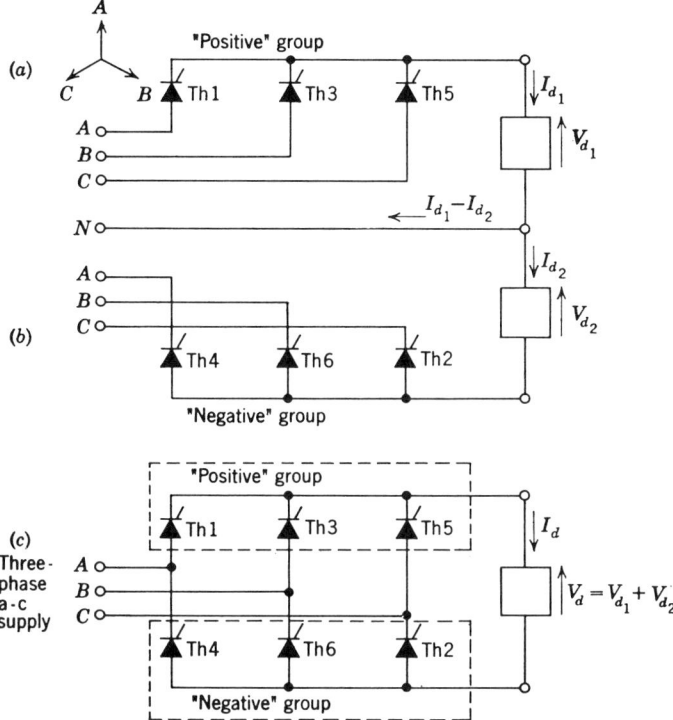

Figure 3.12. Construction of a 6-pulse bridge circuit from two 3-pulse midpoint circuits. (*a*) Three-pulse midpoint converter giving positive d-c terminal voltage at $\alpha = 0°$. (*b*) Three-pulse midpoint converter giving negative d-c terminal voltage at $\alpha = 0°$. (*c*) Six-pulse bridge converter.

bridge circuit is shown at *c*. Typical waveforms associated with this circuit are shown in Fig. 3.13.

Since the bridge circuit is comprised of one "positive" and one "negative" 3-pulse group, the d-c terminals of which are connected in series with one another, it is evident that the mean voltage appearing across the d-c terminals of the bridge is twice the mean d-c terminal voltage of either 3-pulse group on its own. Moreover, since the two groups work with portions of the a-c voltage waves which are displaced by 180°, the waveform of the current in the a-c lines connected to the converter now consists of symmetrically displaced positive and negative rectangular "blocks," each of 120° duration, and there is no net d-c component. For the same reason, the a-c ripple voltage appearing across the d-c terminals of the bridge has a fundamental frequency of $6X$ the input frequency.

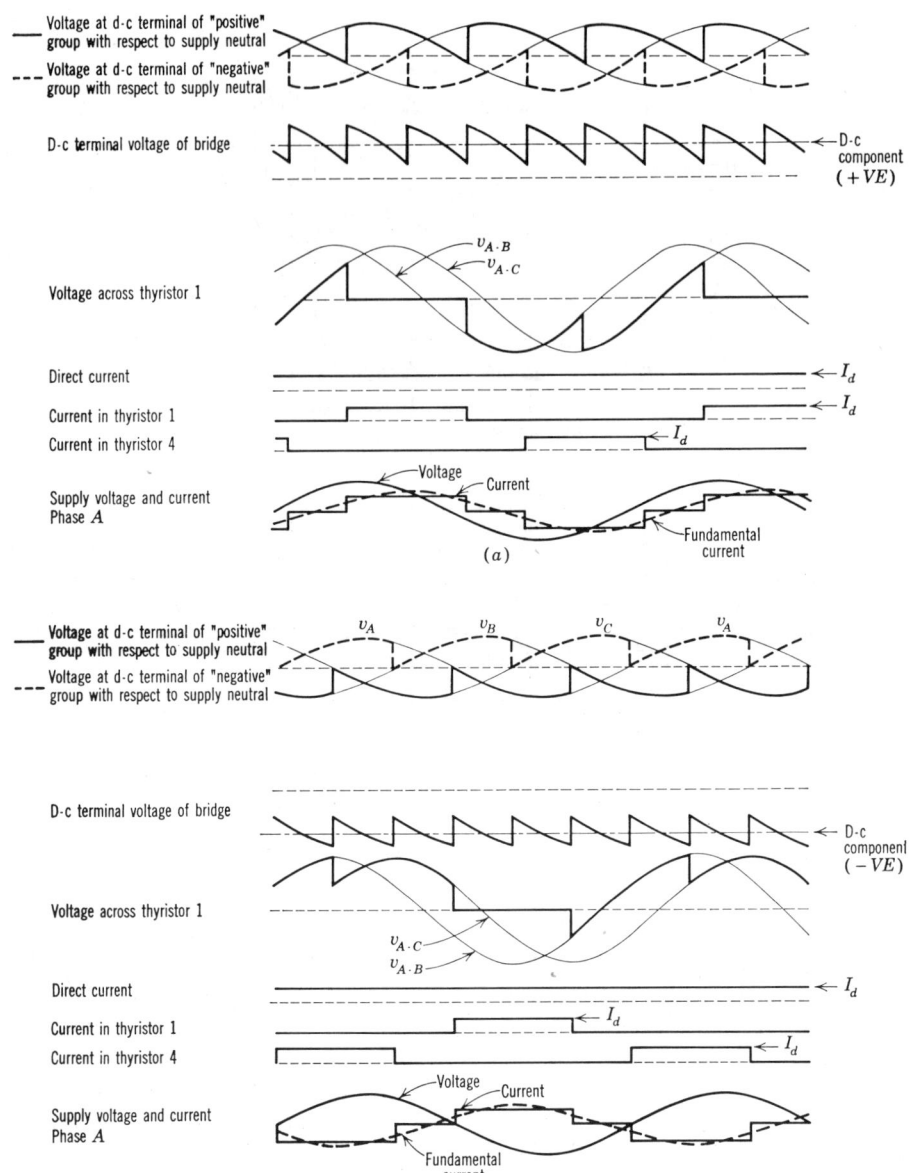

Figure 3.13. Theoretical waveforms associated with the 6-pulse bridge converter. (a) $\alpha = 45°$; and (b) $\alpha = 150°$.

TWO-QUADRANT CONVERTERS 47

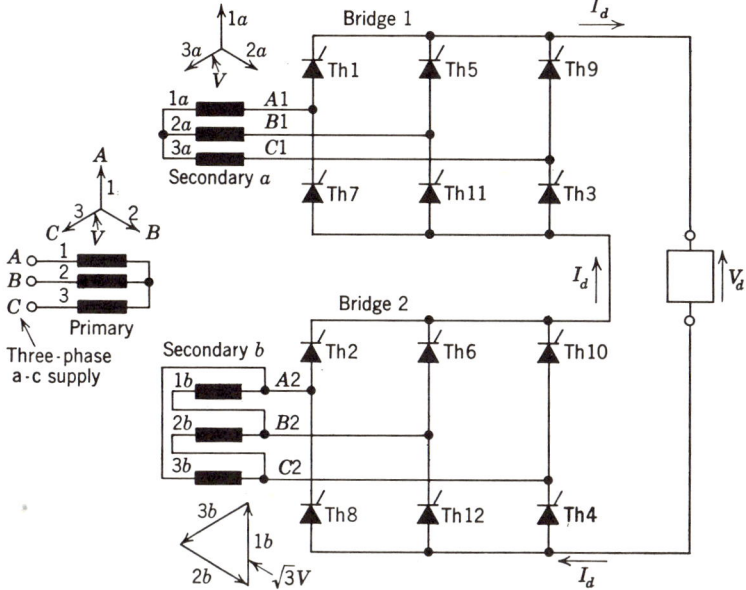

Figure 3.14. Two 6-pulse bridge converter circuits connected in series, to form a 12-pulse bridge converter.

The Twelve-Pulse Bridge Converter

In the same way as it is possible to connect the d-c terminals of two 3-pulse groups in series with one another, to provide a bridge circuit with an overall 6-pulse operation, it is also possible to connect the d-c terminals of two 6-pulse bridge circuits, having appropriately displaced and isolated a-c input voltages, in series with one another, to provide a circuit with an overall 12-pulse operation. The resulting circuit is shown in Fig. 3.14.

The required 30° phase shift, and isolation, between the 2 sets of 3-phase voltages connected to the individual bridges, is achieved by means of the isolated wye and delta connected transformer secondary windings. Typical waveforms associated with this circuit are shown in Fig. 3.15.

Other "Multipulse" Converter Connections

The various converter circuit connections described in the preceding sections do not represent all the practical possibilities. Other circuit configurations can be constructed, by application of the basic principle of connecting the d-c terminals of individual groups, with suitably phase-displaced a-c

48   PHASE-CONTROLLED CONVERTERS

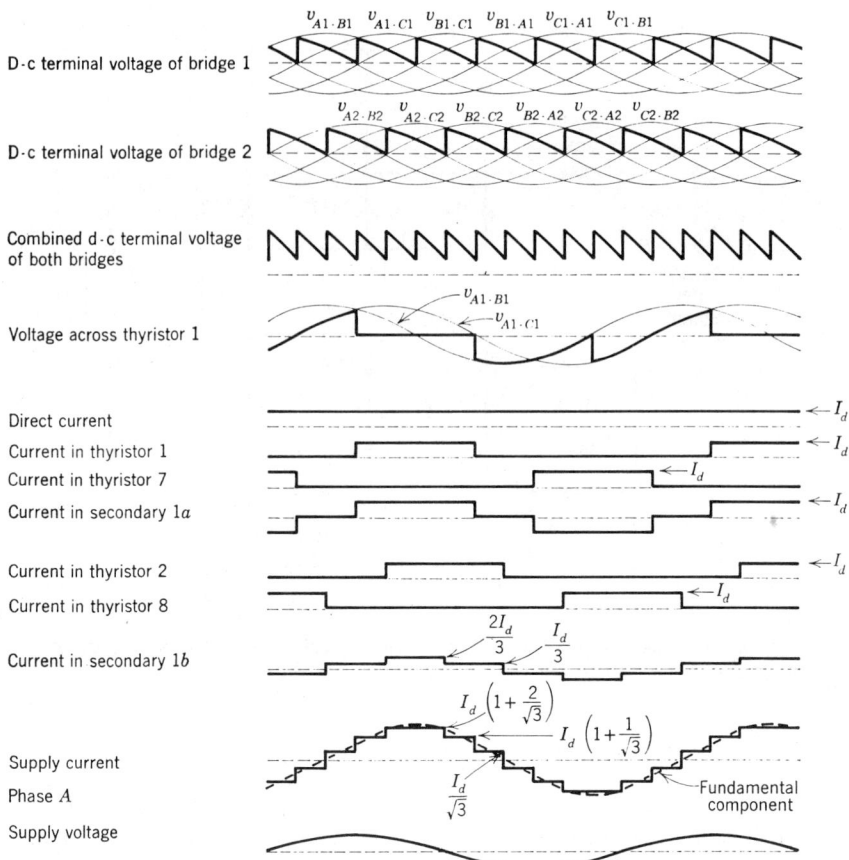

Figure 3.15. Theoretical waveforms associated with the 12-pulse bridge converter of Fig. 3.14. $\alpha = 60°$.

input voltages, in series or parallel with one another, or by combining together series and parallel connections into one system.

In practice, for a "multipulse" converter, a thyristor conduction angle of 120° is almost invariably preferred. In other words, almost all practical multipulse converter circuits are comprised of combinations of the basic 3-pulse commutating group. Each group within the system operates essentially independently of all the other groups, as if it were by itself. Where it is required to connect the d-c terminals of individual groups in parallel with one another, these connections are made through interphase reactors, in order to maintain independent operation of the groups. Connections of groups in series, on the other hand, can be made with "solid" connections at the d-c side. Series connections of *bridges*, however, require isolation

OPERATION OF THE TWO-QUADRANT CONVERTER WITH DISCONTINUOUS CURRENT

It has been assumed so far that the current at the d-c terminals of the converter is continuous and perfectly smooth. This assumption implies that the a-c ripple currents in the d-c circuit are negligible in comparison with the direct component. In practice, of course, this assumption is never entirely valid; and in fact, it may often be the case that the magnitude of the a-c ripple current is comparable with the direct component. However, so long as the peak instantaneous value of the a-c ripple current is less than the direct component, so that the net current is always greater than zero, and thus the converter is kept in continuous conduction, then the d-c terminal voltage waveform at any given firing angle is theoretically the same as with a perfectly smooth current.

On the other hand, if the peak instantaneous "negative" value of the a-c ripple current should tend to exceed the d-c component, then since the net current flow cannot reverse, the direct current waveform becomes discontinuous, and the d-c terminal voltage waveform departs from that obtained with continuous conduction. Moreover, since the voltage waveform is changed, the mean value of this waveform is also changed, and the relationship between the firing angle and the mean d-c terminal voltage is not the same as with continuous conduction.

In practice, the continuity of the load current waveform depends upon the nature of the d-c load circuit, and it is possible for discontinuous conduction to occur either in the rectifying or inverting region of operation. If the load at the d-c terminals is purely "passive," and is capable only of absorbing mean power, then the current waveform inevitably becomes discontinuous as the firing angle is retarded towards $90°$. In order to illustrate this, the operation of a 6-pulse converter with a "passive" series inductance-resistance load will be considered. From an examination of the d-c terminal voltage waveform of a 6-pulse converter (Figs. 3.9 and 3.13), it is clear that if the firing angle lies between $0°$ and $60°$, the instantaneous voltage at the d-c terminals is at all times positive, and hence, over this range of firing angles, the current in a passive load of this type must necessarily be continuous. As the firing angle is retarded beyond this point, however, the voltage at the d-c terminals swings into the negative direction for certain periods of time, and it now depends upon the time constant of the load whether or not this negative voltage excursion can be absorbed, whilst a net positive current flow is maintained.

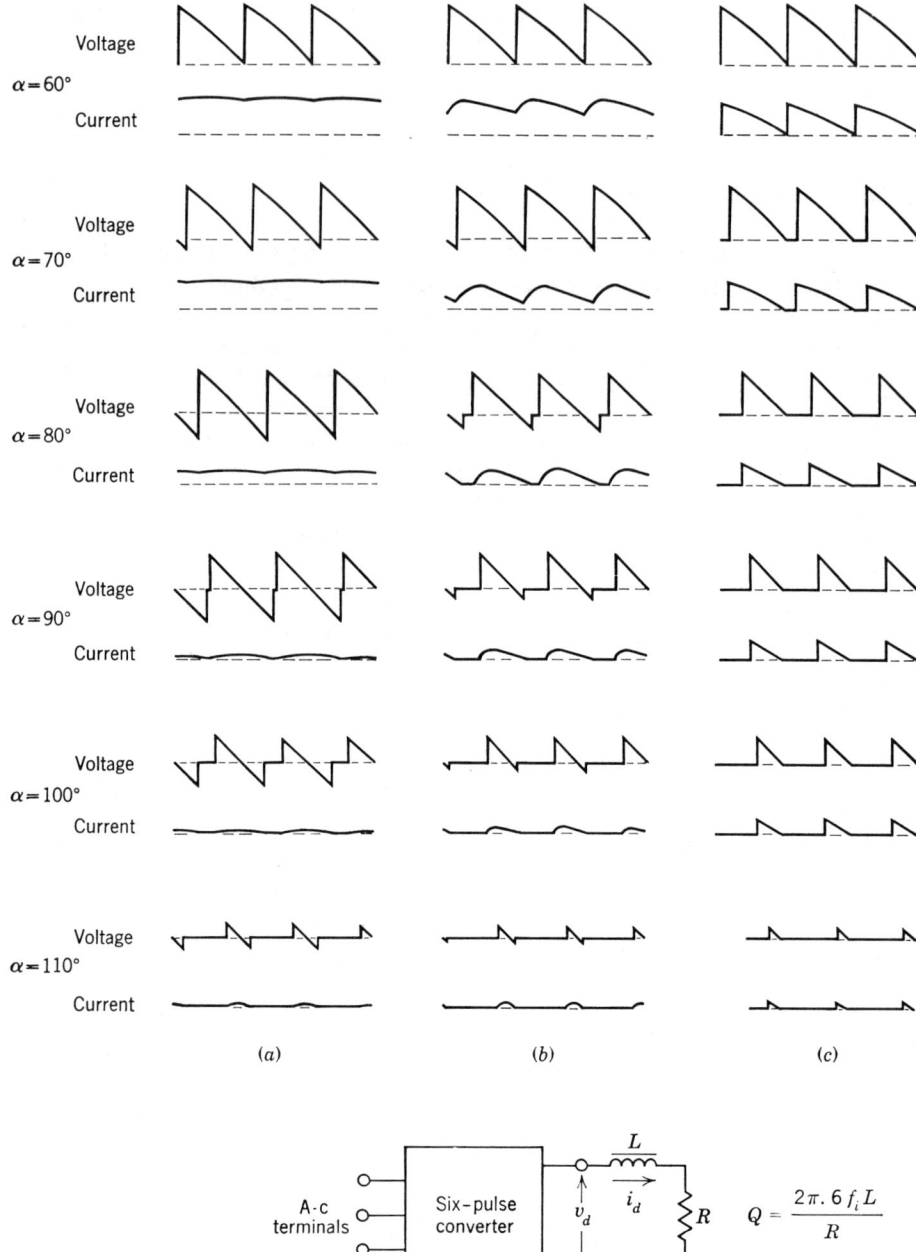

Figure 3.16. D-c terminal voltage and current waveforms obtained with a 6-pulse converter, with a passive series $L - R$ load. (a) $Q \to \infty$; (b) $Q \doteq 3$; and (c) $Q = 0$ (purely resistive load).

OPERATION WITH DISCONTINUOUS CURRENT 51

In Fig. 3.16, typical d-c terminal voltage and current waveforms obtained at various firing angles, for series inductance-resistance loads having various time constants, are shown. At *a*, the load time constant is very large, and in this case the current waveform remains continuous up to a firing angle only slightly in advance of 90°. At $\alpha = 90°$, the current has just become discontinuous, and there is a short period during which there is no voltage or current at the d-c terminals of the converter. It can be seen that the effect of the zero-current period is to "remove" a small area of negative voltage which, with continuous conduction, would otherwise have been present, and hence the mean value of the d-c terminal voltage is not quite zero, but slightly positive, say 0.5% of $V_{d_{max}}$. As the firing angle is retarded beyond 90°, the current waveform becomes more and more discontinuous, and the mean d-c terminal voltage approaches zero more and more closely. Finally, at $\alpha = 120°$, there is no conduction, and the d-c terminal voltage becomes zero. Thus, with this load, within the region of discontinuous conduction, a relatively large movement of the converter firing angle results in only a very small change in the value of the mean d-c terminal voltage.

Examining now the waveforms shown at *b*, which are appropriate to a load Q factor of about 3 (at $6f_i$), it is seen that the current waveform has just become discontinuous at $\alpha = 80°$. Once again, the d-c terminal voltage becomes zero at $\alpha = 120°$. At any intermediate firing angle, the mean value of the d-c terminal voltage waveform is greater than that obtained with the higher Q load.

Finally, at *c* the waveforms are shown for a purely resistive load. In this case, it is clear that the conduction becomes discontinuous at $\alpha = 60°$, and, once again, the d-c terminal voltage becomes zero at $\alpha = 120°$. At any intermediate firing angle, the mean value of the d-c terminal voltage waveform is greater than at either *a* or *b*.

With loads which have the capability for storing voltage, for example a capacitor, or the armature of a d-c machine with an induced back emf, it is possible to obtain operation with discontinuous current at any point within the operating range, if the average load current demand is sufficiently small. Figure 3.17 shows typical waveforms obtained in a 6-pulse converter operating as a "rectifier" with a load producing a "back emf," for various levels of load current. In this particular example, in order to maintain a constant mean d-c terminal voltage, it is seen that it is necessary to advance the converter firing angle by about 25°, between the no load condition with discontinuous current, and the loaded condition with continuous current. By the same token, it can be seen that if the firing angle is kept constant as the load current is varied, a relatively large reduction of mean d-c terminal voltage occurs between no load and full load.

It is also possible to obtain operation with discontinuous current in the

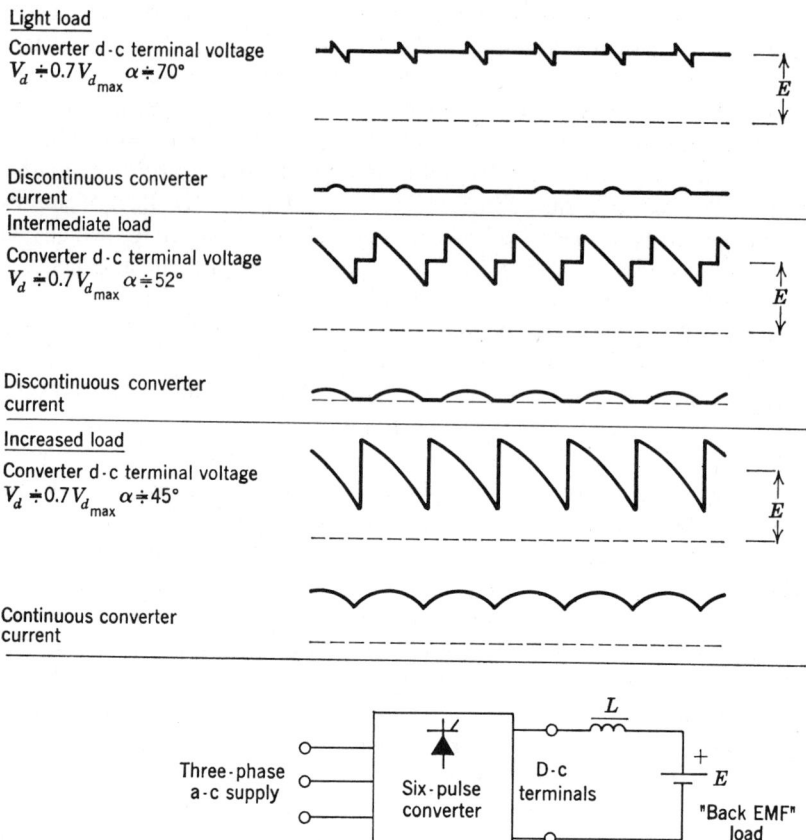

Figure 3.17. D-c terminal voltage and current waveforms obtained in a 6-pulse converter, operating as a rectifier into a "back-EMF" load, with discontinuous conduction at light load, and continuous conduction at increased load.

inverting mode of operation of the converter, if the average load current is sufficiently small. Figure 3.18 shows typical waveforms obtained in a 6-pulse converter, operating in its inverting mode, for various levels of load current. It is seen here that a shift in firing angle of about 20° is required in order to maintain the same mean d-c terminal voltage between operation with discontinuous current at light load, and operation with continuous current at increased load.

The general conclusion to be drawn from the foregoing discussion is that whereas, with continuous conduction, the d-c terminal voltage waveform, and hence also the mean value of this waveform, is clearly defined, and is

OPERATION WITH DISCONTINUOUS CURRENT 53

dependent only upon the converter firing angle, with discontinuous conduction, on the other hand, the d-c terminal voltage waveform, and hence also its mean value, is dependent both upon the firing angle, and upon the load. Thus, at a given firing angle with discontinuous conduction, it is possible for a relatively small change in load current to result in a relatively large change in the mean d-c terminal voltage. Conversely, with a fixed passive load impedance, a relatively large change in firing angle may result in only a small change in the value of the mean d-c terminal voltage.

As will be seen in later chapters, in applications in which two phase-controlled converters are connected back-to-back with one another, the

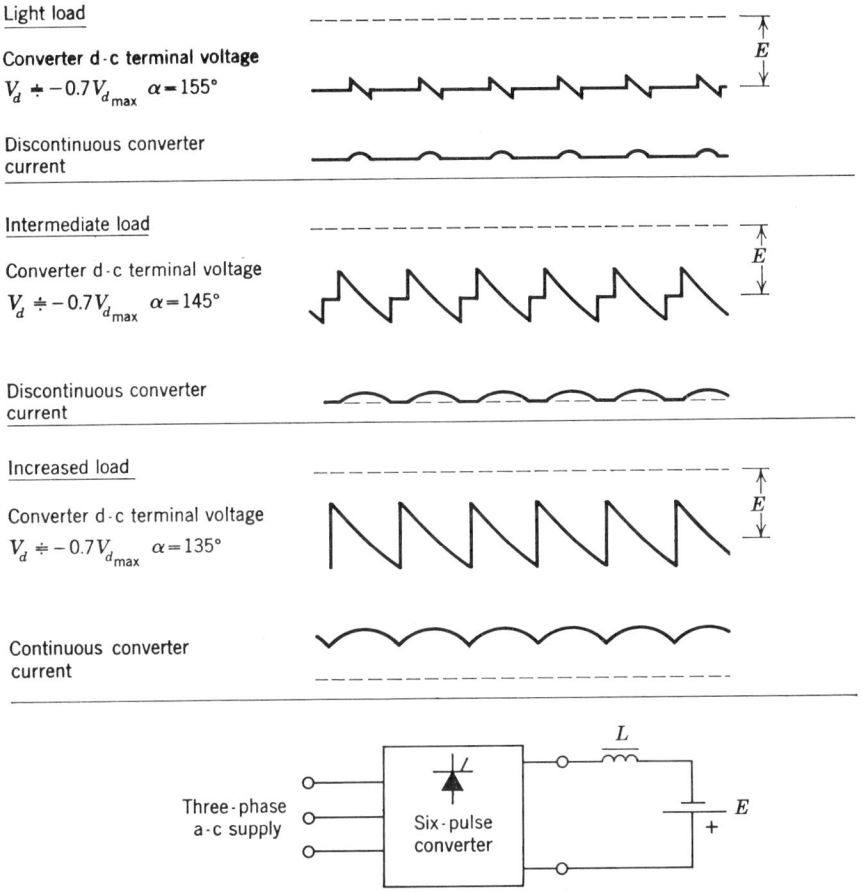

Figure 3.18. D-c terminal voltage and current waveforms obtained in a 6-pulse converter, operating as an inverter, with discontinuous conduction at light load, and continuous conduction at increased load.

phenomena associated with discontinuous conduction may require quite sophisticated control circuitry in order to provide a smooth system operation during transfer of load from one converter to the other.

ONE-QUADRANT CONVERTERS

The phase-controlled converter circuits so far considered are capable of operation with both positive and negative mean voltage at the d-c terminals. Many applications actually require operation only with a positive voltage, that is to say, only in the rectifying mode. In this case, it is generally advantageous to connect uncontrolled diodes into certain parts of the circuit.

In the case of bridge connected circuits, (but not midpoint circuits), uncontrolled diodes, instead of thyristors, can be connected into half the total circuit positions, thus providing a so called "half-controlled" converter. With this type of circuit, it is possible to provide a continuous control of the mean d-c terminal voltage, from maximum to virtually zero, but reversal of the mean voltage is not possible. In other words, only a 1-quadrant operation can be obtained.

Obviously, the half-controlled bridge has economic advantages over the fully-controlled circuit, insofar as it uses uncontrolled diodes, as direct replacements for the thyristors, in half the total number of circuit positions. In addition, as will be seen, the circuit operation is such that the "wattless" components of load at the a-c input side of the converter, obtained at relatively low levels of output voltage, are considerably reduced, as compared to those obtained with the fully-controlled converter. In the case of a "multipulse" converter, however, this advantage is obtained at the expense of a 2 : 1 reduction of ripple frequency at the d-c terminals.

An artifice for obtaining an improvement over the performance characteristics of the 2-quadrant converter, and which, at the same time, renders the circuit capable only of a 1-quadrant operation, is to connect a so-called "freewheel" diode directly across the d-c terminals of the converter.

Connection of a freewheel diode across the d-c terminals is equally applicable to both bridge and midpoint circuits. Its presence has two effects upon the external performance characteristics. First, because it prevents the voltage at the d-c terminals from swinging into the negative direction, it results in a reduction of ripple voltage at the d-c terminals, at reduced levels of output voltage. And second, it results in a reduction of "wattless" power at the a-c side of the converter, at reduced output voltage.

Thus, unlike the 2-quadrant converter, in which the amplitude of the input current theoretically always bears the same relationship to the current at the d-c terminals, regardless of the level of output voltage, both the half controlled converter, and the converter with freewheel diode, have the

ONE-QUADRANT CONVERTERS

common feature that reduction of the output voltage towards zero is accompanied by a corresponding reduction of the ratio between the input current and the direct output current.

In the following sections, various alternative arrangements of 1-quadrant converter, together with associated waveform diagrams, are presented.

The Two-Pulse Half-Controlled Bridge Converter

Two alternative arrangements of the 2-pulse half-controlled bridge converter are shown in Fig. 3.19.

Consider first, the circuit shown at *a*. This can be regarded as comprising two separate midpoint converter circuits, connected in series with one another. The first midpoint circuit, comprising thyristors 1 and 2, is a "positively

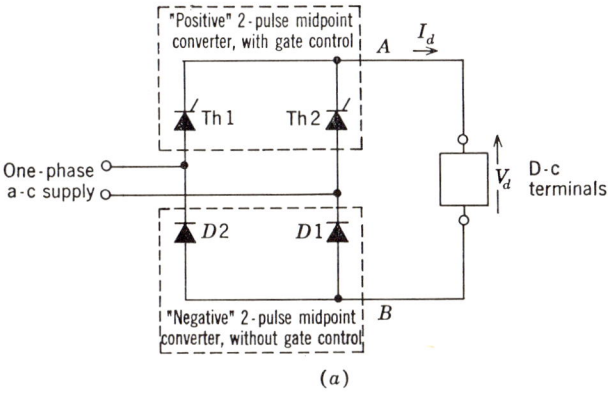

(a)

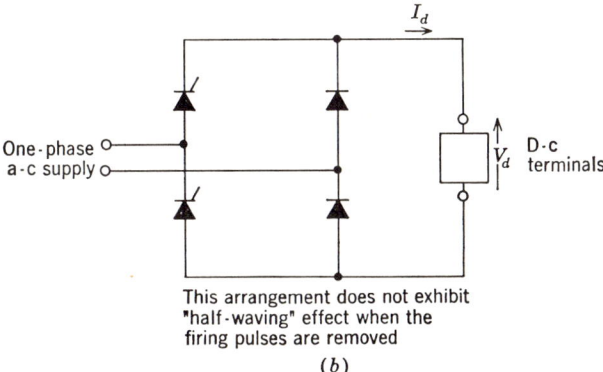

This arrangement does not exhibit "half-waving" effect when the firing pulses are removed

(b)

Figure 3.19. Alternative arrangements of 2-pulse "half-controlled" converters.

poled" 2-pulse 2-quadrant converter, which is capable, through control of its firing angle, of producing a continuously controllable mean voltage of either polarity at its d-c terminal A, with respect to the midpoint potential of the a-c supply. The second midpoint circuit, comprising diodes $D1$ and $D2$, is a "negatively poled," uncontrolled, 2-pulse rectifier, which invariably produces the same maximum mean voltage at its d-c terminal B, which is negative with respect to the supply midpoint potential.

Bearing in mind then, that this half-controlled bridge converter circuit consists essentially of a controlled midpoint converter, connected in series with an uncontrolled midpoint converter, it can be seen that by varying the voltage of the controlled converter from maximum "boosting" to maximum "bucking"—that is to say, from "full rectification" to "full inversion"— so the combined output voltage of the bridge can be controlled from maximum to virtually zero.

The operation of the circuit is illustrated by the waveforms of Fig. 3.20, in which it is assumed that the current at the d-c terminals is perfectly smooth. For each firing angle, the waveforms of the voltage at each of the points A and B, with respect to the midpoint potential of the supply, are shown. The waveform of the voltage appearing across the d-c terminals of the bridge is the difference between these two waves.

At a the firing angle is 0°. In this case, the voltage waves at A and B are exactly equal and opposite to one another, and hence the total voltage across the d-c terminals of the bridge is exactly twice the voltage of either of the individual converters. Under this condition, of course, the operation of the circuit is identical with that of an uncontrolled bridge circuit.

At b, the firing angle is 45°. In this case, for the first 45° period of each input cycle, the voltages at points A and B are exactly equal, and have the same polarity, as one another. Thus the difference between these two voltages is zero, and during this period, the d-c terminal voltage of the bridge is zero. This means to say that the thyristor and diode in the same "leg" of the bridge circuit are simultaneously in conduction. Thus, during this period, the load current merely freewheels through this local path, and there is no current in the input line. At the point at which the firing pulse is applied, the voltage at A reverses, the voltage at the d-c terminals of the bridge now starts to follow the input voltage wave, and the load current starts to flow in the input line. This situation continues until the next input voltage zero is reached, at which point the load current once more commences to "freewheel" through one leg of the bridge, and the current in the input line once again becomes zero.

At c the firing angle is 90°. The mechanism of operation is essentially similar to that already discussed for the 45° firing angle, and requires no further description.

ONE-QUADRANT CONVERTERS

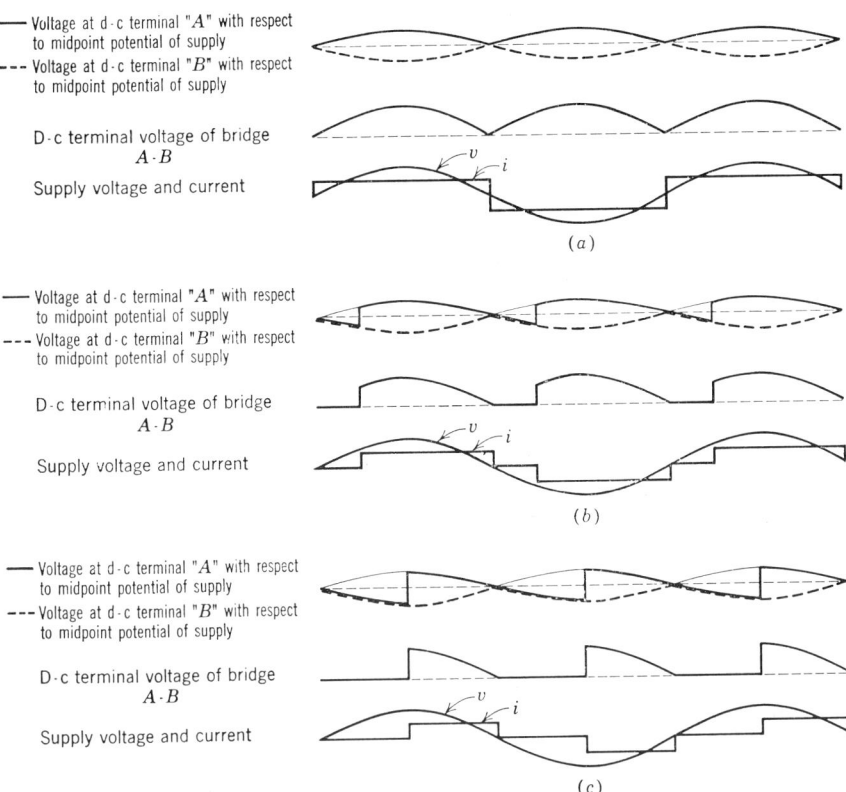

Figure 3.20. Theoretical waveforms associated with the 2-pulse half-controlled converter. (a) $\alpha = 0°$; (b) $\alpha = 45°$; and (c) $\alpha = 90°$.

Since, in practice, it is not permissible to retard the firing angle of a 2-quadrant converter beyond a certain practical limit, which is somewhat in advance of 180°, without the possibility for provoking commutation failures, therefore the maximum possible "bucking" voltage available from the controlled half-bridge is slightly less than the mean voltage of the uncontrolled half bridge. Hence it is not possible, with continuous conduction, to reduce the d-c terminal voltage of the bridge to absolute zero. By the same token, a sudden removal of the firing pulses results in an undesirable "half waving" mode of operation, in which the last thyristor to be fired remains in conduction indefinitely, and every other half-wave of the input voltage appears across the d-c terminals of the bridge.

These difficulties are eliminated in the circuit arrangement shown in Fig. 3.19b. Although the external waveshapes of current and voltage obtained in this circuit are theoretically identical to those of the circuit at *a*, the internal operation of the circuit is different. In this case, the freewheeling current is always carried by the two diodes, instead of by one thyristor and one diode. This means to say that as the firing angle is retarded, so the diode conduction periods, and correspondingly, the thyristor voltage blocking periods, increase their duration. The result is that the period during which reverse voltage is applied across each of the thyristors is invariably 180°, and thus there is no possibility for commutation failures. Thus, with this circuit, the d-c terminal voltage can be safely reduced to zero; and sudden removal of the firing pulses does not result in the undesirable "half-waving" mode of operation associated with the circuit at *a*. In this case, sudden removal of the firing pulses merely results in a natural commutation of the load current into the two diodes, where it continues to "freewheel," for a period depending upon the time constant of the load circuit, eventually decreasing to zero.

In comparing the operation of the half-controlled 2-pulse circuit with that of the fully-controlled circuit, two important points are evident. First, it can be seen that the periods of negative voltage "swing" at the d-c terminals, obtained with the fully-controlled circuit, are, in the half-controlled circuit, replaced by "freewheeling" periods of zero-voltage. The elimination of the negative swings of voltage at the d-c terminals is beneficial, because it results in a reduction of the ripple voltage, with correspondingly reduced filtering requirements. Second, whereas with the fully controlled circuit—for a given smooth load current—the waveform of the current in the input line theoretically remains the same at all firing angles, and only its phase position changes, with the half-controlled circuit, the duration of the input line current pulse is progressively shortened as the output voltage is reduced, until finally, with zero output voltage, the input current is also zero. Thus, the "wattless" load imposed by the half-controlled converter on the a-c supply at reduced output voltage, is considerably less than that due to the fully-controlled converter.

The Two-Pulse Midpoint Converter with Freewheel Diode

The circuit of the 2-pulse midpoint converter with freewheel diode is shown in Fig. 3.21. The presence of the freewheel diode prevents the voltage at the d-c terminals from becoming negative. Thus the periods of negative voltage swing, obtained without the diode, are replaced by periods of load current "freewheeling," during which the voltage at the d-c terminals remains substantially zero, and there is no current at the input side of the converter. Thus the external waveforms of this circuit are the same as those of the 2-pulse half-controlled bridge.

ONE-QUADRANT CONVERTERS 59

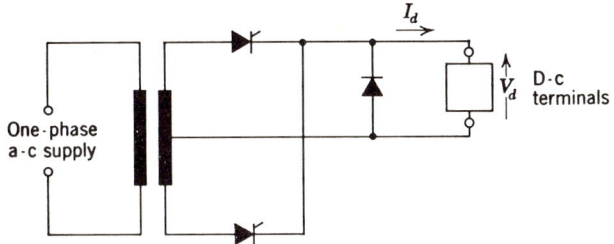

Figure 3.21. Two-pulse midpoint converter with freewheel diode.

The Three-Pulse Half-Controlled Bridge Converter

The circuit of a 3-pulse half-controlled bridge converter is shown in Fig. 3.22.

This circuit comprises two separate 3-pulse midpoint converters, connected in series with one another. The first midpoint circuit, containing the 3 thyristors, is a "positively poled" 2-quadrant converter; the second midpoint circuit, containing the 3 diodes, is a negatively poled uncontrolled 3-pulse rectifier. By controlling the voltage of the 2-quadrant converter from "full rectification" to "full inversion," so it is possible to control the mean voltage at the d-c terminals of the bridge from maximum positive, to virtually zero.

The operation of the circuit is illustrated by the waveforms of Fig. 3.23, in which it is assumed that the current at the d-c terminals is perfectly smooth. For each firing angle, the waveforms are shown of the voltage at the d-c terminal of each 3-pulse group, with respect to the neutral point of the supply. The waveform of the voltage appearing across the d-c terminals of the bridge is the difference between these two waves.

At a, the firing angle is $0°$, and the operation of the circuit is, in appearance, the same as that of an uncontrolled 6-pulse bridge. Thus the fundamental

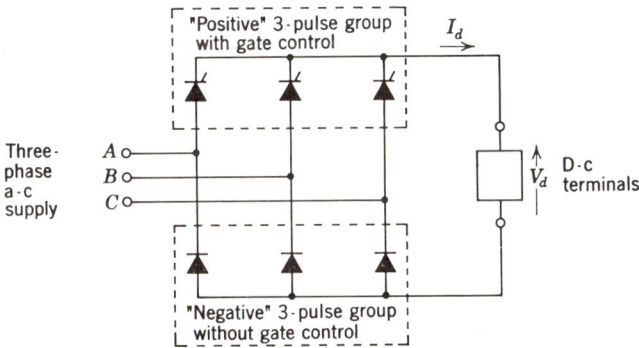

Figure 3.22. Three-pulse "half-controlled" converter.

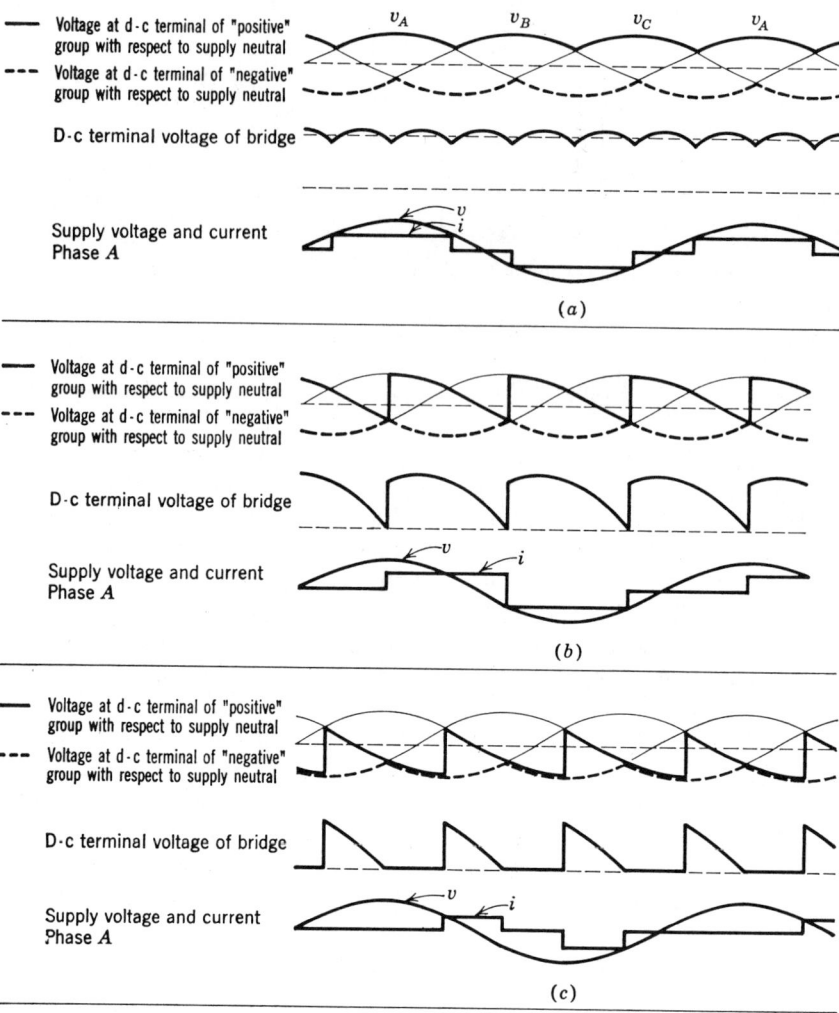

Figure 3.23. Theoretical waveforms associated with the 3-pulse half-controlled converter. (a) $\alpha = 0°$; (b) $\alpha = 60°$; (c) $\alpha = 120°$.

ripple frequency at the d-c terminals is $6\times$ the input frequency, and the waveform of the input current consists of positive and negative rectangular "blocks," each of 120° duration, which are symmetrically displaced from one another.

At b, the firing angle is 60°, and now the fundamental ripple frequency at the d-c terminals is $3\times$ the input frequency; indeed, this is generally the case for any firing angle other than 0°. The input line current waveform still consists of positive and negative rectangular "blocks," each of 120° duration. Now, however, the position of the positive current "block" has been retarded by 60°, whereas the position of the negative current "block" remains unaltered.

Actually, at this particular firing angle, the trailing edge of the positive "block" of current in the input line, due to the thyristor, just meets the leading edge of the "stationary" negative "block" of current, due to the diode. This means that the thyristor and diode in the same leg of the bridge circuit are now just at the point of simultaneous conduction. Thus further retardation of the firing angle beyond this point will result in an "overlapping" of the thyristor and diode current waveforms. In other words, there will be periods of "freewheeling" operation, during which no current flows in the input lines, and the output voltage is substantially zero. This is illustrated by the waveforms at c for a firing angle of 120°.

It is clear that as the output voltage decreases towards zero, so the current carried by the input lines also decreases towards zero. Thus, the 3-pulse half-controlled bridge circuit has the advantage, over its fully-controlled 6-pulse counterpart, that it imposes a reduced "wattless" load on the a-c system, at reduced levels of output voltage. This advantage, however, is obtained at the expense of a 2:1 reduction of ripple frequency at the d-c terminals, as a result of which the filtering requirements are greater than for the fully-controlled 2-quadrant 6-pulse circuit, at least over most of the operating range. At very low levels of output voltage, on the other hand, the ripple situation may actually be in favor of the half-controlled converter.

As in the case of the 2-pulse half-controlled converter circuit of Fig. 3.19a, it is not possible, with continuous current, to reduce the d-c terminal voltage of the 3-pulse half-controlled converter beyond a certain practical minimum. Again, this is because a certain small margin angle must be provided to ensure commutation of the current from one thyristor to the next. In this circuit, unlike the 2-pulse circuit, no rearrangement of diodes and thyristors is possible to overcome this problem. In this case, one method for overcoming the difficulty is to feed the controlled 3-pulse group from a slightly larger a-c voltage than the uncontrolled group. This provides the controlled group with a slightly higher theoretical voltage capability, and enables it to completely "buck out" the voltage of the uncontrolled group, without the risk of commutation failures. If an input transformer is used, the increased voltage for the controlled group can be conveniently obtained from a suitable tapping point on the transformer secondary.

The Three-Pulse Midpoint Converter with Freewheel Diode

The circuit of a 3-pulse midpoint converter with freewheel diode is shown in Fig. 3.24. Theoretical waveforms associated with this circuit, which assume a perfectly smooth load current, are shown in Fig. 3.25.

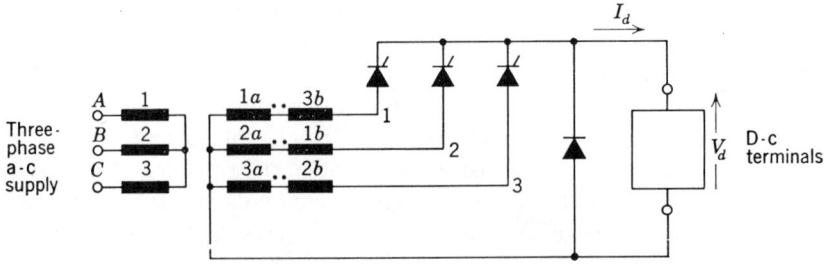

Figure 3.24. Three-pulse midpoint converter with freewheel diode.

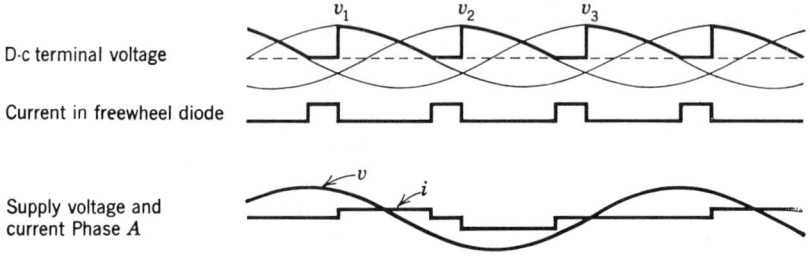

Figure 3.25. Theoretical waveforms associated with the 3-pulse midpoint converter with freewheel diode. $\alpha = 60°$.

For firing angles less than 30°, the d-c terminal voltage of the converter is always positive, and the freewheel diode does not come into operation. As the firing angle is retarded beyond this point, so the load current starts to freewheel through the diode for certain periods, thus cutting off the input line current, and preventing the d-c terminal voltage from swinging into the negative direction. Thus the effect of the freewheel diode is to cause a reduction of ripple voltage at the d-c terminals, and, at the same time, to divert the load current away from the input lines.

The total range of firing angle control required for this circuit is 150°.

The Six-Pulse Half-Controlled Converter

A 6-pulse half-controlled converter can be constructed by connecting a controlled 6-pulse bridge circuit in series with uncontrolled 6-pulse bridge, as shown in Fig. 3.26.

By controlling the voltage of the 2-quadrant bridge from "full rectification" to "full inversion," so it is possible to control the combined voltage of the

ONE-QUADRANT CONVERTERS 63

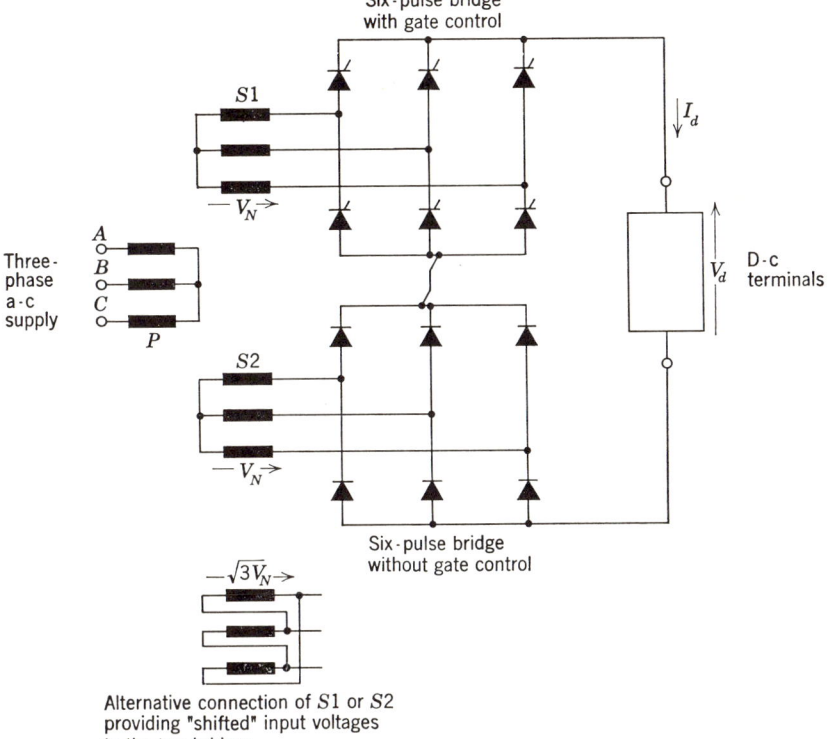

Figure 3.26. Six-pulse "half-controlled" converter.

two bridges from maximum positive to zero. Because of the need for an input transformer to provide isolation between the a-c terminals of the two bridges, it is a simple matter to provide the controlled bridge with a slightly larger a-c voltage than the uncontrolled bridge. This makes it possible to control the combined output voltage down to absolute zero, whilst still preserving a safe commutation angle for the thyristors of the controlled bridge.

The two sets of 3-phase voltages which supply the individual bridges may either have a 30° phase shift between them, or they may be "unshifted." Typical waveforms obtained with both "unshifted" and "shifted" input voltages are shown in Fig. 3.27. These waveforms assume that the current at the d-c terminals is perfectly smooth, and, in either circuit, has the same constant amplitude.

With "shifted" input voltages, the combined voltage waveform of the two bridges has a 12-pulse ripple at $\alpha = 0°$; however, the fundamental

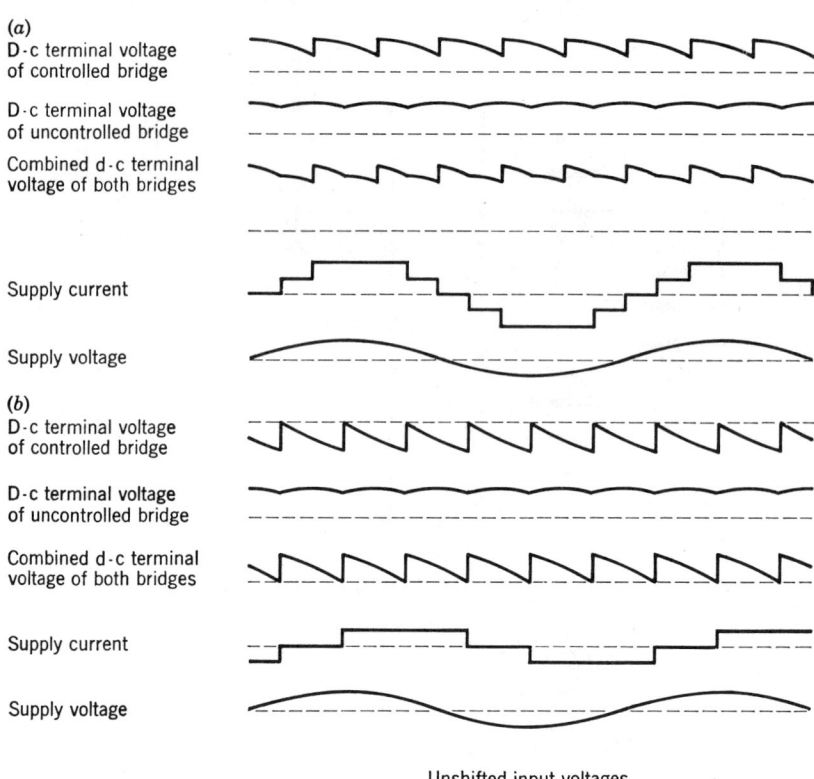

Figure 3.27. Theoretical waveforms associated with the 6-pulse half-controlled converter. (a) $\alpha = 30°$; (b) $\alpha = 120°$.

ripple frequency reverts to $6X$ the input frequency, as soon as the firing angle is moved away from this point. With "unshifted" input voltages, the fundamental ripple frequency is $6X$ the input frequency at all levels of output voltage. As will be seen in Chapter 4, (and as is evident from inspection of the waveforms of Fig. 3.27), there is generally little to choose between the two circuit connections, so far as the amplitude of the ripple voltage at the d-c terminals is concerned.

The waveform of the input line current consists of the addition of the individual waveforms due to the two bridges. Thus it is comprised of the "stationary" current waveform of the uncontrolled bridge, added to the phase-shifted waveform of the controlled bridge. With unshifted input voltages, these two waveforms tend to cancel one another completely as the output voltage is reduced towards zero. With shifted input voltages, on the

ONE-QUADRANT CONVERTERS 65

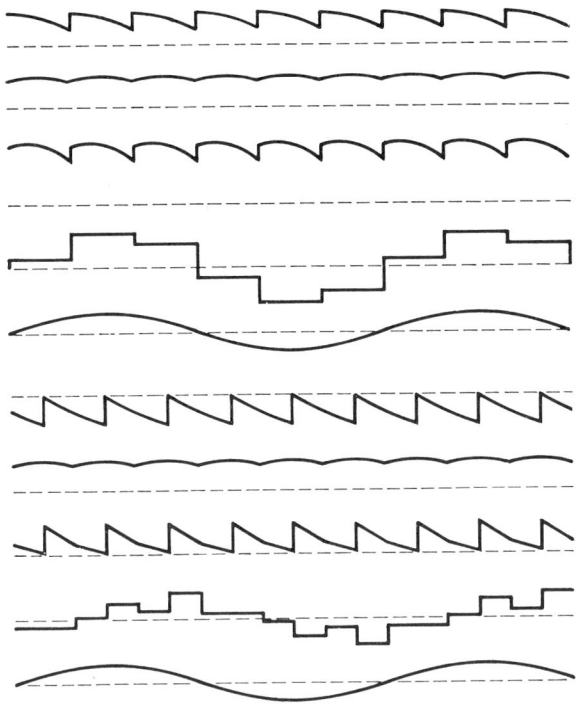

Shifted input voltages

Figure 3.27 Continued

other hand, a complete cancellation of currents does not occur, and some "residual" harmonic input current remains, even with zero output voltage.

The Six-Pulse Converter with One Freewheel Diode

A circuit diagram of a 6-pulse bridge converter with a freewheel diode connected directly across its d-c terminals is shown in Fig. 3.28a. Typical waveforms associated with this circuit, which assume a perfectly smooth load current, are shown in Fig. 3.29.

The effect of the freewheel diode upon the circuit operation is similar to that already discussed for the 3-pulse converter. In this case, however, the diode does not come into operation until a firing angle of 60° is reached. Hence it does not start to exert any influence on the external performance characteristics until a relatively late point in the control range.

The total range of firing angle control required for this circuit is 120°.

66 PHASE-CONTROLLED CONVERTERS

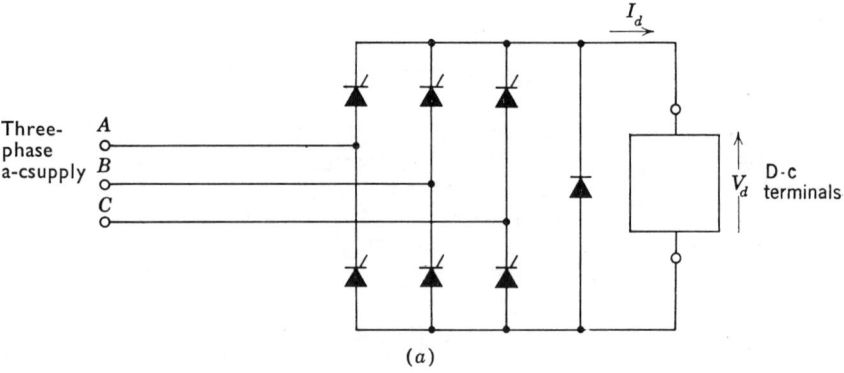

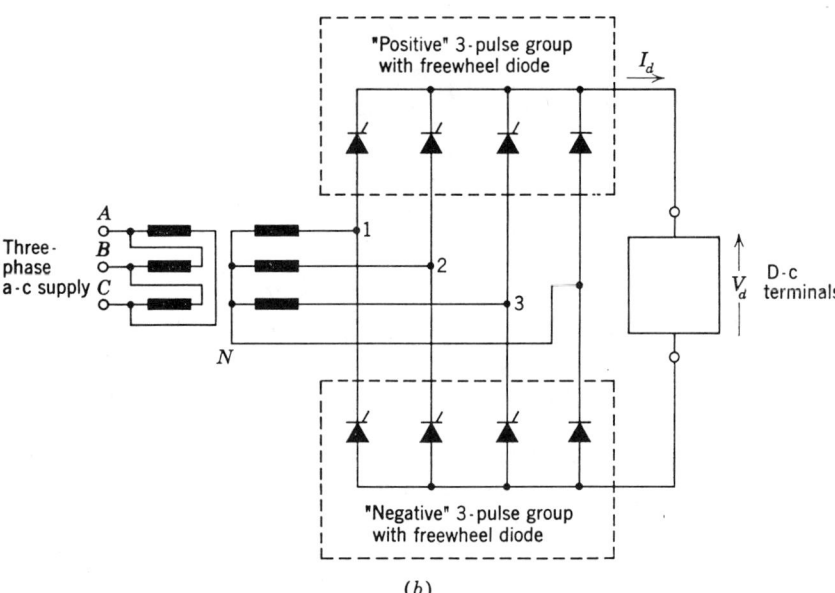

Figure 3.28. Alternative arrangements of 6-pulse bridge converters with freewheel diodes. (a) One freewheel diode. (b) Two freewheel diodes.

ONE-QUADRANT CONVERTERS

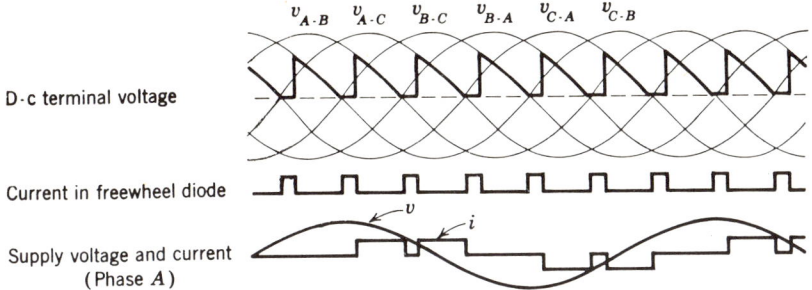

Figure 3.29. Theoretical waveforms associated with the 6-pulse bridge converter with one freewheel diode. $\alpha = 75°$.

The Six-Pulse Converter with Two Freewheel Diodes

It has been seen that the 6-pulse bridge converter consists essentially of two 3-pulse midpoint converters, connected in series with one another. By the same token, it is possible to connect two 3-pulse converters, each with its own freewheel diode, in series with one another. The resulting circuit

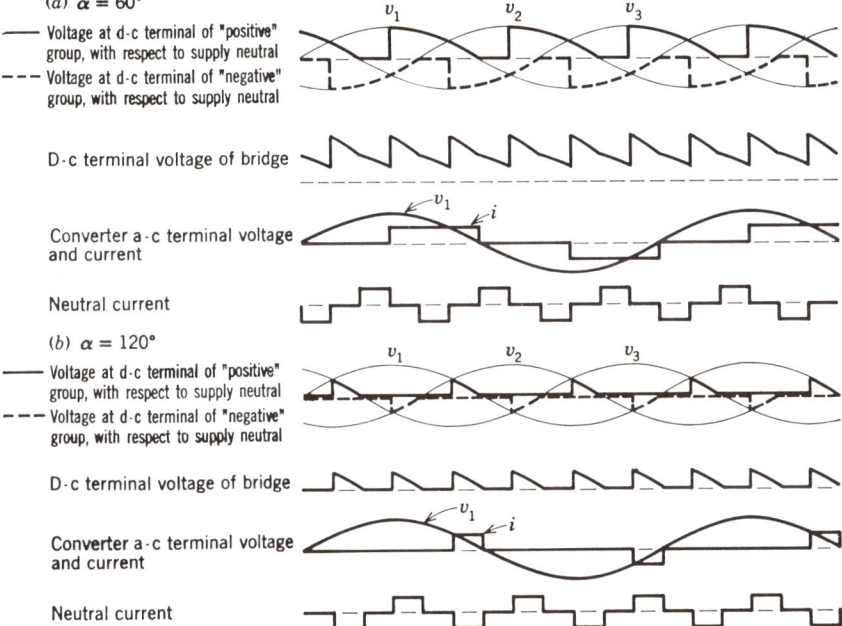

Figure 3.30. Theoretical waveforms associated with the 6-pulse bridge converter with two freewheel diodes. (a) $\alpha = 60°$; (b) $\alpha = 120°$.

connection is shown in Fig. 3.28b. Of course, in order to realize this connection, it is necessary for the neutral point of the 3-phase input voltages feeding the converter to be available; and this implies, furthermore, that a current will flow in the neutral wire.

Each 3-pulse group operates essentially independently of the other, as if it were by itself, so that each freewheel diode now starts to come into operation at a firing angle of 30°. Hence the benefits of the freewheeling operation are obtained at an earlier point in the control range, as compared to the circuit of Fig. 3.28a, with only one freewheel diode.

Typical waveforms associated with this circuit, which assume a perfectly smooth load current, are, shown in Fig. 3.30. For each firing angle, the waveforms of the voltage at the d-c terminal of each 3-pulse group, with respect to the neutral point of the supply, are shown. The waveform of the voltage appearing across the d-c terminals of the bridge is the difference between these two waves.

The total range of firing angle control required for this circuit is 150°.

Chapter Four

The External Performance Characteristics of Phase-Controlled Converters

It has been seen, in Chapter 3, that there are many possible circuit arrangements of the phase-controlled converter. In this chapter, a quantitative analysis is presented of the external performance characteristics of various phase-controlled converters, and it is shown that certain basic relationships exist between the principal performance parameters, which are common to all converters of the same type (i.e., 2-quadrant, or half-controlled 1-quadrant), regardless of the circuit configuration or pulse number. It is also shown that the differences which do exist between the performance characteristics of converters of similar type, with different pulse numbers, can be resolved simply to a question of whether or not given harmonic components of current at the a-c side, and voltage at the d-c side, are present or absent.

DEFINITION OF CONVERTER PERFORMANCE PARAMETERS

It is first necessary to define the important external performance parameters of the phase-controlled converter. These are as follows:

The d-c Voltage Ratio

The mean voltage at the d-c terminals of the converter can be continuously controlled by controlling the firing angle α. The d-c voltage ratio, denoted by r, is defined as the ratio of the mean d-c terminal voltage at a given firing angle α, to the maximum possible mean d-c terminal voltage; that is, that obtained when the firing angle is zero degrees.

The Input Displacement Angle

The waveform of the current in the input a-c lines which feed the phase-controlled converter is not a pure sinusoid. In general, it is comprised of a fundamental component, and a series of superimposed harmonic components. The input displacement angle, denoted by ϕ_i, is defined as the angular displacement between the fundamental component of the a-c line current and the associated line to neutral voltage. In all phase-controlled converter circuits, the fundamental component of current lags, or, in the limiting case, is in phase with, the associated voltage; thus ϕ_i is generally a negative angle.

The Input Displacement Factor

The input displacement factor is defined as the cosine of the input displacement angle.

The Input Power Factor

The input power factor is defined as the ratio of the total mean input power to the total rms input volt-amperes.

The Input Current Distortion Factor

The distortion factor of the current in a given input line is defined as the ratio of the rms amplitude of the fundamental component, to the total rms amplitude.

BASIC RELATIONSHIPS BETWEEN A-C INPUT PARAMETERS

The current at the a-c input terminals of a phase-controlled converter consists of a fundamental component with superimposed harmonic components. On the assumption that the a-c voltage wave is an undistorted sinusoid, certain basic relationships can be shown to exist between the a-c parameters, which are quite independent of the type of converter circuit. Indeed, these identities are generally true in any situation in which fundamental and harmonic currents flow through an undistorted sinusoidal voltage source. These relationships will now be derived.

Relationships Between Input Parameters

Relationship Between Mean a-c Power and Displacement Factor

The current in an a-c line connected to a converter can be represented quite generally by the following expression:

$$i = \hat{I}_1 \sin(\theta_i + \phi_i) + \sum_{n=2}^{n=\infty} \hat{I}_n \sin(n\theta_i + \delta_n)$$

- Fundamental component, having amplitude $\hat{I}_1$, displaced by angle ϕ_i with respect to the associated line-to-neutral voltage
- Series of superimposed harmonic components

The only component in the above series which contributes to the mean a-c input power is the fundamental component. This follows from the following mathematical identity, which shows that the mean power due to each and every harmonic component of current is necessarily zero:

$$\frac{1}{2\pi}\int_0^{2\pi} \hat{V}_N \sin\theta_i \cdot \hat{I}_n \sin(n\theta_i + \delta_n) \cdot d\theta_i = 0$$

where n is any integer except unity. Therefore, the mean a-c input power in a given line is:

P_i = rms line to neutral voltage × rms fundamental current × displacement factor

$$= V_N I_1 \cos\phi_i \qquad (4.1)$$

Relationship Between Displacement, Power, and Distortion Factors

The input a-c power factor is given by

$$\lambda = \frac{P_i}{V_N \times \text{total rms value of alternating current wave}}$$

$$= \frac{P_i}{V_N \times \sqrt{I_1^2 + \sum_{n=2}^{n=\infty} I_n^2}} \qquad (4.2)$$

The input a-c distortion factor is given by

$$\mu = \frac{I_1}{\text{total rms value of alternating current wave}}$$

$$= \frac{I_1}{\sqrt{I_1^2 + \sum_{n=2}^{n=\infty} I_n^2}} \qquad (4.3)$$

72 CONVERTER PERFORMANCE CHARACTERISTICS

From (4.1) through (4.3)

$$\lambda = \mu \times \frac{P_i}{V_N I_1}$$

$$= \mu \cos \phi_i \qquad (4.4)$$

Thus, the power factor is equal to the product of the distortion and displacement factors. Since the distortion factor is always less than unity, the power factor must always be less than the displacement factor. The meaning of this is as follows: Since the a-c supply carries harmonic currents, it must therefore handle wattless harmonic power, in addition to real and reactive fundamental power; therefore the total volt amperes handled by the a-c supply must be greater than the total fundamental volt amperes.

METHOD OF ANALYSIS OF CONVERTER PERFORMANCE CHARACTERISTICS

The approach generally used in the harmonic analysis of the waveforms of phase-controlled converter circuits is to express the complex wave in terms of a general Fourier series, the coefficients of which are then evaluated by normal methods. The basic analytical technique used in this chapter, and generally throughout this book, does not follow precisely this approach. The reason for this is that although the conventional approach can be satisfactorily applied to the analysis of converter waveforms with a steady firing angle, it is not readily applicable to a generalized analysis of cycloconverter waveforms.

The basic analytical technique used here is to represent the waveform of the voltage at the d-c terminals as being the sum of the individual voltage segments generated by each of the thyristors within the converter. Each individual voltage segment is expressed mathematically as the product of the appropriate sinusoidal input voltage, and a "switching function"; this switching function has unity amplitude whenever the associated thyristor is ON, and zero amplitude whenever it is OFF. By expressing each switching function as a harmonic series, so the series for the d-c terminal voltage waveform is obtained in terms of its d-c and a-c harmonic components.

In a similar manner, the current in a given input line is represented as the sum of the individual thyristor currents associated with that line. Each thyristor current is expressed as the product of the d-c terminal current of the associated group, and the thyristor switching function. The resulting expression obtained for the input current waveform is a complete representation of this waveform, in terms of its fundamental and harmonic components. For simplicity, the analysis is restricted to the condition of a

PERFORMANCE OF TWO-QUADRANT CONVERTERS

ripple-free current at the d-c terminals. The same analytical technique, however, could also be applied to an imperfectly smoothed d-c terminal current.

THE PERFORMANCE CHARACTERISTICS OF TWO-QUADRANT CONVERTERS

First, the performance characteristics of the 3-pulse converter will be analyzed. Then, it will be shown how the performance characteristics of other "multipulse" converter circuits, containing combinations of 3-pulse groups, can be obtained directly, without returning to first principles, from the results already derived for the basic 3-pulse group.

The method of harmonic analysis for the 2-pulse converter, as well as the results of such an analysis, are similar in nature to those for the 3-pulse converter. For the sake of conciseness, only the results of such an analysis are given here. This data is presented graphically along with that for the 3- (and multi-) pulse converters.

Analysis of the Performance Characteristics of the Three-Pulse Converter

HARMONIC ANALYSIS OF THE D-C TERMINAL VOLTAGE. The method of analysis is illustrated in Fig. 4.1.

The d-c terminal voltage is represented mathematically by:

$$v_d = \hat{V}_N \sin \theta_i \times F_1(\theta_i - \alpha) + \hat{V}_N \sin \left(\theta_i - \frac{2\pi}{3}\right) \times F_2(\theta_i - \alpha)$$

$$+ \hat{V}_N \sin \left(\theta_i + \frac{2\pi}{3}\right) \times F_3(\theta_i - \alpha) \quad (4.5)$$

$F_1(\theta_i - \alpha)$, $F_2(\theta_i - \alpha)$, and $F_3(\theta_i - \alpha)$ are the "switching" functions associated with thyristors 1, 2, and 3 respectively. Each of these functions has unity amplitude when its associated thyristor is ON, and zero amplitude at all other times.

According to conventional Fourier harmonic analysis, $F_1(\theta_i - \alpha)$, $F_2(\theta_i - \alpha)$, and $F_3(\theta_i - \alpha)$ can be expressed in terms of the following

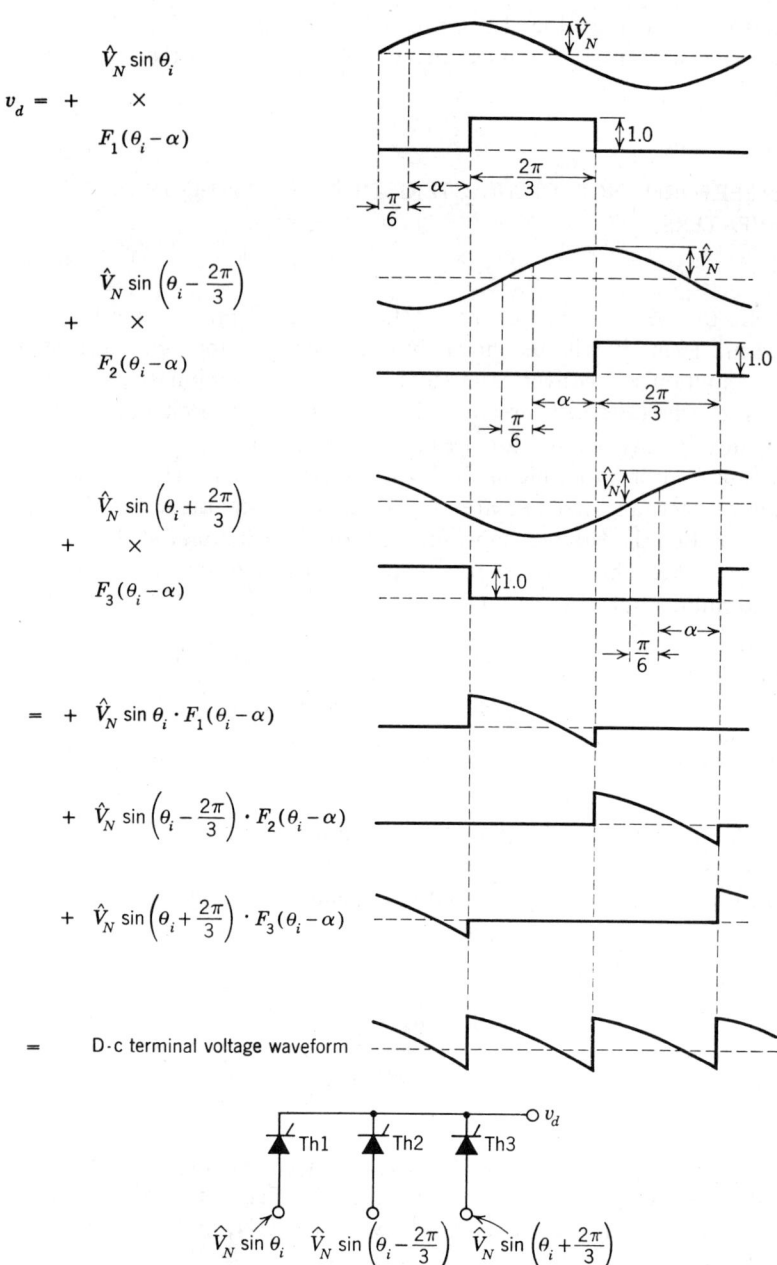

Figure 4.1. Method of analysis of the d-c terminal voltage waveform of the 3-pulse 2-quadrant converter.

PERFORMANCE OF TWO-QUADRANT CONVERTERS

harmonic series:

$$F_1(\theta_i - \alpha) = \frac{1}{3} + \frac{\sqrt{3}}{\pi}[\sin(\theta_i - \alpha) - \tfrac{1}{2}\cos 2(\theta_i - \alpha) - \tfrac{1}{4}\cos 4(\theta_i - \alpha)$$

$$- \tfrac{1}{5}\sin 5(\theta_i - \alpha) - \tfrac{1}{7}\sin 7(\theta_i - \alpha) + \tfrac{1}{8}\cos 8(\theta_i - \alpha)$$

$$+ \tfrac{1}{10}\cos 10(\theta_i - \alpha) + \tfrac{1}{11}\cos 11(\theta_i - \alpha) + \tfrac{1}{13}\cos 13(\theta_i - \alpha) \cdots]$$

(4.6)

$$F_2(\theta_i - \alpha) = \frac{1}{3} + \frac{\sqrt{3}}{\pi}\left[\sin\left(\theta_i - \alpha - \frac{2\pi}{3}\right) - \tfrac{1}{2}\cos 2\left(\theta_i - \alpha - \frac{2\pi}{3}\right)\right.$$

$$\left. - \tfrac{1}{4}\cos 4\left(\theta_i - \alpha - \frac{2\pi}{3}\right)\cdots\right] \quad (4.7)$$

$$F_3(\theta_i - \alpha) = \frac{1}{3} + \frac{\sqrt{3}}{\pi}\left[\sin\left(\theta_i - \alpha + \frac{2\pi}{3}\right) - \tfrac{1}{2}\cos 2\left(\theta_i - \alpha + \frac{2\pi}{3}\right)\right.$$

$$\left. - \tfrac{1}{4}\cos 4\left(\theta_i - \alpha + \frac{2\pi}{3}\right)\cdots\right] \quad (4.8)$$

Therefore, from (4.5)

$$v_d = \hat{V}_N \sin \theta_i \left\{\frac{1}{3} + \frac{\sqrt{3}}{\pi}[\sin(\theta_i - \alpha) - \tfrac{1}{2}\cos 2(\theta_i - \alpha) - \tfrac{1}{4}\cos 4(\theta_i - \alpha)\cdots]\right\}$$

$$+ \hat{V}_N \sin\left(\theta_i - \frac{2\pi}{3}\right)\left\{\frac{1}{3} + \frac{\sqrt{3}}{\pi}\left[\sin\left(\theta_i - \alpha - \frac{2\pi}{3}\right)\right.\right.$$

$$\left.\left. - \tfrac{1}{2}\cos 2\left(\theta_i - \alpha - \frac{2\pi}{3}\right) - \tfrac{1}{4}\cos 4\left(\theta_i - \alpha - \frac{2\pi}{3}\right)\cdots\right]\right\}$$

$$+ \hat{V}_N \sin\left(\theta_i + \frac{2\pi}{3}\right)\left\{\frac{1}{3} + \frac{\sqrt{3}}{\pi}\left[\sin\left(\theta_i - \alpha + \frac{2\pi}{3}\right)\right.\right.$$

$$\left.\left. - \tfrac{1}{2}\cos 2\left(\theta_i - \alpha + \frac{2\pi}{3}\right) - \tfrac{1}{4}\cos 4\left(\theta_i - \alpha + \frac{2\pi}{3}\right)\cdots\right]\right\}$$

76   CONVERTER PERFORMANCE CHARACTERISTICS

By trigonometric manipulation, this reduces to

$$v_d = \frac{3\sqrt{3}\,\hat{V}_N}{2\pi}\left[\cos\alpha + \left(\frac{1}{2^2} + \frac{1}{4^2} - \frac{2}{2\cdot 4}\cos 2\alpha\right)^{1/2}\sin(3\theta_i + \gamma_3)\right.$$
$$+ \left(\frac{1}{5^2} + \frac{1}{7^2} - \frac{2}{5\cdot 7}\cos 2\alpha\right)^{1/2}\sin(6\theta_i + \gamma_6)$$
$$+ \left(\frac{1}{8^2} + \frac{1}{10^2} - \frac{2}{8\cdot 10}\cos 2\alpha\right)^{1/2}\sin(9\theta_i + \gamma_9)$$
$$\left. + \left(\frac{1}{11^2} + \frac{1}{13^2} - \frac{2}{11\cdot 13}\cos 2\alpha\right)^{1/2}\sin(12\theta_i + \gamma_{12}) + \cdots\right]$$

(4.9)

$$= \frac{3\sqrt{3}\,\hat{V}_N}{2\pi}\left\{\underbrace{\cos\alpha}_{\substack{\text{Steady}\\\text{d-c component}}} + \underbrace{\sum_{n=1}^{n=\infty}\left[\frac{1}{(3n-1)^2} + \frac{1}{(3n+1)^2} - \frac{2\cos 2\alpha}{(3n-1)(3n+1)}\right]^{1/2}\sin(3n\theta_i + \gamma_{3n})}_{\substack{\text{Superimposed}\\\text{a-c ripple components}}}\right\}$$

(4.10)

where

$$\gamma_{3n} = -\frac{n\pi}{2} + \tan^{-1}\frac{\dfrac{\cos(3n+1)\alpha}{(3n+1)} - \dfrac{\cos(3n-1)\alpha}{(3n-1)}}{\dfrac{\sin(3n+1)\alpha}{(3n+1)} - \dfrac{\sin(3n-1)\alpha}{(3n-1)}}$$

Thus it is seen that the d-c terminal voltage contains a steady d-c component, and an infinite (but converging) series of a-c ripple components. The frequency of the lowest order term is $3X$ the input frequency; the next highest ripple frequency is $6X$ the input frequency, and so on.

Figure 4.2 shows the peak amplitudes of the a-c ripple components in the d-c terminal voltage of the 3-pulse converter, up to the 24th harmonic of the input frequency, plotted against the firing angle. It is seen here that each curve is symmetrical about the 90° point. Each harmonic has its minimum value at the 2 extremities of the firing angle control range, and its maximum value at $\alpha = 90°$.

THE D-C VOLTAGE RATIO. From expression (4.10), it is seen that the mean component of the d-c terminal voltage is proportional to the cosine of the firing angle. The d-c voltage ratio is given by

$$r = \frac{\dfrac{3\sqrt{3}\,\hat{V}_N}{2\pi}\cos\alpha \;\leftarrow\;\text{Mean d-c terminal voltage at firing angle } \alpha}{\dfrac{3\sqrt{3}\,\hat{V}_N}{2\pi} \;\leftarrow\;\text{Maximum possible mean d-c terminal voltage}}$$

$$= \cos\alpha \qquad (4.11)$$

PERFORMANCE OF TWO-QUADRANT CONVERTERS

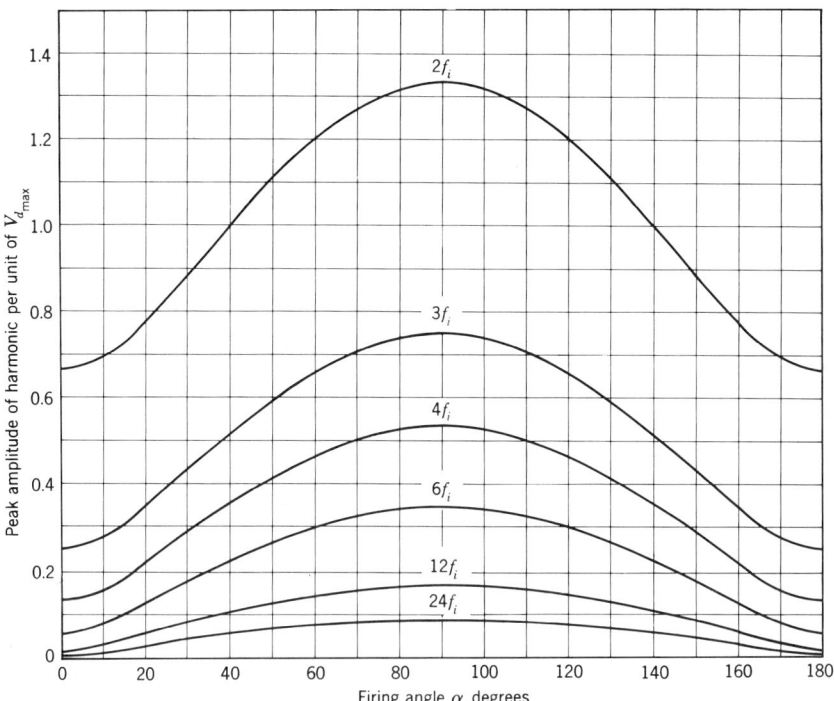

Figure 4.2. Variation with firing angle of the predominant harmonic components present in the d-c terminal voltage of various 2-quadrant converters (with continuous conduction). For a 2-pulse converter, the two lowest order harmonics are the $2f_i$ and $4f_i$ terms (in general, all integer multiples of $2f_i$ are present). For a 3-pulse converter, the two lowest order harmonics are the $3f_i$ and $6f_i$ terms (in general, all integer multiples of $3f_i$ are present). For a 6-pulse converter, the two lowest order harmonics are the $6f_i$ and $12f_i$ terms (in general, all integer multiples of $6f_i$ are present). For a 12-pulse converter, the two lowest order harmonics are the $12f_i$ and $24f_i$ terms (in general, all integer multiples of $12f_i$ are present).

HARMONIC ANALYSIS OF THE INPUT CURRENTS. The method of analysis is illustrated in Fig. 4.3. The current in secondary $S1$ is given by:

$$i_1 = I_d \times F_1(\theta_i - \alpha)$$
$$= \underbrace{\frac{I_d}{3}}_{\text{d-c component}} + \underbrace{\frac{\sqrt{3}\,I_d}{\pi}}_{\text{Fundamental component}} [\sin(\theta_i - \alpha) \underbrace{- \tfrac{1}{2}\cos 2(\theta_i - \alpha) - \tfrac{1}{4}\cos 4(\theta_i - \alpha) - \tfrac{1}{5}\sin 5(\theta_i - \alpha) - \tfrac{1}{7}\sin 7(\theta_i - \alpha) \cdots]}_{\text{Superimposed harmonic components}}$$

(4.12)

The currents in secondaries $S2$ and $S3$ are obtained in a similar manner.

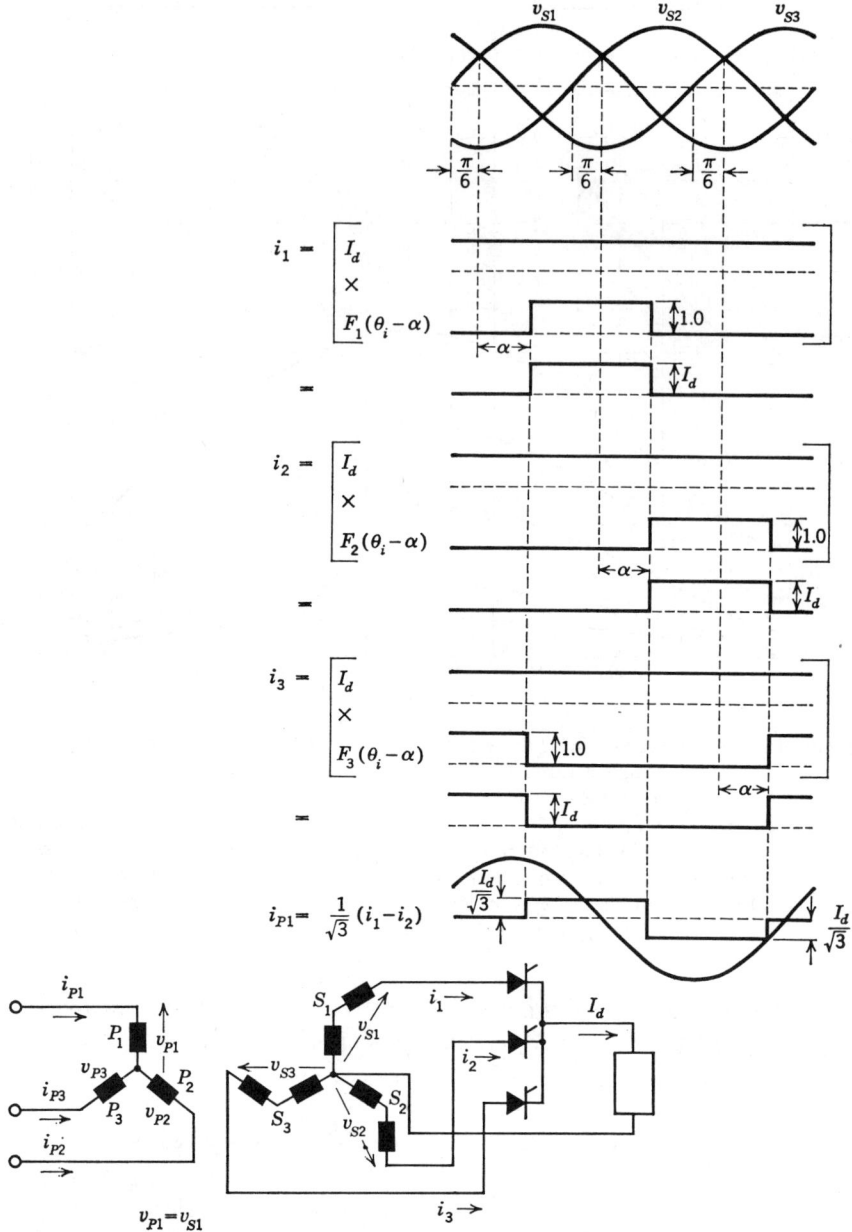

Figure 4.3. Method of analysis of the input currents of the 3-pulse 2-quadrant converter.

PERFORMANCE OF TWO-QUADRANT CONVERTERS 79

Thus the secondary current contains a fundamental component of amplitude $\sqrt{3}I_d/\pi$, which lags the secondary voltage by an angle equal to the converter firing angle α. In addition, it contains a d-c component of amplitude $I_d/3$, and a converging series of superimposed harmonic components. These harmonic components have frequencies of 2, 4, 5, 7, and in general $[3(2n + 1) \pm 1]$ and $[6n \pm 1]$ times the line frequency. The amplitude of any harmonic component, relative to the fundamental, is equal to the reciprocal of the harmonic number. As mentioned in Chapter 3, although there is a d-c component of current in each secondary, there is no net d-c magnetization of the transformer core, because of the zig-zag arrangement of secondary windings.

Assuming the same line-to-neutral voltage at either side of the transformer, the current in the primary $P1$ is given by

$$i_{P1} = \frac{1}{\sqrt{3}}(i_1 - i_2)$$

$$= \frac{\sqrt{3}\,I_d}{\pi}\Bigg[\sin\left(\theta_i - \alpha + \frac{\pi}{6}\right) \quad - \tfrac{1}{2}\cos\left(2\theta_i - 2\alpha - \frac{\pi}{6}\right)$$

$$\underbrace{\phantom{\sin\left(\theta_i - \alpha + \frac{\pi}{6}\right)}}_{\substack{\text{Fundamental}\\ \text{component}}}$$

$$- \tfrac{1}{4}\cos\left(4\theta_i - 4\alpha + \frac{\pi}{6}\right)$$

$$- \tfrac{1}{5}\sin\left(5\theta_i - 5\alpha - \frac{\pi}{6}\right)$$

$$- \tfrac{1}{7}\sin\left(7\theta_i - 7\alpha + \frac{\pi}{6}\right)\cdots\Bigg] \quad (4.13)$$

$$\underbrace{}_{\substack{\text{Superimposed}\\ \text{harmonic}\\ \text{components}}}$$

The associated primary line to neutral voltage is given by:

$$v_{P1} = \hat{V}_N \sin\left(\theta_i + \frac{\pi}{6}\right)$$

Thus the primary line current contains a fundamental component of amplitude $\sqrt{3}\,I_d/\pi$, which lags the primary line to neutral voltage by an angle equal to the converter firing angle α. In addition, it contains a converging series of superimposed harmonic components, the frequencies and relative amplitudes of which are the same as those of the secondary current waveform.

CONVERTER PERFORMANCE CHARACTERISTICS

THE INPUT DISPLACEMENT FACTOR. The input displacement angle is equal to the firing angle, and therefore the input displacement factor is given by:

$$\cos \phi_i = \cos \alpha \qquad (4.14)$$

THE DISTORTION AND POWER FACTORS OF THE TRANSFORMER SECONDARY CURRENT. The distortion factor of the transformer secondary current is given by

$$\mu_S = \frac{1 \quad \leftarrow \text{Relative rms amplitude of fundamental}}{\sqrt{\left(\frac{\sqrt{2}\,\pi}{\sqrt{3}\cdot 3}\right)^2 + \frac{1}{1^2} + \frac{1}{2^2} + \frac{1}{4^2} + \frac{1}{5^2} + \cdots} \;\; \leftarrow \text{Relative total rms amplitude}}$$

↑ Relative rms amplitude of d-c component

$$= 0.675$$

The power factor of the current at the input a-c terminals of the converter is given by

$$\lambda = 0.675 \cos \alpha$$

The power factor as "seen" by the transformer secondary windings is less than this, because the vectorial sum of the zigzag voltages is less than their algebraic sum, by a factor of $\sqrt{3}/2$. Therefore:

$$\lambda_S = 0.675 \times 0.866 \cos \alpha$$
$$= 0.585 \cos \alpha$$

THE DISTORTION AND POWER FACTORS OF THE TRANSFORMER PRIMARY CURRENT. The distortion factor of the transformer primary current is given by:

$$\mu_P = \frac{1 \quad \leftarrow \text{Relative rms amplitude of fundamental}}{\sqrt{1^2 + \frac{1}{2^2} + \frac{1}{4^2} + \frac{1}{5^2} + \cdots} \;\; \leftarrow \text{Relative total rms amplitude}}$$

$$= 0.827$$

And the power factor of the primary current is given by:

$$\lambda_P = 0.827 \cos \alpha$$

IDENTITY BETWEEN D-C AND A-C POWER. Since all losses are assumed to be zero, the d-c power must be equal to the mean a-c input power.

The d-c power is given by

$$P_d = \text{mean d-c terminal voltage} \times \text{direct current}$$
$$= \frac{3\sqrt{3}\,\hat{V}_N}{2\pi} I_d \cos \alpha \qquad (4.15)$$

PERFORMANCE OF TWO-QUADRANT CONVERTERS 81

The total mean a-c input power is given by

$$P_i = 3 \times \underbrace{\frac{\hat{V}_N}{\sqrt{2}}}_{\substack{\text{3-Phases Rms} \\ \text{primary} \\ \text{voltage}}} \times \underbrace{\frac{\sqrt{3}}{\pi} \frac{I_d}{\sqrt{2}}}_{\substack{\text{Rms} \\ \text{fundamental} \\ \text{primary} \\ \text{current}}} \times \underbrace{\cos \alpha}_{\substack{\text{Displacement} \\ \text{factor}}}$$

$$= \frac{3\sqrt{3}\, \hat{V}_N}{2\pi} I_d \cos \alpha \tag{4.16}$$

And hence, from (4.15) and (4.16)

$$P_d = P_i$$

Analysis of the Performance Characteristics of Multipulse Converters, Containing Combinations of the Three-Pulse Group

As has been seen in Chapter 3, most practical "multipulse" converter circuits consist of combinations of the basic 3-pulse commutating group. It will now be seen that the performance characteristics of any multipulse circuit, containing combinations of the basic 3-pulse group, can be deduced quite simply, without returning to first principles, from a knowledge of the performance characteristics of the basic 3-pulse group.

This is not surprising, since each 3-pulse group within a multipulse system operates essentially independently of all the other groups, as if it were by itself, and therefore each group preserves its own d-c terminal voltage and input current waveforms. Thus the mathematical analysis of the external performance characteristics of a multipulse group reduces to a process of addition, in the appropriate phase relationships, of the harmonic series, representing the d-c terminal voltages, or the input currents, as the case may be, of the individual groups.

THE MEAN D-C TERMINAL VOLTAGE OF A MULTIPULSE GROUP. In any circuit configuration which contains combinations of the basic 3-pulse group, the mean d-c terminal voltages of all the basic groups theoretically are equal to one another. Thus, the mean d-c terminal voltage of any number of groups connected in parallel through interphase reactors is equal to the mean voltage of any one group:

$$V_d = \frac{3\sqrt{3}\, \hat{V}_N}{2\pi} \cos \alpha$$

$$= 0.827 \hat{V}_N \cos \alpha \tag{4.17}$$

The mean d-c terminal voltage of a number, s, of series connected groups is given by:

$$V_d = s \cdot \frac{3\sqrt{3}\, V_N}{2\pi} \cos \alpha$$

$$= s \cdot 0.827 \hat{V}_N \cos \alpha \qquad (4.18)$$

Here, it is pointed out that $\hat{V}_N$ is the peak line to neutral voltage of each group. To take a specific example, consider the 6-pulse bridge circuit of Fig. 3.12. This consists of two 3-pulse groups, connected in series with one another. Therefore, for this circuit:

$$V_d = 2 \cdot \frac{3\sqrt{3}\, \hat{V}_N}{2\pi} \cos \alpha$$

$$= 1.654 \hat{V}_N \cos \alpha$$

From the foregoing discussion, it is clear that the d-c voltage ratio of any multipulse group is given by

$$r = \cos \alpha \qquad (4.19)$$

CANCELLATION OF LOWER ORDER HARMONICS IN THE D-C TERMINAL VOLTAGE OF A MULTIPULSE GROUP. Because the a-c input voltages connected to each 3-pulse group are effectively displaced in phase from those of all the other groups, a phase displacement exists between the ripple voltage waveforms at the d-c terminals of the individual groups. This phase displacement results in a cancellation of certain harmonic components in the final d-c terminal voltage of the system, whilst those components which do not cancel remain unaltered, both in phase and relative amplitude. The higher is the overall pulse number of the circuit, the more complete is the cancellation of harmonics in the d-c terminal voltage waveform, and therefore the more "naturally perfect" is the a-c to d-c conversion process.

HARMONIC ANALYSIS OF THE SIX-PULSE VOLTAGE. The d-c terminal voltage of a 6-pulse converter is synthesized from the individual voltages, v_{d3_1} and v_{d3_2}, of two 3-pulse groups:

$$v_d = k(v_{d3_1} + v_{d3_2})$$

where k has a value of 1 for the midpoint circuit, and 2 for the bridge circuit.

The a-c input voltages of the two groups are, in effect, displaced by 180° from each other.

PERFORMANCE OF TWO-QUADRANT CONVERTERS

According to expression (4.9), the d-c terminal voltage of group 1 has been shown to be:

$$v_{d3_1} = \frac{3\sqrt{3}\,\hat{V}_N}{2\pi}\left[\cos\alpha + \left(\frac{1}{2^2} + \frac{1}{4^2} - \frac{2\cos 2\alpha}{2\cdot 4}\right)^{1/2}\sin(3\theta_i + \gamma_3)\right.$$

$$\left. + \left(\frac{1}{5^2} + \frac{1}{7^2} - \frac{2\cos 2\alpha}{5\cdot 7}\right)^{1/2}\sin(6\theta_i + \gamma_6) + \cdots\right]$$

By substituting $(\theta_i + \pi)$ for θ_i into the expression above, the corresponding expression for the d-c terminal voltage of group 2 is obtained:

$$v_{d3_2} = \frac{3\sqrt{3}\,\hat{V}_N}{2\pi}\left[\cos\alpha - \left(\frac{1}{2^2} + \frac{1}{4^2} - \frac{2}{2\cdot 4}\cos 2\alpha\right)^{1/2}\sin(3\theta_i + \gamma_3)\right.$$

$$\left. + \left(\frac{1}{5^2} + \frac{1}{7^2} - \frac{2\cos 2\alpha}{5\cdot 7}\right)^{1/2}\sin(6\theta_i + \gamma_6)\cdots\right]$$

Hence, the final d-c terminal voltage of the 6-pulse group is given by:

$$v_d = k\cdot\frac{3\sqrt{3}\,\hat{V}_N}{2\pi}\left[\cos\alpha + \left(\frac{1}{5^2} + \frac{1}{7^2} - \frac{2\cos 2\alpha}{5\cdot 7}\right)^{1/2}\sin(6\theta_i + \gamma_6)\right.$$

$$\left. + \left(\frac{1}{11^2} + \frac{1}{13^2} - \frac{2\cos 2\alpha}{11\cdot 13}\right)^{1/2}\sin(12\theta_i + \gamma_{12})\cdots\right] \quad (4.20)$$

$$= k\cdot\frac{3\sqrt{3}\,\hat{V}_N}{2\pi}\Bigg\{\underbrace{\cos\alpha}_{\substack{\text{Steady}\\\text{d-c}\\\text{component}}} + \underbrace{\sum_{n=1}^{n=\infty}\left[\frac{1}{(6n-1)^2} + \frac{1}{(6n+1)^2} - \frac{2\cos 2\alpha}{(6n-1)(6n+1)}\right]^{1/2}\sin(6n\theta_i + \gamma_{6n})}_{\text{Superimposed a-c ripple components}}\Bigg\}$$

(4.21)

Thus, those harmonic components in the d-c terminal voltage of group 2 which have frequencies of 3, 9, and in general $3(2n-1) \times$ the line frequency are exactly 180° out of phase with the corresponding components in the d-c terminal voltage of group 1, and therefore each of these components cancel one another in the final d-c terminal voltage of the 6-pulse group. On the other hand, the harmonic components in the d-c terminal voltage of group 2, which have frequencies of 6, 12, and in general $6n \times$ the line frequency, are in-phase with the corresponding components in the d-c terminal voltage of group 1. Therefore these components are retained, with the same relative amplitudes, in the final d-c terminal voltage. The curves

84 CONVERTER PERFORMANCE CHARACTERISTICS

of Fig. 4.2 are therefore applicable to the 6-pulse voltage, it being understood that only the $6n \times$ input frequency components are present.

HARMONIC ANALYSIS OF THE TWELVE-PULSE VOLTAGE. The d-c terminal voltage of a 12-pulse converter is synthesized from the individual voltages, v_{d6_1} and v_{d6_2} of two 6-pulse groups:

$$v_d = C'(v_{d6_1} + v_{d6_2})$$

where C' is an integer whose value depends upon the circuit connection. The a-c input voltages of the two groups are displaced by 30° from each other.

From expression (4.20)

$$v_{d6_1} = k \cdot \frac{3\sqrt{3}\, \hat{V}_N}{2\pi} \left[\cos \alpha + \left(\frac{1}{5^2} + \frac{1}{7^2} - \frac{2 \cos 2\alpha}{5 \cdot 7} \right)^{1/2} \sin(6\theta_i + \gamma_6) \right.$$
$$\left. + \left(\frac{1}{11^2} + \frac{1}{13^2} - \frac{2 \cos 2\alpha}{11 \cdot 13} \right)^{1/2} \sin(12\theta_i + \gamma_{12}) \cdots \right]$$

By substituting $(\theta_i - \pi/6)$ for θ_i into the above expression, the corresponding expression for v_{d6_2} is obtained:

$$v_{d6_2} = k \cdot \frac{3\sqrt{3}\, \hat{V}_N}{2\pi} \left[\cos \alpha - \left(\frac{1}{5^2} + \frac{1}{7^2} - \frac{2 \cos 2\alpha}{5 \cdot 7} \right)^{1/2} \sin(6\theta_i + \gamma_6) \right.$$
$$\left. + \left(\frac{1}{11^2} + \frac{1}{13^2} - \frac{2 \cos 2\alpha}{11 \cdot 13} \right)^{1/2} \sin(12\theta_i + \gamma_{12}) \cdots \right]$$

And hence the final d-c voltage of the 12-pulse group is given by:

$$v_d = C \cdot \frac{3\sqrt{3}\, \hat{V}_N}{2\pi} \left[\cos \alpha + \left(\frac{1}{11^2} + \frac{1}{13^2} - \frac{2 \cos 2\alpha}{11 \cdot 13} \right)^{1/2} \sin(12\theta_i + \gamma_{12}) \right.$$
$$\left. + \left(\frac{1}{23^2} + \frac{1}{25^2} - \frac{2 \cos 2\alpha}{23 \cdot 25} \right)^{1/2} \sin(24\theta_i + \gamma_{24}) \cdots \right] \quad (4.22)$$

$$= C \cdot \frac{3\sqrt{3}\, \hat{V}_N}{2\pi} \left\{ \underbrace{\cos \alpha}_{\text{Steady d-c component}} + \underbrace{\sum_{n=1}^{n=\infty} \left[\frac{1}{(12n-1)^2} + \frac{1}{(12n+1)^2} - \frac{2 \cos 2\alpha}{(12n-1)(12n+1)} \right]^{1/2} \sin(12n\theta_i + \gamma_{12n})}_{\text{Superimposed a-c ripple components}} \right\} \quad (4.23)$$

Thus the 12-pulse d-c terminal voltage waveform contains only those harmonics which have frequencies of 12, 24, and in general $12n \times$ the line

PERFORMANCE OF TWO-QUADRANT CONVERTERS

frequency. The relative amplitudes of these components are the same as in the basic 3-pulse waveform. The curves of Fig. 4.2 are therefore applicable to the 12-pulse voltage, it being understood that only the $12n \times$ line frequency components are present.

THE INPUT DISPLACEMENT FACTOR OF A MULTIPULSE CONVERTER. Each 3-pulse group within a multipulse circuit ideally has the same relative firing angle,* and carries an equal share of the direct current; therefore each group imposes exactly the same amount of real and reactive fundamental loading on the a-c system. Therefore, the phase of the fundamental component of current at any point in the a-c circuit, relative to the associated a-c voltage, invariably is equal to the converter firing angle. Hence the input displacement factor of a multipulse converter is equal to the displacement factor of each of the individual groups. Therefore, for any multipulse converter

$$\cos \phi_i = \cos \alpha \qquad (4.24)$$

CANCELLATION OF LOWER ORDER HARMONICS IN THE INPUT ALTERNATING CURRENTS OF A MULTIPULSE CONVERTER. Because the thyristor conduction periods of each 3-pulse group are displaced in phase from those of all the other groups, a cancellation of certain harmonic components of current occurs in those parts of the input a-c circuit which carry the currents of more than one 3-pulse group. At the same time, those harmonic currents which do not cancel, remain with the same relative amplitudes as in the 3-pulse waveform.

The pulse number of a given alternating current waveform is defined as being equal to the number of individual 3-pulse groups which contribute to that waveform, multiplied by 3. Since the primary input line current of a multipulse converter is necessarily constituted from the currents of each 3-pulse group within the system, it follows that the pulse number of the primary input line current is equal to the overall pulse number of the converter. In general, the higher is the pulse number of the current waveform, the more complete is the cancellation of harmonics, and therefore the less is the harmonic load carried by the a-c system. In other words, the higher is the converter pulse number, the closer the input power factor approaches the input displacement factor, and the more "nearly sinusoidal" is the load imposed by the converter on the a-c system.

HARMONIC ANALYSIS OF THE SIX-PULSE CURRENT. In Figs. 4.4 and 4.5, the current waveforms (along with other data) obtained in 6-pulse circuits having various transformer connections are shown.

---

* This, of course, assumes "concurrent" firing of all groups. See p. 110.

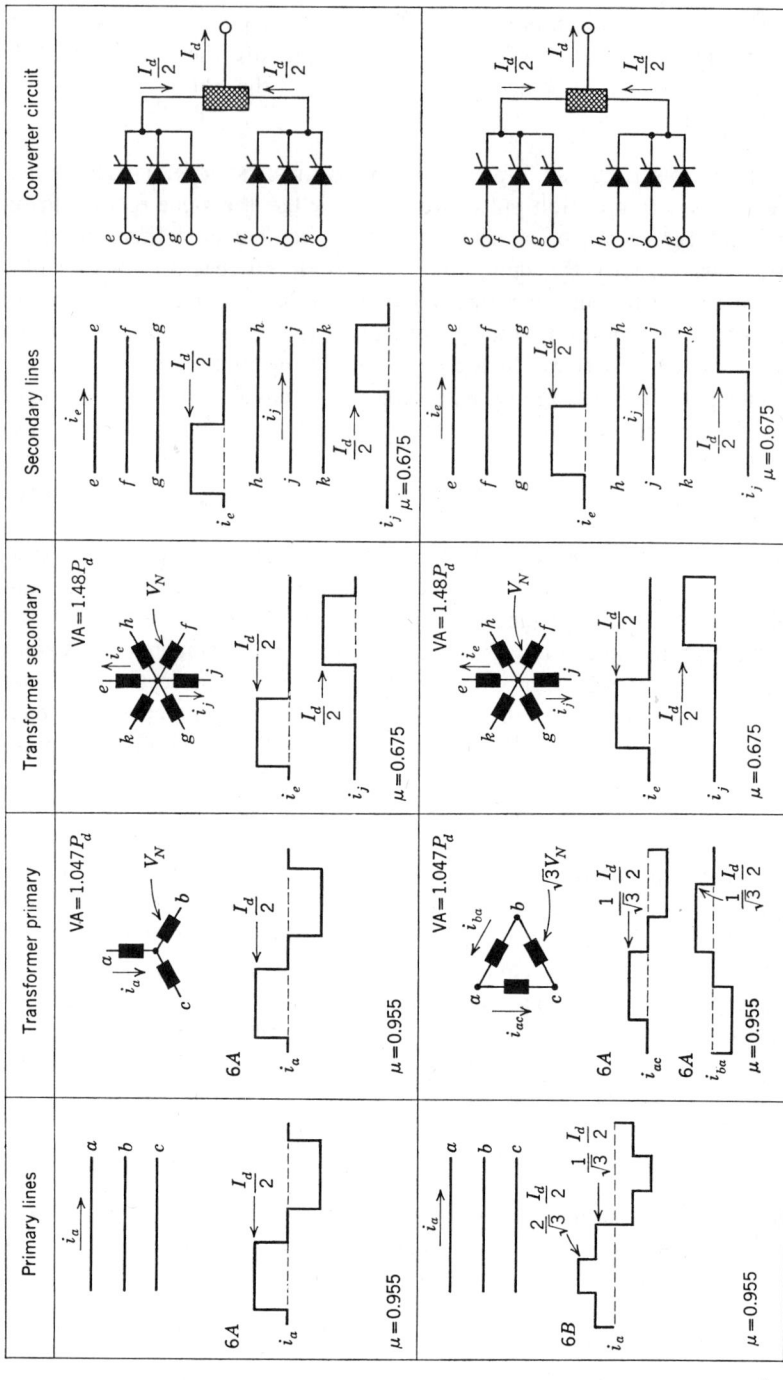

Figure 4.4. Current waveforms and related data for 2-quadrant 6-pulse converter circuits. Transformer volt-amperes are in terms of d-c power at $\alpha = 0°$.

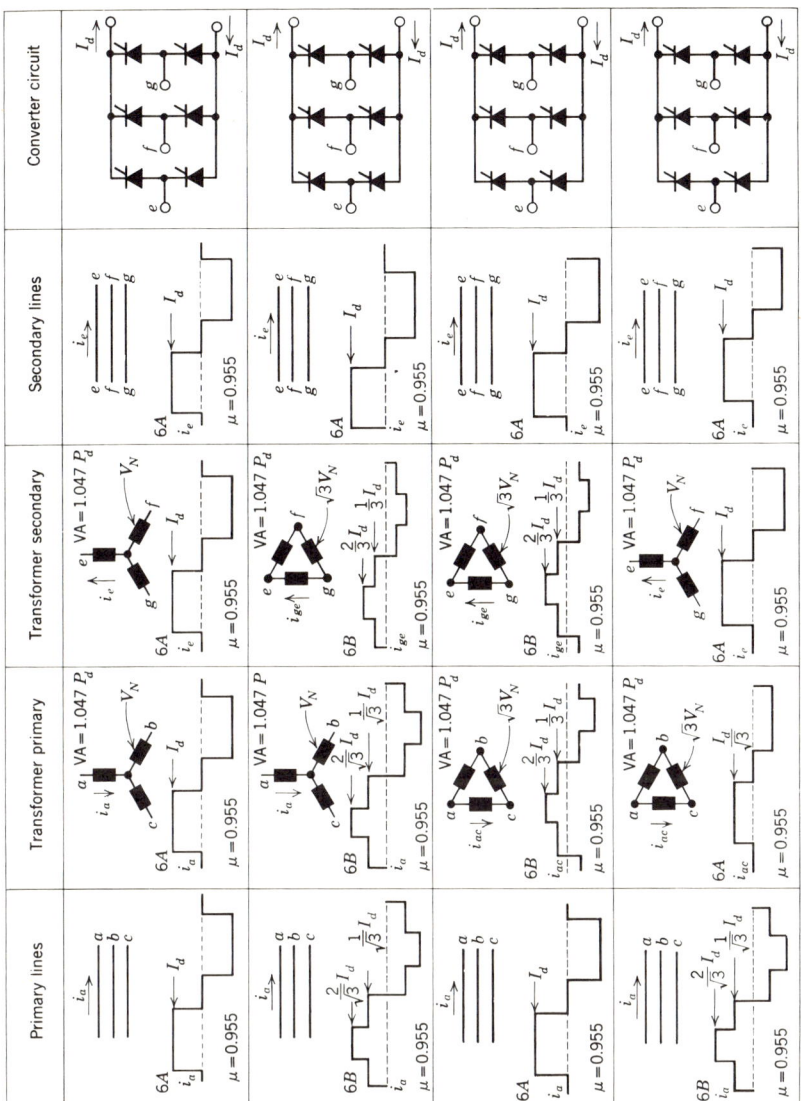

Figure 4.5. Current waveforms and related data for 2-quadrant 6-pulse converter circuits. Transformer volt-amperes are in terms of d-c power at $\alpha = 0°$.

It is seen from these illustrations that the waveform of the 6-pulse current in any of these circuits invariably has one or other of two alternative shapes, labeled 6A and 6B.

If the peak value of waveform 6A is defined as having unit amplitude, then this waveform is given by

$$i_{6A} = F_1(\theta_i - \alpha) - F_1(\theta_i - \alpha + \pi)$$

$$= \frac{2\sqrt{3}}{\pi}[\sin(\theta_i - \alpha) - \tfrac{1}{5}\sin 5(\theta_i - \alpha) - \tfrac{1}{7}\sin 7(\theta_i - \alpha)$$

$$+ \tfrac{1}{11}\sin 11(\theta_i - \alpha) + \tfrac{1}{13}\sin 13(\theta_i - \alpha) \cdots] \quad (4.25)$$

Thus the amplitude of the fundamental component is $2\sqrt{3}/\pi \times$ the peak value. Each of the even harmonic components of the 3-pulse waveform completely cancel one another in the 6-pulse waveform, whereas the remaining harmonic components do not cancel, and these are retained with the same relative amplitudes. These components have frequencies of 5, 7, and in general $(6n \pm 1) \times$ the fundamental frequency, and the amplitude of any given harmonic, relative to the fundamental, is equal to the reciprocal of the harmonic number.

If the peak value of waveform 6B is defined as having unit amplitude, then this waveform is given by

$$i_{6B} = 0.5 F_1\left(\theta_i - \alpha + \frac{\pi}{6}\right) - 0.5 F_1\left(\theta_i - \alpha - \frac{5\pi}{6}\right)$$

$$+ 0.5 F_1\left(\theta_i - \alpha - \frac{\pi}{6}\right) - 0.5 F_1\left(\theta_i - \alpha + \frac{5\pi}{6}\right)$$

$$= \frac{3}{\pi}[\sin(\theta_i - \alpha) + \tfrac{1}{5}\sin 5(\theta_i - \alpha) + \tfrac{1}{7}\sin 7(\theta_i - \alpha)$$

$$+ \tfrac{1}{11}\sin 11(\theta_i - \alpha) + \tfrac{1}{13}\sin 13(\theta_i - \alpha) \cdots] \quad (4.26)$$

Thus the amplitude of the fundamental component of waveform 6B is $3/\pi \times$ the peak value, and the harmonic components have the *same* frequencies and relative amplitudes as those of waveform 6A.

The explanation for the result that both these waves have the same relative harmonic content, but different shapes, resides in the fact that certain of the harmonic components have a different phase relationship with respect to the fundamental. This can be seen from an inspection of expressions (4.25) and (4.26).

HARMONIC ANALYSIS OF THE TWELVE-PULSE CURRENT. In Figs. 4.6 and 4.7 the current waveforms (along with other data) obtained in 12-pulse circuits having various transformer connections are shown.

PERFORMANCE OF TWO-QUADRANT CONVERTERS

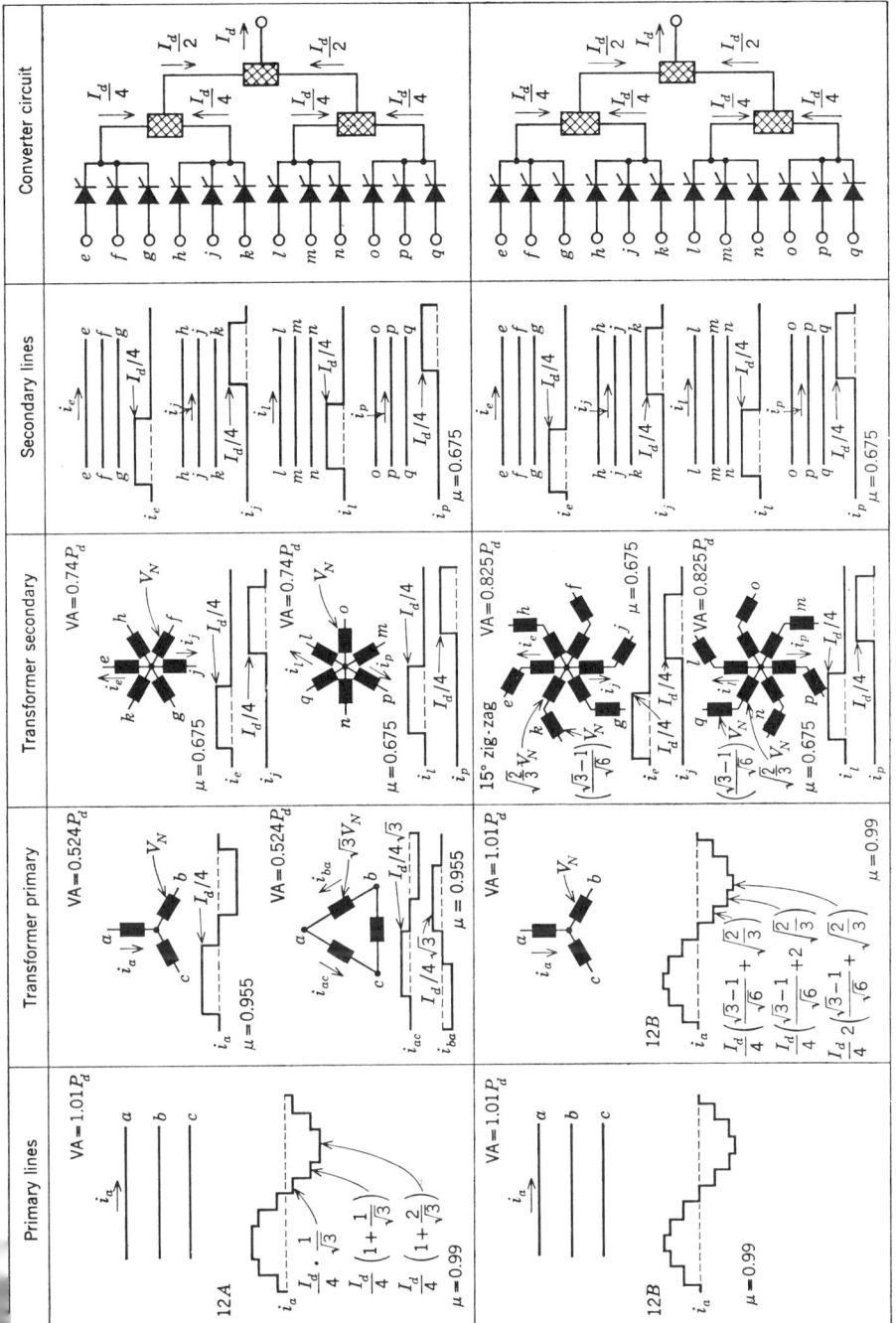

Figure 4.6. Current waveforms and related data for 2-quadrant 12-pulse converter circuits. Transformer and input volt-amperes are in terms of d-c power at $\alpha = 0°$.

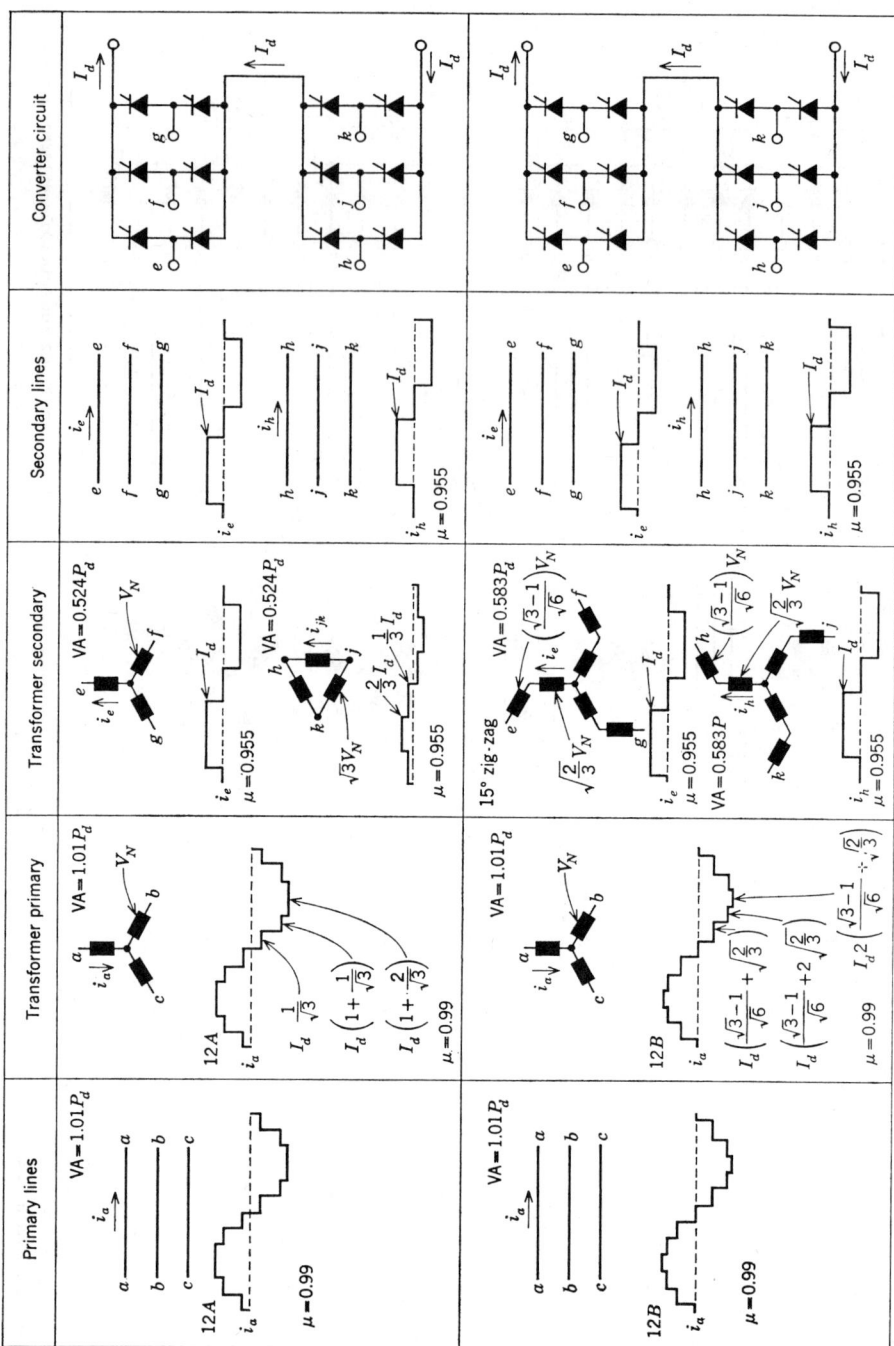

Figure 4.7. Current waveforms and related data for 2-quadrant 12-pulse converter circuits. Transformer and input volt-amperes are in terms of d-c power at $\alpha = 0°$.

PERFORMANCE OF TWO-QUADRANT CONVERTERS

It is seen from these illustrations that the waveform of the 12-pulse current in any of these circuits invariably has one of two shapes. These have been labeled 12A and 12B.

If the peak value of waveform 12A is defined as having unit amplitude, then this waveform is given by

$$i_{12A} = 0.464F_1(\theta_i - \alpha) - 0.464F_1(\theta_i - \alpha + \pi)$$

$$+ 0.268F_1\left(\theta_i - \alpha - \frac{\pi}{6}\right) - 0.268F_1\left(\theta_i - \alpha + \frac{5\pi}{6}\right)$$

$$+ 0.268F_1\left(\theta_i - \alpha + \frac{\pi}{6}\right) - 0.268F_1\left(\theta_i - \alpha - \frac{5\pi}{6}\right)$$

$$= 1.024[\sin(\theta_i - \alpha) + \tfrac{1}{11}\sin 11(\theta_i - \alpha) + \tfrac{1}{13}\sin 13(\theta_i - \alpha)$$

$$+ \tfrac{1}{23}\sin 23(\theta_i - \alpha) + \tfrac{1}{25}\sin 25(\theta_i - \alpha) \cdots] \quad (4.27)$$

Thus the amplitude of the fundamental component is $1.024 \times$ the peak value. The harmonic components have frequencies of 11, 13 and in general $(12n \pm 1) \times$ the fundamental frequency. Once again, the amplitude of any given harmonic, relative to the fundamental, is equal to the reciprocal of the harmonic number.

If the peak value of waveform 12B is defined as having unit amplitude, then this waveform is given by

$$i_{12B} = 0.366F_1\left(\theta_i - \alpha + \frac{\pi}{12}\right) - 0.366F_1\left(\theta_i - \alpha - \frac{11\pi}{12}\right)$$

$$+ 0.366F_1\left(\theta_i - \alpha - \frac{\pi}{12}\right) - 0.366F_1\left(\theta_i - \alpha + \frac{11\pi}{12}\right)$$

$$+ 0.134F_1\left(\theta_i - \alpha - \frac{\pi}{4}\right) - 0.134F_1\left(\theta_i - \alpha + \frac{3\pi}{4}\right)$$

$$+ 0.134F_1\left(\theta_i - \alpha + \frac{\pi}{4}\right) - 0.134F_1\left(\theta_i - \alpha - \frac{3\pi}{4}\right)$$

$$= 0.99[\sin(\theta_i - \alpha) - \tfrac{1}{11}\sin 11(\theta_i - \alpha) - \tfrac{1}{13}\sin 13(\theta_i - \alpha)$$

$$+ \tfrac{1}{23}\sin 23(\theta_i - \alpha) + \tfrac{1}{25}\sin 25(\theta_i - \alpha) \cdots] \quad (4.28)$$

Thus the amplitude of the fundamental component is $0.99 \times$ the peak value, and the harmonic components have the same frequencies and relative amplitudes as those of waveform 12A.

Again, the explanation for the result that both waves have the same harmonic content, but different shapes, is that certain of the harmonic components have a different phase relationship with respect to the fundamental. This can be seen from inspection of expressions (4.27) and (4.28).

The Universal Relationships Between Firing Angle, d-c Voltage Ratio, and Input Displacement Factor, for All Two-Quadrant Converters

For all 2-quardrant phase-controlled converters, regardless of pulse number, (with smooth current at the d-c terminals), the same basic relationship exists between the firing angle, the d-c voltage ratio, and the input displacement factor. This relationship is evident from the preceding analytical work, and it is stated mathematically as follows:

$$\cos \alpha = r = \cos \phi_i \tag{4.29}$$

These relationships are illustrated graphically in Figs. 4.8 and 4.9.

Expression (4.29) is a mathematical statement of the characteristic operation of all 2-quadrant phase-controlled converters. It indicates that a continuous reduction of the d-c terminal voltage, from maximum to zero, is brought about by a continuous phase-retardation of the firing angle, from 0 to 90°. This reduction in voltage is accompanied by a proportional reduction in the input displacement factor. By implication, this means that the

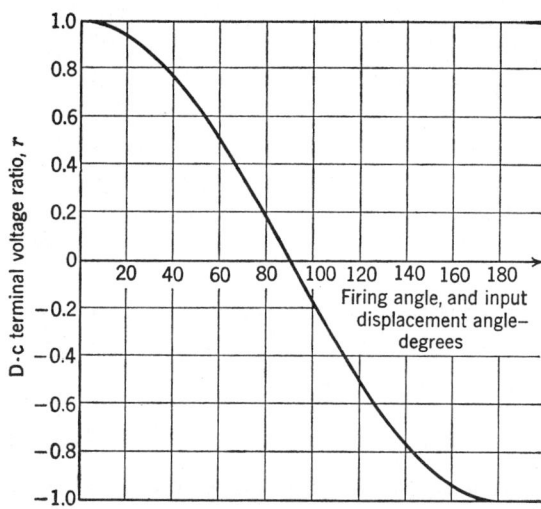

Figure 4.8. Universal relationship between the d-c terminal voltage ratio, the firing angle, and the input displacement angle, for all 2-quadrant phase-controlled converters (with smooth current at the d-c terminals).

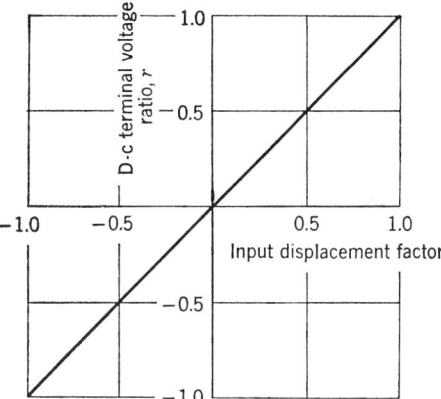

Figure 4.9. Universal relationship between the d-c terminal voltage ratio and the input displacement factor, for all 2-quadrant phase-controlled converters (with smooth current at the d-c terminals).

amplitude of the fundamental component of current at the input side invariably bears the same relationship to the output current, and only its phase position changes as the output voltage is changed. Over this part of the control range, the converter operates as a rectifier, and the power flow is from the a-c to the d-c side.

Expression (4.29) indicates that further phase delay of the firing angle, beyond 90°, causes the d-c terminal voltage to reverse its polarity; and increasing phase delay now produces an increasing negative voltage. This increasing negative voltage is accompanied by a proportionally increasing negative displacement factor at the input side. Over this part of the operating range, the converter is operating as a "line commutated" inverter, and the power flow is from the d-c to the a-c side of the system.

The characteristic operation, described by expression (4.29), is applicable to all 2-quadrant converter circuits, regardless of the circuit configuration or pulse number. Thus, the differences which do exist between the external performance characteristics of circuits with different pulse numbers, are related entirely to the matter of the harmonic distortion at the input and output sides of the circuit. Moreover, from the foregoing analysis, it is clear that the differences of harmonic distortion which exist between circuits of different pulse numbers can be resolved, quite simply, into a question of whether or not given harmonic components are present or absent; furthermore, those harmonic components which are present invariably have the same relative amplitude.

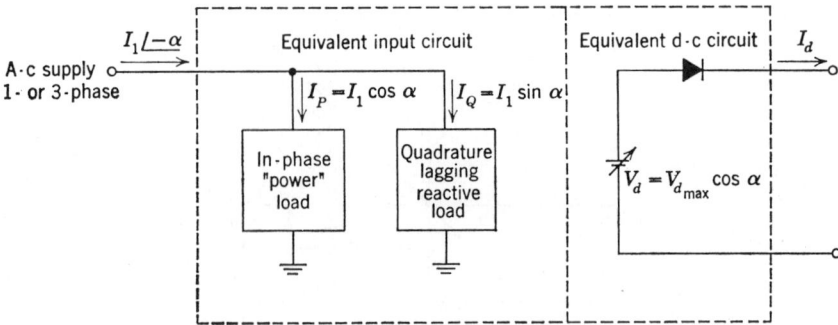

Figure 4.10. Simplified equivalent circuit representation of the 2-quadrant phase-controlled converter.

From expression (4.29), it is possible to derive the simple equivalent circuit, shown in Fig. 4.10, which is universally applicable to all 2-quadrant converters. This equivalent circuit neglects the harmonic currents at the a-c input side, and the harmonic voltages at the d-c side, which are a function of the circuit pulse number, and includes only the fundamental component of current at the input side, and the direct component of voltage at the output side, which are independent of the circuit pulse number. At the input side, the load presented by the converter to the a-c system is represented as an in-phase component of current, $I_P = I_1 \cos \alpha$, and a quadrature component of current, $I_Q = I_1 \sin \alpha$. At the d-c side, the converter is represented as a controllable direct voltage source, $V_d = V_{d_{max}} \cos \alpha$, connected in series with a diode, which represents the condition of unidirectional current flow through the converter.

The Universal Harmonic Relationships for All Two-Quadrant Converters

For all 2-quadrant converters, regardless of pulse number, a simple relationship exists between the circuit pulse number, the harmonic frequencies present in the input line current, and the frequencies present in the d-c terminal voltage. This relationship is evident from the preceding analytical work, and it can be stated as follows:

The harmonic frequencies present in the d-c terminal voltage have orders which are integer multiples of the circuit pulse number. For every harmonic frequency present in the d-c terminal voltage, there are two associated harmonic frequencies present in the converter input line current; the first of these frequencies has one lower order, and the second, one higher order, than the frequency of the associated d-c terminal voltage harmonic.

A specific example will serve to illustrate this rule. For a 6-pulse converter, the harmonic frequencies present in the d-c voltage are 6, 12, and in general

PERFORMANCE OF ONE-QUADRANT CONVERTERS 95

$6n \times$ the input frequency. The corresponding harmonic frequencies present in the alternating input currents are 5, 7, 11, 13, and in general $(6n \pm 1) \times$ the line frequency.

The following simple rules apply for determining the amplitudes of any voltage or current harmonic:

The peak amplitude, relative to the maximum d-c component, of a given voltage harmonic of order n at a firing angle α, is given by

Amplitude of voltage harmonic of order n, relative to the maxmium dc terminal voltage
$$= \left[\frac{1}{(n-1)^2} + \frac{1}{(n+1)^2} - \frac{2\cos 2\alpha}{(n+1)(n-1)}\right]^{1/2} \quad (4.30)$$

The amplitude of any given harmonic component of current, relative to the fundamental, is, quite simply, equal to the reciprocal of the harmonic number. Thus

$$\text{relative amplitude of current harmonic of order } n = 1/n \quad (4.31)$$

This result (which assumes a smooth current at the d-c terminals of the converter) is quite independent of the converter firing angle.

THE PERFORMANCE CHARACTERISTICS OF ONE-QUADRANT CONVERTERS

In Chapter 3 it was seen that a converter which is capable of only a 1-quadrant operation can be constructed either by connecting uncontrolled diodes in place of the thyristors in certain positions of the converter circuit, or by connecting a freewheel diode, or diodes, at the d-c terminals. In either case, the result is to modify the external performance characteristics, as compared with those of the 2-quadrant converter.

At the d-c side, the harmonic distortion of the voltage, and the relationship between the mean voltage and the firing angle, are both altered. At the a-c side, the harmonic spectrum of the input current is modified; and, of particular importance, the amplitude of the quadrature component of current, as well as the total rms value of the current wave, are considerably reduced at low values of d-c terminal voltage ratio, as compared with the corresponding currents obtained with the 2-quadrant converter.

These effects are briefly examined in the following sections.

Harmonic Content of the d-c Terminal Voltage, and of the a-c Input Current, of Various One-Quadrant Converters

Mathematical expressions for the d-c terminal voltage, and the a-c input current, of any 1-quadrant converter circuit can be deduced, without returning to first principles, from the analytical data already derived for 2-quadrant converters. For the sake of conciseness, however, the derivation of these harmonic series, for 1-quadrant converters, is not presented here; rather, just the final results of a harmonic analysis of the d-c terminal voltage, and of the a-c input current, of various 1-quadrant converter circuits, are given in graphical form.

The following is a summary of the specific 1-quadrant converter circuits considered, together with the figure numbers of the related graphical information:

Circuit	*Figure Numbers*
2-Pulse Half-Controlled Converter	4.11, 4.12, 4.21, 4.24, 4.25 and 4.26
2-Pulse Midpoint Converter with Freewheel Diode	4.11, 4.12, 4.22, 4.27, 4.28 and 4.29
3-Pulse Half-Controlled Bridge Converter	4.13, 4.14, 4.21, 4.24, 4.25 and 4.26
3-Pulse Converter with Freewheel Diode	4.15, 4.16, 4.22, 4.27, 4.28 and 4.29
6-Pulse Half-Controlled Converter	4.17, 4.18, 4.21, 4.24, 4.25 and 4.26
6-Pulse Converter with One Freewheel Diode	4.19, 4.20, 4.22, 4.27, 4.28 and 4.29
6-Pulse Converter with Two Freewheel Diodes	4.19, 4.20, 4.22, 4.27, 4.28 and 4.29

Relationships Between Firing Angle, d-c Voltage Ratio, and Input Displacement Factor for One-Quadrant Converters

UNIVERSAL RELATIONSHIPS FOR ALL HALF-CONTROLLED CONVERTERS. Regardless of the circuit pulse number, the half-controlled converter essentially consists of a controlled 2-quadrant converter, connected in series with an uncontrolled converter. Thus, just as for the 2-quadrant converter, the half-controlled converter has associated with it a set of "universal"

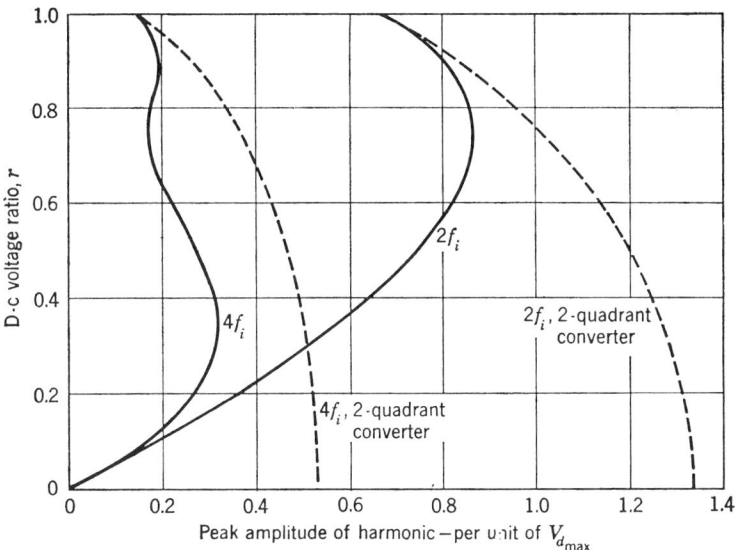

Figure 4.11. Curves showing the variation, with d-c voltage ratio, of the amplitudes of the predominant harmonic components in the d-c terminal voltage of the 2-pulse half-controlled bridge converter, and the 2-pulse midpoint converter with freewheel diode.

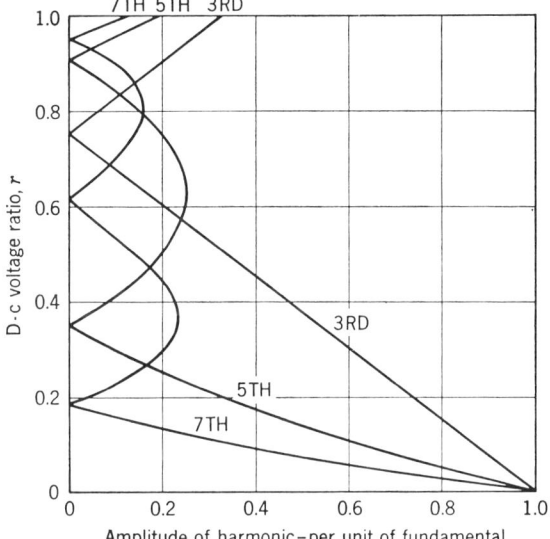

Figure 4.12. Curves showing the variation, with d-c voltage ratio, of the amplitudes of the predominant harmonic components of input current of the 2-pulse half-controlled bridge converter, and the 2-pulse midpoint converter with freewheel diode.

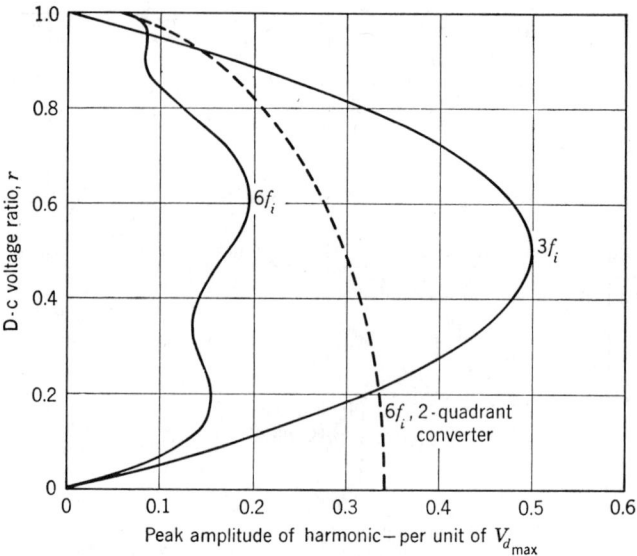

Figure 4.13. Curves showing the variation, with d-c voltage ratio, of the amplitudes of the predominant harmonic components in the d-c terminal voltage of the 3-pulse half-controlled bridge converter.

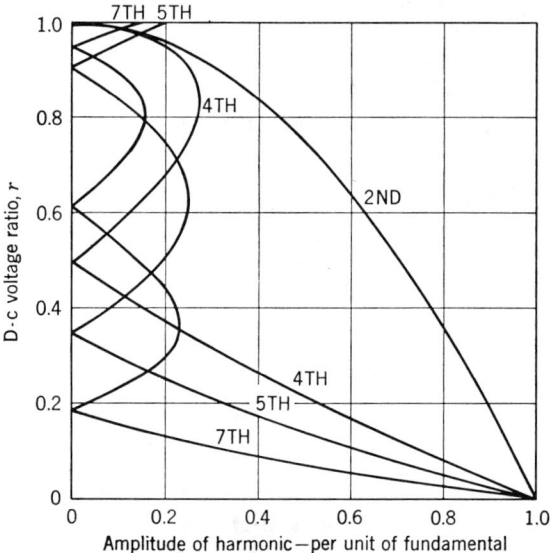

Figure 4.14. Curves showing the variation, with d-c voltage ratio, of the amplitudes of the predominant harmonic components of input current of the 3-pulse half-controlled bridge converter.

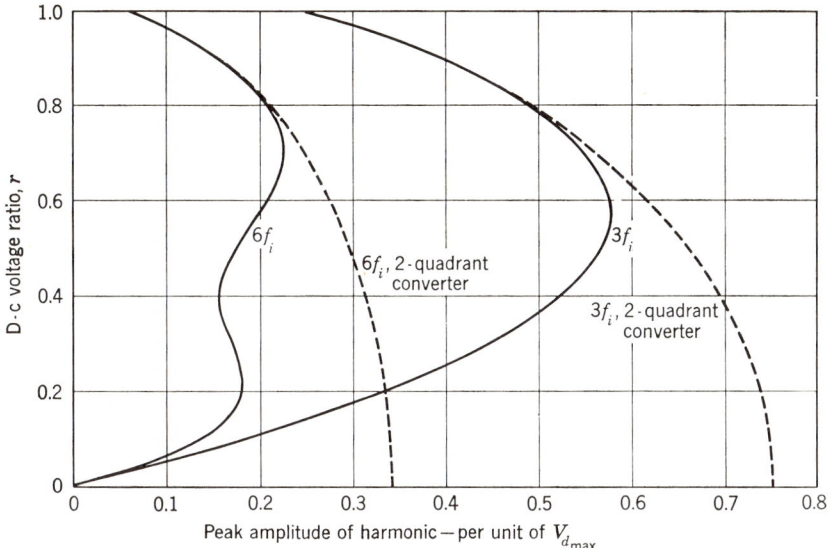

Figure 4.15. Curves showing the variation, with d-c voltage ratio, of the amplitudes of the predominant harmonic components in the d-c terminal voltage of the 3-pulse converter with freewheel diode.

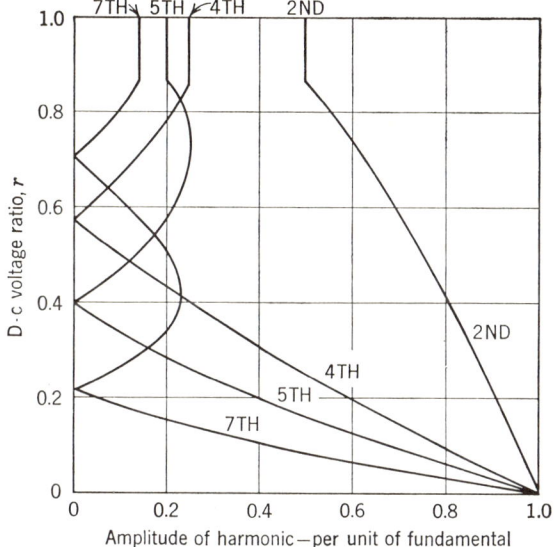

Figure 4.16. Curves showing the variation, with d-c voltage ratio, of the amplitudes of the predominant harmonic components of input current of the 3-pulse converter with freewheel diode.

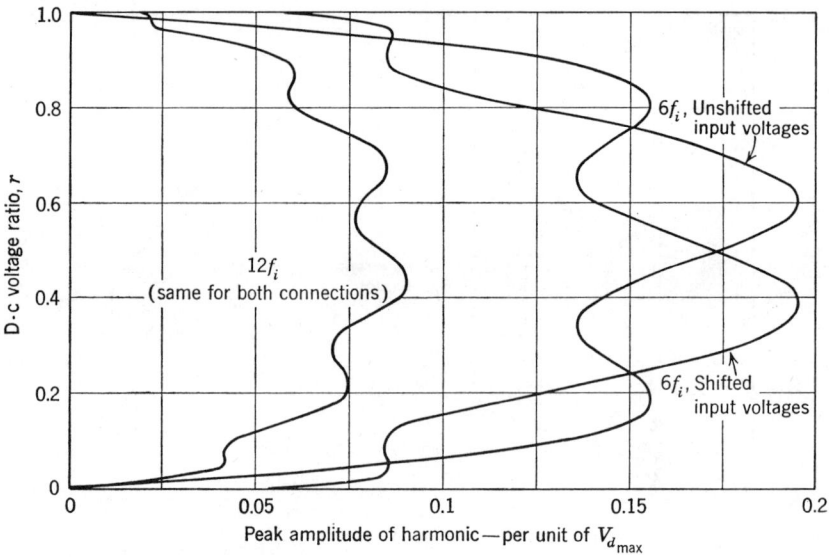

Figure 4.17. Curves showing the variation, with d-c voltage ratio, of the amplitudes of the predominant harmonic components in the d-c terminal voltage of the 6-pulse half-controlled converter, with shifted and unshifted input voltages.

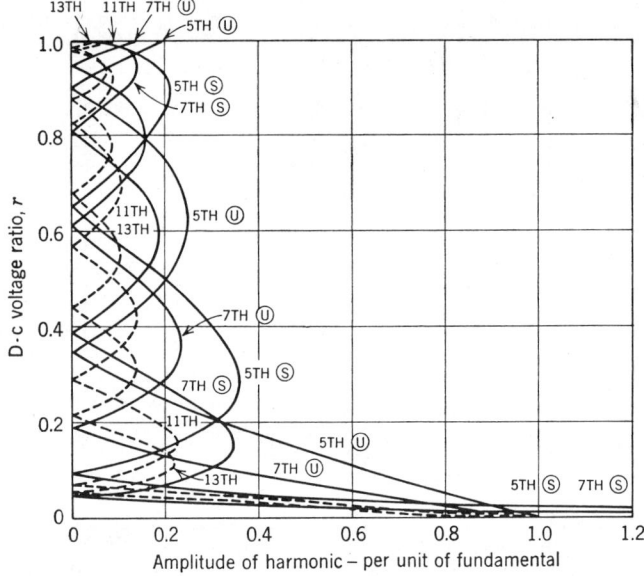

Figure 4.18. Curves showing the variation, with d-c voltage ratio, of the amplitudes of the predominant harmonic components of input current of the 6-pulse half-controlled converter, with shifted and unshifted input voltages. U Signifies unshifted input voltages; S signifies shifted input voltages. Eleventh and thirteenth harmonics are the same for both circuit connections.

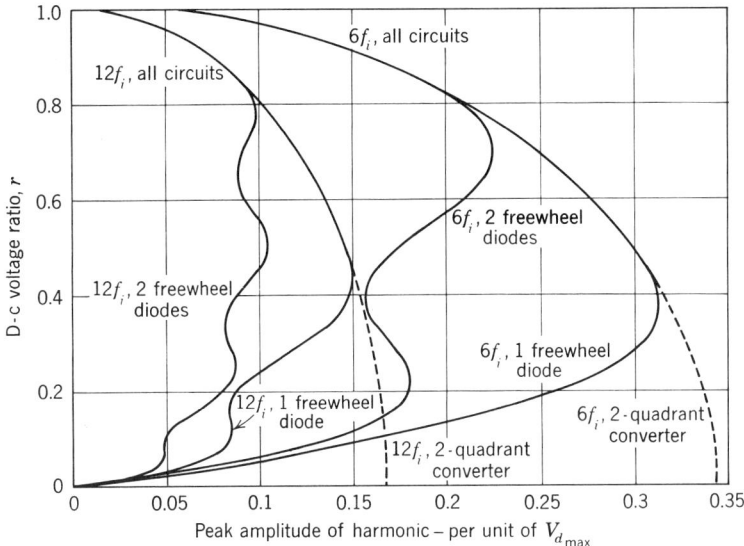

Figure 4.19. Curves showing the variation, with d-c voltage ratio, of the amplitudes of the predominant harmonic components in the d-c terminal voltage of the 6-pulse converter, with one and two freewheel diodes.

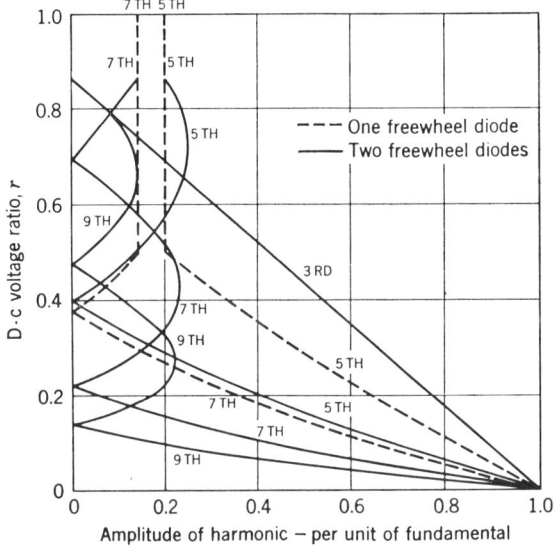

Figure 4.20. Curves showing the variation, with d-c voltage ratio, of the amplitudes of the predominant harmonic components of input current of the 6-pulse converter, with one and two freewheel diodes.

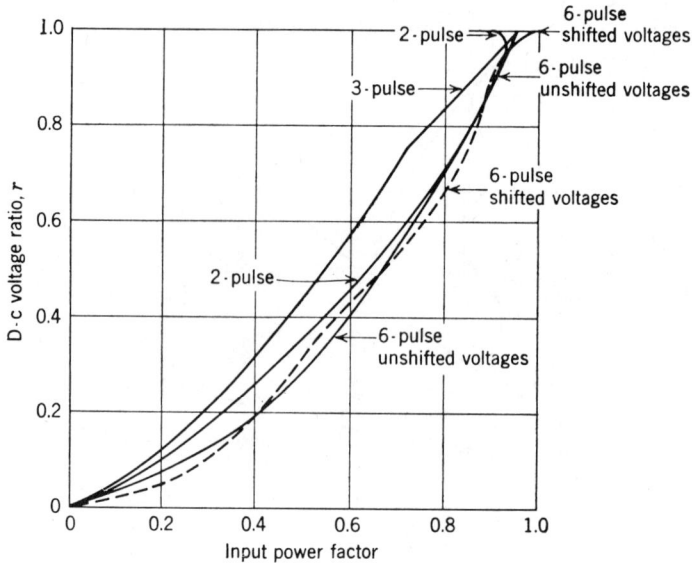

Figure 4.21. Relationships between the d-c voltage ratio and the input power factor, for 2-, 3-, and 6-pulse half-controlled converters.

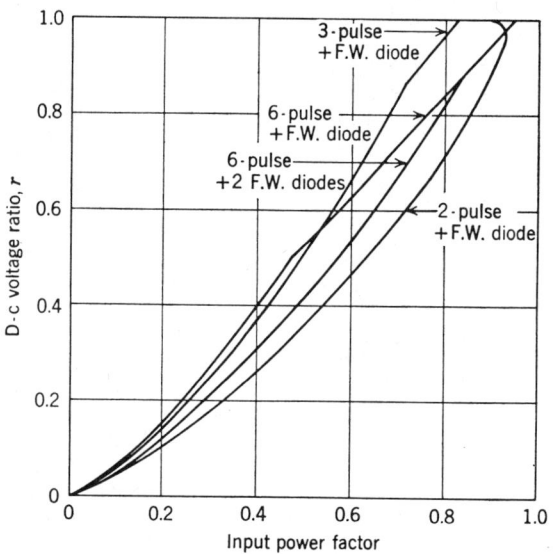

Figure 4.22. Relationships between the d-c voltage ratio and the input power factor, for 2-, 3-, and 6-pulse converters with freewheel diodes.

PERFORMANCE OF ONE-QUADRANT CONVERTERS

relationships—between the firing angle, d-c voltage ratio, and input displacement factor—which are independent of the circuit pulse number. These relationships can be deduced, without resort to detailed waveform analysis, from the corresponding universal relationships, already known, for the 2-quadrant converter.

Firstly, it is helpful to derive the simplified equivalent circuit diagram for the half-controlled converter, from the corresponding circuit diagram already deduced for the 2-quadrant converter (Fig. 4.10). The derivation of this equivalent circuit is illustrated in Fig. 4.23.

The equivalent circuit at the d-c side consists of a controllable direct voltage source, $V_{d_{\max}}/2 \cos \alpha$, representing the voltage of the controlled half of the converter, connected in series with a fixed direct voltage source, $V_{d_{\max}}/2$, representing the voltage of the uncontrolled half of the converter.

Thus, the d-c terminal voltage ratio of the half-controlled converter is given by

$$r = \frac{V_d}{V_{d_{\max}}} = \tfrac{1}{2}(1 + \cos \alpha) \tag{4.32}$$

This relationship is illustrated graphically in Fig. 4.24.

The fundamental current at the a-c side consists of an in-phase component, $I_1/2 \cos \alpha$, and a quadrature component, $I_1/2 \sin \alpha$, due to the controlled half-converter, and a second, fixed amplitude, in-phase component, $I_1/2$, due to the uncontrolled half-converter.

Thus, the total in-phase component of current is given by

$$I_P = \frac{I_1}{2}(1 + \cos \alpha) \tag{4.33}$$

$$= I_1 \cdot r \tag{4.34}$$

And the total quadrature component of current is given by

$$I_Q = \frac{I_1}{2} \sin \alpha \tag{4.35}$$

$$= I_1 \sqrt{r - r^2} \tag{4.36}$$

The input displacement factor is given by

$$\cos \phi_i = \frac{I_P}{\sqrt{I_P^2 + I_Q^2}}$$

Hence

$$\cos \phi_i = \sqrt{r} = \cos \frac{\alpha}{2} \tag{4.37}$$

This relationship is illustrated graphically in Fig. 4.25.

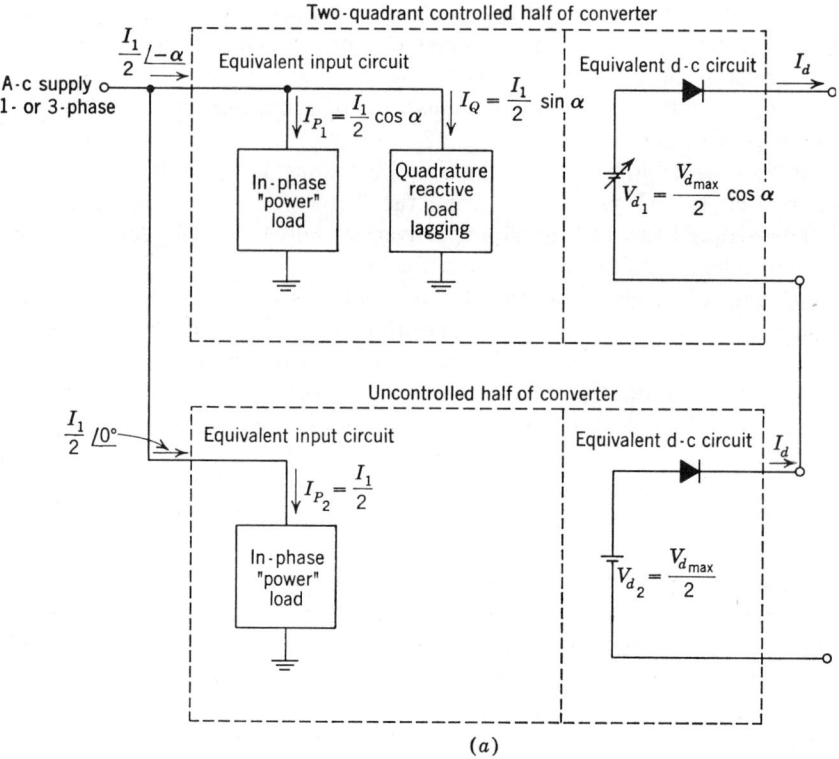

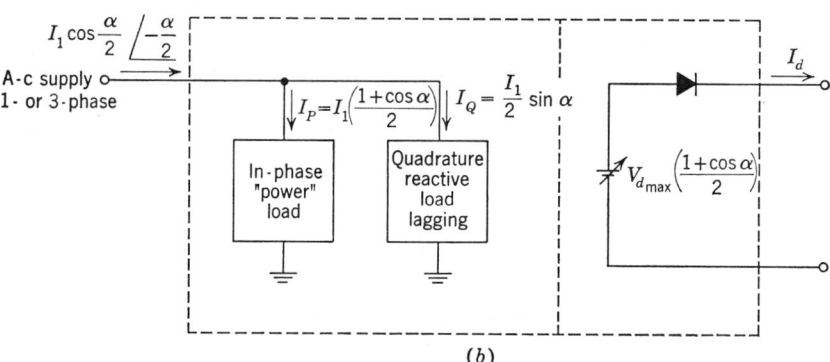

Figure 4.23. Equivalent circuit representation of a half-controlled converter. (*a*) Equivalent circuit representation of a half-controlled converter, consisting of a controlled converter connected in series with an uncontrolled converter. (*b*) Composite equivalent circuit representation of a half-controlled converter.

PERFORMANCE OF ONE-QUADRANT CONVERTERS

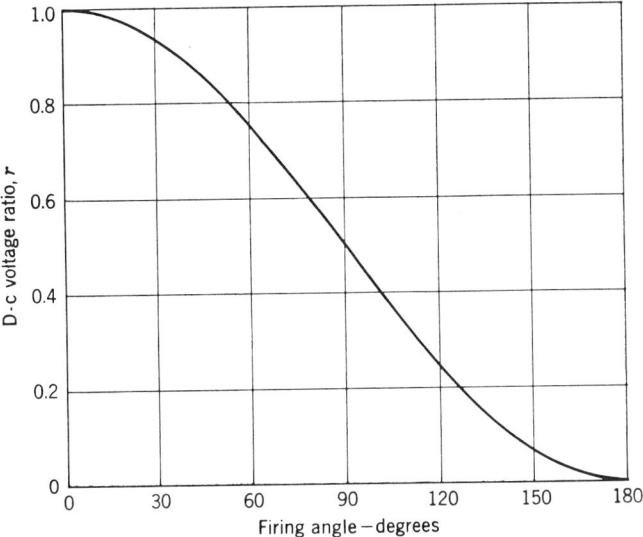

Figure 4.24. Universal relationship between the d-c terminal voltage ratio and the firing angle, for all half-controlled converters.

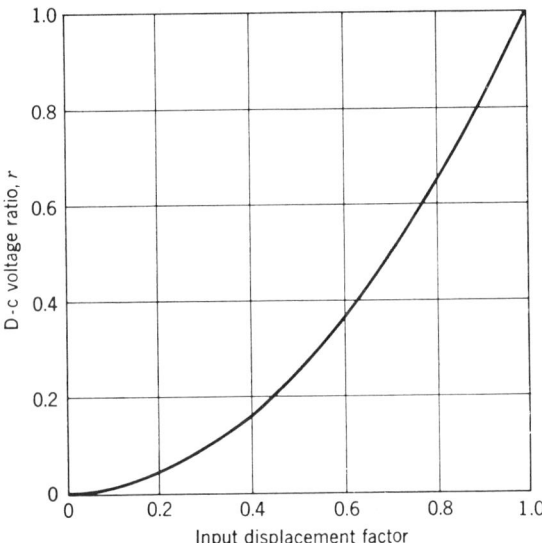

Figure 4.25. Universal relationship between the d-c terminal voltage ratio and the input displacement factor, for all half-controlled converters.

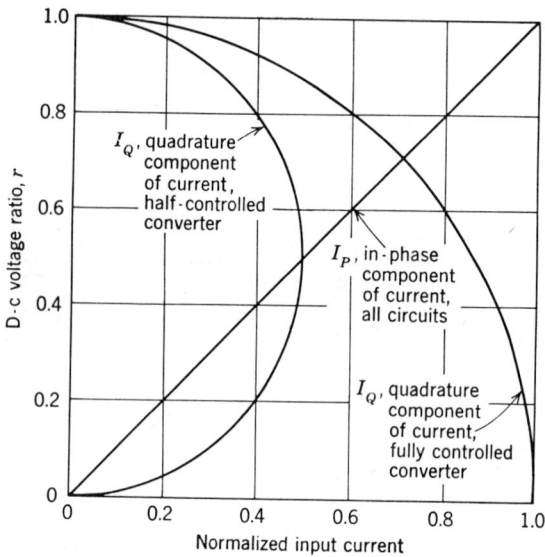

Figure 4.26. Universal relationships between the d-c terminal voltage ratio, and the in-phase and quadrature components of input current, for all half-controlled and fully controlled (2-quadrant) converters.

$$\text{Normalized input current} = \frac{\text{amplitude of input current component with } I_d \text{ at output}}{\text{amplitude of fundamental input current with } r=1, I_d \text{ at output}}.$$

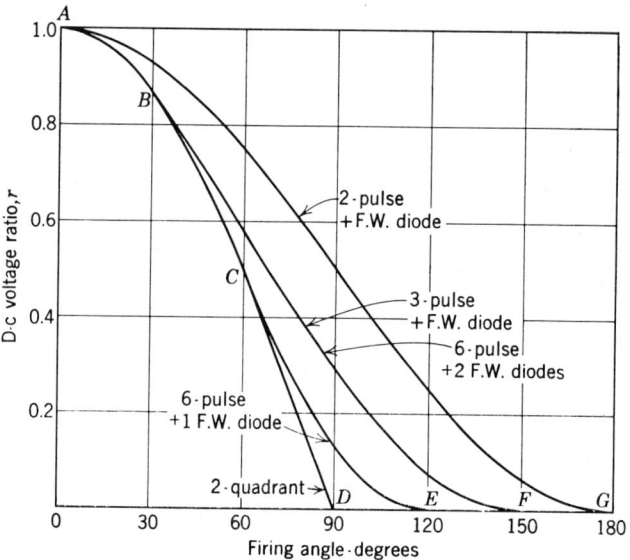

Figure 4.27. Relationships between the d-c terminal voltage ratio and the firing angle for 2-, 3-, and 6-pulse converters with freewheel diodes. *A-G*—2-pulse converter with freewheel diode; *A-B-F*—3-pulse converter with freewheel diode; *A-B-F*—6-pulse converter with two freewheel diodes; *A-B-C-E*—6-pulse converter with one freewheel diode; and *A-B-C-D*—2-quadrant converter.

PERFORMANCE OF ONE-QUADRANT CONVERTERS 107

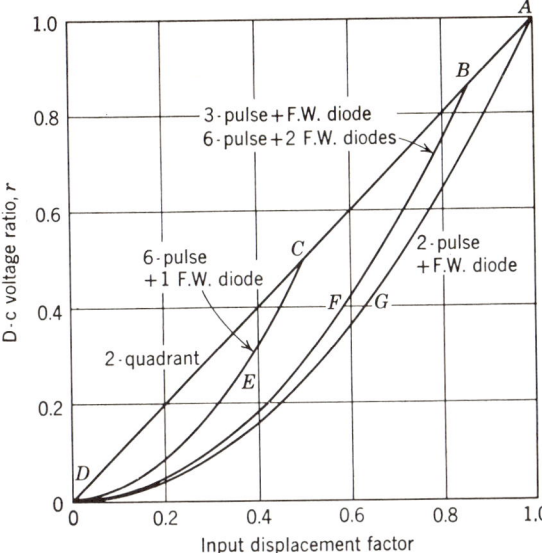

Figure 4.28. Relationships between the d-c terminal voltage ratio and the input displacement factor, for 2-, 3-, and 6-pulse converters with freewheel diodes. *A-G-D*—2-pulse converter with freewheel diode; *A-B-F-D*—3-pulse converter with freewheel diode; *A-B-F-D*—6-pulse converter with two freewheel diodes; *A-B-C-E-D*—6-pulse converter with one freewheel diode; and *A-B-C-D*—2-quadrant converter.

The curves of Fig. 4.26 illustrate the universal relationships between the d-c voltage ratio, and the normalized in-phase and quadrature components of current at the input of the half-controlled converter. Also shown are the corresponding relationships for the 2-quadrant converter. For both types of converter, the same linear relationship exists between the normalized in-phase component of current and the d-c voltage ratio. This is necessarily so, since this relationship is implicit in the fact that the input and output powers (theoretically) are equal to one another. For the half-controlled converter, however, the maximum quadrature component of current is only a half of that of the fully-controlled converter, and this is obtained at half maximum output voltage. At zero output voltage, the half-controlled converter consumes no fundamental input current, whereas the fully-controlled converter consumes its maximum (normalized) quadrature current. Thus, so far as the loading of the supply at reduced voltage ratio is concerned, the half-controlled converter has considerably superior characteristics to the fully-controlled converter.

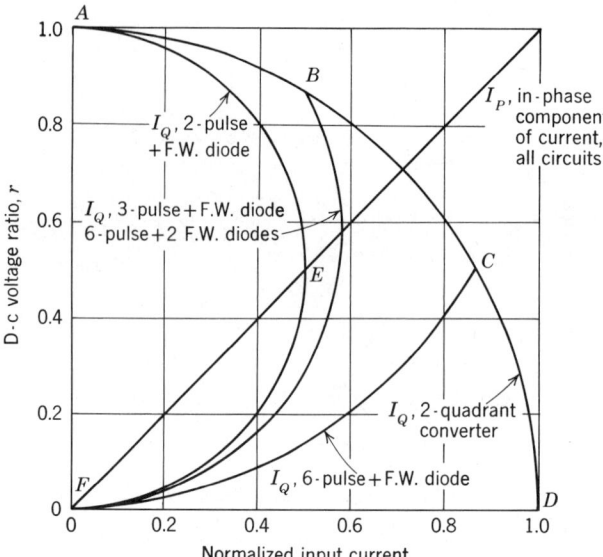

Figure 4.29. Relationships between the d-c terminal voltage ratio, and the in-phase and quadrature components of input current, for 2-, 3-, and 6-pulse converters with freewheel diodes. *A-E-F*—quadrature current: 2-pulse converter with freewheel diode; *A-B-F*—quadrature current: 3-pulse converter with freewheel diode; *A-B-F*—quadrature current: 6-pulse converter with two freewheel diodes; *A-B-C-F*—quadrature current: 6-pulse converter with one freewheel diode; and *A-B-C-D*—quadrature current: 2-quadrant converter.

$$\text{Normalized input current} = \frac{\text{amplitude of input current component with } I_d \text{ at output}}{\text{amplitude of fundamental current with } r = 1, I_d \text{ at output}}$$

RELATIONSHIPS FOR CONVERTERS WITH FREEWHEEL DIODES. For converters with freewheel diodes, the relationships between the firing angle, the d-c voltage ratio, and the input displacement factor, depend upon the point in the control range at which the freewheel diode starts to come into operation, which itself depends upon the pulse number of the converter.

The curves in Figs. 4.27 and 4.28 illustrate respectively the relationships between the d-c terminal voltage ratio and the firing angle, and between the d-c terminal voltage ratio and the input displacement factor, for converters of various pulse numbers, with freewheel diodes.

The curves of Fig. 4.29 show the relationships between the d-c terminal voltage ratio, and the normalized in-phase and quadrature components of current at the input, for converters of various pulse numbers, with freewheel diodes. These curves illustrate the fact that the later in the control range the freewheel diode starts to come into operation, the greater is the maximum quadrature component of current at the input side. Nevertheless, in all cases, the total fundamental input current decreases to zero, as the output voltage is reduced to zero.

REDUCTION OF REACTIVE LOADING OF THE SUPPLY BY THE TWO-QUADRANT CONVERTER, BY MEANS OF CONSECUTIVE FIRING ANGLE CONTROL

To complete this chapter, brief mention should be made of the fact, which actually is evident from the analysis of the performance characteristics of the half-controlled converter, that the lagging component of input current normally associated with the 2-quadrant converter can be reduced, by means of a process of "consecutive" firing angle control.

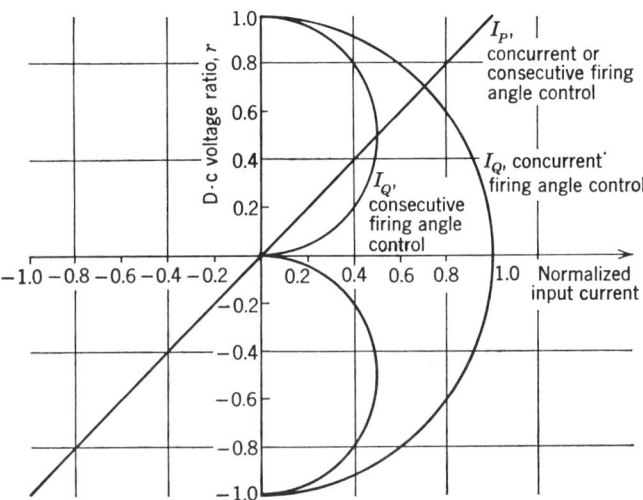

Figure 4.30. Relationships between the d-c terminal voltage ratio, and the in-phase and quadrature components of input current, for a 2-quadrant converter with 2-stage "consecutive firing angle control." Relationships for "concurrent firing angle control" also shown.

Normalized input current = $\dfrac{\text{amplitude of input current component with } I_d \text{ at output}}{\text{amplitude of fundamental input current with } r=1, I_d \text{ at output}}$.

I_P = In-phase component of current; I_Q = quadrature component of current. *Note:* Practical I_Q characteristic with consecutive firing angle control departs slightly from the theoretical characteristic shown because of commutation recovery angle required for inverter operation of first-controlled converter stage.

It has already been seen that the half-controlled converter produces a reduced quadrature component of load at the input side, as compared with the fully controlled converter. Thus, by controlling the firing angles of two series connected groups of thyristors, within a 2-quadrant converter, in such a way that a half-controlled-like operation is obtained, it is possible to reduce the quadrature load at the input side. The basic principle is to control the voltage of one group from "full rectification" to "full inversion," whilst the other stays at "full rectification." Over this part of the control range, the combined d-c terminal voltage is reduced from maximum to virtually zero. In order to operate in the second quadrant, the voltage of the second group is controlled from "full rectification" to "full inversion," whilst the first group stays at "full inversion."

With this "consecutive" firing angle control technique, the maximum normalized quadrature component of load occurs at half maximum voltage, in both the positive and negative voltage quadrants, and has an amplitude of half that obtained with a conventional "concurrent" firing angle control. To offset this advantage, however, the fundamental ripple frequency at the d-c terminals of the converter is reduced by a factor of 2:1 (except for the 2-pulse converter).

The curves of Fig. 4.30 show the relationships between the d-c terminal voltage ratio, and the normalized in-phase and quadrature components of current at the input, for a 2-quadrant converter with "consecutive" firing angle control. For purposes of comparison, the corresponding curve for conventional "concurrent" firing angle control, is also shown.

Of course, the principle of "consecutive" firing angle control can be extended to circuits containing more than 2 groups of thyristors connected in series, with further corresponding reductions in the quadrature components of current at the input side.

Chapter Five

The Dual-Converter

It has been seen that the phase-controlled converter can provide a continuously controllable d-c voltage of either polarity at its output terminals. However, due to the unidirectional current carrying property of the thyristors within the converter, the current at the d-c terminals can flow in only one direction. Thus a single converter is capable of providing only a so-called "2-quadrant" operation.

A converting system which gives a complete "4-quadrant" operation, that is to say, which operates with both polarities of voltage and current at the d-c terminals, can be constructed by connecting the d-c terminals of two "oppositely poled" phase-controlled converters in parallel with one another, thus providing a bidirectional path for the current. The resulting circuit connection is known as a *dual-converter*.

In this chapter, the basic principles of operation and control of the dual-converter are discussed and explained.

BASIC PRINCIPLE OF THE DUAL-CONVERTER

A dual-converter consists of two similar 2-quadrant converter circuits, with the thyristors "facing in opposite directions," the d-c terminals of which are connected in parallel with one another. With this arrangement, the current can flow in either direction at the d-c terminals, "positive" load current being carried by the "positive" converter, and "negative" current by the "negative" converter. Figure 5.1 shows diagrams of typical dual-converter circuits.

The basic principle of operation of the dual-converter can be explained by reference to the simplified equivalent diagram of the d-c circuit, shown in Fig. 5.2. In this simple representation, the a-c ripple voltage components which appear at the d-c terminals of the converters are neglected, and the equivalent circuit for each 2-quadrant converter is assumed to be a controllable direct voltage source, connected in series with a diode, which represents the condition of unidirectional current flow through the converter.

112 THE DUAL-CONVERTER

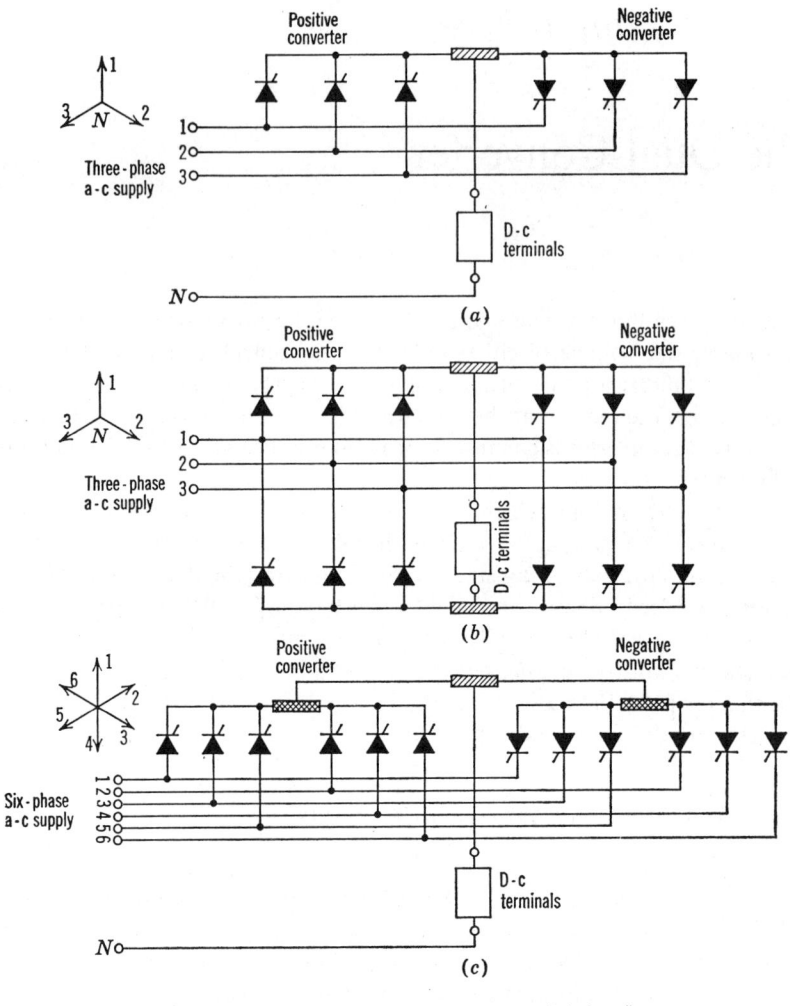

▨ = Circulating current reactor. This may or may not be required, depending upon whether dual-converter operates with or without circulating current
▧ = Interphase reactor

Figure 5.1. Three- and six-pulse dual-converter circuits. (*a*) 3-pulse midpoint dual-converter; (*b*) 6-pulse bridge dual-converter; and (*c*) 6-pulse midpoint dual-converter.

BASIC PRINCIPLE OF THE DUAL-CONVERTER

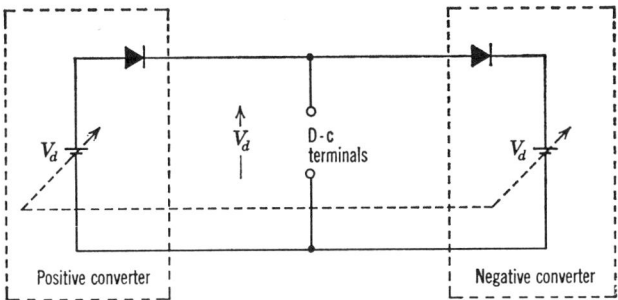

Figure 5.2. Idealized equivalent d-c circuit for the dual-converter.

The basic control principle of the ideal dual-converter is to regulate the firing angles of the individual converters so that their d-c voltages are always exactly equal, and of the same circuit polarity, as one another. Thus, when one converter operates as a rectifier, having a given d-c terminal voltage, the other operates as an inverter, having exactly the same "counter" voltage, and vice versa. Since the d-c terminal voltages of the individual converters are equal, and have the same circuit polarity as one another, the voltage at the d-c output terminals of the dual-converter is equal to the voltage of either converter, and, in this ideal circuit, the current has an equal freedom to flow either through the "positive" or the "negative" converter.

In Fig. 5.3 the firing angle—d-c terminal voltage relationships of the

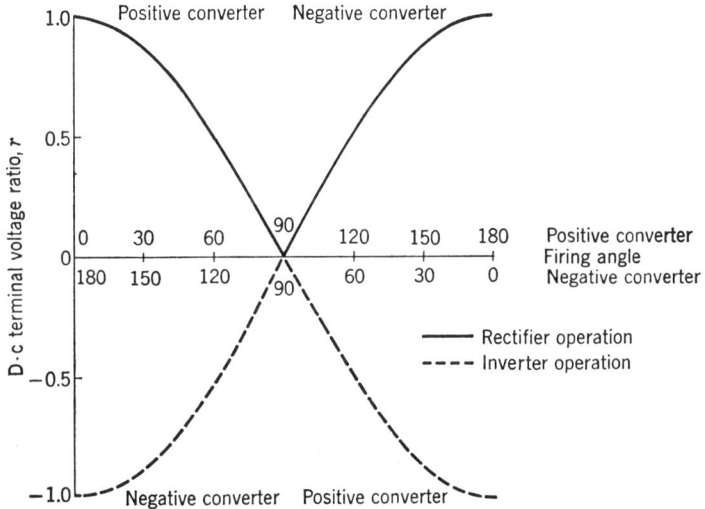

Figure 5.3. Firing angle—d-c terminal voltage relationships for the two converters of the dual-converter. (Continuous conduction assumed.

individual 2-quadrant converters of the dual-converter are illustrated. It is evident that if the d-c terminal voltages of the individual converters are to be controlled so that they are always equal to one another, then the firing angles of the two converters must be related to one another by the following expression:

$$\alpha_P + \alpha_N = 180°$$

where α_P = firing angle of the "positive" converter

α_N = firing angle of the "negative" converter

In practice, if the firing angles of the converters are controlled in this manner, then although the mean d-c terminal voltages of the 2 converters are equal to one another, there are inevitably instantaneous inequalities between the a-c ripple voltages appearing at the d-c terminals of the two converters. Therefore, with this simple control method, it is not permissible to make a "solid" connection between the d-c terminals of the two converters, since this would result in a theoretically infinite circulating ripple current. It is necessary therefore in practice to use some means for controlling the flow of circulating current between the two converters. Basically, there are two alternative methods for doing this.

The first method is simply to inhibit completely the flow of circulating current, through appropriate automatic control of the firing pulses, so that only that converter which carries the load current is in conduction, and the other, temporarily "idle," converter is "blocked." This is the so-called "circulating current-free" mode of operation.

Sometimes, as will be seen, it may be desirable to operate the dual-converter with a controlled amount of circulating current. In this case, the firing angles of the individual converters are controlled in basically the same manner as for the ideal circuit, and the amplitude of the circulating current is limited to an acceptable level by means of a "circulating current reactor," which is connected between the d-c terminals of the two converters. This is the so-called "circulating current" mode of operation.

Both of these operating modes are discussed more fully in the following sections.

THE CIRCULATING CURRENT-FREE MODE OF OPERATION

The Basic Principle

Since only one converter of the dual-converter actually carries load current at any given time, it is not necessary for the "idle" converter to be kept in conduction at all, and this converter can be "blocked," through suitable control of its firing pulses, thus preventing the flow of circulating current.

CIRCULATING CURRENT-FREE OPERATION

Blocking of the idle converter can be accomplished in one of two ways:

(i) The firing pulses can be completely removed.

(ii) The phase of the firing pulses can be adjusted so that there is no tendency for circulating current to flow.

This second method involves a retardation of the firing angle of the idle converter, by at least a certain critical angle, dependent upon the firing angle of the active converter, and the circuit configuration, away from the point at which $\alpha_P + \alpha_N = 180°$. In other words, the firing angle of the idle converter must be shifted in such a direction as to tend to make its d-c terminal voltage oppose the flow of circulating current, and thus it becomes biased off.

This method for blocking the idle converter can be applied through the action of the normal phase control circuits, and it is therefore sometimes more convenient to implement than method (i), in which the firing pulses are completely inhibited. It does, however, necessarily result in the application of the firing pulses to the thyristor gates whilst their anode voltages are in the reverse direction—a practice which may be undesirable, since it results in increased reverse leakage current through the thyristors.

With the circulating current-free mode of operation, then, the need arises for a control system which automatically "blocks" and "deblocks" the individual converters in accordance with the direction of load current. Clearly, if an external performance is to be obtained which is as good as that of the "ideal" dual-converter, then the automatic converter "bank selection" control must be such that there is always, apparently, a freedom for the external load current to flow in the direction of its choice. At the same time, of course, the control should never permit both converters to be simultaneously in conduction, since this could result in a large and undesirable circulating current. The requirements of the control system can be summarized as follows:

(a) The control of the firing pulses should be such that only that converter which carries the load current is kept in conduction, and the temporarily idle converter should be blocked, so that there is no tendency for circulating current to flow.

(b) Any desired changes in the direction of load current should be anticipated, and the converter firing pulses should be manipulated in such a way that the load current is free to reverse its direction at the desired instant.

(c) In order to achieve a smooth changeover of current from one converter to the other, irregular "jumps" in the level of the d-c terminal voltage at the point of current reversal must be avoided. Thus the firing pulse control should, ideally, be such that the mean d-c terminal voltage of the incoming converter, at the instant of current reversal, is the same as that of the outgoing converter.

116 THE DUAL-CONVERTER

In practice, these functions can be achieved relatively easily, if the application is such that the load current waveform is always continuous. If, however, the steady state-load current should become discontinuous under certain conditions, as is often the case, then a more sophisticated control becomes necessary, especially if a fast system response is required, with no appreciable "dead zone" at the current "crossover" point.

A Simple "Open-Loop" Control Scheme for a Dual-Converter Operating with no Circulating Current

Figure 5.4 shows, in simplified diagrammatic form, a control scheme for a circulating current-free dual-converter. Such a control scheme is satisfactory, provided that the load current is always continuous.

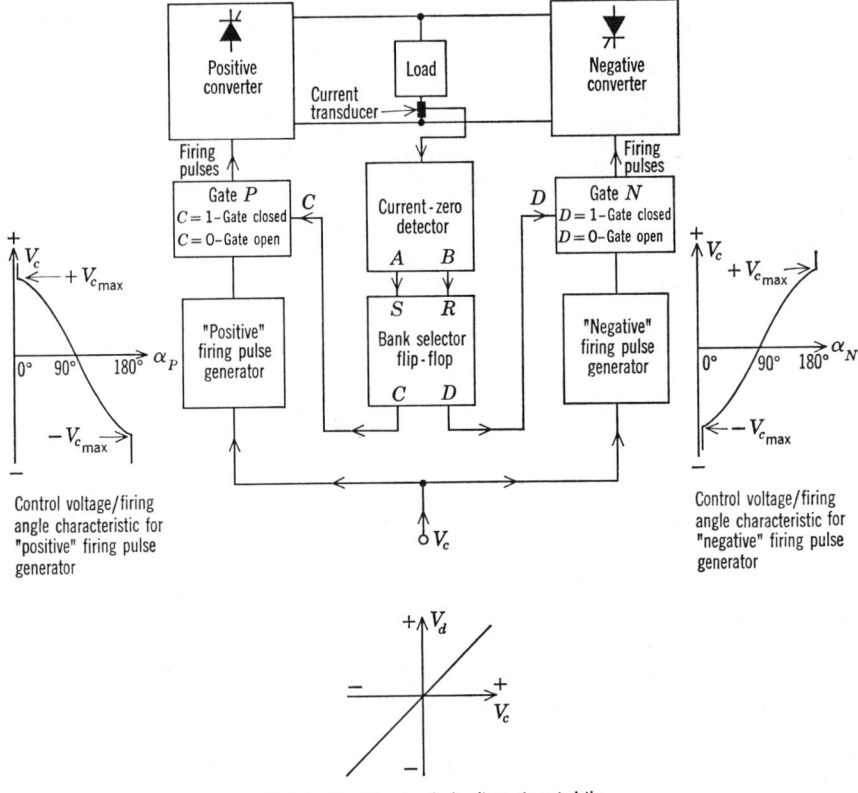

Figure 5.4. Diagrammatic representation of a simple control scheme for a circulating current-free dual-converter with continuous load current.

CIRCULATING CURRENT-FREE OPERATION

The function of each of the firing pulse generators is to supply firing pulses to the thyristors within the associated converter, provided that the associated gate, "P" or "N," is closed.* Since these gates are controlled by the complementary outputs, C and D, of the bank selector flip-flop, it is inherent that one of these must always be closed, and the other open, and hence only one of the two converters has firing pulses applied at any one time.

The phase of the firing pulses produced by the firing pulse generators is controlled in accordance with a low level d-c analog control voltage, V_c. The relationships between the firing angle and the control voltage for the two converters are given by:

$$\frac{V_c}{V_{c_{max}}} = \cos \alpha_P \text{—for the positive converter}$$

$$\frac{V_c}{V_{c_{max}}} = -\cos \alpha_N \text{—for the negative converter}$$

Thus, for any value of V_c, the sum of the converter firing angles, $(\alpha_P + \alpha_N)$, is always 180°, and therefore, in the—hypothetical—event that the firing pulses were simultaneously applied to both converters, each would produce the same mean d-c terminal voltage, with the same circuit polarity.

Moreover, since the d-c terminal voltage ratio of each converter is equal to the cosine of the firing angle, it follows that the mean d-c terminal voltage is proportional to the analog control voltage V_c, and therefore each converter essentially behaves as a power amplifier with a linear voltage transfer characteristic:

$$\frac{V_d}{V_{d_{max}}} = \cos \alpha_P = \frac{V_c}{V_{c_{max}}}$$

$$\therefore V_d = V_c \frac{V_{d_{max}}}{V_{c_{max}}}$$

In practice, the control voltage V_c might be the "error signal" derived from a feedback control loop. For example, in an application in which it is required to control the speed of a d-c machine, V_c might be the difference between the speed reference, and a signal proportional to the actual machine speed.

The "set" and "reset" input signals to the bank selector flip-flop are obtained from the output of the current-zero detector circuit. The function of this latter circuit is to provide a pulse at its output A whenever the load current falls to zero from a negative direction, and a pulse at its output B whenever the load current falls to zero from a positive direction.

---

* According to the terminology employed throughout this book, a "closed" gate permits the signal to pass through it (i.e., it acts as a closed switch); an "open" gate blocks the passage of the signal (i.e., it acts as an open switch).

In order to explain the operation of this control system, assume, for example, that the dual-converter supplies the armature current of a separately excited d-c machine. Suppose that a situation exists in which the positive converter supplies current to the machine at a given positive voltage level. Under this condition, gate N is open, and therefore the firing pulses for the negative converter are inhibited.

Suppose now that an overhauling torque is applied to the shaft of the machine, which tends to make it accelerate, thus increasing its induced emf, and consequently cutting off the armature current supplied by the positive converter. At the instant at which the positive converter current reaches zero, the current-zero detector produces a pulse at its output B, thus changing the state of the bank selector flip-flop. This results in gate P being opened, and the firing pulses to the positive converter thereby are inhibited. At the same time, gate N is closed, and firing pulses now are applied to the negative converter. Since $\alpha_P + \alpha_N = 180°$, the mean voltage which appears at the d-c terminals of this converter is the same as that which previously appeared at the d-c terminals of the positive converter. Thus reverse armature current now starts to increase smoothly through the negative converter, until an equilibrium point is reached, at which the counter torque of the machine becomes equal and opposite to the applied overhauling torque. The d-c machine is of course now operating as a generator, and is delivering power into the d-c terminals of the negative converter.

It can be seen that the external operating characteristics of this scheme are essentially those of the "ideal" dual-converter. That is to say, so far as the external d-c load is concerned, it "sees" a theoretically zero impedance voltage source, for currents of either polarity, and the fact that one of the two converters is always blocked is not externally apparent.

Of course, it will be appreciated that this simple control scheme would not operate properly with discontinuous load current, nor indeed under any circumstances in which a current-zero is not followed by a reversal in the direction of the load current.

Control Difficulties due to Discontinuous Load Current

The simple "open loop" control system described in the previous section operates in the proper manner, so long as the load current is continuous. If the steady state load current should become discontinuous, then, for proper operation, a more sophisticated control system is necessary.

One immediate difficulty which arises is in the operation of the converter bank selection control. Clearly, with discontinuous conduction, it is not possible to use indiscriminately the zero values of the current waveform as

CIRCULATING CURRENT-FREE OPERATION

the logic condition for switching the firing pulses from one converter to the other, as this would result in an erratic and haphazard operation of the system.

A less obvious, but perhaps more fundamental difficulty, arises from the fact that with discontinuous conduction, as explained in Chapter 3, the mean d-c terminal voltage of the phase-controlled converter depends not only upon the firing angle, but also upon the extent of the discontinuity of the load current waveform. Thus with a load which, under given conditions, is capable only of accepting mean power from the d-c terminals of the converter, it is clearly not possible for the steady state mean d-c terminal voltage to become negative, and the current inevitably becomes discontinuous at some firing angle in advance of 90°. At this point, the relationship between the firing angle and mean d-c terminal voltage departs considerably from the cosine relationship obtained with continuous conduction.

A typical relationship between the mean d-c terminal voltage and the firing angle of a 6-pulse converter, with a passive inductance-resistance load, is illustrated in Fig. 5.5a; the corresponding relationship between the control input voltage to the firing pulse generator, and the mean d-c terminal voltage of the converter, assuming that the firing angle is controlled in accordance with the inverse cosine of the control voltage, is shown in Fig. 5.5b.

A similar type of steady state relationship might be obtained, for example, with a load comprising the armature of a d-c machine, which is being driven from the converter. (Although, actually, with this type of load, due to the induced back emf, it would also be possible for the load current to become discontinuous at virtually any output voltage level, depending upon the value of the series inductance of the armature circuit, and the minimum load on the machine.)

A disadvantage of the characteristic of Fig. 5.5 is that, in the region of discontinuous conduction, the mean d-c terminal voltage is not directly proportional to the input control voltage to the firing pulse generator. At the same time, the dynamic voltage gain of the system is much less than with continuous load current. Since, in most practical applications, it is required to control the output of the converter in at least fairly close correspondence with an input reference, this non-linear and load dependent transfer characteristic of the converter, in the discontinuous current region of operation, results in the necessity for using some form of closed loop control of the converter output. It should be mentioned, of course, that this particular phenomenon is fundamental to the 2-quadrant phase-controlled converter with discontinuous conduction, and is not associated specifically with the circulating current-free dual-converter.

A particular control difficulty with the circulating current-free dual-converter, due to operation with discontinuous load current, is associated with obtaining a smooth transfer of current from one converter to the other.

120 THE DUAL-CONVERTER

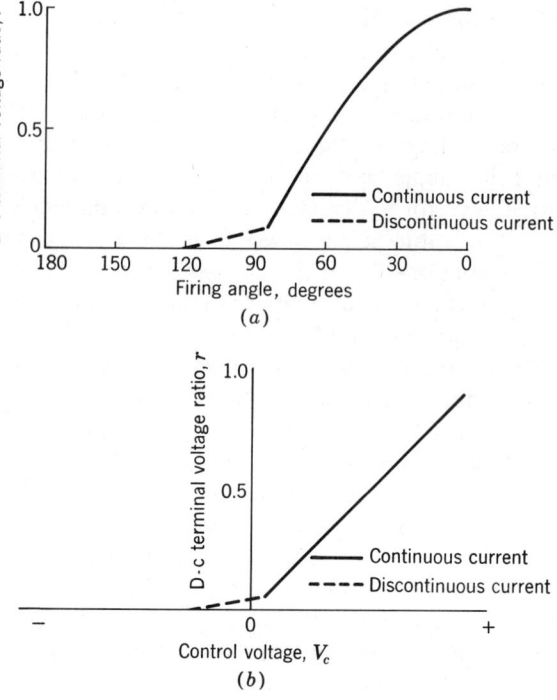

Figure 5.5. Typical relationships between the firing angle, the mean d-c terminal voltage, and the control voltage, for a 6-pulse 2-quadrant converter with a passive L-R load. (*a*) Relationship between firing angle and mean d-c terminal voltage with passive L-R load.

$$\frac{2\pi \cdot 6f_i L}{R} \doteq 6.$$

(*b*) Resulting relationship between mean d-c terminal voltage and control voltage input to firing pulse generator.

$$\frac{V_c}{V_{c\max}} = \cos \alpha.$$

The basic problem is that since, with discontinuous conduction, the mean d-c terminal voltage is determined by the load, as well as by the converter firing angle, it is no longer required that the sum of the converter firing angles be invariably 180°, and, ideally, the control system should predict the firing angle for the incoming converter which results in the same mean d-c terminal voltage as that of the outgoing converter.

In order to illustrate this point, consider an application in which the dual-converter supplies the armature of a d-c machine. Assume that a situation exists in which the positive converter supplies a discontinuous current, at a relatively low positive voltage, say 5% of maximum, with a corresponding firing angle of 110°. If now it is required to transfer the current to the negative converter, in order, say, to reverse the direction of rotation of the machine, the required firing angle for this converter, which would result in the same 5% voltage level, assuming continuous conduction, would be about 93°. On the other hand, if the sum of the converter firing angles is rigidly maintained at 180°, then the firing angle of this converter would be initially 70°. That is to say, this converter initially would be operating as a rectifier, rather than as an inverter, with a mean d-c terminal voltage of 34% of maximum. This sudden large change in the voltage level applied to the machine could result in a very large and unacceptable transient overcurrent, and the system operation would be far from ideal.

In practice, in order to avoid a transient "jump" in the voltage level which would result from a too far advanced firing angle for the incoming converter, and would thus be in such a direction as possibly to result in a large transient overcurrent, the control system may often be arranged so that the firing angle of the incoming converter is initially too far retarded (commonly to 180°), and is subsequently advanced, by the action of a closed loop control, to the final required operating point. This eliminates the possibility of current "jerk" at the crossover point, but it may result in a generally less than ideal system response, with a small dead-time zone during which there is no load current in either direction. However, except in special cases where the performance requirements are particularly stringent, the control system can usually be "tailored" to fit the application, so that the practical consequences of this less than perfect system response are often inconsequential.

Another approach, which is effective so long as the application is such that the incoming converter is always initially in continuous conduction, is to arrange the control scheme so that the firing angle of this converter is automatically preadjusted to the value which would result in the same mean d-c terminal voltage as that last appearing at the outgoing converter, regardless of whether the current of this latter converter was previously continuous or discontinuous. This technique can provide a very smooth change over, with virtually no "dead" time.

A control scheme of this general type is described in more detail in the following section.

A Closed-Loop Control Scheme for a Dual-Converter Operating with no Circulating Current

The design of a control scheme for a circulating current-free dual-converter depends largely upon the circumstances and requirements of the particular application. In order to illustrate the general type of control scheme which may be used, a control system for a circulating current-free dual-converter,

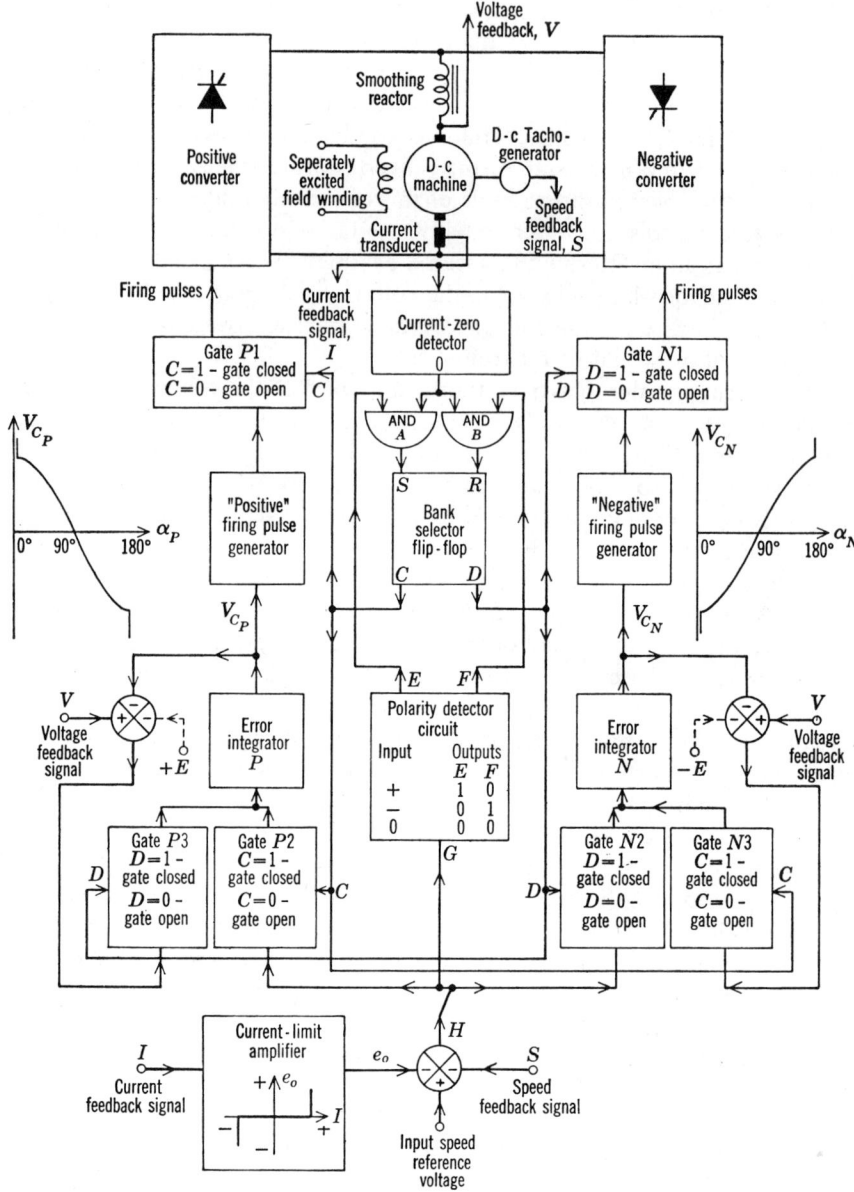

Figure 5.6. Simplified diagrammatic representation of a closed-loop speed control scheme for a separately excited d-c machine, with armature voltage controlled by means of a dual-converter operating in the circulating current-free mode. System provides forward and reverse speed control, with automatic regenerative braking and armature current limiting.

which provides closed-loop speed control of a separately excited d-c machine, will be described.

The scheme is shown in simplified diagrammatic form in Fig. 5.6. This scheme provides continuous closed loop control of the speed of the machine, by means of controlling its applied armature voltage, from full forward to full reverse, with automatic regenerative braking and current limiting.

The "positive" and "negative" firing pulse generators are functionally the same as those used in the simple control scheme of Fig. 5.4. Thus, with continuous conduction, the relationship between the analog control input voltage of the firing pulse generator, and the mean d-c terminal voltage of the converter, is linear, and the same control voltage applied to either firing pulse generator results in the same d-c terminal voltage at either converter, with the same polarity in the power circuit.

The output pulses of the positive and negative firing pulse generators are applied to the thyristors of the associated converters through the gates $P1$ and $N1$ respectively. These gates are controlled from the complementary outputs C and D of the bank selector flip-flop, and hence it is inherent that only one of the two converters has firing pulses applied at any one time.

The function of the current-zero detector is to produce a logic "1" signal at its output O whenever the load current is zero, and a logic "0" signal at its output whenever the absolute load current is greater than zero.

The polarity detector circuit produces a logic signal "1" at its output E whenever the polarity of the analog input signal at G, which represents the error between the demanded speed and the actual speed, is positive, and a logic signal "1" at its output F, whenever the polarity of the signal at G is negative. With "zero" input signal at G (more precisely, with input signals less than a small threshold level) both outputs E and F have a logic state "0." The purpose of this circuit is to "steer" the output signal of the current zero detector to the appropriate input terminal of the bank selector flip-flop, and to enable the bank selection control to distinguish between current zero values which occur, under steady state conditions, during the course of discontinuous conduction, under which circumstances it is not required to switch the firing pulses from one converter to the other, and current zero values which occur under dynamic conditions, as a result of a change in the set conditions, under which circumstances it is required to switch the firing pulses from one converter to the other. Thus, under steady state conditions, with no error between the speed reference and the speed feedback signal, the voltage at G is zero, and hence both E and F have logic zero values. As a result both the AND gates A and B are open, and therefore it is not possible for the bank selector flip-flop to receive any input signals from the current-zero detector. Under transient conditions, however, during which an error voltage exists between the speed reference and the speed feedback signal, this voltage appears at G, and, depending upon its polarity, it results in a logic "1" output at either E or F. This results in the closing of the appropriate AND gate, A or B, in readiness for the possibility of a logic "1" signal from the current-zero detector, indicating that the load current has fallen to zero, and that the firing pulses should be transferred from one converter to the other. (Of course, this scheme would not operate satisfactorily if the armature current becomes discontinuous *during* the process of acceleration or deceleration—in practice, however, this possibility is remote.)

The function of each of the error integrators P and N is to provide the analog control voltages V_{C_P} and V_{C_N}, for the positive and negative firing pulse generators respectively. These voltages are obtained as a result of the "error" voltages appearing at the inputs of the integrators.

The primary function of the system as a whole is to provide a closed loop control of the speed of the machine being driven. The function of the speed control loop is to compare a d-c analog input reference voltage, representing the desired speed, with the speed feedback signal, obtained from a d-c tachogenerator, representing the actual speed. Any error between the demanded speed and the actual speed is applied to the input of the appropriate error integrator, in such a direction as to eliminate the error.

There are, in addition, two auxiliary feedback control loops. The first of these provides an automatic armature current-limiting facility. In this control loop, the current feedback signal is compared with a preset threshold reference. If the current is less than this threshold value, then there is no output from the current limit amplifier, and under these circumstances this control loop has no effect upon the operation of the system. If however the armature current should tend to exceed the preset threshold level, then the current limit amplifier produces a steeply rising output voltage signal. This signal is injected into the speed control loop, and it has the effect of overriding the speed control and producing an essentially constant current mode of operation. This continues so long as there is a tendency for the armature current to exceed the preset limiting value. As soon as the current falls below this value, the operation of the system automatically reverts to the speed control loop.

The second auxiliary control loop uses a feedback from the output voltage, and its function is to adjust automatically the analog control input voltage to the firing pulse generator associated with the idle converter, so that the d-c terminal voltage of this converter, in the event that its firing pulses should be applied, will be as close as possible to the armature voltage of the machine. The principle of this control is to compare the analog control voltage at the input of the firing pulse generator with the sum of a fixed bias voltage E (which may or may not be zero) and a suitably scaled analog signal, proportional to the mean voltage actually existing across the armature of the machine. Any error is applied, through the appropriate gate, to the input of the error integrator, in such a direction as to eliminate the error.

If the application is such that it can be relied upon that the incoming converter is always initially in continuous conduction (regardless of whether the current of the outgoing converter was previously continuous or discontinuous), then the fixed bias voltages E can be set to zero, and in this event the voltage of the incoming converter automatically will be preadjusted to the voltage existing on the machine. On the other hand, if the possibility exists that the incoming converter may not always be in continuous conduction, then the bias voltages E can be set to some appropriate value to prevent current jerk at the cross-over point. (Of course, this introduces some "dead time" for some operating conditions—but not so much as would result from, say, an incoming converter firing angle which is invariably 180°.)

The function of gates $P2$ and $P3$, and $N2$ and $N3$, is to switch the inputs to the error integrators, P and N, from the speed control loop, to the voltage control loop, and vice versa, depending upon which converter is in conduction. These gates are controlled from the complementary outputs C and D of the bank selector flip-flop. Thus, when C has a logic "1" value, in which event the positive converter is in conduction, gates $P2$ and $N3$ are closed, thereby connecting the speed control loop to the positive error integrator, and the voltage control loop to the negative error integrator. Likewise, when the negative converter is in conduction, the speed control loop is connected to the negative error integrator, and the voltage control loop to the positive error integrator.

Thus it can be seen that the arrangement is such that the speed control loop always operates on the firing angle of the converter which actively controls the output voltage, and, in the meantime, the voltage control loop unobtrusively operates on the firing angle of the idle converter, thus, as nearly as possible, matching its, for the time being "fictitious," terminal voltage to the voltage actually existing at the load.

In order to further explain the operation of this system, consider firstly a steady state condition, with the positive converter supplying current which is less than the limiting value. Under these circumstances, the output C of the bank selector flip-flop has a "1" value, the speed of the machine is equal to the demanded speed, and the error input voltage to the integrator P is zero. The output voltage of this integrator has exactly the value which produces the voltage at the d-c terminals of the positive converter required to maintain the set speed. That this is so can be seen by considering the effect of, say, a momentary reduction in speed. This would result in the appearance of a momentary positive error signal at the input of the error integrator P, which would cause an increase in the output voltage of this integrator, such that the output voltage of the positive converter would also be increased, thus compensating against the original error.

Next, consider the effect of a reversal in the speed reference voltage. In this event, the speed error signal initially becomes negative, since the speed of the machine cannot change instantaneously, and thus a negative voltage is applied at the input of the error integrator P. Thus the output voltage of this integrator decreases, the firing angle of the positive converter is retarded, and the armature current of the machine decreases towards zero. In the meantime, the negative speed error voltage appears at the input of the polarity detector circuit, and this results in a logic signal "1" at the output F of this circuit. Thus AND gate B is closed, in readiness for a logic "1" signal from the output of the current zero detector, indicating that the armature current has fallen to zero. When this occurs, the bank selector flip-flop changes its state, and the firing pulses are transferred from the positive to the negative converter. At the same time, gates $N2$ and $P3$ are closed, thereby applying the speed control loop to the negative converter, and the voltage control loop to the positive converter. As previously explained, the firing angle of the incoming negative converter has already been automatically preset to the value which results in its d-c terminal voltage being (as nearly as possible) equal to the armature voltage of the machine. Thus the induced voltage of the machine now drives a "regenerative" current into the d-c terminals of the negative converter. Since the speed error signal is relatively large, this now causes the firing angle of the negative converter to advance to the point at which the current reaches the preset limit value. At this point, the current limit amplifier produces a steeply rising negative output voltage, which is subtracted from the speed error signal, such that the error voltage appearing at the input of the error integrator N automatically assumes a level which restrains the rate of advance of the firing angle to that which results in an essentially constant current, equal to the set limit value. This situation exists so long as there is a tendency for the current to exceed the limit value; that is to say, almost to the point at which the speed of the machine reaches the "reversed" reference setting. Thus, throughout the period during which the machine decelerates to standstill, and then accelerates in the "reverse" direction to the new speed setting, the armature current automatically is held constant at the prescribed limit level. Finally, when the new set speed is reached, the armature current falls below the limit value, and the speed control loop reestablishes control.

THE CIRCULATING CURRENT MODE OF OPERATION

The Basic Principle

With the dual-converter operating in the circulating current-free mode, it has been seen that certain control problems arise, as a result of operation with discontinuous load current. Depending upon the application requirements, these problems may necessitate the use of a quite sophisticated control scheme, in order to achieve the desired external system performance.

An alternative operating technique, with which the difficulties associated with the circulating current-free dual-converter in discontinuous conduction do not arise, is to connect a current-limiting reactor between the d-c terminals of the two converters, and to regulate the firing angles of the individual converters in such a way that a controlled amount of current is allowed to circulate between them. This circulating current, by itself, keeps both converters in virtually continuous conduction, over the whole control range.

This has two important effects upon the system operation. Firstly, a natural freedom is provided in the power circuit for the load current to flow in either direction at any time. Thus a reversal of load current is inherently a "natural" and smooth procedure, which is completely unhampered by any control functions otherwise required to detect the direction of the load current, and to "block" and "deblock" the appropriate converters. Secondly, a simple cosine relationship is preserved between the mean d-c terminal voltage and the firing angle, independently of whether the external load current is continuous or discontinuous, or indeed, or whether there is any load current at all.

The firing angle control principle for the dual-converter in circulating current operation is basically the same as that of the idealized dual-converter circuit of Fig. 5.2. That is to say, the firing angles of the individual converters are controlled so that each produces the same mean d-c terminal voltage, with the same circuit polarity. Thus, when one converter operates as a rectifier, producing a given mean voltage at its d-c terminals, the other operates as an inverter, producing the same mean "counter voltage." As has been seen, in theory, this requires that the sum of the firing angles of the two converters should be invariably 180°.

It is possible, for example, to arrange the firing angle control in such a way that the dual converter behaves as a voltage amplifier having a linear open-loop transfer characteristic, over the whole control range (irrespective of the external load). This is accomplished by employing a cosine relationship between the control voltage and the firing angle, as illustrated in the simplified functional diagram of Fig. 5.7.

With this mode of operation, then, the natural external operating characteristics of the dual-converter are virtually equivalent to those of the idealized dual converter, and thus it is possible to obtain a near perfect external system performance.

The need for the circulating current reactor arises from the fact that although it is possible to exercise a continuous control over the mean value of the voltage which appears at the d-c terminals of each converter, no

128 THE DUAL-CONVERTER

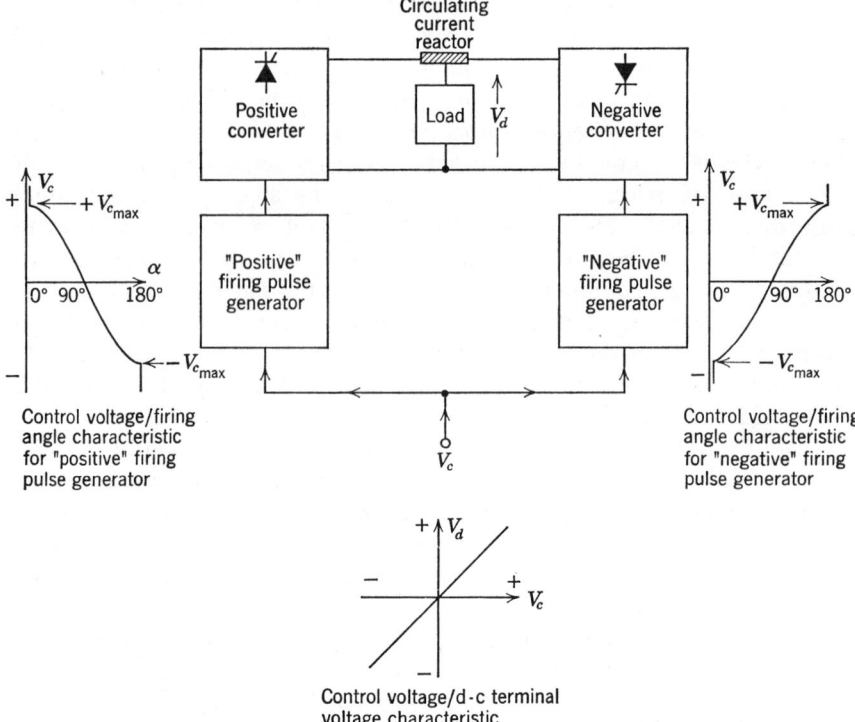

Figure 5.7. Diagrammatic representation of a simple open-loop control scheme for a dual-converter operating with a circulating current.

control is possible over the magnitude and phase of the accompanying superimposed ripple voltage. In circulating current operation, it inevitably occurs that certain components of the ripple voltages appearing at the d-c terminals of the individual converters are in phase opposition with one another around the circuit, and thus it is necessary to connect a reactor between the d-c terminals of the two converters, in order to support this circulating ripple voltage, and, in so doing, to limit the amplitude of the circulating ripple current. This, then, is perhaps the major disadvantage of this mode of operation—the fact that it requires an additional power circuit component, in the form of a circulating current reactor, the size and cost of which may be quite significant, especially at high power levels.

Details of Operation

In order to explain the details of the operation of the dual-converter in the circulating current mode, the operation of the 3-pulse midpoint circuit will

OPERATION WITH CIRCULATING CURRENT

be considered. It will be assumed that the circulating current reactor has zero resistance, and that the firing angles of the two converters are controlled so that their sum is 180°. It will also be assumed—as is virtually mandatory in practice for proper circuit operation—that the firing pulses applied to the gates of the thyristors extend for a full period of 120°.

NO LOAD OPERATION. Firstly, the operation of the circuit with no external load will be considered. This case is of interest, since it represents the extreme condition under which the only current which flows in the circuit is the circulating current. As will be seen, even in this case, for all practical purposes, both converters are kept in a state of just-continuous conduction, and hence the waveforms of the voltage in the circuit are clearly defined.

Figure 5.8 illustrates the operation of the circuit at various firing angles. The waveforms at a are appropriate to the case when the firing angles of both the converters are 90°. Consider the time period $t_1 - t_2$, during which the voltages of both converters are instantaneously "positive." The voltage of the positive converter is more positive than that of the negative converter, and therefore the voltage across the circulating current reactor is in a "positive" direction, and this results in an increasing circulating current. During the subsequent period t_2 to t_3, the voltages of both converters are still positive, but now the voltage of the negative converter is greater than that of the positive converter. Thus the voltage across the circulating current reactor is in a "negative" direction, and the circulating current decreases towards zero. At time t_3, the net integral of the voltage waveform appearing across the circulating current reactor during the elapsed period since time t_1 is zero, and hence the circulating current is also instantaneously zero. At this time, however, the next thyristor in sequence of the negative converter is fired. During the period t_3 to t_4, both converter voltages are now negative with respect to neutral, but the voltage of the negative converter is more negative than that of the positive converter. Thus the circulating voltage is again in the positive direction, and the circulating current once more starts to increase. At time t_4, the circulating voltage becomes negative, and subsequently, at t_5 its integral over the elapsed period since time t_3 becomes zero, and the circulating current therefore also becomes instantaneously zero once more.

It is seen, then, that the alternating ripple voltage which appears between the d-c terminals of the two converters, by itself, maintains a just-continuous flow of circulating current, and the circulating voltage and current waveforms are mutually self sustaining. Thus, both converters are kept in a state of just-continuous conduction, even though there is no external load current.

It is of course inherent that the circulating current flows in one direction only, that is to say, from the positive to the negative converter. It should be

130   THE DUAL-CONVERTER

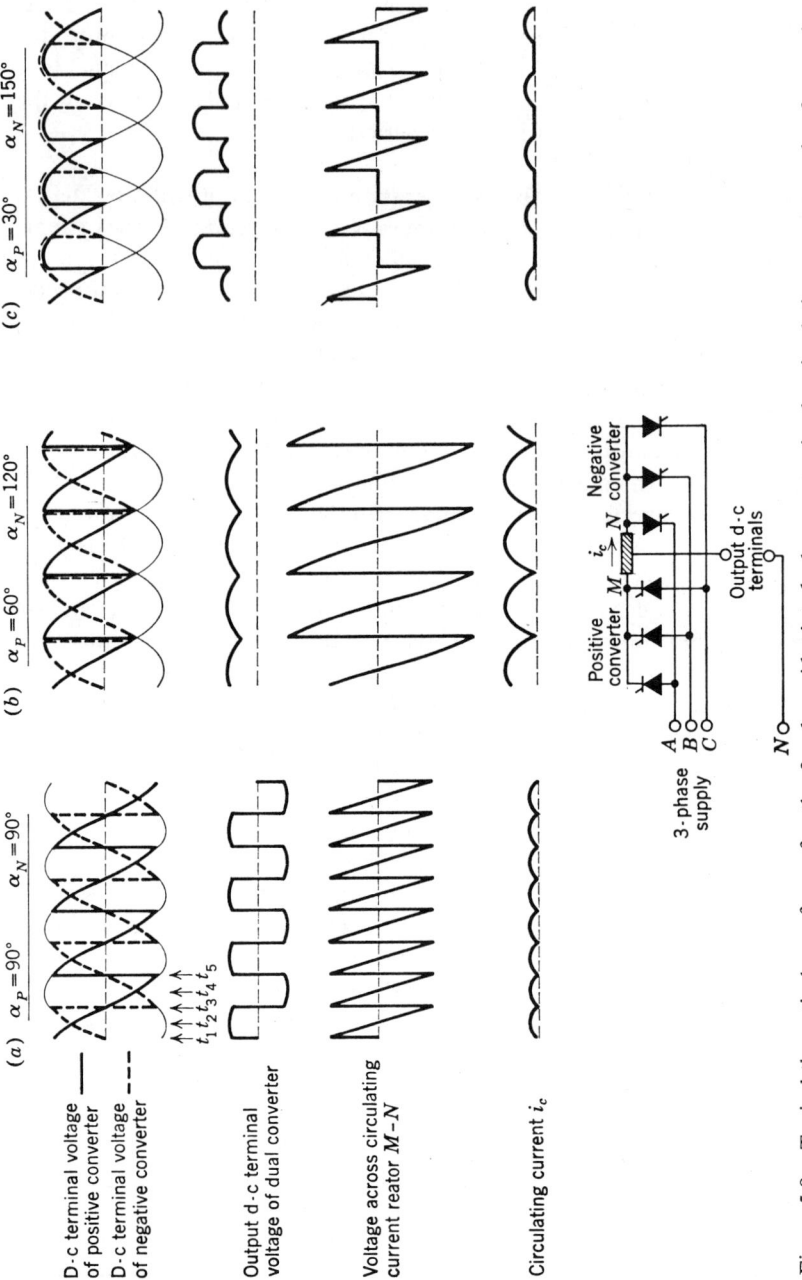

Figure 5.8. Typical theoretical waveforms for the 3-pulse midpoint dual-converter in the circulating current mode of operation, with no external load.

OPERATION WITH CIRCULATING CURRENT

appreciated also, however, that it is equally inherent that the voltage which causes the circulating current to flow constitutes a pure alternating wave, with no d-c component (at least in this idealized case).

The voltage at the midpoint of the circulating current reactor, which is the output terminal of the dual-converter, is the instantaneous average of the voltages of the individual converters. It can be seen that the waveform of this voltage has a different general appearance to that of either of the constituent voltages; (indeed, at some particular firing angles, it also has a different fundamental ripple frequency). The mean value of the output voltage wave, however, is, of necessity, equal to the mean value of each of the constituent voltage waveforms. Of course, for the particular case of a 90° firing angle, the mean d-c terminal voltage of each converter is zero, and hence the mean output voltage of the dual-converter is also zero. This is confirmed from an inspection of the waveform of the output voltage, which shows that it contains no direct component.

The waveforms of Fig. 5.8b are appropriate to firing angles for the positive and negative converters of 60 and 120° respectively. Under this condition, the positive converter is operating as a rectifier, producing half its maximum voltage, and the negative converter is operating as an inverter, producing exactly the same "counter" voltage. It is seen, once again, that the circulating voltage and current waveforms mutually sustain one another, and the result is that both converters are kept in a state of just continuous conduction. The output voltage waveform, which is the average of the individual converter voltage waveforms, once again is different in appearance to either of the constituent waveforms, but its mean value is necessarily the same as that of each of the individual converters.

The waveforms of Fig. 5.8c are appropriate to firing angles for the positive and negative converters of 30 and 150° respectively. In this particular case, for the first 30° period of the firing pulse applied to a given thyristor of the positive converter, the thyristor which is connected to the same a-c input line in the negative converter also has a firing pulse applied to it, and hence there is no difference of voltage between the d-c terminals of the two converters during this time, and no voltage is impressed across the circulating current reactor. Thus, in this period, there is theoretically no circulating current. Since, however, both the positive and negative thyristors are continuously "fired," a low impedance path exists for current of either polarity between the d-c terminals of the dual-converter and the associated input line, and hence the voltage which appears at the d-c terminals of each of the converters, as well as at the midpoint of the circulating current reactor, is the voltage of the input line. Thus, even though there is theoretically no circulating current during this period, the output voltage waveform is nevertheless the same as if the converters were in conduction. During the second 30° period of each firing pulse for the positive converter, the next thyristor in sequence of the negative converter is fired, and now a voltage is impressed across the circulating current reactor. This voltage is initially in the "positive" direction, and therefore it causes an increasing circulating current to flow. Subsequently, the circulating voltage reverses its polarity, and the circulating current then starts to decrease, reaching a zero value at the instant at which the next positive thyristor is fired. Once again, of course, the voltage waveform which appears at the output terminals of the dual-converter is the instantaneous average of the constituent waveforms, and its mean value is the same as the mean value of either of the individual converter voltages.

Summarizing the foregoing description, it is seen that the result of applying firing pulses to both converters, so that the sum of the firing angles is 180°, is to produce a flow of circulating ripple current between the two converters, as a result of which a clearly defined voltage waveform is produced at the d-c terminals of each converter. This voltage waveform is the same as that obtained with continuous conduction, even though there is no external load current, and even though, over part of the firing angle control range, the circulating current itself may not be continuous.

Although only three specific firing angles have been considered here, it can be seen quite easily that this same result is obtained over the whole firing angle control range. Furthermore, it is applicable not only to the 3-pulse midpoint dual-converter considered here, but, more generally, to all dual-converter circuits.

OPERATION WITH A SMOOTH LOAD CURRENT. The next case which will be considered briefly is that of a perfectly smooth current at the d-c terminals. In this simple case, the waveforms of the circulating ripple current are theoretically the same as with no load, and hence the net waveform of the current in the "active" converter is the addition of the no load circulating ripple current, and the external load current, whereas the waveform of the current in the "idle" converter is simply the circulating ripple current.

Theoretical waveforms for a 3-pulse dual-converter operating at half-maximum d-c terminal voltage, with a smooth load current, are shown in Fig. 5.9.

At *a* the load current is in the positive direction, and therefore this current is carried by the positive converter. The total current in this converter is thus the sum of the load current and the circulating ripple current. The current in the negative converter, on the other hand, is simply the circulating ripple current.

At *b*, the load current is in the negative direction. (Still, however, the mean d-c terminal voltage is positive, which implies that a source of voltage

OPERATION WITH CIRCULATING CURRENT 133

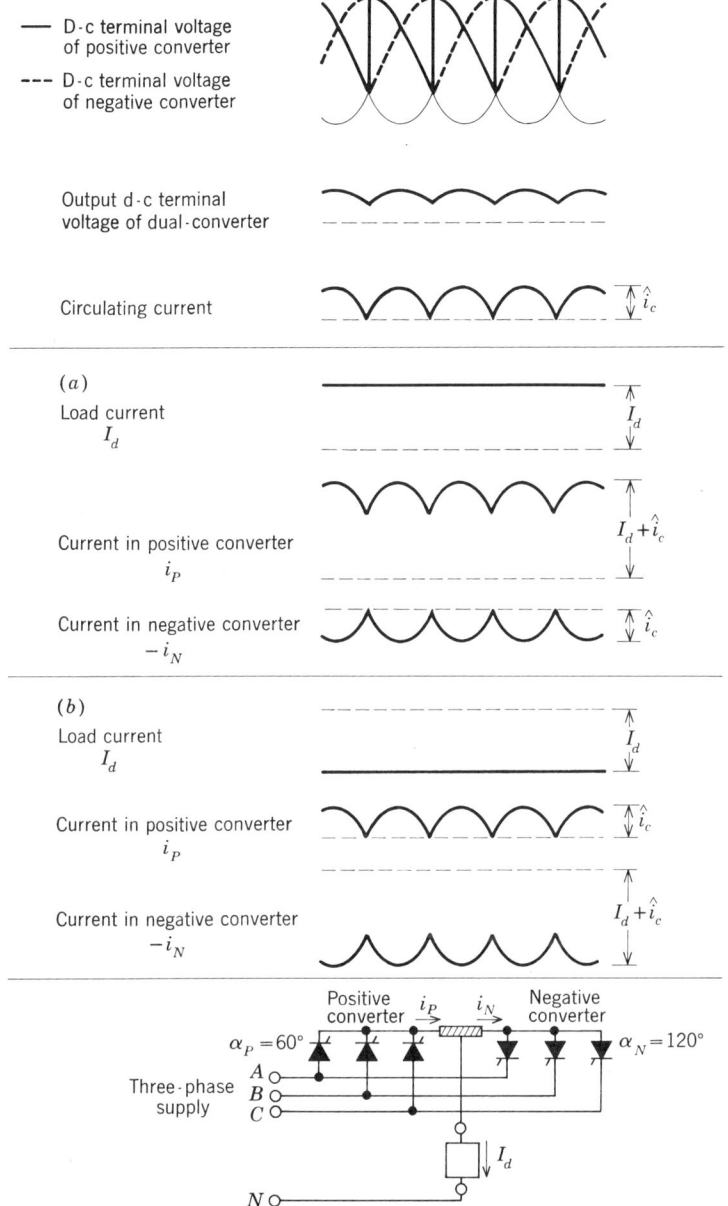

Figure 5.9. Typical theoretical waveforms for the 3-pulse midpoint dual-converter in the circulating current mode of operation, with a smooth load current. (a) Positive load current; (b) negative load current.

134   THE DUAL-CONVERTER

is present in the d-c circuit, which drives this current against the "counter" voltage of the negative converter.) The current in the negative converter is now the sum of the load current and the circulating ripple current, whereas the current in the positive converter is just the circulating current.

OPERATION WITH AN UNSMOOTHED LOAD CURRENT. In practice, of course, the output current of the dual-converter is not perfectly smooth, but invariably it contains ripple components, due to the ripple voltage at the d-c terminals of the dual-converter. (In fact, as has already been explained, the effects of these ripple voltages are, in the first place, the basic motive for operating the dual-converter with a circulating current.)

It has already been seen with no load, as well as with a smooth load current, that both converters are kept in virtually continuous conduction, and the voltage waveforms in the circuit are therefore the same in either case. It is to be expected, therefore, that the same voltage waveforms would also be obtained with an unsmoothed or discontinuous load current. This is indeed the case, but it requires some explanation, since at first sight it is not obvious that the presence of alternating components of output current in the windings of the circulating current reactor do not (under "steady" conditions) produce induced voltages across the reactor.

In order to explain the details of the circuit operation with an unsmoothed output current, it is convenient, firstly, to neglect the presence of the ripple voltages (even though it is these voltages which in practice produce the ripple current), and to consider the operation of the idealized equivalent circuit of Fig. 5.10a, in which each converter is represented as a direct voltage source connected in series with a diode. An assumed waveform of the output current is shown in Fig. 5.10b. This consists of a steady positive component, with a superimposed sinusoidal ripple component, of such a magnitude that the net current becomes negative for certain periods. The corresponding waveforms of the individual converter currents are shown in Fig. 5.10c and d. Consider the conditions at time t_1. It is assumed that, previous to this time, there has been no load current, and that the current now starts to increase in a positive direction. This current is drawn through the positive converter, and it results in the voltage waveform shown at f being induced across the circulating current reactor. The polarity of this voltage is such that M is positive with respect to N, and it is, therefore, in such a direction that it appears as a reverse voltage across the diode associated with the negative converter. This converter is therefore biased off during this initial period.

At time t_2 the output current attains its peak positive value. The mmf of the circulating current reactor at this time is given by

$$\text{mmf} = \hat{\imath}_o \frac{n}{2} \qquad (5.1)$$

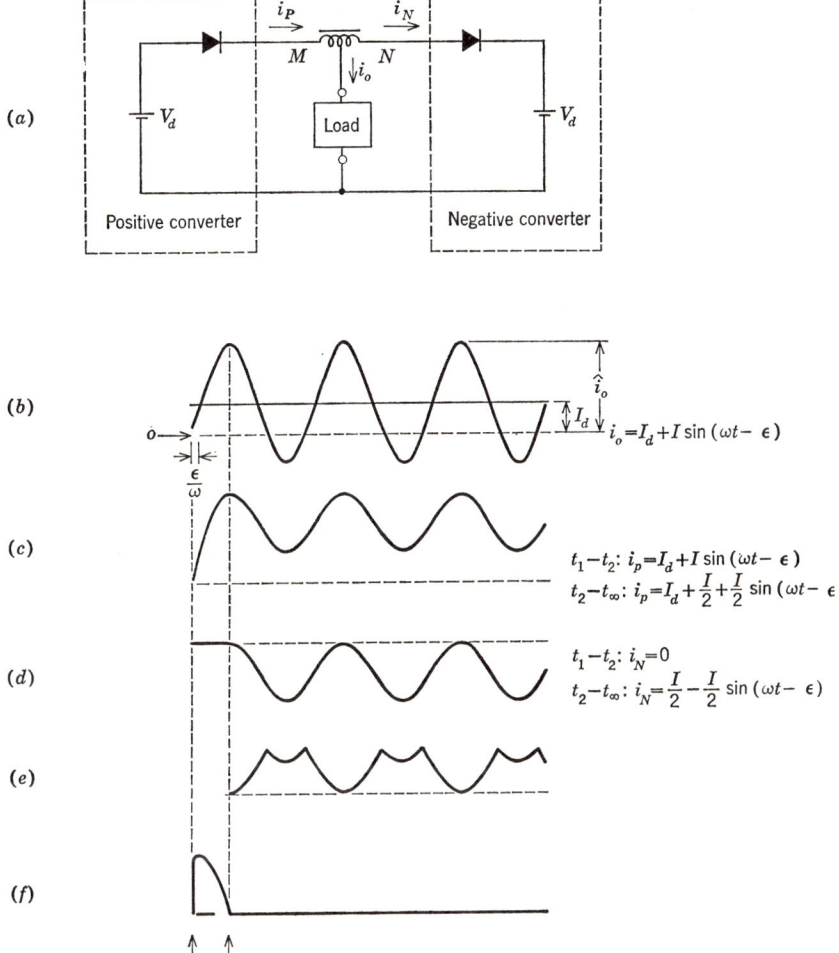

Figure 5.10. Operation of an "idealized" dual-converter in the circulating current mode, with an unsmoothed load current. (a) Idealized equivalent d-c circuit of the dual-converter with circulating current reactor; (b) output current, i_o; (c) current in positive converter, i_P; (d) current in negative converter, $-i_N$; (e) "self-induced" circulating current; (f) voltage M-N across circulating current reactor.

136 THE DUAL-CONVERTER

where n is the total number of turns on the circulating current reactor. This mmf produces a corresponding flux in the core of this reactor, which, in itself, represents a given amount of stored energy.

The load current now starts to decrease, thus creating a tendency for a reduction of the mmf, and hence of the stored energy, of the circulating current reactor. However, in order to extract this stored energy from the reactor, it is necessary for a voltage to be developed in the opposite direction across its winding; that is, with M negative with respect to N. A voltage of this polarity is not possible, however, so long as the voltages of the two converters are exactly equal to one another (assuming "perfect" diodes). This is because any tendency for a voltage in this direction immediately drives the diodes of both converters into forward conduction, thus "clamping" the voltage to zero.

The conclusion is reached, then, that the energy stored in the circulating current reactor as a result of the peak output current in its winding cannot be recuperated, but remains permanently "trapped" in the reactor, so long as the voltages of both converters remain exactly equal to one another. The practical implication of this is that once the output current has attained its peak value, any subsequent reduction in the output current necessarily results in a surplus of "trapped" current in the reactor. This excess current can only "freewheel" between the two converters, and it is, therefore, a circulating current. Moreover, any changes of load current, below the peak value of $\hat{\imath}_o$, are free to take place instantaneously, and do not result in any induced voltage across the circulating current reactor, because the load is always free to draw upon the "reservoir" of current already circulating between the converters, up to this limit value. (Of course, in practice, due to the losses of the reactor, as well as to other circuit imperfections, the trapped energy would "leak away" at a rate dependent upon the losses, and this would result in a slight departure from the theoretically "ideal" mode of operation considered here.)

Expressing this phenomenon in mathematical terms, the basic criterion is that the total mmf of the circulating current reactor remains constant at its peak value:

$$i_P \frac{n}{2} + i_N \frac{n}{2} = \hat{\imath}_o \frac{n}{2} \tag{5.2}$$

where i_P = instantaneous current in the positive converter

i_N = instantaneous current in the negative converter

$\hat{\imath}_o$ = peak output current

This means to say that if the output current contains a ripple component, then this ripple component by itself must produce a virtually continuous

OPERATION WITH CIRCULATING CURRENT   137

"self-induced" circulating current. This "self-induced" component of circulating current is, of course, quite separate from, and additional to, the circulating current produced by the circulating ripple voltage which, in practice, appears across the reactor.

Returning now to the conditions at time t_2 (Fig. 5.10), the mmf of the circulating current reactor at this time has just attained its maximum steady value, and therefore from this time onwards, the relationship between the positive and negative converter currents is given by expression (5.2).

In addition, of course, the following basic relationship exists between i_P, i_N, and i_o:

$$i_P - i_N = i_o \qquad (5.3)$$

From expressions (5.2) and (5.3), the instantaneous positive and negative converter currents i_P and i_N can each be expressed in terms of the instantaneous output current, and the maximum output current $\hat{i}_o$, as follows:

$$i_P = \frac{\hat{i}_o + i_o}{2} \qquad (5.4)$$

$$i_N = \frac{\hat{i}_o - i_o}{2} \qquad (5.5)$$

For the case under consideration, the instantaneous output current is given by:

$$i_o = I_d + I \sin(\omega t - \varepsilon)$$

Therefore:

$$\hat{i}_o = I_d + I$$

Thus, expressions (5.4) and (5.5) become, more specifically, for this case:

$$i_P = I_d + \frac{I}{2} + \frac{I}{2} \sin(\omega t - \varepsilon) \qquad (5.6)$$

$$i_N = \frac{I}{2} - \frac{I}{2} \sin(\omega t - \varepsilon) \qquad (5.7)$$

From these expressions, the waveforms of the positive and negative converter currents, i_P and i_N respectively, can be deduced immediately. An inspection of these waveforms confirms that with any level of output current below the maximum value of $\hat{i}_o$, a "self-induced" component of circulating current is forced to flow between the two converters. The waveform of this self-induced circulating current, shown in Fig. 5.10e, is arrived at as follows: Whenever the output current is positive, this current must be drawn from the positive converter; hence the current carried by the negative converter is the circulating current. Conversely, whenever the output current is negative, the current carried by the positive converter is the circulating current.

The waveforms of current can now be deduced for a practical case, in which the effect of the circulating ripple voltage, as well as that of the output ripple current, is taken into account. For this case, the currents flowing in each converter, due to the output current, are given by expressions (5.4) and (5.5), but now there is an additional superimposed component of circulating current, i_c, due to the circulating ripple voltage. Thus

$$i_P = \frac{\hat{i}_o}{2} + \frac{i_o}{2} + i_c \qquad (5.8)$$

$$i_N = \frac{\hat{i}_o}{2} - \frac{i_o}{2} + i_c \qquad (5.9)$$

In theory, the waveform of i_c could be assumed to be always unidirectional—as is the case in the waveforms of Figs. 5.8 and 5.9. Under this assumption, however, the total circulating current may (depending on the firing angle) remain permanently "above zero," due to a phase displacement between the zero points on the waveform of i_c and the zero points on the waveform of the "self-induced" component of circulating current. This is a somewhat theoretical state of affairs, which does not materialize in practice (with $\alpha_P + \alpha_N = 180°$), due to the inevitable presence of resistance in the circuit, as well as to the finite forward voltage drops across the thyristors. Thus, in reality, due to these "imperfections," the waveform of the circulating current is always "just discontinuous" (with $\alpha_P + \alpha_N = 180°$). Thus, a valid approximation to the practical situation is to assume that the waveform of i_c has a superimposed d-c bias, of such a level that the waveform of the total circulating current just "touches zero."

In order to illustrate an example, it is convenient, for simplicity, to consider the particular case of a pure resistance load connected at the d-c terminals of the dual converter. In this case, the waveform of the current at the output terminals must be the same as that of the voltage.

Figure 5.11 shows theoretical steady state waveforms for a 3-pulse dual-converter, operating in the circulating current mode, with a pure resistance load.

At a the firing angles of the two converters are each 90°, and the mean d-c terminal voltage is zero. However, an alternating ripple component of voltage appears at the d-c terminals, and this produces a corresponding output ripple current, shown at (4). In this case, this current has the same peak value, $\hat{i}_o$, in either direction. The waveforms of the positive and negative converter currents resulting from this output current waveform can be derived from expressions (5.4) and (5.5), and are shown at (5) and (6) respectively. The waveform of the component of circulating current due to the circulating ripple voltage is shown at (3). This current, when added to each of the currents at (5) and (6), gives the waveforms of the total current in the positive and negative converters respectively. These waveforms are shown at (7) and (8). Finally, the waveform of the total circulating current is shown at (9).

The waveforms at b and c are appropriate to firing angles for the positive converter of 60° and 30° respectively. These waveforms are derived by application of exactly the same "rules," and therefore do not require any further explanation.

Operation Under Dynamic Load Changing Conditions

It has been assumed, so far, that the current which circulates between the converters is, in essence, simply a ripple current, the waveform of which is, as a rule, only just continuous. This, of course, is all that is necessary merely to keep both converters in continuous conduction, and, thereby, to maintain a well defined steady state system operation.

If, however, the ultimate in system response is required under dynamic load changing conditions, then it may be necessary purposefully to produce a well defined level of circulating current, which is considerably in excess of the minimum level required for continuous conduction, at least prior to any anticipated increase of output current. (This can be achieved by means of appropriate control of the firing angle of the "idle" converter, as discussed later.)

The reason for this is that if the circulating current reactor does not already have a sufficient "reservoir" of circulating current in its winding, then this reactor places some restriction on the rate at which the absolute level of the output current can be increased. This is because an increase in the level of the output current (which may or may not also involve a change of polarity) necessarily requires an increase in the mmf of the half winding of the reactor which carries this current. Thus, if the reactor is not precharged with the requisite mmf (in the form of a circulating current), then an increase in the level of the load current necessarily results in the production of an induced voltage across the reactor, which thereby presents a transient impedance to the increasing current. Thus, for increasing levels of load current, unless the circulating current reactor is precharged with a sufficient circulating current, the effective transient output inductance of the dual converter is equal to the inductance of the half-winding of the circulating current reactor. That is to say, the transient output inductance is $L/4$, where L is the inductance of the total winding.

140 THE DUAL-CONVERTER

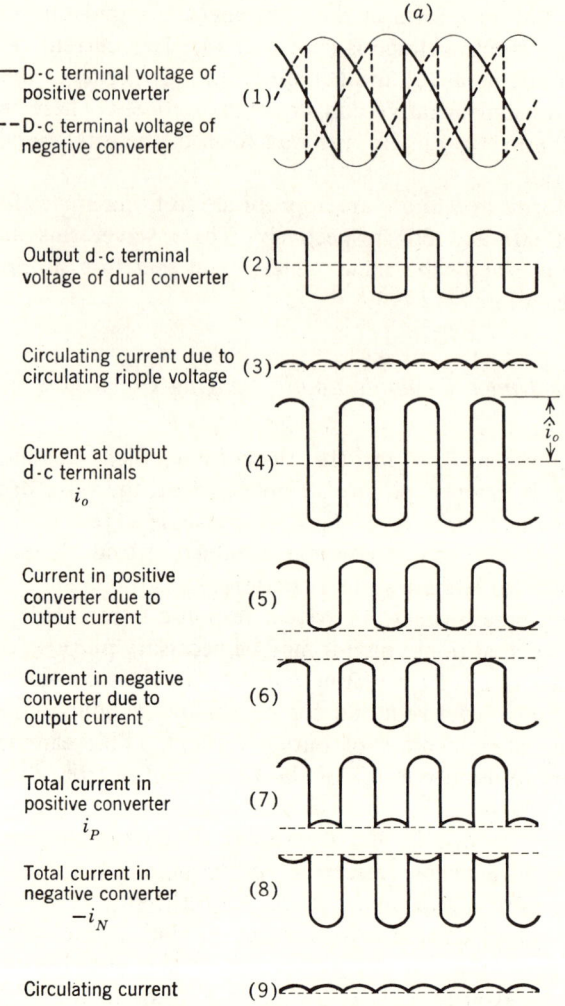

Figure 5.11. Typical theoretical waveforms for the 3-pulse mid- with a pure resistance load. (a) $\alpha_P = 90°$, $\alpha_N = 90°$; (b) $\alpha_P = 60°$,

OPERATION WITH CIRCULATING CURRENT

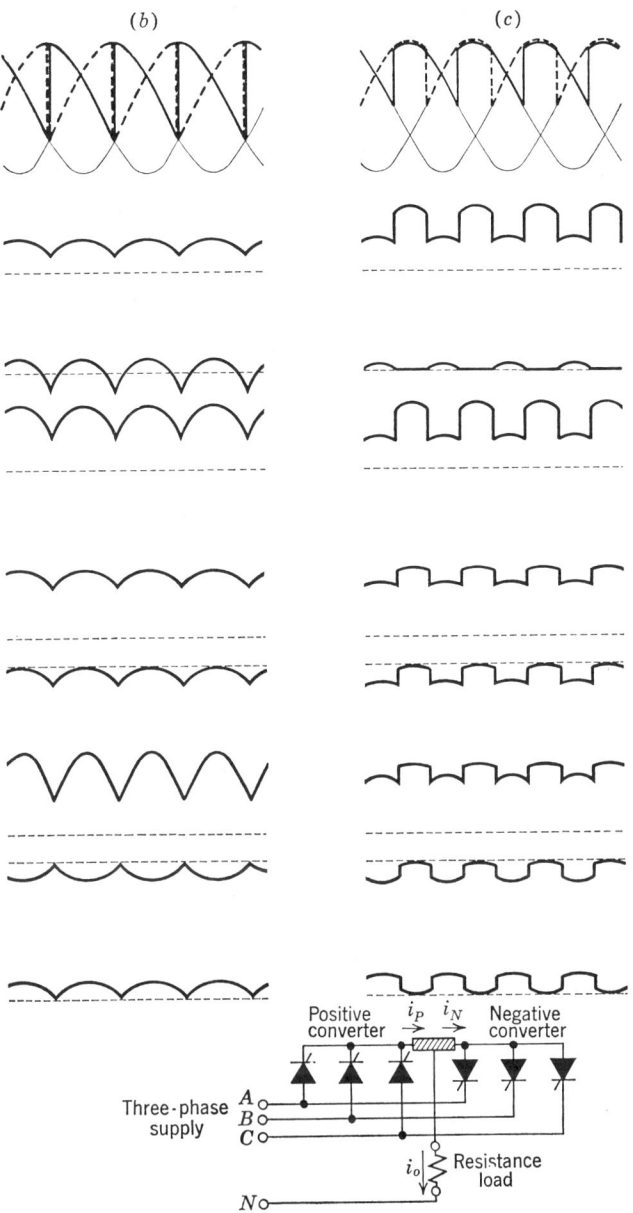

point dual-converter in the circulating current mode of operation, $\alpha_N = 120°$; (c) $\alpha_P = 30°$, $\alpha_N = 150°$.

Of course, this transient inductance may not necessarily be objectionable, in which case "precharging" of the circulating current reactor, with the requisite level of circulating current, is unnecessary. Moreover, reductions in the absolute level of the output current are not restricted in this way. This is because, with a reduction in load current, an excess of circulating current is momentarily left in the reactor. This excess current simply decays away, after a time depending upon the losses of the circuit, unless it is purposefully maintained by the action of the control system.

In order to produce a definite controlled level of circulating current, it is necessary to apply a small, but well defined, difference of mean voltage across the circulating current reactor. This voltage difference can be produced by means of a small adjustment to the relative converter firing angles, so that their sum becomes slightly less than 180°. Since the resistance of the circulating current reactor is, ideally, relatively small, a small shift in the relative converter firing angles, once the circulating current is already continuous, results in a relatively large increase in the direct component of circulating current. Thus it can be seen that it is essential to keep a precise control over the converter firing angles, relative to one another, in order to maintain a proper control over the circulating current. In practice, if the system requirements are such that a definite level of circulating current is required, rather than just a circulating ripple current, due to the circulating ripple voltage, then this can best be achieved by means of a closed loop control scheme, which responds to current feedback signals obtained from the individual converters, and applies an appropriate regulation to the converter firing angles so as to produce the prescribed circulating current.

The Waveform of the d-c Terminal Voltage of the Dual-Converter Operating in the Circulating Current Mode

The d-c terminal voltage of the dual-converter operating with a circulating current has a waveshape which results from the average of the d-c terminal voltage waveshapes of each of the constituent 2-quadrant converters, and the appearance of this waveshape is generally different to that of either of the constituent voltage-waveshapes. This is evident from the waveshapes of Figs. 5.8, 5.9, and 5.11.

Because of this "voltage averaging" effect, the harmonic content of the d-c terminal voltage of a dual-converter operating with circulating current is generally different to that of either of the individual converters.

It can be shown that the peak amplitude of any given voltage harmonic of order n, relative to $V_{d_{\max}}$, is given by

$$\left| \left\{ \frac{\cos(n-1)\alpha_P}{n-1} - \frac{\cos(n+1)\alpha_P}{n+1} \right\} \right|$$

OPERATION WITH CIRCULATING CURRENT 143

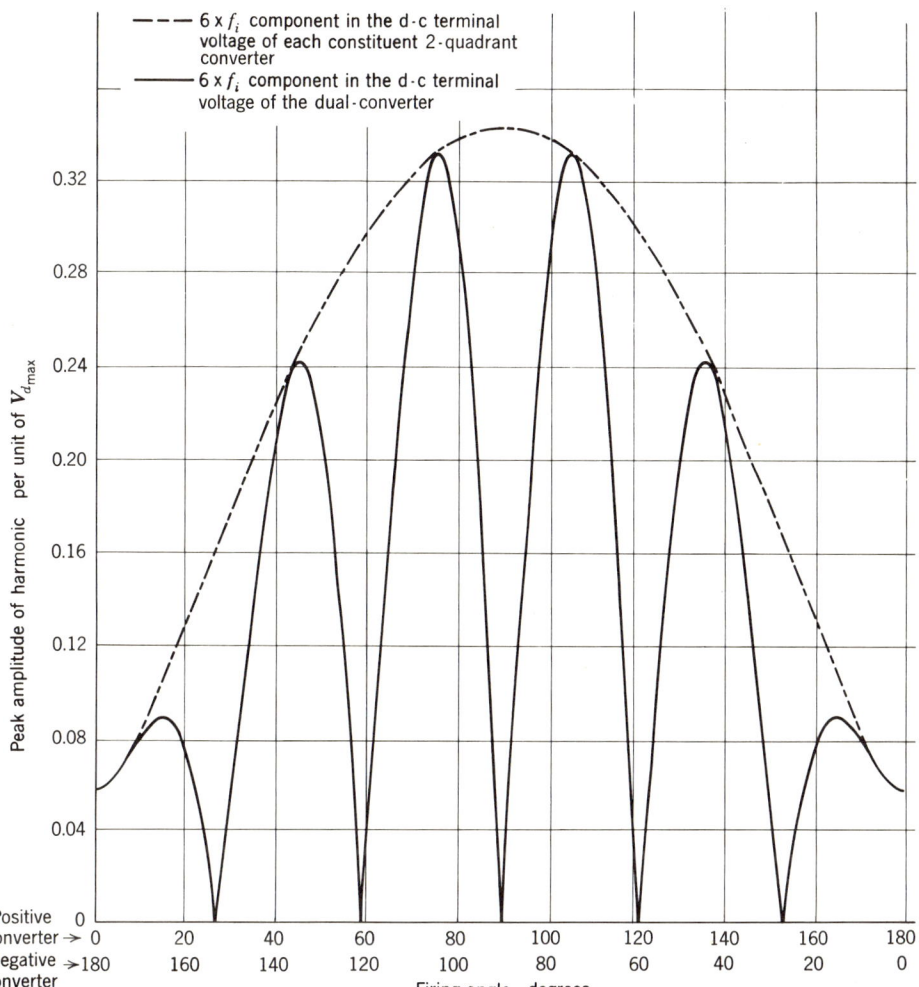

Figure 5.12. Chart showing the variation with firing angle of the lowest harmonic component (i.e., $6 \times f_i$) in the d-c terminal voltage of a 6-pulse dual-converter in the circulating current mode of operation.

This general result is applicable to circuits of any pulse number, it being understood, of course, that the only harmonic frequencies present in the d-c terminal voltage of a converter of a given pulse number are those which are integer multiples of that pulse number.

In Fig. 5.12 the amplitude of the lowest order harmonic present in the d-c terminal voltage of the 6-pulse dual-converter, operating in the circulating

144   THE DUAL-CONVERTER

current mode, is plotted against the converter firing angles. For purposes of comparison, the amplitude of the same harmonic in the d-c terminal voltage of each of the individual 2-quadrant converters is plotted alongside.

It is seen that the harmonic curve for the circulating current mode of operation undergoes several fluctuations as the firing angle is changed through 180°. At certain operating points, a quite large reduction, or complete suppression, of this particular harmonic voltage takes place. At other operating points, however, the reduction is small, or nonexistent.

In general, it can be seen that, unless the required firing angle operating range is small, and happens also to be favorably situated, the circulating current mode of operation has little to offer from the viewpoint of providing a reduced distortion of the d-c terminal voltage.

Chapter Six

The Phase-Controlled Cycloconverter

It has been seen that the dual-converter is able to carry current in both directions at its d-c terminals, and that by controlling the phase of its firing pulses with respect to the a-c supply voltage, it can be made to produce a continuously controllable mean d-c terminal voltage, of either polarity.

It would be expected, therefore, that by a process of continuous "to and fro" phase modulation of the converter firing angles, the dual-converter could be made to produce a continuously varying mean voltage level, of first one, and then the other polarity, at its output terminals. In other words, the wanted output voltage of the converter would now be an alternating, rather than a direct, component. This is indeed the case, and the dual-converter now operates as a direct a-c to a-c frequency changer, or "cycloconverter," the function of which is to convert directly the incoming supply frequency to some different output frequency.

In essence, then, the phase-controlled cycloconverter consists of a dual-converter, which is controlled, through the timing of its firing pulses, so that it produces an alternating, rather than a direct, output voltage. By controlling the frequency and "depth" of phase modulation of the firing angles of the converters, so it is possible to control the frequency and amplitude of the "wanted" component of output voltage. Thus the cycloconverter has the facility for continuous and independent control of both its output frequency and voltage.

Just as the dual-converter provides a full "4-quadrant" operation, so the cycloconverter has the inherent property of being able to handle loads of any power factor, and power is free to flow in either direction through the cycloconverter; thus operation is possible with both "passive" and "regenerative" loads.

Of course, the output voltage waveshape inevitably contains harmonic distortion components, in addition to the wanted sinusoidal component. These distortion terms are produced as a necessary outcome of the basic mechanism of the cycloconverter, whereby the output voltage is fabricated from segments of the input voltage waves. These distortion terms generally can be readily filtered off, if necessary, so that the distortion content of the filtered output voltage of the cycloconverter can be reduced to any desired extent. Thus, by means of a filter, it is possible to obtain a high quality output voltage wave from the cycloconverter.

As would be expected, the distortion of the output voltage becomes increasingly objectionable with increasing ratio between output and input frequency. As would be expected, also, the relative distortion of the output voltage decreases with increasing circuit pulse number; nevertheless, due to the necessity for always producing a natural commutation of current from one input phase to the next, the maximum attainable useful output frequency is, for most practical purposes, *less* than the input frequency, irrespective of the pulse number of the converter. This, then, is a fundamental limitation of the phase-controlled cycloconverter.

In this chapter, the basic principles of operation of the phase-controlled cycloconverter are explained. The nature of the harmonic distortion of the raw output voltage, and of the load presented by the cycloconverter to the input a-c system, are also discussed.

THE BASIC PRINCIPLE OF OPERATION OF THE CYCLOCONVERTER

In Chapter 5, the basic principle of operation of the dual-converter was explained by reference to the simplified equivalent diagram of the d-c circuit, shown in Fig. 5.2. In this simple representation, the ripple voltages are neglected, and each 2-quadrant converter is represented as a controllable direct voltage source, connected in series with a diode. The firing angle control is such that the d-c voltages of the 2 converters are equal, and have the same circuit polarity as one another.

The basic principle of operation of the cycloconverter can also be explained by reference to a similar simplified equivalent output circuit, shown in Fig. 6.1. Each 2-quadrant converter is now represented as an alternating voltage source, which corresponds to the "fundamental" or "wanted" voltage component generated at its output terminals, connected in series with a diode, which represents the condition of unidirectional current flow through the converter. Once again, the ripple voltage components are neglected, and therefore, in this simple representation, there is no "circulating" ripple voltage, and hence no requirement for a circulating current reactor. Thus, for the time being, the practical tendency for a circulating current to flow will be neglected.

THE BASIC PRINCIPLE OF OPERATION

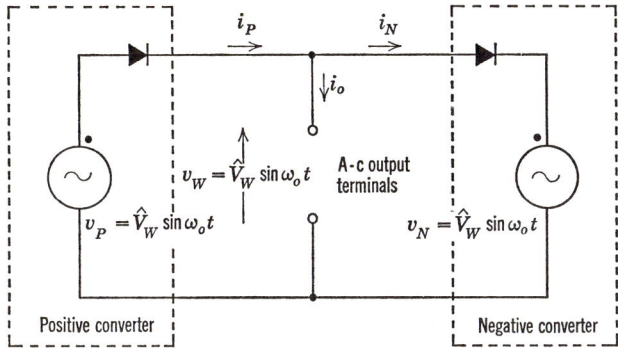

Figure 6.1. Idealized equivalent output circuit of the cycloconverter.

The basic control principle of the ideal cycloconverter is to continuously modulate the firing angles of the individual converters, so that each produces the same sinusoidal a-c voltage at its output terminals. Thus the voltages of the two generators in Fig. 6.1 have the same amplitude, frequency, and phase as one another, and the voltage at the output terminals of the cycloconverter is equal to the voltage of either of these generators. It is clear, so far as the load is concerned, that it "sees" a theoretically zero impedance a-c generator "behind" the output terminals of the cycloconverter, and therefore the load current can flow in either direction at any time, regardless of the instantaneous polarity of the voltage. In other words, it is possible for the mean power to flow either "to" or "from" the output terminals, and the cycloconverter is inherently capable of operation with "loads" of any phase angle, within a complete spectrum of 360°.

Of course, because of the unidirectional current-carrying property of the individual converters, it is inherent that the positive half cycle of load current must always be carried by the positive converter, and the negative half cycle, by the negative converter, regardless of the phase of the current with respect to the voltage. This means, in the general case of a non-unity displacement factor load, that during the course of a given half cycle of load current, the associated 2-quadrant converter produces both "positive" and "negative" portions of the load voltage wave, for given periods of time. In other words, each 2-quadrant converter operates both in its *rectifying* and in its *inverting* region, during the course of its associated half-cycle of current.

This is illustrated by the waveforms of Fig. 6.2. These waveforms are appropriate to the idealized cycloconverter circuit of Fig. 6.1, for loads of various displacement angles.

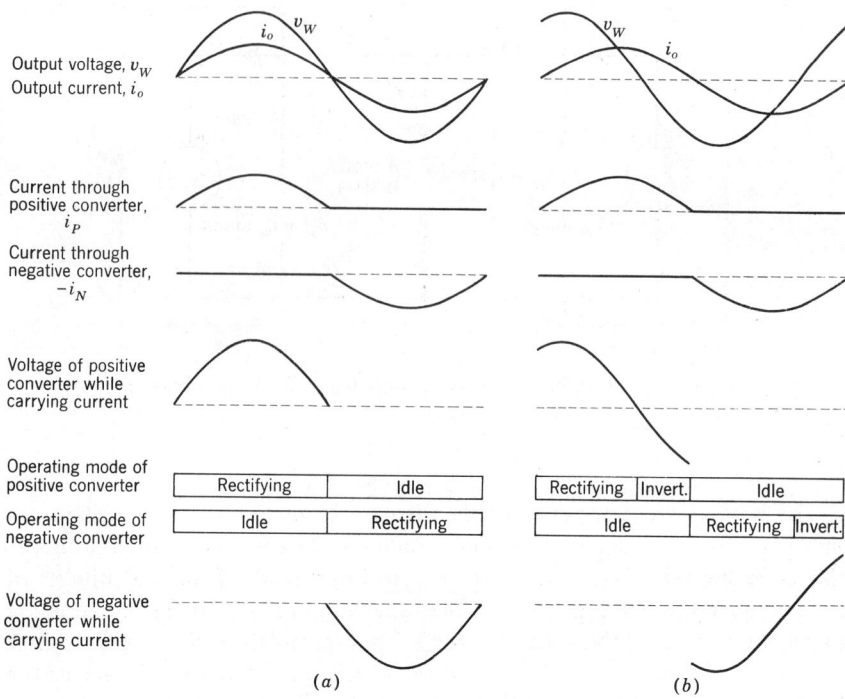

Figure 6.2. Waveforms illustrating the operation of the idealized cycloconverter circuit of Fig. 6.1, with loads of various displacement angles. (a) $\phi_0 = 0°$; (b) $\phi_0 = -60°$; (c) $\phi_0 = +60°$; (d) $\phi_0 = 180°$. (ϕ_0 = Load displacement angle.)

At *a* the displacement angle of the load is 0°. In this case each converter carries the load current only while it operates in its rectifying region, and it remains idle throughout the whole period in which its terminal voltage is in the inverting region of operation.

At *b*, the displacement angle of the load is 60° lagging. During the first 120° period of each half cycle of load current, the associated converter operates in its rectifying region, and delivers power to the load. During the latter 60° period of each half cycle of load current, on the other hand, the associated converter operates in its inverting region, and under this condition the load is "regenerating" power back into the cycloconverter output terminals, and thence into the a-c system at the input side.

At *c*, the displacement angle of the load is 60° leading. In this case, during the first 60° period of the load current half cycle, the associated converter operates in its inverting region; and during the latter 120° period, in its rectifying region.

THE BASIC PRINCIPLE OF OPERATION 149

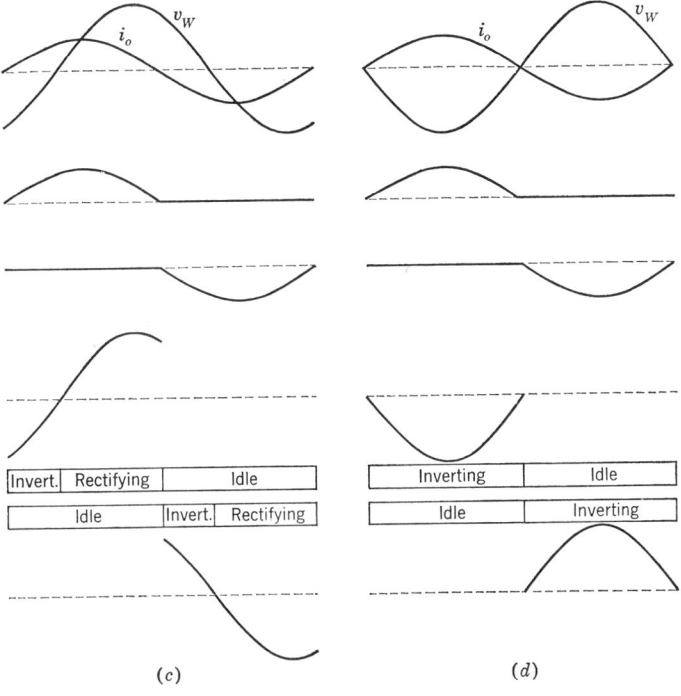

Figure 6.2 Continued

Finally, at *d* the displacement angle of the load is 180°. In this case the load is "fully regenerative," and it continuously delivers power into the output terminals of the cycloconverter, over the whole period of each output cycle. Thus, during each half cycle of current, the associated converter operates permanently in its inverting region.

It has been said that the basic control principle of the cycloconverter is to regulate the firing angles of the individual converters so that they produce identical wanted voltage components. The principles of the firing angle control of a cycloconverter are discussed in Chapter 9; for the time being, however, a brief qualitative examination will be made of the general manner in which the firing angles of the converters of the ideal cycloconverter must be controlled.

The "quiescent" condition for a cycloconverter can be considered to be when both converter firing angles are 90°, and under this condition the mean output voltage is zero. This is the condition which exists in the absence of any input stimulus to the firing pulse timing controls, which would make the cycloconverter produce a "useful" component of alternating output voltage. In order to produce a mean positive output voltage, the firing angle of the positive converter must be advanced, and that of the negative converter retarded, by an equal amount, away from the 90° "quiescent" point. Conversely, for a mean negative output voltage, the firing angle of the positive converter must be retarded, and that of the negative converter advanced, by an equal amount, away from the "quiescent" point. Thus, in general terms, it can be seen that in order to produce a varying mean voltage level of first one, and then the other, polarity at the output terminals of the cycloconverter, it is necessary to make the firing angle of each converter undergo a process of continuous "to and fro" oscillation about the 90° "quiescent" point. The oscillations of the two converter firing angles must be in phase opposition to one another, so that, as with the dual-converter providing a steady d-c output, (with continuous conduction) the sum of the firing angles should always be 180°. By controlling the frequency and amplitude of the oscillation of the converter firing angles, so the frequency and amplitude of the wanted output voltage component can be controlled. It should, of course, be appreciated that the firing angle cannot in fact be continuously controlled with respect to time, since it is possible to produce the firing pulses only at discrete intervals.

In practice, as with a steady d-c output, if the control is such that the firing pulses are applied simultaneously to both converters, and the sum of the converter firing angles is 180°, then although the wanted output voltage components of the converters are equal to one another, there are inevitably instantaneous inequalities between their ripple voltages. In the absence of any circulating current-limiting impedance, this circulating ripple voltage would give rise to a theoretically infinite circulating ripple current. It is therefore necessary to use some means for controlling the amplitude of, or completely suppressing, the circulating current. As with the dual-converter providing a steady d-c output, this can be achieved either by using a circulating current limiting reactor, or by suitably controlling the converter firing pulses in accordance with the direction of the load current, so as to "block" the idle converter; or, quite commonly, by a combination of both of these techniques.

The two basic alternative modes of operation of the cycloconverter, that is to say, operation without, and with, a circulating current, are described in the following sections. It is assumed that the operation is such that, in the first case, no circulating current is allowed to flow at any time; and, in the second, that circulating current is allowed to flow all the time. The circumstances under which the cycloconverter is operated with or without a circulating current are discussed in Chapter 7. It will be seen that although each of these basic operating modes may be used under certain conditions, it is often the case that a "partial" circulating current is permitted to flow, just within certain specified periods of each output cycle.

THE CIRCULATING CURRENT-FREE MODE OF OPERATION

In the circulating current-free mode of operation, each 2-quadrant converter fabricates a voltage waveform at its output terminals, and is allowed to conduct, only during its associated half cycle of load current. During the "idle" half cycle, the converter is completely blocked, through suitable control of its firing pulses. Thus, only one converter is in conduction at any one time, and no current circulates between the converters.

The operation of the cycloconverter with no circulating current is illustrated by the waveforms of Figs. 6.3 through 6.6, which are appropriate to a 6-pulse circuit with loads of various displacement angles. It is assumed here that the output current is sinusoidal, and shifted from the wanted component of the output voltage by the displacement angle of the load. In practice, of course, the output current waveform inevitably has ripple components superimposed upon it, due to the ripple voltages at the output of the cycloconverter. The magnitude of these ripple current components depends upon the impedance of the load circuit to the harmonic voltages.

The waveforms of Figs. 6.3 through 6.6 clearly illustrate how the converters operate both in their rectifying and inverting regions during specified periods of each output cycle, according to the displacement angle of the load. Also illustrated is the facility for controlling the amplitude of the wanted output voltage component of the cycloconverter, by means of appropriate control of the "depth" of modulation of the converter firing angles.

THE "NATURAL" CIRCULATING CURRENT MODE OF OPERATION

In practice, depending upon the circumstances of the application, there may be a tendency for the load current waveform to become discontinuous, at least under some conditions. As is explained in detail in Chapter 7, a discontinuous load current waveform gives rise to certain control difficulties, which sometimes can be best overcome by operating the cycloconverter with a circulating current.

It has been seen, in the case of a dual-converter providing a steady d-c output, that the basic principle of the circulating current mode of operation is to apply firing pulses continuously to both converters, without regard to the direction of the load current, so that each converter produces exactly the same mean voltage at its d-c terminals. This same control principle, when applied to a cycloconverter, would result in each converter producing exactly the same "wanted" alternating voltage component at its output terminals. This control principle, and the resulting manner of operation

152 THE PHASE-CONTROLLED CYCLOCONVERTER

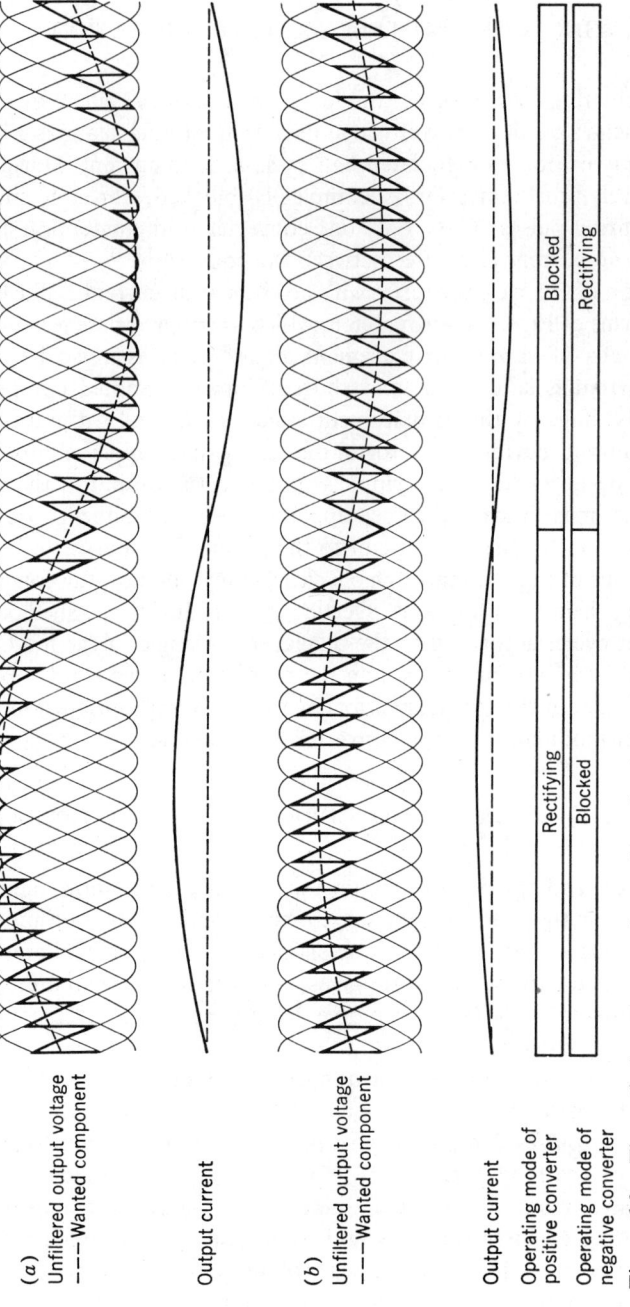

Figure 6.3. Theoretical output waveforms for a 6-pulse cycloconverter operating with no circulating current. (a) Maximum output voltage (i.e., $r = 1.0$); (b) half-maximum output voltage (i.e., $r = 0.5$). Load displacement angle $\phi_0 = 0°$; and output frequency = $\frac{1}{6} \times$ input frequency.

CIRCULATING CURRENT-FREE OPERATION 153

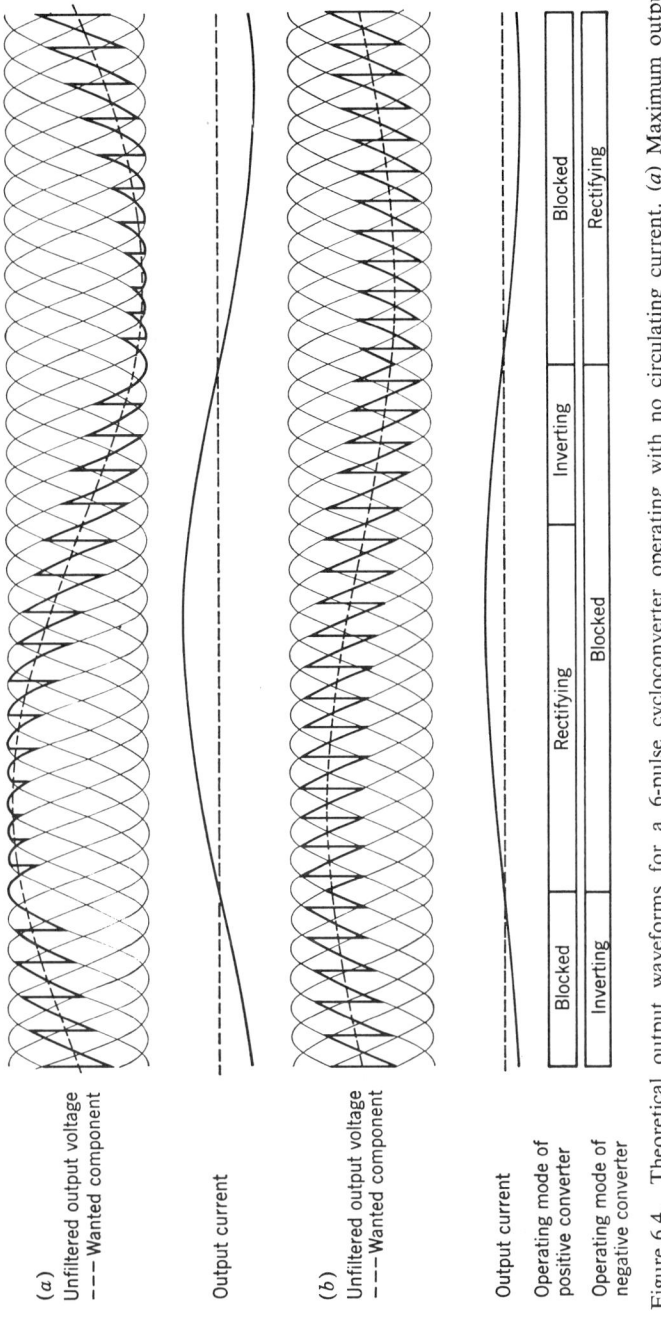

Figure 6.4. Theoretical output waveforms for a 6-pulse cycloconverter operating with no circulating current. (a) Maximum output voltage (i.e., $r = 1.0$); (b) half-maximum output voltage (i.e., $r = 0.5$). Load displacement angle $\phi_0 = 60°$ lagging; and output frequency = $\frac{1}{6} \times$ input frequency.

154 THE PHASE-CONTROLLED CYCLOCONVERTER

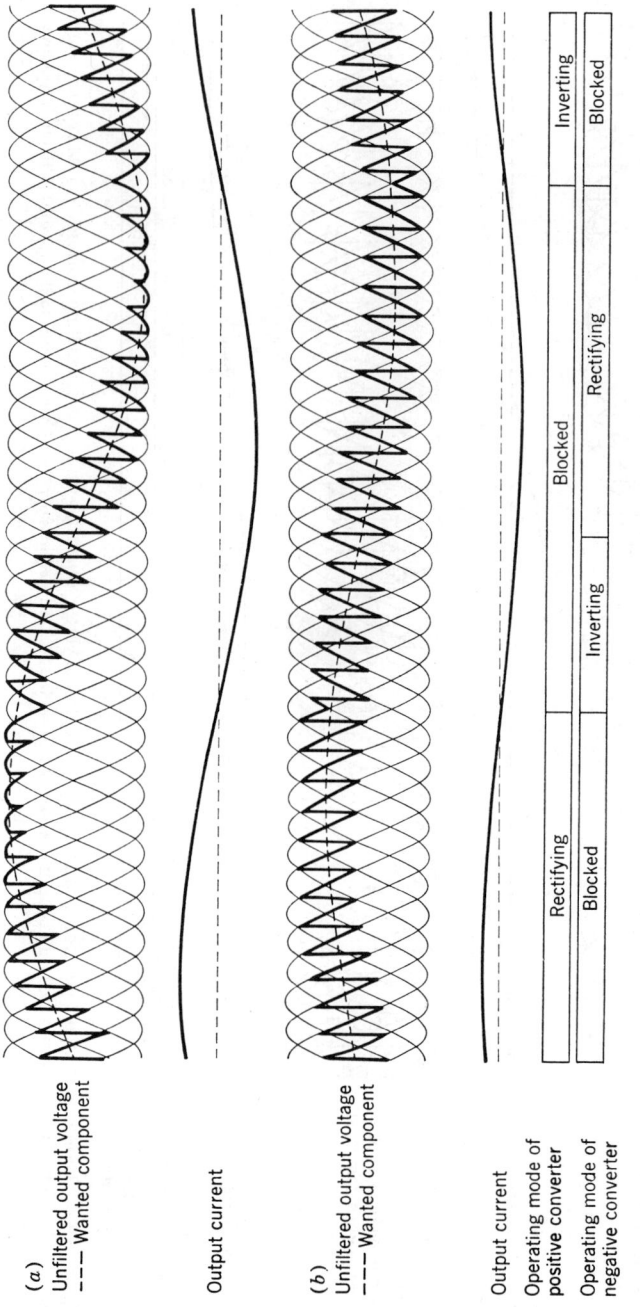

Figure 6.5. Theoretical output waveforms for a 6-pulse cycloconverter operating with no circulating current. (a) Maximum output voltage (i.e., $r = 1.0$); (b) half-maximum output voltage (i.e., $r = 0.5$). Load displacement angle $\phi_0 = 60°$ leading; and output frequency $= \frac{1}{6} \times$ input frequency.

CIRCULATING CURRENT-FREE OPERATION 155

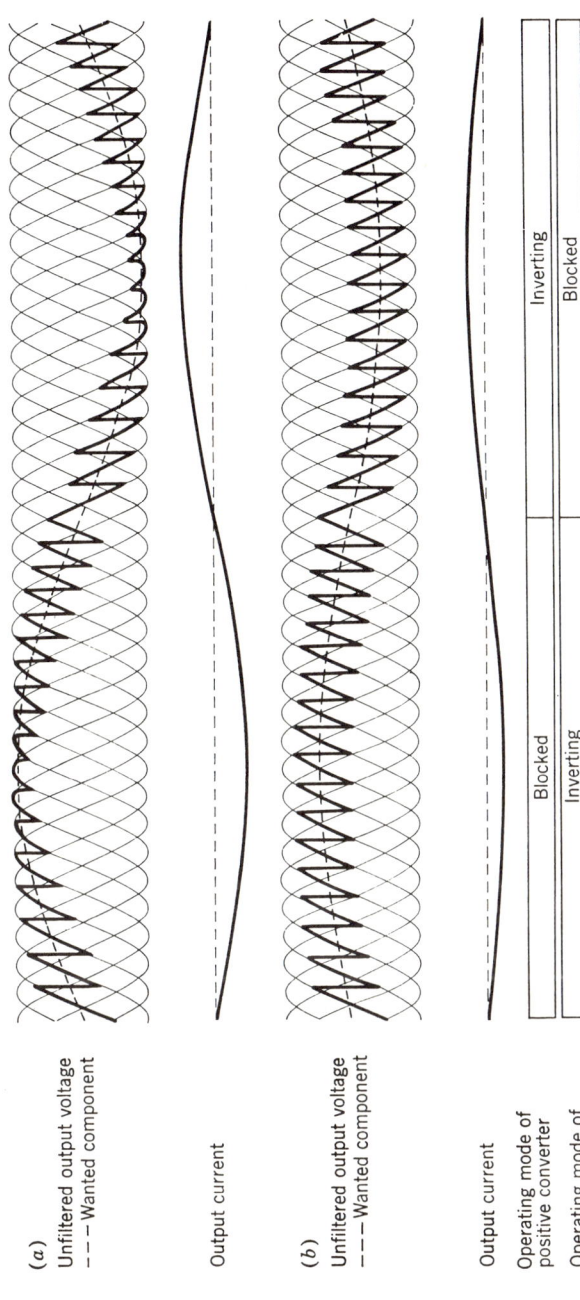

Figure 6.6. Theoretical output waveforms for a 6-pulse cycloconverter operating with no circulating current. (a) Maximum output voltage (i.e., $r = 1.0$); (b) half-maximum output voltage (i.e., $r = 0.5$). Load displacement angle $\phi_0 = 180°$ (i.e., "fully regenerative load"); and output frequency $= \frac{1}{6} \times$ input frequency.

of the power circuit, can be regarded as constituting the "natural" circulating current mode of operation of the cycloconverter, and this is the mode of operation which will now be considered. As will be seen, the operation of the power circuit is such that a relatively large amount of current, in addition to the circulating ripple current, circulates between the two converters. For this reason, this "natural" mode of operation is used in practice only under special circumstances, in which its use is virtually obligatory. Its consideration here is useful, however, since this provides a clear basis for a later understanding of the more commonly used practical modes of operation, in which a "partial" circulating current is permitted to flow, just within certain specified periods of each output cycle.

In Fig. 6.7, voltage and current waveforms are shown for a 6-pulse cycloconverter, which illustrate this "natural" circulating current mode of operation. Since the wanted voltage components generated at the output terminals of each converter are equal to one another, there can be no difference of voltage, at the wanted output frequency, developed across the circulating current reactor. Thus the amplitude of the wanted voltage component, at the midpoint of this reactor, must be the same as that of either of the individual converters. However, as can be seen, the general appearance of the raw voltage waveform at this point is different to that of the voltage waveforms of either of the constituent converters. This is because this waveform is the instantaneous average of the constituent waveforms, and certain harmonic components contained in the individual converter waveforms cancel one another, and therefore do not appear at the output terminals.

In Chapter 5 it was seen that the presence of an alternating ripple current at the output terminals of the dual-converter operating in the circulating current mode results in a "trapped" mmf in the circulating current reactor, which gives rise to a so-called "self-induced" component of circulating current. In the case of the cycloconverter, ideally the output current is comprised entirely of an alternating component, having the wanted output frequency. Once more, therefore, the phenomenon of a "self-induced" component of circulating current arises. In this case, however, this "self-induced" component of circulating current is produced as a result of the wanted component of output current, rather than as a result of a generally relatively small unwanted ripple component. Its relative magnitude is therefore considerably larger, and as a result, its presence is generally considerably more objectionable.

To the reader who is familiar with the mechanism, described in Chapter 5, whereby a "self-induced" component of circulating current arises as a result of the ripple current at the output terminals of the dual-converter, the waveforms of current for a cycloconverter operating in its "natural" circulating current mode are immediately obvious. For the sake of completeness,

THE "NATURAL" CIRCULATING CURRENT MODE

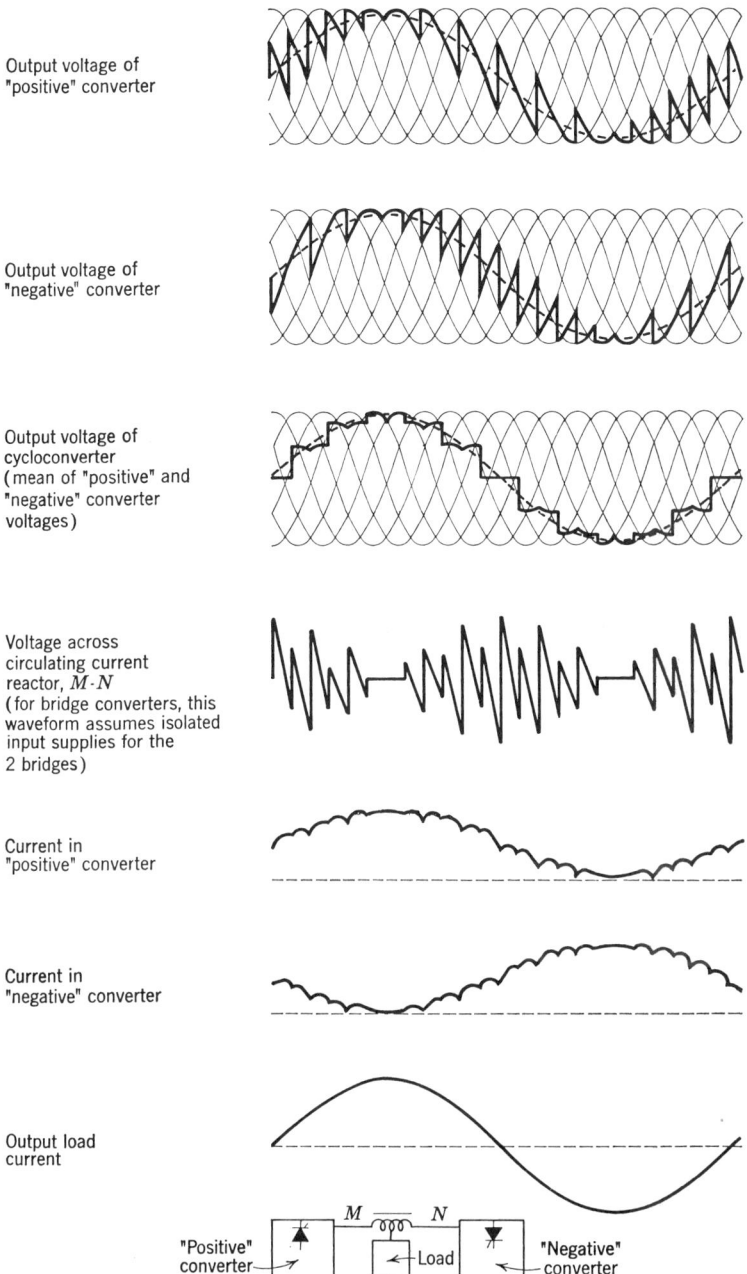

Figure 6.7. Waveforms illustrating the operation of a 6-pulse cycloconverter in the "natural" circulating current mode.

however, the derivation of these waveforms will be briefly explained here. In order to do this, it is helpful to consider the simple equivalent output circuit of the cycloconverter, in which the ripple voltages are neglected, with a circulating current reactor connected between the output terminals of the two converters. This simplified circuit is shown in Fig. 6.8.

The operation of the circuit is explained by reference to the waveforms shown in Fig. 6.9. It is assumed that the load is switched on at time $t_1 = 0$, and that a current, $i_o = \hat{I}_o \sin \omega_o t$, starts to flow. During the first quarter cycle, this current is drawn through the "positive" converter, and it results in the voltage waveform shown at e being developed across the circulating current reactor. The polarity of this voltage is clearly in such a direction that the diode associated with the "negative" converter is reverse biased, which confirms that this converter is shut off during this period. At time $t_2 = \pi/2\omega_o$, the load current starts to decrease, thus creating a tendency for the voltage across the reactor to reverse. This is prevented, however, due to the clamping action of the diode associated with the negative converter, and hence this converter, as well as the positive converter, comes into conduction at this time. With both converters in conduction, there can be no voltage across the circulating current reactor, since points M and N are now necessarily at the same potential. So long as this situation exists, the total mmf of the circulating current reactor must remain constant, and equal to the peak value existing at time t_2

$$i_P \frac{N}{2} + i_N \frac{N}{2} = \hat{I}_o \frac{N}{2} \tag{6.1}$$

$$\therefore \quad i_P + i_N = \hat{I}_o \tag{6.2}$$

Now $$i_P - i_N = \hat{I}_o \sin \omega_o t \tag{6.3}$$

Hence $$i_P = \frac{\hat{I}_o}{2} + \frac{\hat{I}_o}{2} \sin \omega_o t \tag{6.4}$$

and $$i_N = \frac{\hat{I}_o}{2} - \frac{\hat{I}_o}{2} \sin \omega_o t \tag{6.5}$$

Equations (6.4) and (6.5) describe the operation of the circuit so long as both converters remain in continuous conduction; that is to say, so long as i_P and i_N, as given by these equations, both remain positive. In fact, it can be seen that the right-hand side of each of these equations is always positive, except just instantaneously once each cycle at the appropriate peaks of the load current wave. Thus, once the current waveforms described by equations (6.4) and (6.5) are in existence, theoretically there is never any

THE "NATURAL" CIRCULATING CURRENT MODE

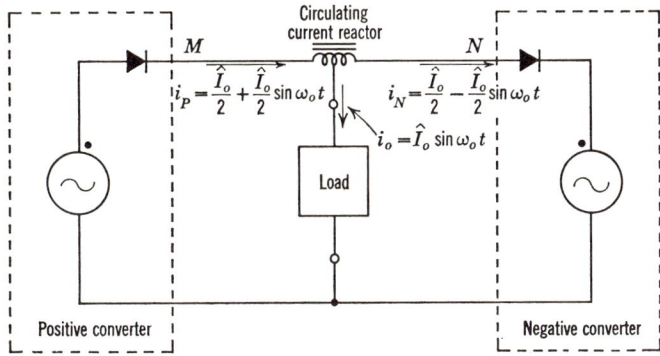

Figure 6.8. Idealized equivalent output circuit of the cycloconverter with circulating current reactor.

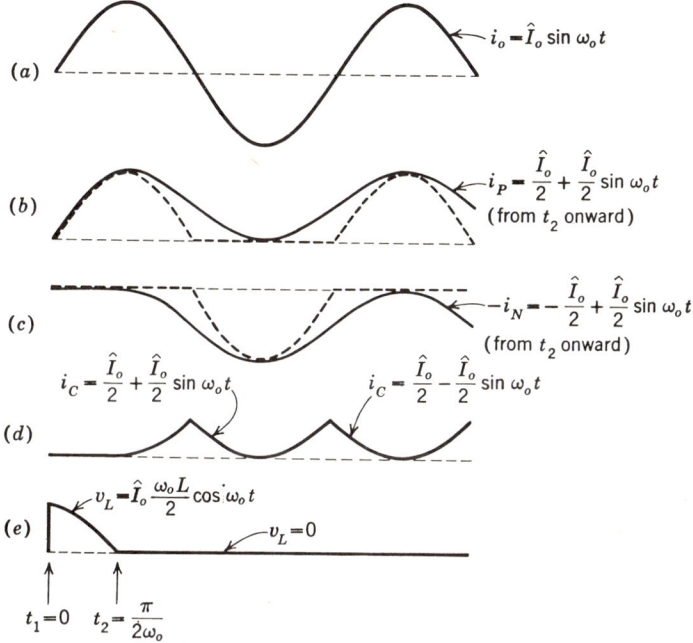

Figure 6.9. Waveforms associated with the idealized cycloconverter circuit of Fig. 6.8 illustrating the operation in the "natural" circulating current mode, with each converter producing the same wanted voltage component. (a) Output load current; (b) ———total current carried by positive converter, - - -output load current carried by positive converter; (c) ———total current carried by negative converter, - - -output load current carried by negative converter; (d) "self-induced" circulating current; (e) voltage across circulating current reactor, M-N.

more than a momentary tendency for either converter to drop out of conduction, and these equations therefore describe the final steady state operation of this theoretically "ideal" circuit.

Thus it is seen that, with the wanted voltage components generated at the output terminals of the two converters exactly equal to one another, unavoidably energy becomes trapped in the circulating current reactor, and this forces both converters into continuous conduction, with the result that a continuous "self-induced" component of circulating current is produced. This is in contrast to the case of the dual-converter providing a smooth direct output current, in which case, if the control is such that the mean d-c terminal voltages of the individual converters are exactly equal to one another, then the only circulating current is a relatively small ripple component due to the "circulating" ripple voltage.

This "self-induced" component of circulating current is illustrated in Fig. 6.9d. This waveform is arrived at as follows: During the positive half cycle of output current, all of the output current must be drawn from the "positive" converter; therefore the current carried by the "negative" converter during this time does not flow at the output terminals, and is purely "circulating current." Conversely, during the negative half cycle of output current, the current carried by the positive converter is purely circulating current.

The self-induced component of circulating current is, of course, quite independent of any additional ripple components of circulating current which are produced in practice due to ripple voltages appearing at the outputs of the converters. Furthermore, it is theoretically independent of the inductance value of the circulating current reactor, assuming that this has a relatively high Q factor at the wanted output frequency. In practice, if the output frequency is relatively low in comparison with the input frequency, this assumption may not be valid, and in this case the self-induced component of circulating current may be reduced, or even almost nonexistent.

Returning now to the "practical" converter current waveforms associated with the voltage waveforms in Fig. 6.7, but still assuming that the load current is sinusoidal, and the circuit losses are zero, it can be seen that these have the same general shape as the waveforms of the simple equivalent circuit, but with superimposed circulating ripple components, due to the circulating ripple voltage, which in practice appears across the circulating current reactor.

It can be shown that the average value of the "self-induced" component of circulating current is $0.57 \times$ the average output load current carried by the converters. Clearly, this imposes a substantial "wattless" load on the converters, in addition to the "useful" load. For this reason, this mode of operation, in which a continuous circulating current is permitted to flow, and in which the wanted voltage components generated by the two converters are exactly equal to one another, is used in practice only under special

DISTORTION OF THE OUTPUT VOLTAGE 161

circumstances, and, more particularly, at relatively light levels of output load, under which conditions the relatively large circulating current can be tolerated.

DISCUSSION ON THE HARMONIC DISTORTION OF THE CYCLOCONVERTER OUTPUT VOLTAGE

General Considerations

A detailed theoretical analysis of the predominant harmonic distortion terms theoretically present in the output voltage waveform of the cycloconverter is made in Chapter 11. At this stage, in order to provide a general understanding of this aspect of the cycloconverter performance, the following qualitative discussion is presented.

As has been seen, with a steady d-c output, the lowest harmonic frequency present in the voltage waveform of the phase-controlled converter is $P \times$ the input frequency, where P is the pulse number of the converter; and, in general, the harmonic series of the distortion terms follows a regular pattern, the frequency of each term being an exact integer multiple of the "lowest" term.

With an alternating output voltage, however, due to the process of continuous "to and fro" phase modulation of the converter firing angles, as well as to certain practical effects, which are of no relevance with a steady d-c output, the harmonic spectrum of the waveform of the output voltage is more complex. The harmonic distortion terms which appear in the output voltage of the cycloconverter may be classified into three categories. These are discussed separately in the following sections.

Necessary Distortion Terms

The necessary distortion terms contained in the output voltage of the cycloconverter are akin to the ripple distortion terms obtained with a steady d-c output, in as much as they are produced as a necessary outcome of the basic mechanism of the converter, whereby the output voltage waveform is "pieced together" from segments of the input voltage waves.

It is usually assumed in the literature that the frequency spectrum of the necessary distortion terms is substantially the same as with a steady firing angle. This is a reasonable approximation so long as the output frequency is low in comparison with the input frequency; however, it becomes progressively less valid as the ratio of the output to input frequency is raised. In fact, it is the progressive departure of the actual frequency spectrum of the necessary harmonic distortion terms, from that obtained with a steady d-c output, which is the basic limiting factor determining the maximum attainable useful output to input frequency ratio of the cycloconverter. Thus, consideration of the actual spectrum of the necessary harmonic frequencies contained in the output voltage is a key factor in determining the limits of performance of the cycloconverter.

As will be seen in Chapter 11, because of the direct process of waveform conversion through the cycloconverter, the frequency spectrum of the necessary distortion terms contained in the output voltage is related to both the input frequency and to the output frequency. Thus, the main class of necessary distortion terms constitute "beat frequency" components, having frequencies which are both sums and differences of multiples of both the input and output frequencies.

Because of the production of these beat frequency components, it is necessary, when considering the harmonic spectrum of the output voltage of the cycloconverter, to dissociate oneself from the common conception of classical Fourier harmonic analysis, according to which a complex waveform can be resolved into a fundamental component, the frequency of which is equal to the fundamental repetition frequency of the wave, and a series of superimposed harmonic components, having frequencies which are integer multiples of the fundamental frequency. This is because, for the output voltage waveform of the cycloconverter, the necessary "harmonic" frequencies are not generally integer multiples of the wanted output frequency; nor, for that matter, are they necessarily integer multiples of the fundamental repetition frequency of the waveform, which, as will be seen, is generally less than the wanted output frequency.* Furthermore, there may not be necessarily a "harmonic" component in the output voltage waveform having the fundamental repetition frequency. This, at first sight, may be an unacceptable proposition to the reader familiar with the classical concept of Fourier harmonic analysis. Nonetheless, it is a simple truth that a complex waveform need not necessarily contain a component having the fundamental repetition frequency.

The production of beat frequency distortion components in the output voltage of the cycloconverter arises from the fact that it is not possible for the waveform of the output voltage during one output cycle to be precisely identical with the waveform during the next output cycle, unless the output frequency happens to be an exact integer submultiple of the product of the input frequency and the pulse number of the converter. This is because the

* Strictly speaking, the term "harmonic" applies only to a frequency which is an integer multiple of a fundamental frequency. The term is used in a more general sense in this book, to apply to all *distortion* components generated at the output (or input) of the cycloconverter.

DISTORTION OF THE OUTPUT VOLTAGE 163

basic "building pieces" for the output voltage waveform are the input voltage waves, and since like points on successive input voltage waves occur only at discrete intervals of time, the resolution possible in the fabrication of the output waveform is discrete, rather than continuous. This, however, does not imply that it is possible to obtain only discrete output frequencies from the cycloconverter. This is not the case, and, in fact, the facility for continuous control of the output to input frequency ratio of the cycloconverter is the essential feature of most of its practical applications. Nor does the limited resolution possible in the fabrication of the output waveform necessarily imply that subharmonic* frequency distortion components are present. In point of fact, however, with an open loop firing angle control, small amounts of these components are usually present; but, for all practical purposes, these can be eliminated by means of suitable feedback control techniques, as discussed in Chapter 9. What is implied, more generally, is simply that distortion components are present having frequencies which are not integer multiples of the wanted output frequency; in other words, that "beat frequency" components are present.

As a demonstration of the fact that only a limited resolution is possible in the fabrication of the output voltage waveform, consider, for example, the waveform of a 3-pulse cycloconverter. In this case, it is possible for the waveform of the voltage during one output cycle to be exactly the same as the next, only if the output frequency is related to the input frequency by the following expression:

$$f_o = \frac{3f_i}{n}$$

where n is a positive integer.

This is illustrated by the waveforms shown in Fig. 6.10. At a, the output frequency is exactly $\frac{1}{3}$ of the input frequency, and the waveform of the output voltage during one output cycle is the same as in the next. At b, the output frequency is a little higher than $\frac{1}{3}$ of the input frequency; in this case, the waveform of the output voltage during one output cycle is not precisely the same as in the next, but it repeats itself exactly only every third cycle. More generally, it can be seen that the fundamental repetition frequency of the output waveform may be much lower than the wanted output frequency.

A second, usually less significant, class of "necessary" distortion, which should be mentioned, arises as a result of the practical presence of internal impedance of the input source. The presence of this impedance causes

---

* In this book, the term "subharmonic frequency," when applied to the output voltage, is taken to mean any frequency below the wanted output frequency; when applied to the input current, it is taken to mean any frequency below the input line frequency.

164 THE PHASE-CONTROLLED CYCLOCONVERTER

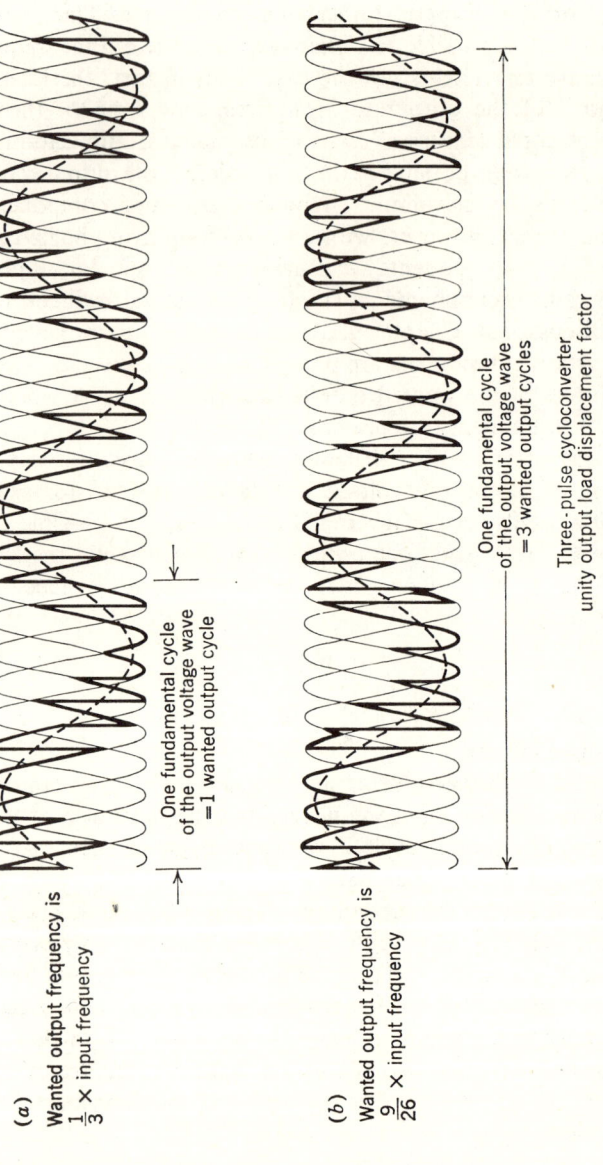

(a) Wanted output frequency is $\frac{1}{3} \times$ input frequency

One fundamental cycle of the output voltage wave = 1 wanted output cycle

(b) Wanted output frequency is $\frac{9}{26} \times$ input frequency

One fundamental cycle of the output voltage wave = 3 wanted output cycles

Three-pulse cycloconverter unity output load displacement factor

Figure 6.10. Waveforms demonstrating that successive wanted cycles of the output voltage of the cycloconverter are not necessarily precisely the same as one another.

DISTORTION OF THE OUTPUT VOLTAGE 165

"notches" to be produced in the output voltage waveform of the cycloconverter, and these "notches" give rise to small amounts of odd harmonic distortion. This will not be elaborated upon here, since it is discussed in Chapter 13.

Unnecessary Distortion Terms

If the firing angle is controlled in response to an analog reference voltage, then by suitable design of the relationship between the firing angle and this reference voltage, the phase-controlled converter can be made to behave as an amplifier with a linear voltage transfer characteristic. Thus, if the reference voltage is a sinusoid, representing the desired wanted component of output voltage, the "mean" component of the output voltage is also a corresponding sinusoid. Under these circumstances, the output voltage waveform produced is the theoretically "ideal" one, inasmuch as the only distortion terms produced are those which are an inevitable by-product of the basic waveform fabrication technique. In other words, these are the necessary distortion terms, which have already been discussed.

In practice, it may be, as a matter of practical convenience in the design of the control system, that either the analog input reference voltage is not a "perfect" sinusoid, or that the relationship between the analog input reference voltage and the mean output voltage is not linear, or, possibly, that both of these factors may be simultaneously applicable. In this event, as would be expected, the output voltage waveform of the cycloconverter departs from the ideal one. In other words, it contains so-called unnecessary distortion terms, in addition to the necessary ones.

Thus, the unnecessary distortion terms can be defined as those components which are generated as a result of a firing angle modulation process which does not control the timing of the firing pulses in the theoretically ideal manner. Since the unnecessary distortion terms are produced as a result of a nonlinear voltage transfer characteristic, or as a result of an imperfect reference voltage, or both, the spectrum of frequencies of these terms generally consists only of integer multiples of the wanted output frequency. This is in contrast to the frequency spectrum of the necessary distortion terms, which, as has been stated, is related to both the input and output frequencies.

"Practical" Distortion Terms

The types of distortion so far considered can be regarded as having theoretical origins, insofar as they are produced as an outcome of the basic waveform fabrication process, possibly in conjunction with theoretical inadequacies which may be designed into the firing pulse timing circuits.

In addition to these "theoretical" distortion terms, further "nontheoretical" distortion components can be produced in the output voltage waveform of the cycloconverter, as a result of various practical effects. These distortion components are referred to here as "practical" distortion terms.

These terms may be produced, for example, as a result of practical imperfections in the control circuits. Thus, quite small practical errors in the timing of the firing pulses can, under some conditions (as explained in Chapter 9), produce a relatively large nontheoretical distortion of the output voltage wave.

Another source of practical distortion is due to the nonlinear conduction voltage characteristic of the thyristors themselves. This effect is not generally significant, so long as the amplitude of the wanted component of output voltage is high, in relation to the thyristor forward conduction voltage. However, it may become quite significant at relatively low levels of output voltage.

DISCUSSION ON THE LOADING EFFECT OF THE CYCLOCONVERTER ON THE INPUT SYSTEM

General Considerations

The cycloconverter, in its basic form, consists merely of a collection of static switches connected directly between the input a-c system and the load circuit, and the basic principle of power conversion is to fabricate an output voltage waveform having the desired frequency, simply by opening and closing the switches according to a predetermined program. Thus, unlike other types of frequency converting equipment—for example, a motor-generator, or a rectifier-inverter (with an intermediate d-c filter)—there are, basically, no energy storage elements connected between the input system and the "raw" output terminals of the cycloconverter.

Thus, the process of energy transfer through the cycloconverter is a very direct one, and, of necessity, the input system always directly "sees" the load at the output terminals. Thus, with a single cycloconverter, supplying a single phase load, the instantaneous fluctuations of power at the output terminals, inherent in the production of the alternating output, are transmitted directly to the input system. This gives rise to harmonic components of input line current having frequencies which are "beat" components between the output and input frequencies. These harmonic components of current are not, therefore, "locked" to the input frequency, except at certain discrete output-to-input frequency ratios, but generally "drift" with respect to the input voltage wave. Moreover, the presence of these harmonic components of current is quite independent of the pulse number of the converter, since they are inherent in the basic process of instantaneous power transfer from the input to the output of the cycloconverter.

Of course, in addition to the harmonic currents which flow in the input system as a result of the fluctuations of the output power, there are also harmonic currents which flow as a result of the basic rectifier-like operation of the cycloconverter. The presence or absence of given families of these latter harmonic components, as with a steady d-c output, is determined by the pulse number of the converter.

For a cycloconverter supplying a balanced 3-phase output, the harmonic load "seen" by the input system is much reduced, as compared to the case of a single phase load. This is because although, of course, the power at each of the individual output phases still fluctuates, the load which is now "seen" by the input system is the total instantaneous power of all three output phases, and this remains constant. Thus the beat frequency components of current which supply the individual fluctuating power components at the three output phases, now merely circulate between the three cycloconverters, and do not flow in the input system.

Of course, the only component of current at the input of the cycloconverter, which is capable of supplying a mean component of power at the output, is the fundamental* "in-phase" component, because the mean power due to any quadrature or harmonic components of current is necessarily zero. Thus the fundamental "in-phase" component of current drawn from the supply necessarily assumes a value appropriate to the mean power at the output of the cycloconverter.

In addition to the in-phase component of current, the cycloconverter also consumes a lagging quadrature component. The presence of this component is inherent in the basic control mechanism of the cycloconverter, whereby an output voltage waveform having a sinusoidal envelope is fabricated by means of a process of firing angle phase delay.

It has already been seen, in the case of a steady d-c output, that the phase control process gives rise to a lagging displacement angle between the fundamental component of the input line current and the associated voltage. Since the firing angle of the cycloconverter is made to undergo a process of continuous phase modulation away from, and then back towards, the position of zero phase retard, the hypothetical "mean firing angle," averaged over a complete output cycle, is necessarily in retard of 0°. As a result, a lagging reactive component of current is produced at the input, even if the power factor of the load at the output is unity. Indeed, as is shown in Chapter

---

* The "fundamental" component of input current is here designated as being the component having input line frequency.

12, the minimum possible theoretical displacement angle between the input voltage and the fundamental component of the input current of the phase-controlled cycloconverter is 32.5°. This occurs when the cycloconverter generates its maximum possible output voltage, into a load of unity displacement factor.

With reactive loads at the output, the input displacement angle is greater than with a unity power factor load. Moreover, it is a fundamental property of the phase-controlled cycloconverter, that either a lagging or a leading load at the output appears as a lagging load at the input. This, again, is due to the basic phase-control mechanism, which is inherently such that the input line current must lag the input voltage, in order to produce a natural commutation of current from one phase to the next.

Just as, with the phase-controlled converter operating at a steady firing angle, the displacement angle of the fundamental component of the input line current is determined only by the relative level of the output voltage, and is independent of the converter pulse number, so, in the case of the cycloconverter, the displacement angle of the input line current is dependent only upon the relative level of the output voltage, and the displacement factor of the load, and is independent of the converter pulse number, and of the number of output phases, and of the ratio between the output and input frequencies.

Figures 6.11 through 6.14 are illustrations of theoretical input line current waveforms obtained with a phase-controlled cycloconverter supplying balanced 3-phase loads of various displacement angles. In each case, it is assumed, for convenience, that the output frequency is exactly $\frac{1}{3}$ of the input frequency. It should be realized, however, that unless there is a direct and purposeful synchronization between the output and input frequencies, then in practice a slow "drifting" between the output and input waves inevitably occurs, even if the output frequency is nominally an exact integer fraction of the input frequency. Thus, in practice, the detailed appearance of the waveforms of the currents in the individual input lines would not, over a long term period, be different to one another, as they are with a rigid synchronization between the output and input frequencies; but rather there would be a slow "drifting" effect, during the course of which the detailed appearance of the waveform of the current in each input line would undergo a slow, but continuous, change. This, of course, is simply another way of repeating what has already been said—namely, that harmonic distortion components are generally present in the waveform of the input line current, which have "beat" frequencies, related to both the input and output frequencies of the cycloconverter.

In Figs. 6.11 through 6.14 the waveforms of the input line current due to the current of each output phase are shown separately, in addition to the

LOADING EFFECT ON THE INPUT SYSTEM 169

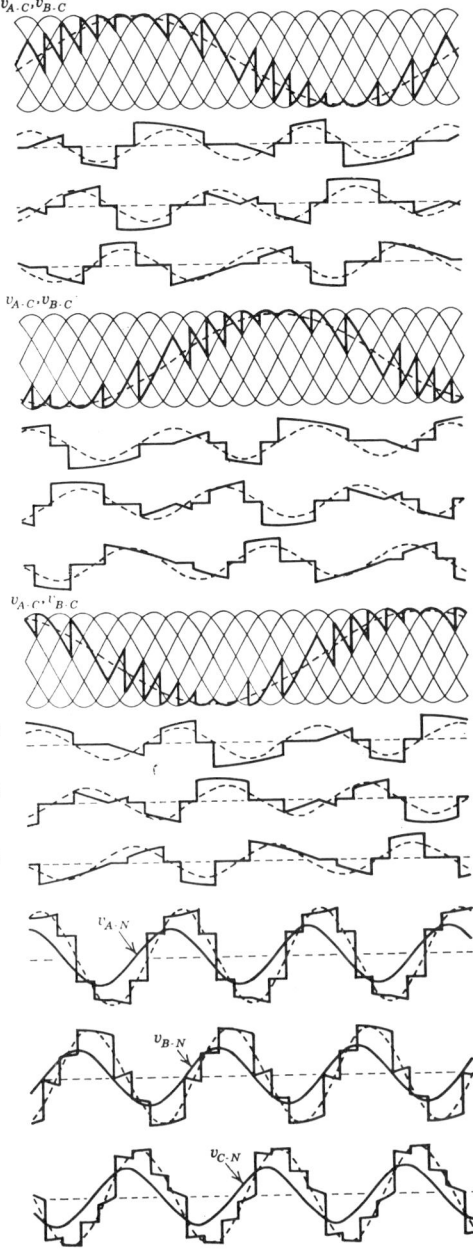

Figure 6.11. Theoretical waveforms for a 3-phase to 3-phase, 6-pulse bridge cycloconverter. Output load displacement angle $= 0°$; output frequency $= \frac{1}{3} \times$ input frequency; and maximum output voltage, that is, $r = 1.0$.

170 THE PHASE-CONTROLLED CYCLOCONVERTER

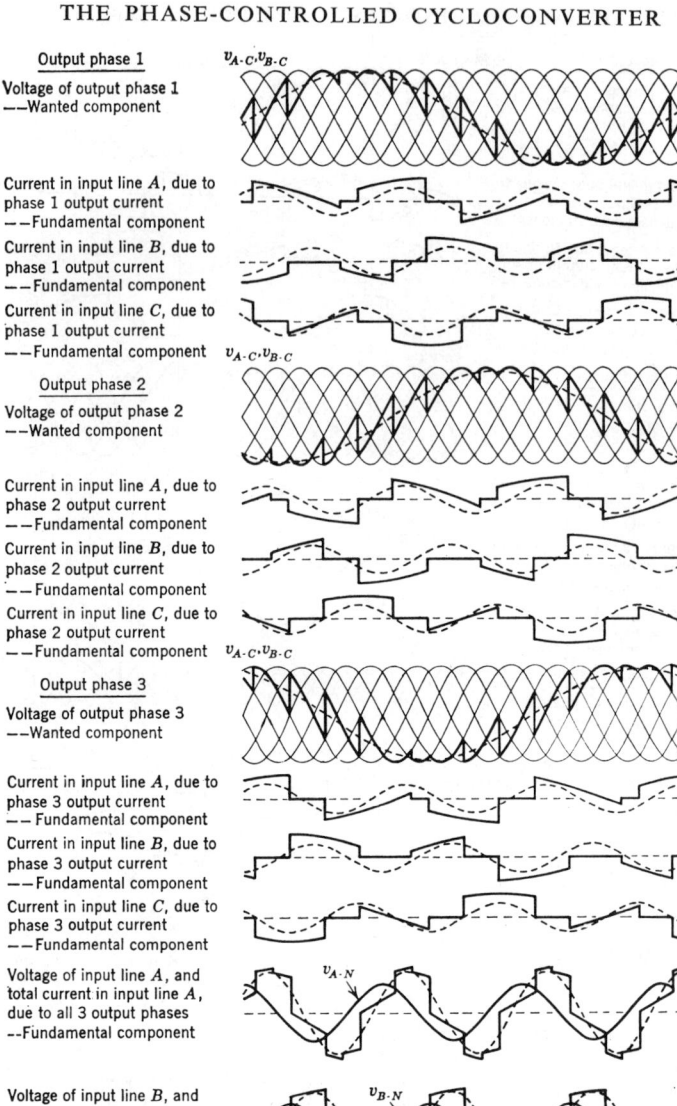

Figure 6.12. Theoretical waveforms for a 3-phase to 3-phase, 6-pulse bridge cycloconverter. Output load displacement angle = 60° lagging; output frequency = $\frac{1}{3}$ × input frequency; and maximum output voltage, that is, $r = 1.0$.

LOADING EFFECT ON THE INPUT SYSTEM

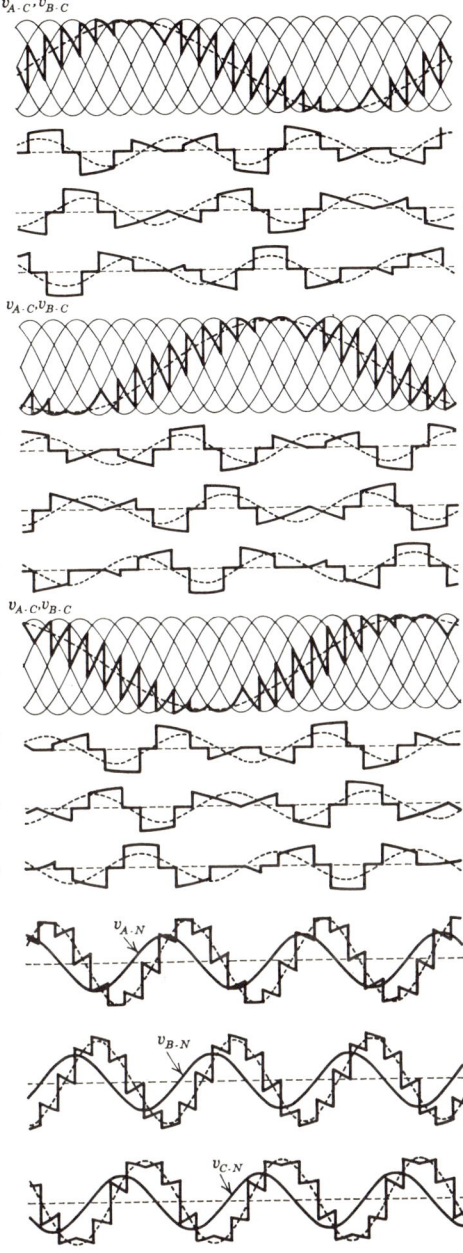

Output phase 1

Voltage of output phase 1
— —Wanted component

Current in input line A, due to phase 1 output current
— —Fundamental component

Current in input line B, due to phase 1 output current
— —Fundamental component

Current in input line C, due to phase 1 output current
— —Fundamental component

Output phase 2

Voltage of output phase 2
— —Wanted component

Current in input line A, due to phase 2 output current
— — Fundamental component

Current in input line B, due to phase 2 output current
— — Fundamental component

Current in input line C, due to phase 2 output current
— — Fundamental component

Output phase 3

Voltage of output phase 3
— —Wanted component

Current in input line A, due to phase 3 output current
— — Fundamental component

Current in input line B, due to phase 3 output current
— — Fundamental component

Current in input line C, due to phase 3 output current
— — Fundamental component

Voltage of input line A, and total current in input line A, due to all 3 output phases
— —Fundamental component

Voltage of input line B, and total current in input line B, due to all 3 output phases
—Fundamental component

Voltage of input line C, and total current in input line C, due to all 3 output phases
—Fundamental component

Figure 6.13. Theoretical waveforms for a 3-phase to 3-phase, 6-pulse bridge cycloconverter. Output load displacement angle = 60° leading; output frequency = $\frac{1}{3}$ × input frequency; and maximum output voltage, that is, $r = 1.0$.

172 THE PHASE-CONTROLLED CYCLOCONVERTER

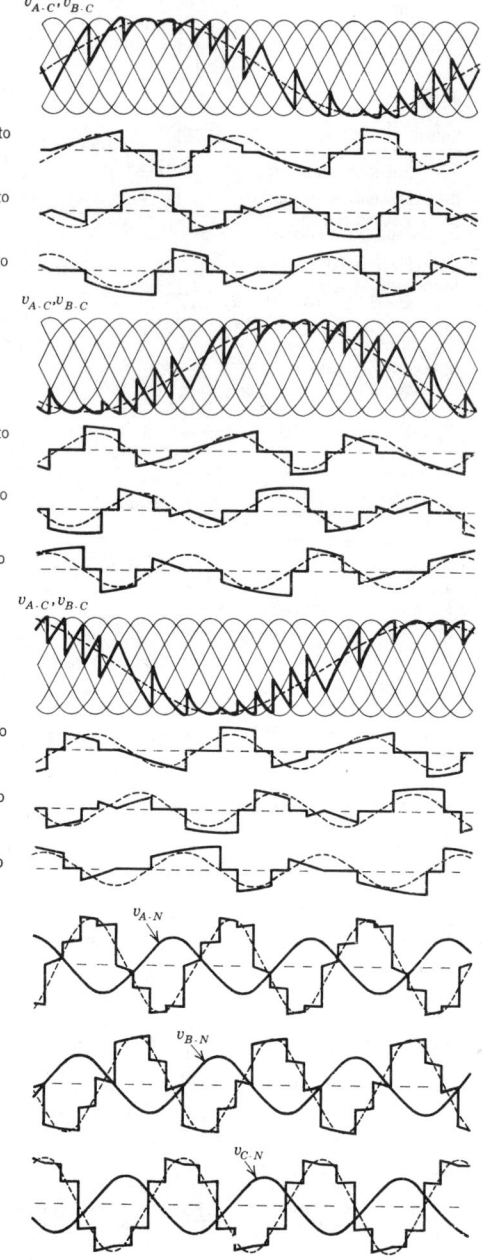

Output phase 1

Voltage of output phase 1
— — Wanted component

Current in input line A, due to phase 1 output current
— — Fundamental component

Current in input line B, due to phase 1 output current
— — Fundamental component

Current in input line C, due to phase 1 output current
— — Fundamental component

Output phase 2

Voltage of output phase 2
— — Wanted component

Current in input line A, due to phase 2 output current
— — Fundamental component

Current in input line B, due to phase 2 output current
— — Fundamental component

Current in input line C, due to phase 2 output current
— — Fundamental component

Output phase 3

Voltage of output phase 3
— — Wanted component

Current in input line A, due to phase 3 output current
— — Fundamental component

Current in input line B, due to phase 3 output current
— — Fundamental component

Current in input line C, due to phase 3 output current
— — Fundamental component

Voltage of input line A, and total current in input line A, due to all 3 output phases
— — Fundamental component

Voltage of input line B, and total current in input line B, due to all 3 output phases
— — Fundamental component

Voltage of input line C, and total current in input line C, due to all 3 output phases
— — Fundamental component

Figure 6.14. Theoretical waveforms for a 3-phase to 3-phase, 6-pulse bridge cycloconverter. Output load displacement angle $= 180°$ (i.e., "fully regenerative load"); output frequency $= \frac{1}{3} \times$ input frequency; and maximum output voltage, that is $r = 1.0$.

LOADING EFFECT ON THE INPUT SYSTEM 173

waveforms of the net current in each input line, due to the currents of all three output phases. These illustrations demonstrate the fact that the distortion of the net waveform of the input line current, due to all three output phases, is much less than the distortion of the waveform of the input line current due to one output phase. It is also seen, for a load of a given displacement factor, that the fundamental component of the input line current due to one output phase bears the same phase relationship to the associated input voltage, as does the fundamental component of the input line current due to all three output phases.

These waveforms also demonstrate the fact that the fundamental component of the input line current invariably lags the input voltage, regardless of the displacement angle of the load. Thus, in the case of a unity displacement factor load, as illustrated by the waveforms in Fig. 6.11, the fundamental component of the input line current lags the associated input voltage by approximately 32.5°.

The waveforms of Figs. 6.12 and 6.13 are for load displacement angles of 60° lagging and 60° leading respectively. Here, it is interesting to see, in either case, that the fundamental component of the input line current lags the voltage by the same angle, of approx. 60°. Thus, these waveforms confirm the fact that both lagging and leading loads at the output appear as a lagging load at the input of the phase-controlled cycloconverter.

The waveforms of Fig. 6.14 are for a load phase angle of 180°. In this case, the load is "fully regenerative," and power is delivered into the output terminals of the cycloconverter at unity displacement factor. It is seen that the fundamental component of the input line current lags the input voltage by approx. 147.5°. This, of course, confirms the fact that power is being delivered into the input a-c system; but still with the lagging displacement factor characteristic of the phase-controlled cycloconverter.

Estimation of the Approximate In-Phase and Quadrature Composition of the Input Current

The exact composition of the load presented by the cycloconverter to the input system can be calculated by means of a detailed harmonic analysis of the waveform of the input current, and this analysis is presented in Chapter 12. For the time being, however, it is useful to obtain some approximate quantitative results, without resort to detailed mathematics, and, in so doing, to obtain a further insight into the physical nature of the loading effect of the cycloconverter on the input system.

In order to do this, consider the simplified equivalent representation of an "ideal" cycloconverter supplying a single phase load, as illustrated in Fig. 6.15b. In this representation it is assumed that the components of current drawn by the cycloconverter from the 3-phase a-c system are only those which result from the basic nature of the load presented by the cycloconverter—in other words, it is assumed that this ideal cycloconverter has an infinite pulse number, and thus the theoretically unnecessary imperfections of a practical cycloconverter with a finite pulse number are neglected. Thus this ideal cycloconverter, when operated as a phase-controlled dual-converter with a steady d-c output voltage, as represented in Fig. 6.15a, would draw a pure sinusoid of current from the input system.

In order to estimate the input power factor and displacement factor of the "ideal" cycloconverter, consider firstly the situation with the converter producing a steady d-c output voltage, $V_{d_{max}}$, and a steady direct current, I_d. Assume, with this output current, that the rms amplitude of the input line current is I_1, as indicated in Fig. 6.15a. With maximum output voltage, the firing angle of the converter is 0°, and the input line current—which, for this ideal converter, with an infinite pulse number, is a pure sinusoid— is in phase with its associated input voltage. The power output is given by

$$P_d = V_{d_{max}} I_d \tag{6.6}$$

When operated as a cycloconverter, into a unity displacement factor load, so as to produce a sinusoidal output voltage waveform having a peak value of $V_{d_{max}}$, and a sinusoidal current waveform having a rms value of I_d, the instantaneous output voltage and current are given by

$$v_o = \underset{\underset{\text{Peak value of voltage wave}}{\uparrow}}{V_{d_{max}}} \sin \theta_o \tag{6.7}$$

$$i_o = \sqrt{2}\, \underset{\underset{\text{Peak value of current wave}}{\uparrow}}{I_d} \sin \theta_o \tag{6.8}$$

The instantaneous power output is given by

$$P_o = V_{d_{max}} \sin \theta_o \cdot \sqrt{2}\, I_d \sin \theta_o$$

$$= \underset{\underset{\substack{\text{Steady component} \\ \text{of output power}}}{\uparrow}}{\frac{V_{d_{max}} I_d}{\sqrt{2}}} - \underset{\underset{\substack{\text{Oscillatory component} \\ \text{of output power}}}{\uparrow}}{\frac{V_{d_{max}} I_d}{\sqrt{2}}} \cos 2\theta_o \tag{6.9}$$

Thus the output power is comprised of a steady mean component, onto which is superimposed an oscillatory component, having the same amplitude as the steady component, and a frequency of twice the output frequency.

LOADING EFFECT ON THE INPUT SYSTEM

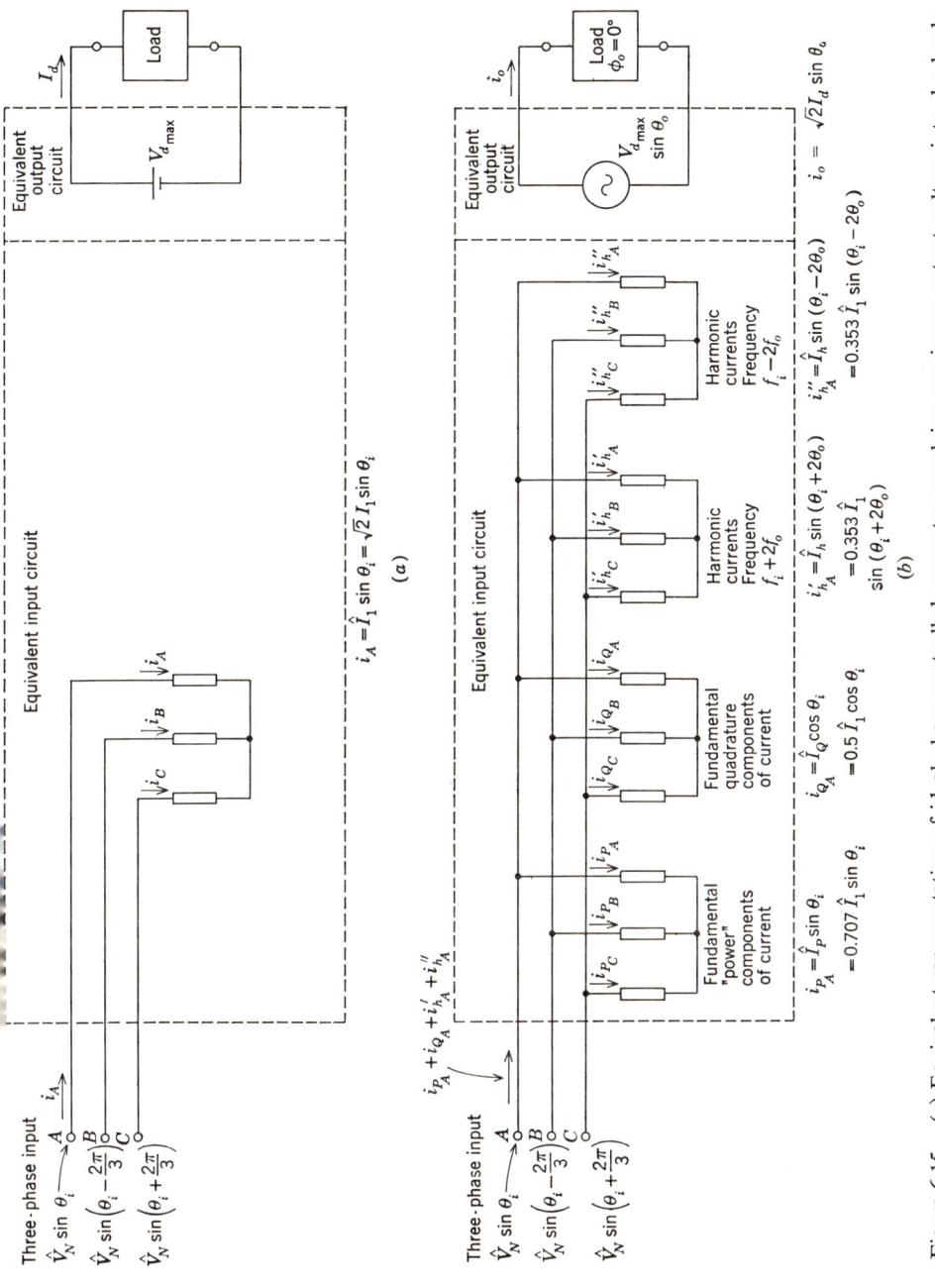

Figure 6.15. (*a*) Equivalent representation of ideal phase-controlled converter supplying maximum output voltage into d-c load; (*b*) approximate equivalent representation of ideal cycloconverter with single-phase output supplying maximum voltage to unity displacement factor load.

The first point to be made is that with the same peak output voltage and rms output current, the mean output power obtained in the case of an alternating output is only $1/\sqrt{2}$ of that obtained in the case of a steady d-c output voltage. Since the ratio of the rms input current to the rms output current is independent of whether the converter is controlled so as to produce a d-c or an a-c output, it follows that, with an a-c output, a given value of rms input line current produces only 0.707 of the output power obtained with a steady d-c output. Thus since, in this ideal converter, the power output is equal to the power input, it follows that the maximum input power factor (as defined at the beginning of Chapter 4), with a single phase unity displacement factor load at the output, is 0.707, as compared to 1.0 with a steady d-c output. And, in fact, for a cycloconverter of any pulse number with a single phase output, the maximum input power factor is 0.707 of the maximum power factor of the converter with a steady d-c output. Thus, for example, the maximum theoretical input power factor of a 3-pulse cycloconverter, with a unity displacement factor load at its output, is $0.707 \times 0.675 = 0.477$. This inferior input power factor is, of course, a direct outcome of the process of phase modulation of the converter firing angles, inherent in the basic control mechanism of the cycloconverter.

The relative magnitudes of the in-phase, quadrature, and harmonic components of current at the input of the cycloconverter can be estimated, from a consideration of the fact that the total instantaneous power at the 3-phase input terminals of the converter must be equal to the total instantaneous power at the output terminals.

Considering firstly the case of a steady d-c output, the instantaneous power input is given by

$$p_i = \hat{V}_N \sin \theta_i \cdot \hat{I}_1 \sin \theta_i + \hat{V}_N \sin\left(\theta_i - \frac{2\pi}{3}\right) \cdot \hat{I}_1 \sin\left(\theta_i - \frac{2\pi}{3}\right)$$

$$+ \hat{V}_N \sin\left(\theta_i + \frac{2\pi}{3}\right) \cdot \hat{I}_1 \sin\left(\theta_i + \frac{2\pi}{3}\right)$$

$$= \frac{3\hat{V}_N \hat{I}_1}{2} \tag{6.10}$$

Hence

$$V_{d\max} I_d = \frac{3\hat{V}_N \hat{I}_1}{2} \tag{6.11}$$

With an alternating output, both the steady and oscillatory components of output power, given by equation (6.9), must be balanced by corresponding components of power at the input.

LOADING EFFECT ON THE INPUT SYSTEM 177

The net instantaneous power at the 3-phase input due to the quadrature components of line current is zero:

$$p_Q = \hat{V}_N \sin \theta_i \cdot \hat{I}_Q \cos \theta_i + \hat{V}_N \sin\left(\theta_i - \frac{2\pi}{3}\right) \cdot \hat{I}_Q \cos\left(\theta_i - \frac{2\pi}{3}\right)$$

$$+ \hat{V}_N \sin\left(\theta_i + \frac{2\pi}{3}\right) \cdot \hat{I}_Q \cos\left(\theta_i + \frac{2\pi}{3}\right)$$

$$= 0 \quad (6.12)$$

The net instantaneous power at the 3-phase input due to the in-phase component of current has a steady value, and it is this component of input power which balances the steady power component at the output:

$$p_i = \hat{V}_N \sin \theta_i \cdot \hat{I}_P \sin \theta_i + \hat{V}_N \sin\left(\theta_i - \frac{2\pi}{3}\right) \cdot \hat{I}_P \sin\left(\theta_i - \frac{2\pi}{3}\right)$$

$$+ \hat{V}_N \sin\left(\theta_i + \frac{2\pi}{3}\right) \cdot \hat{I}_P \sin\left(\theta_i + \frac{2\pi}{3}\right)$$

$$= \frac{3\hat{V}_N \hat{I}_P}{2} \quad (6.13)$$

Hence, from (6.9)

$$\frac{V_{d\max} I_d}{\sqrt{2}} = \frac{3\hat{V}_N \hat{I}_P}{2} \quad (6.14)$$

Combining equations (6.11) and (6.14) gives

$$I_P = \frac{I_1}{\sqrt{2}} \quad (6.15)$$

The oscillatory component of power at the output of the cycloconverter must be supplied by harmonic currents at the input. These harmonic currents could theoretically have frequencies of either $(f_i + 2f_o)$, or $(f_i - 2f_o)$, since, for each of these frequencies, multiplication of the harmonic current component by the input voltage, having frequency f_i, yields a power term having a frequency of $2f_o$. It would be reasonable to assume—as is in fact the case in practice (for an output load with unity displacement factor)—that harmonic components of current having each of these frequencies are present, with equal amplitudes.

Thus, the harmonic currents carried by the input lines are assumed to be

$$i_{hA} = \hat{I}_h \sin(\theta_i + 2\theta_o) + \hat{I}_h \sin(\theta_i - 2\theta_o)$$

$$i_{hB} = \hat{I}_h \sin\left(\theta_i + 2\theta_o - \frac{2\pi}{3}\right) + \hat{I}_h \sin\left(\theta_i - 2\theta_o - \frac{2\pi}{3}\right)$$

$$i_{hC} = \hat{I}_h \sin\left(\theta_i + 2\theta_o + \frac{2\pi}{3}\right) + \hat{I}_h \sin\left(\theta_i - 2\theta_o + \frac{2\pi}{3}\right)$$

Thus, the total instantaneous harmonic power at the input is given by

$$p_h = \hat{V}_N \sin \theta_i \cdot \hat{I}_h[\sin(\theta_i + 2\theta_o) + \sin(\theta_i - 2\theta_o)]$$

$$+ \hat{V}_N \sin\left(\theta_i - \frac{2\pi}{3}\right) \cdot \hat{I}_h\left[\sin\left(\theta_i + 2\theta_o - \frac{2\pi}{3}\right) + \sin\left(\theta_i - 2\theta_o - \frac{2\pi}{3}\right)\right]$$

$$+ \hat{V}_N \sin\left(\theta_i + \frac{2\pi}{3}\right) \cdot \hat{I}_h\left[\sin\left(\theta_i + 2\theta_o + \frac{2\pi}{3}\right) + \sin\left(\theta_i - 2\theta_o + \frac{2\pi}{3}\right)\right]$$

$$= 3\hat{V}_N \hat{I}_h \cos 2\theta_o \tag{6.16}$$

Hence, from (6.9)

$$\underbrace{\frac{V_{d\max} I_d \cos 2\theta_o}{\sqrt{2}}}_{\substack{\text{Oscillatory component} \\ \text{of power at output}}} = \underbrace{3\hat{V}_N \hat{I}_h \cos 2\theta_o}_{\substack{\text{Oscillatory component} \\ \text{of power at input}}} \tag{6.17}$$

Combining equations (6.11) and (6.17) gives

$$I_h = \frac{I_1}{2\sqrt{2}} \tag{6.18}$$

Now, for the cases of the d-c and a-c output considered, the total rms input current is the same:

$$\underbrace{(I_P^2 + I_h^2 + I_h^2 + I_Q^2)^{1/2}}_{\substack{\text{Total rms input line current} \\ \text{with a-c output}}} = \underbrace{I_1}_{\substack{\text{Rms input line current} \\ \text{with d-c output}}} \tag{6.19}$$

Substituting for I_P and I_h, from expressions (6.15) and (6.18) respectively, it is now possible to find the magnitude of the quadrature component of input current of the "ideal" cycloconverter:

$$\frac{I_1^2}{2} + \frac{I_1^2}{8} + \frac{I_1^2}{8} + I_Q^2 = I_1^2$$

Therefore

$$I_Q^2 = \frac{I_1^2}{4}$$

and

$$I_Q = \frac{I_1}{2} \tag{6.20}$$

Finally, combining equations (6.15) and (6.20) gives

$$I_Q = \frac{I_P}{\sqrt{2}} \tag{6.21}$$

LOADING EFFECT ON THE INPUT SYSTEM 179

Hence the input displacement factor of this "ideal" cycloconverter is given by

$$\cos \phi_i = \frac{I_P}{(I_P^2 + I_Q^2)^{1/2}}$$

$$= \frac{I_P}{\left(I_P^2 + \frac{I_P^2}{2}\right)^{1/2}}$$

$$= 0.817 \tag{6.22}$$

Actually, as is shown in Chapter 12, this result is *not* quite correct, and the theoretical displacement factor for the case under consideration is 0.843. This implies that the relative amplitude of the quadrature component of current is somewhat less than that deduced here from a knowledge of the total rms current, and the in-phase and assumed harmonic currents. The reason for the discrepancy is that there are actually additional, relatively low amplitude, harmonic components of current present in the input of the ideal cycloconverter, which have frequencies of $(f_i \pm 2nf_o)$. Each conjugate pair of these harmonic currents have such a phase relationship that the total instantaneous power due to the pair is zero, and hence the presence of these harmonics would not be suspected purely from a consideration of the equality of the instantaneous input and output powers.

With the realization, then, that the foregoing representation of the ideal cycloconverter is, in fact, an oversimplification of the true situation, it is still, nevertheless, of interest to carry this simplified treatment through to the case of a cycloconverter supplying a balanced 3-phase load, since this leads to the important conclusion that the harmonic currents associated with the pulsating output power of the single phase load no longer appear in the input system.

For the case when each output phase delivers maximum voltage, into a unity displacement factor load, the total instantaneous output power of all three phases is given by

$$p_o = V_{d\max} \sin \theta_o \cdot \sqrt{2} I_d \sin \theta_o$$

$$+ V_{d\max} \sin \left(\theta_o - \frac{2\pi}{3}\right) \cdot \sqrt{2} I_d \sin \left(\theta_o - \frac{2\pi}{3}\right)$$

$$+ V_{d\max} \sin \left(\theta_o + \frac{2\pi}{3}\right) \cdot \sqrt{2} I_d \sin \left(\theta_o + \frac{2\pi}{3}\right)$$

$$= \frac{3 V_{d\max} \sqrt{2} I_d}{2} \tag{6.23}$$

Thus, the total instantaneous output power is constant, and there is no net oscillatory component. Thus, from a consideration of the equality of the total instantaneous input and output powers, there can be no harmonic currents which would give rise to a net instantaneous harmonic power at the 3-phase input. Thus, according to this simplified treatment, which assumes that the presence or absence of harmonic currents at the input is dictated solely by the presence or absence of oscillatory power at the output, it can be concluded that there are no harmonic currents at the input of the ideal cycloconverter supplying a balanced 3-phase load. The equivalent circuit representation for this case therefore reduces to that shown in Fig. 6.16. The amplitudes of the in-phase and quadrature components of current, relative to one another, are the same as with the single phase load, since the displacement factor of the cycloconverter is a basic property, which is unrelated to the number of output phases. (This same conclusion is reached, of course, simply by adding together the input currents of 3 "ideal" cycloconverters, each supplying single phase loads displaced by 120° from one another.)

Actually, as will be seen in Chapter 12, with a balanced 3-phase load, the input currents of the cycloconverter do contain certain relatively low amplitude harmonic components (that is, other than those attributable to the finite pulse number) the total instantaneous power contribution of which is zero. Nevertheless, the conclusion reached here is valid, that the a-c system does not "see" the pulsating loads of the individual output phases. As a result of this, the input line current waveforms are relatively undistorted, as compared to the case with a single phase load at the output.

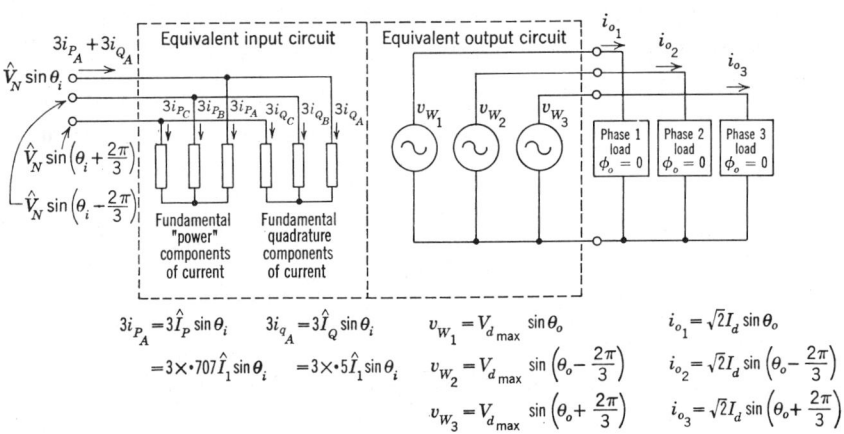

Figure 6.16. Approximate equivalent representation of ideal cycloconverter with 3-phase output, supplying maximum voltage to unity displacement factor load.

Chapter Seven

Cycloconverter Control Problems and Techniques Associated with Discontinuous Current

In Chapter 6, it was assumed that the waveform of the output current of the cycloconverter is a pure sinusoid. In practice, of course, there are ripple components superimposed onto the output current waveform, due to the ripple voltage at the output of the cycloconverter. So long as these superimposed distortion components are sufficiently small that the net current waveform contains only a single current-zero at the end of each output half cycle, as a result of which each converter is kept in continuous conduction throughout its half cycle, then the requirements of the control scheme for the cycloconverter are relatively straightforward. This is because, with continuous conduction, the relationship between the converter firing angle and the mean output voltage is independent of the load, and hence a prearranged firing pulse timing pattern results in the production of the "proper" output voltage waveform, independently of the load.* In addition, the converter "bank selection" control can be implemented simply by switching the firing pulses from one converter to the other at each load current zero.

In practice, the application often may be such, under some conditions at least, that the converter current tends to become discontinuous during the course of the output half cycle. In this case, certain basic difficulties arise in controlling the operation of the cycloconverter.

In this chapter, the basic control problems associated with discontinuous load current are explained, and various control methods are discussed, involving operation both with and without circulating current, which provide solutions to these problems.

---

* But see the remarks in Chapter 9, concerning the production of objectionable distortion components with an open loop firing angle control.

BASIC CONTROL DIFFICULTIES ASSOCIATED WITH DISCONTINUOUS CURRENT

Basically, two difficulties arise with discontinuous load current. Firstly, the mean terminal voltage of the cycloconverter does not depend only upon the firing angle, as it does with continuous conduction, but it depends also upon the degree of discontinuity of the load current wave. Thus, with an open-loop firing pulse timing control, which, with continuous conduction, is designed to produce the "correct" output voltage waveform, with discontinuous conduction a degree of "unnecessary" distortion of the output voltage is produced. Secondly, since the current waveform is discontinuous during the course of the wanted output half-cycle, it is not possible to use indiscriminately the zero values of this waveform as the logic decision making condition for switching the firing pulses from one converter to the other.

In order to demonstrate these points, consider firstly, for the sake of discussion, the operation of a cycloconverter with a load consisting of a pure resistance,* with which discontinuous conduction inevitably occurs. This is illustrated by the cycloconverter output voltage waveform in Fig. 7.1b. In this case, the corresponding current wave has the same shape as the voltage wave. It is assumed here that the firing angle control is such as to produce the proper output voltage waveform, shown in Fig. 7.1a, with continuous conduction. It can be seen that in the region of discontinuous load current, the "mean" output voltage is greater than it would be with continuous current, due to the inability of the converter output voltage to swing into the reverse direction with a resistance load. The effect of this is to produce an "unnecessary" distortion of the mean wave of output voltage. It can be appreciated also, in this case, that the zero values of the current waveform do not generally represent the logic condition for switching the firing pulses from one converter to the other.

Another, more severe case, which accentuates the difficulties of operation with discontinuous current, occurs with a predominantly capacitive load circuit at the output of the cycloconverter. The voltage waveforms of Fig. 7.2a illustrate an extreme theoretical case in which the "load" connected to the output of the cycloconverter simply consists of a capacitor. In this case, the capacitor charges to the instantaneous value of the output voltage at each firing instant, and then retains this voltage until the next firing instant,

---

* The reader should be careful to distinguish between a load consisting of a pure resistance, and a load which has unity displacement factor at the wanted output frequency. The latter often consists of a complex load containing reactive filter components, and does not necessarily draw a discontinuous current.

DIFFICULTIES WITH DISCONTINUOUS CURRENT 183

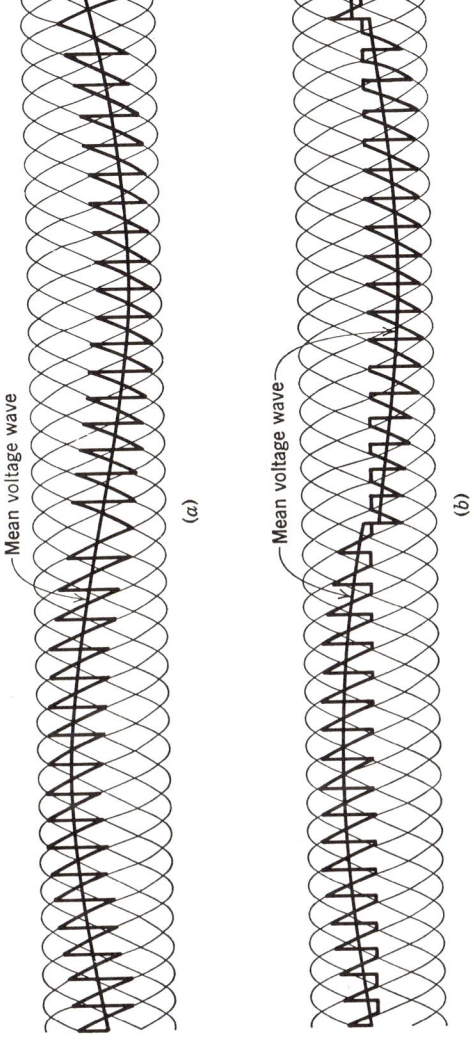

Figure 7.1. Cycloconverter output voltage waveforms showing how discontinuous conduction gives rise to unnecessary distortion of the "mean" output voltage wave. The firing angle control is such as to produce the "correct" voltage wave with continuous conduction. (a) Output voltage wave with continuous conduction. Load displacement angle = 0°; (b) output voltage wave with pure resistance load.

184 CYCLOCONVERTER CONTROL PROBLEMS

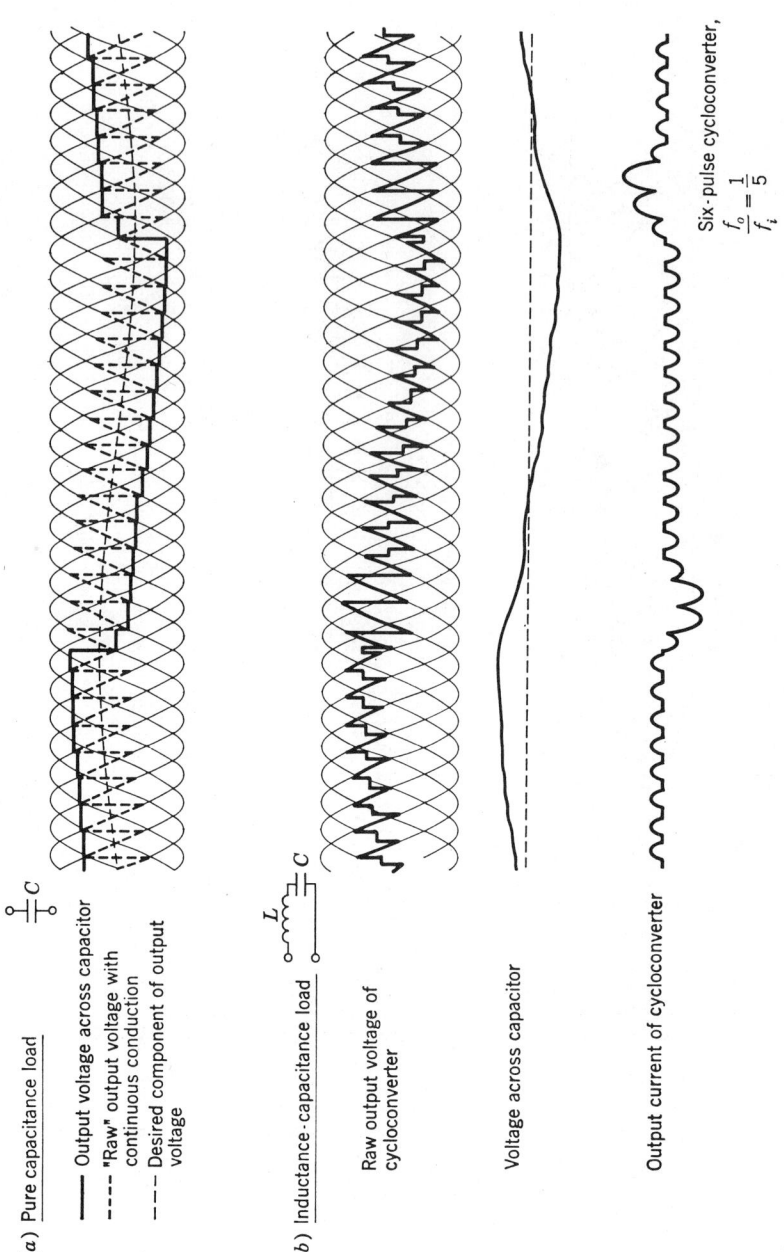

Figure 7.2. Theoretical cycloconverter waveforms for predominantly capacitive loads with no circulating current, showing how discontinuous conduction gives rise to "unnecessary" distortion of the output voltage wave.

DIFFICULTIES WITH DISCONTINUOUS CURRENT

at which point its voltage is increased further. The corresponding current waveform (not shown) consists of a train of very short duration, high amplitude "spikes," which occur at each firing point. At the peak of the wanted output voltage wave, which, for this pure capacitance load, is the point at which the "fundamental" half cycle of current terminates, the firing pulses are switched from one converter to the other. This causes a sharp change in the level of the capacitor voltage, due to the reversal of phase, with respect to the firing instants, of the raw ripple voltage of the cycloconverter.

Thus it is seen that because the current wave is discontinuous to a large degree, there is a considerable shift of the mean output voltage wave, from the desired wave; that is, the wave which would be obtained with continuous conduction. This shift is in the "positive" direction during the "positive" half cycle of current, and in the "negative" during the "negative" half cycle, and the result is that the output voltage waveform contains a relatively high degree of distortion.

Of course, the foregoing example represents a theoretical extreme condition which would not arise in practice. A more practical example is illustrated by the waveforms of Fig. 7.2b, in which case an inductor is connected in series with the capacitor. The relative values of inductance and capacitance are such that under the particular conditions considered, the output current is discontinuous. As in the case of the pure capacitance load, it can be seen that the discontinuous current results in a deviation of the mean output voltage from the desired wave, and, as previously, this results in a relatively high degree of distortion of the voltage developed across the capacitor.

This type of L-C circuit is commonly used as a filter for the output voltage of the cycloconverter, the external load being connected across the capacitor. Thus the waveforms of Fig. 7.2b illustrate the conditions which could arise with an unloaded L-C filter, the inductance of which is insufficient to maintain continuous current under certain conditions.

Of course, under other conditions, this same L-C circuit might not produce a discontinuous current waveform, and, in this event, the type of distortion illustrated by the waveforms of Fig. 7.2 would no longer be present. Thus, for example, if either the relative output voltage level, or frequency, is raised, the result is to increase the fundamental component of the capacitor current, relative to the ripple current, and the tendency for discontinuous current is diminished. Alternatively, of course, if an external load is connected across the capacitor, so as to produce a continuous current, then the distortion associated with the unloaded condition automatically disappears.

In practice, the solutions which may be adopted to the control problems associated with discontinuous conduction, depend largely upon the circumstances and requirements of the particular application, and, in particular, upon the nature of the load, and the desired quality of the output voltage waveform.

In the following sections, approaches to the "voltage distortion" and "bank-selection" problems arising as a result of the tendency for operation of the cycloconverter with discontinuous conduction, are discussed under separate headings. The presentation is made in this way largely as a matter of convenience in providing an ordered sequence of discussion. As will be seen, the approaches to the two problems may not necessarily be divorced from one another.

APPROACHES TO THE VOLTAGE DISTORTION PROBLEM

Closed-Loop Control of the Output Voltage

The cycloconverter, taken in conjunction with its firing angle control circuitry, can be arranged to function as an amplifier, the "mean" output voltage of which is controlled in accordance with an analog input reference voltage. With continuous current at the output of the converter, a linear voltage transfer characteristic can be obtained. With discontinuous current, on the other hand, the voltage transfer characteristic departs from the linear relationship, and in general the mean output voltage tends to become "more positive" (for the "positive" converter) than with continuous current.

In theory, a method for eliminating distortion, due to the nonlinear voltage transfer characteristic associated with discontinuous current, is to incorporate a negative feedback control loop, the action of which is automatically to compensate the firing angle, so as to "force" the wanted component of the output voltage to follow the input reference voltage. This can be accomplished by means of producing a signal which is the integral of the error between the output voltage and the sinusoidal analog reference voltage, and applying this error signal, suitably amplified, and in the proper sense, so as to compensate the firing angle control, and thus minimize the error.

With this approach, it is quite possible, for the most part, to eliminate the distortion of the output voltage wave which would otherwise occur with discontinuous current. However, it is difficult to avoid some distortion of the output voltage wave, in the region of the current crossover. The basic control problem is exactly the same as that for the dual-converter operating with discontinuous current; namely, that in order to avoid irregular jumps in the level of the output voltage at the point of change-over of the current from one converter to the other, ideally it is necessary to preadjust the firing angle of the incoming converter so that this immediately delivers the correct level of voltage at its output. Whereas with continuous conduction, the sum of the converter firing angles should be invariably 180°, this is not so with discontinuous conduction. In this case, the required firing angle of the incoming converter is not directly related to that of the outgoing converter, but it depends upon the particular characteristics of the load.

Thus, for loads which show a tendency to draw a discontinuous current during the period immediately following the bank selection point, it is virtually impossible to devise a firing angle control scheme which inherently completely avoids the possibility for "unnecessary" distortion of the output voltage in this region. This is not to say, however, that it is not possible to "tailor" the firing angle control according to the requirements of the particular application, so as to provide a virtually distortion-free crossover, especially if the application is such that the range of variation of load, voltage, frequency ratio, and so on, is not too great.

In practice, in any case, the application may be such that some "crossover" distortion of the output voltage wave (i.e., in the region of the *current* crossover) may not be objectionable. This could be the case where the output frequency is much lower than the input frequency, in which event the "crossover" distortion need occupy only a relatively small portion of the total output cycle, and thus, as it were, can pass almost unnoticed.

On the other hand, if the application is one in which it is required to produce a high quality output voltage wave, at relatively high output to input frequency ratios, and, at the same time, a possibility exists for operation with discontinuous output current, then the best practical approach may be to operate the cycloconverter with a circulating current, at least during the periods when the output current would otherwise be discontinuous. This approach avoids the crossover distortion problem associated with discontinuous conduction, simply by avoiding discontinuous conduction itself. Thus, it is ensured that at least one converter is always in continuous conduction, and hence the open-loop voltage transfer characteristic is at all times well defined, irrespective of the load conditions. This is discussed further in the following sections.

Operation with Circulating Current

Consider the particular example illustrated by the waveforms of Fig. 7.2b, in which the current into an unloaded *L-C* circuit is discontinuous for relatively large portions of each output half cycle. As a result of the discontinuous current waveform, the output voltage, with an open-loop firing angle control, is highly distorted. Although the use of a closed-loop negative voltage feedback might improve this waveform, it would still be difficult to avoid some "crossover" distortion.

Clearly, if instead of operating the cycloconverter with no circulating current, it is operated with a continuous circulating current, then the voltage distortion difficulties are avoided, since with both converters in continuous conduction, the "mean" wave of the voltage presented to the filter circuit necessarily corresponds at all times to the sinusoidal reference wave demanded by the firing pulse timing control. Putting this another way, since now there is always a freedom for the load current to flow in either direction, it is always possible for the capacitor current to cross freely to and fro about zero, and hence the previous tendency, for the capacitor to charge to the peaks of the raw ripple voltage wave, does not exist. Thus it is possible to produce a voltage waveform across the capacitor which is completely free from the distortion associated with discontinuous load current.

The practical oscillograms of Fig. 7.3a and b, appropriate to a 6-pulse circuit, are a typical illustration of the differences between the operation of the cycloconverter, with an open-loop firing angle control, with and without a circulating current, with this type of load. Both oscillograms correspond to the same unloaded L-C filter circuit, with the same output to input frequency ratio (about 1:5) and the same sinusoidal input reference voltage applied to the firing angle control circuits.

At b, the cycloconverter operates without a circulating current. The distortion of the output voltage, due to the discontinuous load current, is readily apparent.

At a, the firing pulses are continuously applied to both converters, and the cycloconverter operates with a continuous circulating current. As can be seen, the output voltage waveform does not contain the distortion associated with the discontinuous load current, and the output current has a complete freedom to cross "to and fro" about zero.

It is also seen that the raw voltage waveform presented to the filter circuit has an entirely different appearance to that obtained in the circulating current-free mode. This voltage waveform is the instantaneous average of the positive and negative converter voltage waveforms. As will be seen in Chapter 11, the raw voltage waveform of the cycloconverter operating with a continuous circulating current has a lower harmonic content than that of the cycloconverter operating with no circulating current. However, it should be appreciated that the improvement in the quality of the filtered output voltage waveform of Fig. 7.3a is not basically due to the reduced ripple voltage presented to the filter. Rather, it is due to the more fundamental reason that because the converter current is continuous, the mean value of the "raw" voltage wave generated at the output terminals of the cycloconverter necessarily corresponds with the sinusoidal reference wave.

Of course, as has been seen in Chapter 6, if the cycloconverter is operated with a continuous circulating current, then the current carried by each

THE VOLTAGE DISTORTION PROBLEM 189

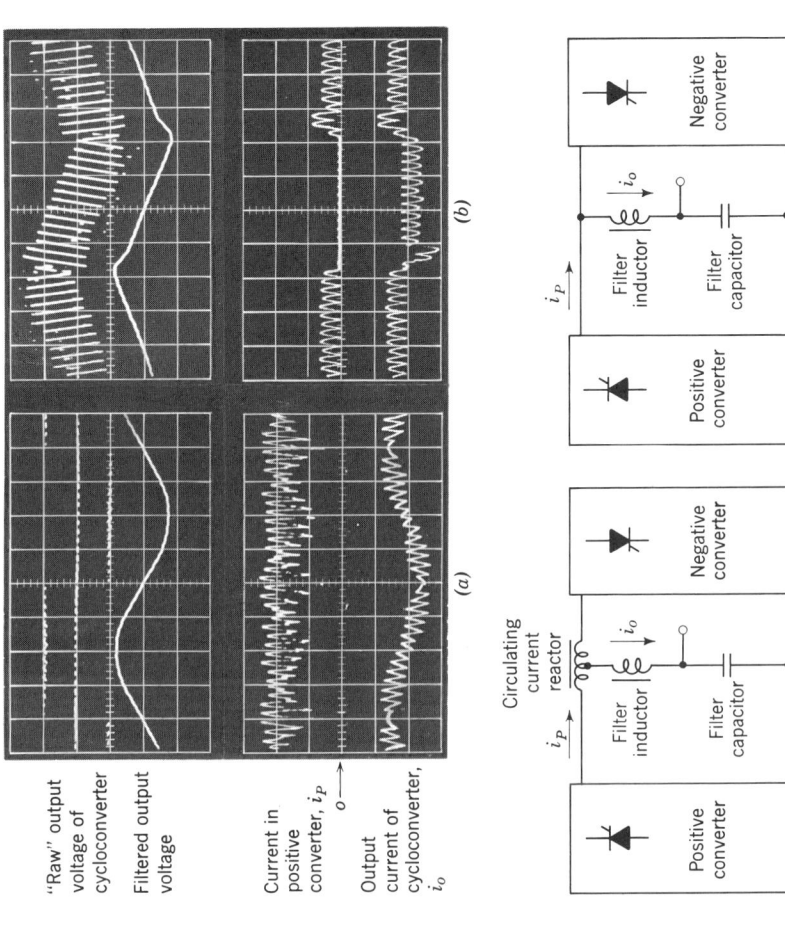

Figure 7.3. Oscillograms illustrating the difference between operation of the cycloconverter with and without circulating current, with an unloaded L-C filter circuit. (a) Operation with continuous circulating current; (b) operation with no circulating current. *Notes:* 1. Six-pulse cycloconverter, $f_o/f_i \simeq \frac{1}{5}$. 2. Open-loop firing angle control.

converter is considerably larger than that due to the output current alone, due to the production of a quite substantial component of "self-induced" circulating current.

Actually, for the particular case of the unloaded L-C filter under consideration, the presence of a relatively large component of circulating current is not really objectionable, since, with no external load, the loading of the converters due to the circulating current is not excessive, at least not in terms of the full-load current.

More generally, however, the presence of a circulating current becomes undesirable, if it results in a total loading of the converters which is substantially in excess of that due to the maximum external load. Fortunately, in practice, the "need" to operate with a circulating current arises only when there is a tendency for the output current otherwise to become discontinuous. Under these particular circumstances, the amplitude of the output current is invariably relatively small.

This leads to the concept of operating the cycloconverter with a circulating current, only when there is otherwise a tendency for the load current to become discontinuous, and to operate the cycloconverter without a circulating current, whenever the load current is continuous. In this way, it is possible to take advantage of the benefits to be obtained from operation with a circulating current, only when these are really required. At the same time, the disadvantages of operation with a circulating current can be avoided; that is to say, the average loading of the converters due to the circulating current, relative to the maximum output load current, can be kept so small as to be insignificant.

A scheme which implements the type of control principle under discussion is presented in the following section.

CONTROLLED FIRING PULSE OVERLAP. A functional control scheme is shown in simplified form in Fig. 7.4.

The function of the "positive" and "negative" firing pulse generators is to produce firing pulses for the thyristors of the associated converters. The phase of these firing pulses is modulated in response to the analog sinusoidal input reference voltage, in such a manner that (with continuous conduction) the wanted component of voltage generated at the output of the converter is directly proportional to this reference. In this simplified diagrammatic representation, the pulses from the positive and negative firing pulse generators are transmitted to the positive and negative converters through the gates P and N respectively. The control signals for these gates are respectively the output signals from the "positive" and "negative" current level detectors.

The analog input voltage to the positive and negative level detectors is a replica of the output current waveform of the cycloconverter. The positive

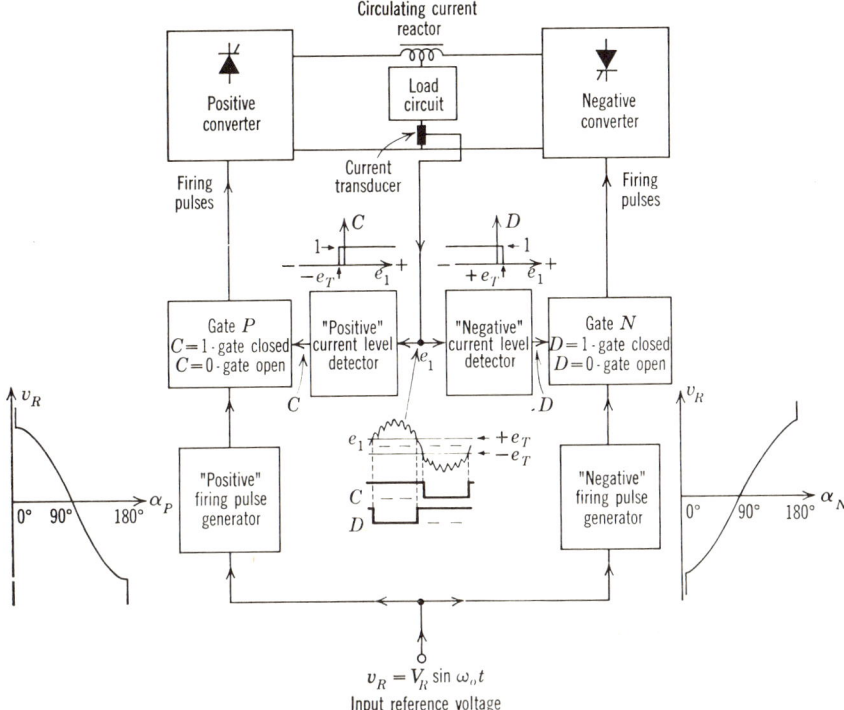

Figure 7.4. Diagrammatic representation of a control scheme for a cycloconverter using "controlled firing pulse overlap."

level detector produces a logic signal "1" at its output, so long as the input voltage e_1 is positive, or greater than a prescribed negative level, $-e_T$. Whenever e_1 falls below this level, the output signal of the positive level detector reverts to the logic state "0," thus blocking the firing pulses from the positive converter. The function of the negative level detector is similar, but in this case, a positive input signal, in excess of a prescribed level, $+e_T$, results in a logic signal "0" at its output, which blocks the firing pulses from the negative converter.

It can be seen that the action of this control scheme is to apply the firing pulses to both the positive and negative converters, thus permitting a circulating current to flow, and thereby keeping the converters in continuous conduction, so long as the load current is instantaneously less than a pre-scribed level, (below which there would otherwise be a possibility for discontinuous current). Whenever the load current exceeds this "threshold" level, the firing pulses are automatically removed from the converter which

192 CYCLOCONVERTER CONTROL PROBLEMS

does not carry the load current, and the circulating current thereby is inhibited.

With this type of control scheme, the required size of the circulating current reactor is not in fact too substantial, since this can be designed to support the circulating ripple voltage only under the circumstances when the load current in its windings is relatively small. Thus the core of this reactor can be permitted to saturate at higher values of load current, at which the action of the control ensures that only one converter is in conduction. Under this latter condition, of course, the circulating current-limiting function of the reactor is not needed, and hence it is of no concern that its core is, for the time being, saturated.

The operation of the power circuit with this type of control is illustrated by the idealized waveforms shown in Fig. 7.5. In order to demonstrate the basic principle, it is assumed here that the load current waveform is a sinusoid, and the ripple voltages at the output terminals of the converters are neglected. Also, for simplicity, it is assumed that the circulating current reactor does not saturate. The waveforms at a and b are appropriate to 2 levels of load current, designated "full load" and "light load" respectively. These waveforms show how the period of the firing pulse overlap increases as the amplitude of the load current decreases. Eventually, of course, if the peak amplitude of the load current decreases below the threshold level I_T, then the result is that the firing pulses are continuously applied to both converters. This, of course, is the special case considered in Chapter 6, in which it was seen that a continuous "self-induced" component of circulating current is produced, due to "trapped" energy in the circulating current reactor, which manifests itself as a continuous circulating current.

In the more general case under consideration here, in which the periods of firing pulse overlap are not continuous, there is still, nevertheless, a tendency for a "self-induced" component of circulating current to flow during the pulse overlap periods (as distinct from circulating ripple current, due to the circulating ripple voltage, which, of course, also flows). This is explained by reference to the waveforms of Fig. 7.5. Consider, for example, the conditions at time t_1, (Fig. 7.5a), when only the positive converter is in conduction. At this time the voltage induced across the circulating current reactor due to the load current is in such a direction as to create a "forward bias" across the terminals of the negative converter. Thus when this converter is "de-blocked" at time t_2, it immediately starts to conduct, and the result is that both converters carry a self-induced component of circulating current, as illustrated, during the period of firing pulse overlap.

By means of a similar analysis to that presented in Chapter 6, for the special case of continuous firing pulse overlap, it can be shown in this case, that the positive and negative converter currents, during the pulse overlap period,

THE VOLTAGE DISTORTION PROBLEM 193

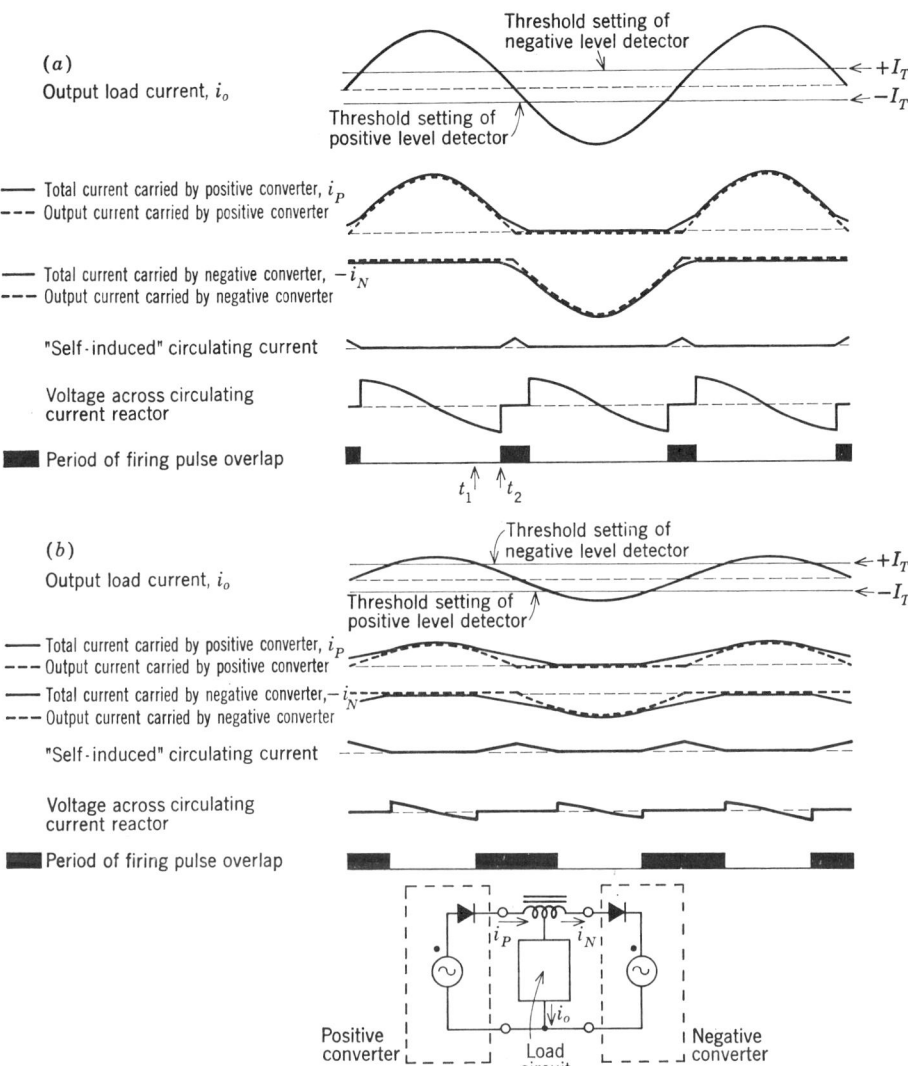

Figure 7.5. Idealized waveforms associated with the cycloconverter control scheme of Fig. 7.4. (a) Full load; (b) light load.

194 CYCLOCONVERTER CONTROL PROBLEMS

are given by the following equations:

$$i_P = \frac{\hat{I}_o}{2} \sin \theta_o + \frac{I_T}{2} \qquad (7.1)$$

$$i_N = \frac{\hat{I}_o}{2} \sin \theta_o - \frac{I_T}{2} \qquad (7.2)$$

where $\hat{I}_o$ = peak output current
I_T = threshold current below which firing pulses are applied to both converters ($I_T \leqslant \hat{I}_o$)

It can also be shown that the current carried by each converter, averaged over the complete output cycle, is given by

$$I_{av} = \frac{\hat{I}_o}{\pi}\left[\sin \Delta + \left(\frac{\pi}{2} - \Delta\right)\cos \Delta\right] \qquad (7.3)$$

where $\Delta = \cos^{-1} I_T/\hat{I}_o$

Now, with no circulating current, that is, with no firing pulse overlap, the average current I'_{av} carried by each converter is found by substituting $\Delta = 90°$ (i.e., $I_T = 0$) into expression (7.3):

$$I'_{av} = \frac{\hat{I}_o}{\pi}[1 + 0]$$

$$= \frac{\hat{I}_o}{\pi} \qquad (7.4)$$

(This, of course, is the well known result for a half wave rectified sinusoid). Dividing (7.3) by (7.4), and substituting $\Delta = \cos^{-1} I_T/\hat{I}_o$ gives:

$$\frac{I_{av}}{I'_{av}} = \left\{\left[1 - \left(\frac{I_T}{\hat{I}_o}\right)^2\right]^{1/2} + \left[\frac{\pi}{2} - \cos^{-1}\left(\frac{I_T}{\hat{I}_o}\right)\right]\left[\frac{I_T}{\hat{I}_o}\right]\right\} \qquad (7.5)$$

The curve of Fig. 7.6 shows the ratio I_{av}/I'_{av} plotted against the ratio $I_T/\hat{I}_o$. From this curve it is evident, as would be expected, that if the threshold level of current, I_T, at which firing pulse overlap is made to occur, is small by comparison with the peak load current, then the average current carried by each converter is hardly greater than with no circulating current. That is to say, the average value of the self-induced component of circulating current is comparatively insignificant. Consideration of a specific example will confirm this:

Assume

$$\frac{I_T}{\hat{I}_o} = 0.2$$

THE VOLTAGE DISTORTION PROBLEM

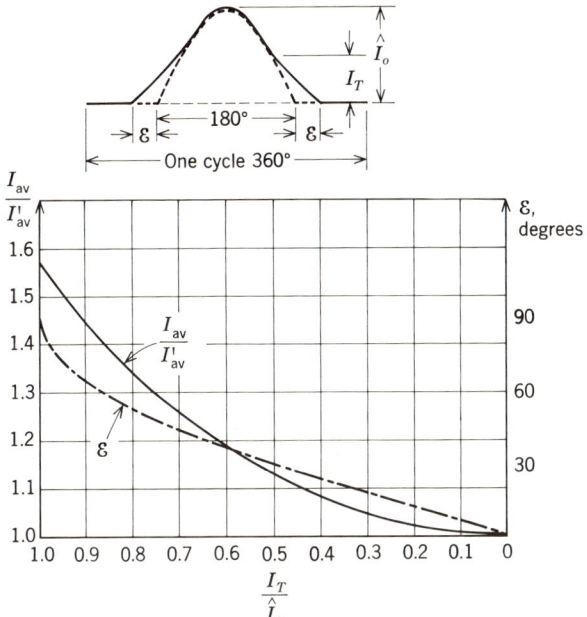

Figure 7.6. Normalized curves showing the variation of the average current in each 2-quadrant converter, and of the overlap angle ε, with the threshold load current I_T, below which the cycloconverter operates with a circulating current. I_{av} = Average value of total current in one converter; proportional to area under "full" wave——; and I'_{av} = average value of half sine wave of output current; proportional to area under dashed wave – – –.

Then, from the chart of Fig. 7.6

$$\frac{I_{av}}{I'_{av}} = 1.02$$

Thus, in this case, the average current carried by each converter is 2% in excess of that which would be obtained with no circulating current.

It is mentioned again here that the foregoing simplified theoretical treatment of the operation of the power circuit, together with the waveforms of Fig. 7.5 and the chart of Fig. 7.6, neglect the presence of the ripple voltages at the output of the cycloconverter, and assume a sinusoidal load current. As has been explained, the reason for operating the cycloconverter in the manner under discussion stems, in the first place, from the fact that the load current is not a perfect sinusoid, but has superimposed ripple components,

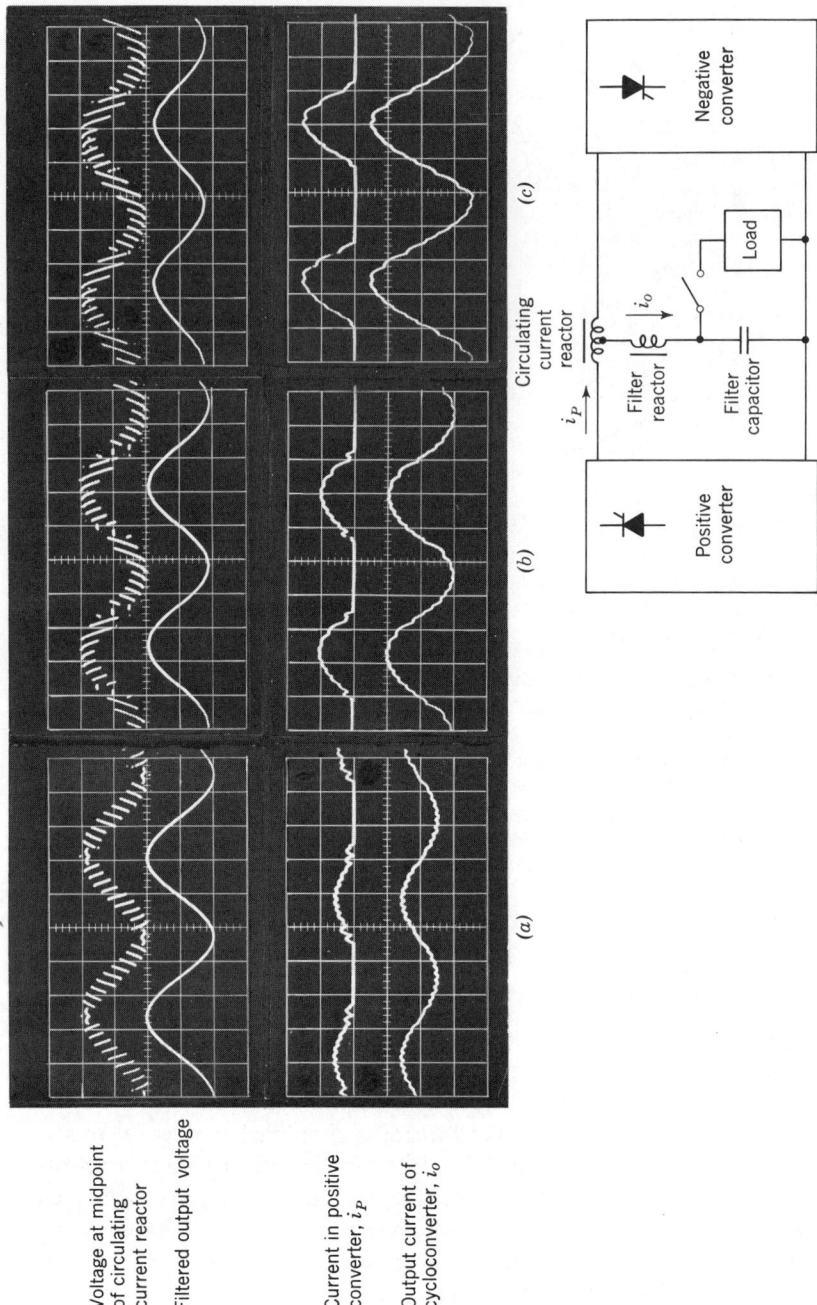

Figure 7.7. Typical cycloconverter oscillograms obtained with controlled firing pulse overlap—L-C-R output circuit. (a) No-load; (b) half-maximum load (resistive); (c) maximum load (resistive). Notes: 1. Six-pulse cycloconverter, $f_o/f_i \simeq \frac{1}{3}$. 2. Circulating current reactor designed to saturate in the circulating current-free region of operation.

THE VOLTAGE DISTORTION PROBLEM

caused by the ripple voltages at the output of the cycloconverter. Thus, in practice, the waveforms of Fig. 7.5, and the qualitative information of Fig. 7.6, would be modified to some extent by the presence of the superimposed ripple components at the output of the cycloconverter. Nevertheless, this simplified analytical treatment does provide a sound basis for an understanding of the power circuit operation with this control method.

A further point to be mentioned, which, in fact, is evident from the waveforms, shown in Fig. 7.5, of the voltage across the circulating current reactor, is that this mode of operation does, in itself, produce some distortion of the output voltage wave. This is due to the fact that the function of the control in switching the conduction from one converter to both converters, and vice versa, is not essentially a linear process. Thus, with only one converter in conduction, the effective "internal" inductance of the cycloconverter, due

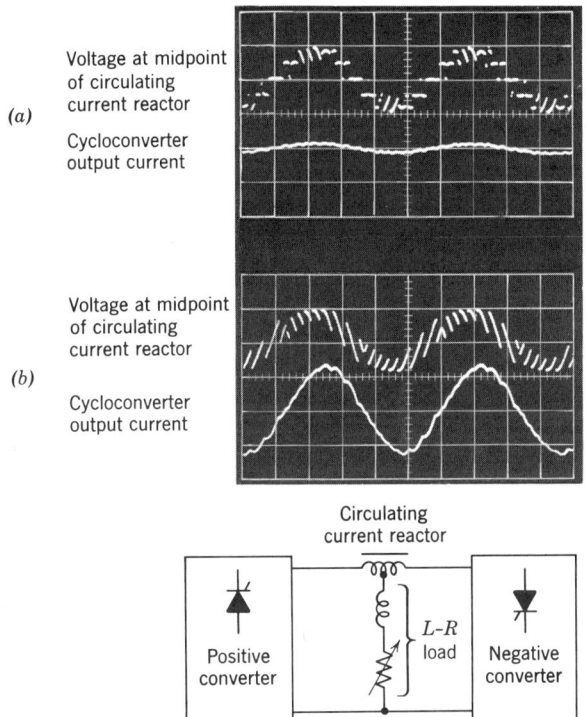

Figure 7.8. Typical cycloconverter oscillograms obtained with controlled firing pulse overlap—*L-R* output circuit. (*a*) Light load; (*b*) full load. *Notes: 1*. Six-pulse cycloconverter $f_o/f_i \simeq \frac{1}{3}$. *2*. Circulating current reactor designed to saturate in the circulating current-free region of operation.

to the circulating current reactor, as seen from the mid-point of this reactor, is $L/4$; whereas, with both converters in conduction, it is zero. The theoretical effect of this sudden change of effective output inductance is to cause a "step" change in the voltage at the mid-point of the circulating current reactor. Thus the waveform of the "internal" voltage drop of the cycloconverter theoretically constitutes a "chopped" sinusoid, consisting, basically, of a fundamental component with superimposed odd harmonics. (Actually, since the circulating current reactor would normally be designed so that it saturates above a certain level of load current, the situation is more complex than this.)

At all events, the distortion of the output voltage caused by this effect is generally relatively small, so that it does not constitute a practical problem. Certainly this type of distortion is negligible by comparison with the "cross-over" distortion which could otherwise result from operation with no circulating current.

The oscillograms of Figs. 7.7 and 7.8 are typical illustrations of the practical operation of this type of control scheme. In these oscillograms the firing pulse overlap periods can be clearly discerned, and it can be seen that the general appearance of the converter current waveforms is little different from that to be expected with no circulating current; this confirms that the additional loading of the converters, due to the circulating current, is relatively slight.

APPROACHES TO THE BANK SELECTION PROBLEM

With the control methods, discussed in the previous section, in which a circulating current is permitted to flow, it is evident that the output current has a "natural" freedom to cross through zero at the instant (or instants) of its choice. Thus, no additional control function is required in order to handle the load current crossover, and the bank-selection problem, as such, does not exist. This type of control scheme, then, simultaneously offers a solution both to the "voltage distortion" and "bank selection" problems.

In practice, the application may be such that the distortion of the output voltage, produced as a result of operation with discontinuous current, is not, in itself, objectionable. Or, it may be that the tendency for discontinuous current is so slight that the resulting distortion of the output voltage is unnoticeable, at least for all practical purposes. In either event, since no practical objection exists so far as the voltage distortion is concerned, it may be deemed unjustifiable to operate the cycloconverter with a circulating current, just for the purpose of providing a solution to the bank-selection problem.

THE BANK SELECTION PROBLEM 199

The practical control problem to be considered, then, is how to automatically and reliably switch the current from one converter to the other at the proper instants, in the face of a current waveform which may have several zero values during the course of each output half-cycle.

Various alternative approaches to this problem are discussed in the following sections.

"First Current-Zero" Bank Selection

Clearly, with a discontinuous load current, the zero values of the current waveform cannot be used indiscriminately as the logic condition for switching the firing pulses from one converter to the other, since this would result in an erratic and unacceptable system operation. Thus, with a discontinuous current waveform at the beginning of the output half cycle, the firing pulses could be prematurely switched away from the "incoming" converter, with the possibility that complete half cycles of output current would be "missed."

This kind of erratic operation can be avoided quite easily, albeit at the expense of the introduction of a further "unnecessary" (but possibly unobjectionable) distortion of the cycloconverter output waveforms, simply by arranging the control so that the first current-zero which occurs at the "end" of the output half cycle is taken to be the point at which to switch the firing pulses from one converter to the other. That is to say, the "end" of the half cycle is taken to be the point at which the "first current-zero" is reached. Subsequent premature transfer of the firing pulses back to the previously conducting converter, during the initial period of discontinuous current in the "incoming" converter, is then prevented, simply by arranging the control to ignore the zero values of the load current wave for a fixed time period.

A variation of the principle of using a fixed time period, during which a premature "bank selection" is prevented, is to prevent a "bank selection" until after the current in the incoming converter has reached a prescribed instantaneous level. This level would be chosen to occur only once the current has, for the time being, entered a "safe" period of continuous conduction.

Another possibility presents itself in the case of a cycloconverter supplying a balanced 3-phase output. The principle relies upon the 120° displacement between the currents in the 3 output phases, and utilizes the fact that during the "dangerous" start-up period of discontinuous current in a given output phase, the current in the next phase has a certain given polarity, whereas at the end of the half cycle, the polarity of this current has reversed. Thus, for the control of each output phase, it is possible to use the polarity of the current in the next output phase as the logic condition for deciding whether the "bank selection" should or should not be initiated.

This principle has the advantage, over those described previously, that it is not time or amplitude-dependent, and for this reason it is well suited to applications in which it is required to control the output frequency of the cycloconverter over a wide range, and/or in which the loading condition may be variable over a wide range.

LIMITATIONS OF "FIRST CURRENT-ZERO" BANK SELECTION. The main attraction of the first current-zero bank selection control method is that it is capable of providing a "reliable," though not theoretically ideal, circulating current-free system operation, with the minimum of complexity. However, the method is not endowed with "intrinsic" intelligence, inasmuch as its basic action, in effect, is to "blindly" switch the conduction from one converter to the other, at a point which may or may not correspond with the theoretical ideal.

The propensity for this type of bank selection control to make "theoretical mistakes" is illustrated by the practical oscillograms of Figs. 7.9 and 7.10.

The waveforms in Fig. 7.9 are for a load consisting of series inductance and resistance. The oscillograms at *a* are appropriate to a first current-zero bank selection control. By contrast, the oscillograms at *b* are appropriate

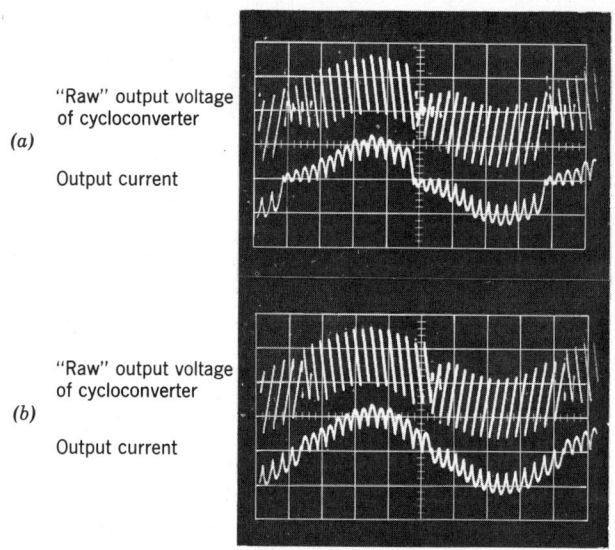

Series *L-R* load

Figure 7.9. Oscillograms illustrating the difference between the operation of "first current-zero" and "fundamental current-zero" bank selection control schemes. (*a*) "first current-zero" bank selection; (*b*) "fundamental current-zero" bank selection.

THE BANK SELECTION PROBLEM 201

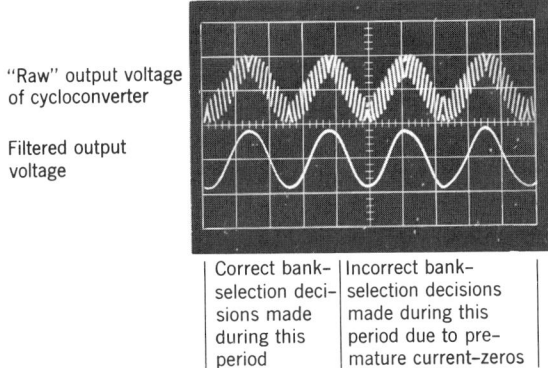

Figure 7.10. Oscillogram illustrating the unsatisfactory operation of a "first current-zero" control method for a cycloconverter with an unloaded L-C filter.

to a bank selection control (of the type discussed in the following section) which contrives to switch converter banks at the theoretically correct current-zeros.

An inspection of the latter oscillograms shows that the theoretically "correct" current waveform has a "premature" zero-value in advance of the bank-selection point. It is clear, then, that if the converter banks are switched at his first current-zero, the theoretically correct waveforms cannot be produced. This is confirmed by the oscillograms at *a*, for which all conditions, other than the bank-selection control method, are identical with those at *b*. It is evident that the premature switching of the firing pulses from one converter to the other, "forcibly" imposed by the first current-zero control method, results in an appreciable (though not catastrophic, and therefore conceivably tolerable) distortion of the output waveforms.

The oscillogram of Fig. 7.10 is applicable to an unloaded L-C filter circuit connected at the output of the cycloconverter, with a "first current zero" bank-selection control. In this particular instance, the output current waveform (not shown) has a tendency for occasional "premature" current-zeros, ahead of the "correct" bank-selection point. The tendency can clearly be seen for the control sometimes to make a correct bank-selection decision, in which case a good output voltage waveform is obtained; and sometimes, due to a premature current zero, to make an incorrect bank-selection decision, as a result of which a quite distorted output voltage wave is produced.

"Fundamental Current-Zero" Bank Selection

With a perfectly smooth output current waveform, which consists only of the wanted sinusoidal output component, with no superimposed ripple components, it is clear that the proper (and only) point at which to switch converter banks is at the zero-crossing of this "fundamental" waveform. It would be reasonable to expect, then, (at least with a linear load) that the presence of superimposed ripple components does not alter the basic concept that the theoretically correct point at which to switch converter banks is at (or as close as possible to) the zero value of the "fundamental" component of output current.

An inspection of the waveforms of Fig. 7.9 supports this supposition. Clearly, the zero-crossing of the "fundamental" component of the current waveform at a lags the first-current zero bank-selection point. Thus, if the control circuit is arranged to filter out the fundamental component of the load current waveform, and the bank-selection is made to occur at the zero-crossing of this fundamental component, then the converter switching point will be delayed, and, it can be postulated, the envelope distortion of the output current waveform will be reduced. This, in fact, is confirmed from the oscillograms shown at b, which are appropriate to this "fundamental current-zero" type of bank-selection control method. It is evident, in this case, that the "fundamental current-zero" bank-selection principle results in a complete elimination of the unnecessary waveform distortion produced by the first-current zero control method.

Of course, in practice, if it is required, at all costs, to positively prevent the flow circulating current, then it is not generally permissible to switch the firing pulses from one converter to the other precisely at the zero-crossing of the fundamental component of current, since this does not necessarily correspond with a zero value of the actual current waveform. Thus, it would seem that the best compromise would be to determine the bank-selection point from the actual current zero, "nearest" to the zero-crossing of the fundamental. Provided that the "nearest" actual current-zero occurs after the fundamental current-zero, this is quite easy to accomplish; in this case, the control requirement is simply one of memorizing that the fundamental current-zero has occurred, and switching banks at the actual current-zero immediately following this. If on the other hand, the nearest actual current-zero occurs before the fundamental current-zero, (and there is no inherent reason why this should not be so) the basic control requirement is now to recognize that this is so, and act accordingly. The recognition of this fact is difficult, if not impossible, to implement, at least in such a way that it inherently provides the proper operation, since, essentially, the requirement is to make an infallible prediction of "future events."

A further practical problem which may arise in the implementation of a fundamental current-zero bank selection principle is that it is necessary, for control purposes, to filter off a signal representing the "fundamental" component of the output current, with virtually no phase shift in the filter circuit. This may be difficult to achieve in applications in which the output frequency and load conditions are variable over a wide range.

Finally, another factor to be mentioned is that if the load connected to the cycloconverter is non-linear, so that it draws harmonic components of current, with a sinusoidal voltage applied to it, then clearly the fundamental current-zero bank-selection principle, in its basic form, does not hold good.

In conclusion, then, although the "fundamental current-zero" bank-selection principle is capable of providing the theoretically correct system operation under some conditions, with certain types of load, nonetheless, for the kinds of reasons discussed, it does not have the inherent ability to make invariably "correct" bank-selection decisions, at least not if a complete freedom from circulating current is to be preserved.

As a corollary to this discussion, it can reasonably be concluded that the only control principle which provides an "inherently correct" solution to the bank-selection problem, for all conceivable load conditions, is the type (discussed earlier in this chapter) in which circulating current is permitted to flow, for a certain "overlap" period, in the vicinity of current crossover, thus allowing a completely natural "freedom of movement" for the load current wave during this critical time. Thus this type of control approach provides an "inherent" solution not only to the "voltage distortion," but also to the "bank-selection" problem.

Voltage-Sensing Bank-Selection

A fundamentally different approach to the bank selection is to employ a closed-loop control of the output voltage, arranged in such a way that it automatically "selects" the converter banks. The firing pulses are permanently applied to both converters, but a bias voltage is introduced between their voltage transfer characteristics, so that there is no possibility for appreciable circulating current to flow. The action of the closed-loop voltage control is such that it automatically counteracts the bias voltage of whichever converter, for the time being, fabricates the output voltage wave. In this way, the current carrying converter is automatically "pulled" into operation, whilst, in the meantime, the idle converter is "pushed to one side." At the current crossover point, the action of the voltage feedback loop is inherently such that the error signal "seeks" the incoming converter, and thus the bank selection function is furnished automatically. A control scheme of this type is shown in simplified functional form in Fig. 7.11.

The function of each of the "positive" and "negative" firing pulse generators is to produce firing pulses for the thyristors of the associated converters. The phase of these firing pulses is controlled in accordance with the analog

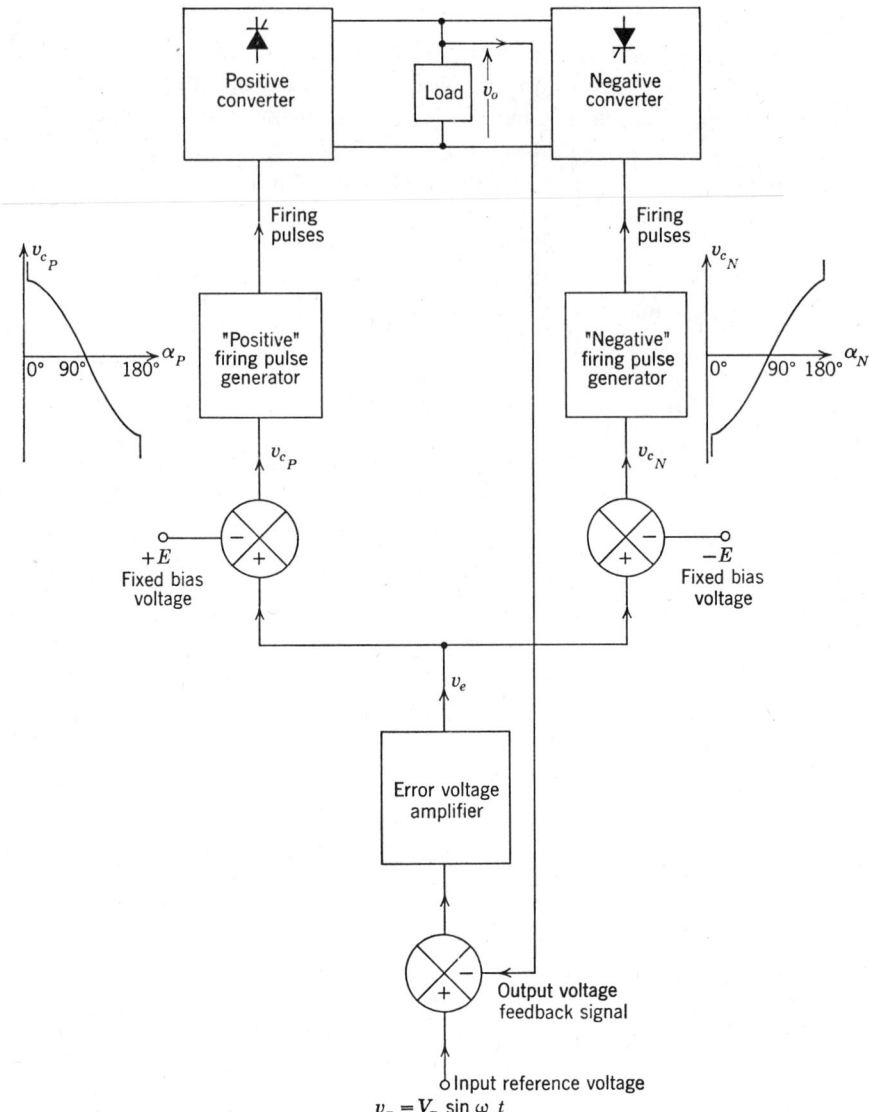

Figure 7.11. Diagrammatic representation of control scheme for a cycloconverter using the "voltage-sensing" principle of converter bank selection.

input voltages v_{c_P} and v_{c_N}, so that (with continuous conduction) the mean output voltage of the associated converter is directly proportional to the respective input voltage.

Each of the firing pulse generators have equal and opposite fixed bias voltages E applied to their inputs. The polarities of these voltages are such as to tend to retard the converter firing angles, and thus to "bias off" the converters, with respect to one another, thereby preventing the flow of circulating current. A second, common control voltage is applied to the inputs of both firing pulse generators; this is the output signal of the error voltage amplifier.

The operation of the scheme is explained by reference to the waveforms shown in Fig. 7.12. For simplicity, the presence of ripple voltage at the output of the cycloconverter is neglected, and continuous conduction (except at the current crossover) into a load of unity displacement factor, is assumed.

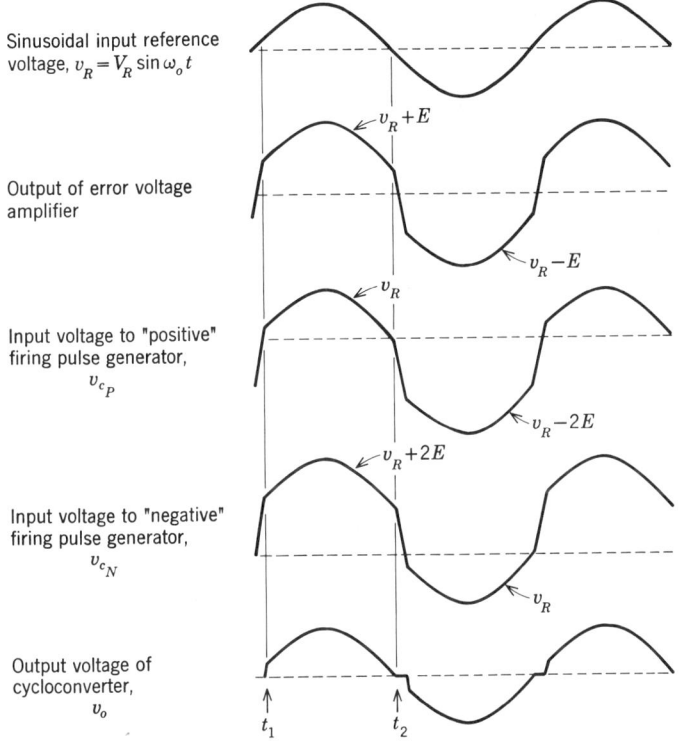

Figure 7.12. Theoretical waveforms associated with the cycloconverter control scheme of Fig. 7.11. Ripple voltage at output of cycloconverter neglected. Unity load displacement factor. Continuous conduction assumed.

Consider the operation during the time period $t_1 - t_2$, during which the positive converter is in conduction. The action of the voltage feedback loop during this time is automatically such as to "force" the output voltage of the positive converter to follow the reference input voltage. This necessarily implies that the error voltage amplifier produces at its output, in addition to a sinusoidal modulating signal, a steady bias voltage of $+E$. The presence of this component of voltage is necessary in order to counterbalance the steady bias voltage applied to the positive firing pulse generator, thereby reducing the net bias voltage at the input of this firing pulse generator to zero. In the meantime, the total bias voltage appearing at the input of the negative firing pulse generator is $2E$, and therefore this converter is "biased off."

At time t_2, the current in the positive converter falls to zero. The output voltage now remains momentarily at zero, (since, with no current, it cannot reverse), and hence the polarity of the actual voltage error becomes negative. The output signal of the error voltage amplifier now moves in a negative-going direction, thus "seeking" the negative converter, and "pulling" this into conduction, whilst, at the same time, the positive converter is automatically "pushed" to one side.

This type of control scheme is attractive because of its simplicity, and because it provides a naturally robust and reliable mode of system operation. It should be appreciated, however, that the principle relies for its operation upon the deliberate production of an error between the input reference voltage and the mean output voltage, in the region of current crossover. It is this error which forces the output signal of the error voltage amplifier to "seek" the incoming converter.

The degree of "crossover distortion" introduced into the output voltage waveform depends largely upon the amount of bias voltage separation between the transfer characteristics of the two converters, necessary to prevent the flow of substantial amounts of circulating current. This, in turn, depends upon the configuration and pulse number of the converter circuit (and generally becomes smaller with increasing circuit pulse number). Additionally, of course, the use of a circulating current reactor would enable the voltage transfer characteristics of the two converters to be brought more closely towards one another, thereby reducing the "crossover distortion," without the penalty of introducing high amplitude circulating ripple currents.

Chapter Eight

Cycloconverter Circuits

Many alternative arrangements of cycloconverter circuit, having varying degrees of complexity, and providing either single or multiphase outputs, are feasible.

As in the case of the rectifier or phase-controlled converter circuit, from the viewpoint of reducing the external harmonic voltages and currents to a minimum, the pulse number of the cycloconverter circuit should be as high as possible. Of course, this necessarily implies that a relatively large number of thyristors be employed in the circuit, and therefore this requirement generally cannot be met economically, unless the application is such that a large number of thyristors are required in any case, purely from the viewpoint of realizing the necessary output power.

Fortunately, in practice, a less than ideal external performance characteristic can usually be tolerated, and thus it is often possible to employ relatively simple circuitry.

In this chapter, the most common cycloconverter circuits are presented. This treatment is by no means exhaustive; however, an understanding of these circuits should enable the reader to arrive at alternative configurations, according to the requirements of any particular application.

It is generally assumed that the input is 3-phase, and that either a single- or three-phase output is required. Cycloconverter circuits operating from a single phase supply are, of course, also feasible. However, these circuits are considered to be relatively trivial, and in any case have limited application, because of their relatively poor performance characteristics. They are not, therefore, considered to be worthy of inclusion here.

For the convenience of the reader in making comparisons between the rating requirements of the various circuits, certain basic quantitative information is included in association with each circuit diagram. This data is obtained largely from the results of the analytical work of Chapters 11 and 12, and, therefore, at this stage, it is offered without explanation as to its derivation.

In each circuit diagram, circulating current reactors, and interphase reactors, where appropriate, are included. As discussed in Chapter 7, circulating current reactors may or may not be required, depending upon the method of control of the converters, and the performance requirements of the application. Of course, it is also true that interphase reactors are not essential. However, since considerable benefits arise from using these reactors, in terms of thyristor and transformer utilization, from a practical viewpoint these can be assumed to be essential circuit ingredients.

"SYMMETRICAL" CYCLOCONVERTER CIRCUITS

In practical applications, the cycloconverter is commonly required to deliver a 3-phase output from a 3-phase input. The basic cycloconverter module essentially consists of a dual-converter, which, of course, produces only a single phase output. Thus, the most logical—but, as will be seen, not the only—method for providing a 3-phase output, is to use three similar, essentially independent, dual-converters, one for each output phase. Such an arrangement for providing a 3-phase output is referred to in this book as a "symmetrical" circuit.

Various "symmetrical" circuits are presented in the following sections. Where only a single phase output is required, only one of the three sections of the circuit is used. Similarly, for a 2-phase output, only two of the three sections of the circuit are used.

"Symmetrical" Three-Pulse Circuit

A diagram of a symmetrical 3-pulse midpoint cycloconverter circuit is shown in Fig. 8.1. This comprises 3 identical 4-quadrant 3-pulse converters, one for each output phase. The common transformer secondary feeds the input terminals of all 3 converters. As will be seen in Chapter 12, for a balanced 3-phase output, theoretically, there are no zero-sequence components of current in the input lines. Thus there is no zero-sequence magnetization of the transformer core, and the zig-zag transformer secondary winding arrangement, normally used for a 3-pulse converter, is not necessary. In practice, in order to positively prevent the flow of zero-sequence currents, the connection between the neutral point of the load, and that of the transformer secondary, can be omitted. For a single phase load, of course, the production of zero-sequence currents is inevitable, and in this case, a zig-zag connection of the transformer secondary would normally be required.

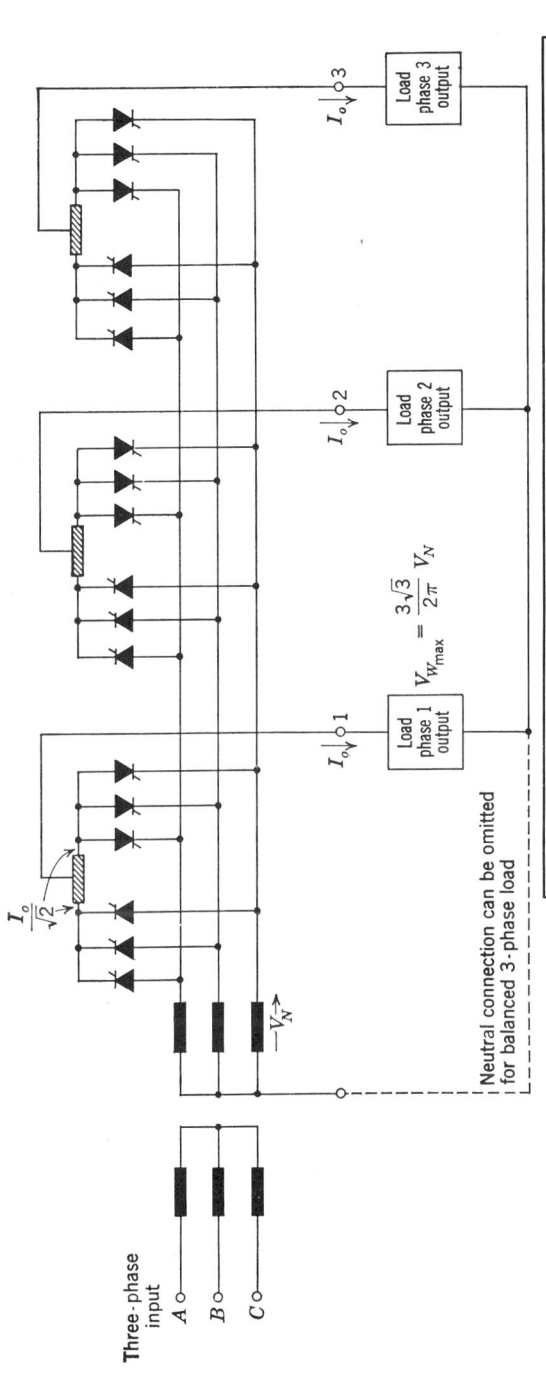

Figure 8.1. Three-pulse midpoint cycloconverter circuit.

"Symmetrical" Six-Pulse Midpoint Circuit

A diagram of a "symmetrical" 6-pulse midpoint cycloconverter circuit is shown in Fig. 8.2. This comprises three identical 4-quadrant 6-pulse midpoint converters, one for each output phase. The input terminals of all 3 converters are connected to the common transformer secondary.

"Symmetrical" Twelve-Pulse Midpoint Circuit

A diagram of a "symmetrical" 12-pulse midpoint cycloconverter circuit is shown in Fig. 8.3. This comprises three identical 4-quadrant 12-pulse midpoint converters, one for each output phase. Again, all three converters share common secondary windings on the input transformer.

"Symmetrical" Six-Pulse Bridge Circuit with Isolated Loads

With 6-pulse bridge cycloconverter circuits, supplying 3-phase loads, it is necessary to provide electrical isolation either between the inputs to the individual bridges, or between the output load circuits. This is because there is no point of common connection—nor is one permissible—between the input and output sides of the circuit.

In practice, the most economical approach, if feasible, is to isolate the 3-phase loads from one another, thus avoiding the requirement for an input transformer with three isolated secondary windings. Actually, with isolated loads, the need for an input transformer is avoided altogether, and it is possible to operate the cycloconverter directly from the input line—assuming, of course, that the voltage is of an appropriate level for the converter.

A diagram of a "symmetrical" 6-pulse bridge cycloconverter circuit with isolated loads is shown in Fig. 8.4. This comprises three identical 4-quadrant 6-pulse bridge converters, one for each output phase.

This type of circuit is commonly used for 3-phase a-c machine loads, since it is usually a simple matter to electrically isolate the 3-phase windings of the machine from one another.

"Symmetrical" Six-Pulse Bridge Circuit with Non-Isolated Loads

A diagram of a "symmetrical" 6-pulse bridge cycloconverter circuit, with non-isolated loads, is shown in Fig. 8.5. This comprises three identical 4-quadrant 6-pulse bridge converters, one for each output phase. The input

"SYMMETRICAL" CYCLOCONVERTER CIRCUITS

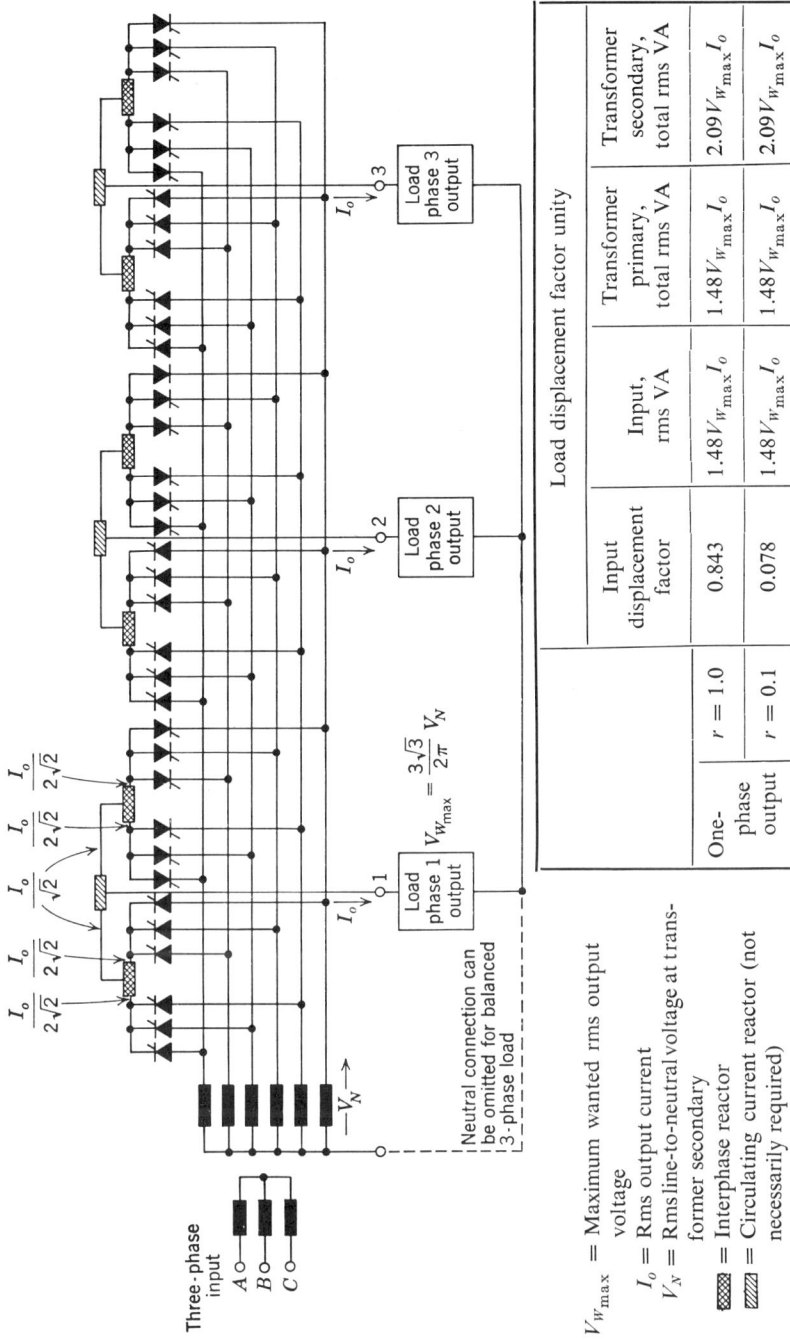

Figure 8.2. Six-pulse midpoint cycloconverter circuit.

$V_{W\max}$ = Maximum wanted rms output voltage
I_o = Rms output current
V_N = Rms line-to-neutral voltage at transformer secondary
▨ = Interphase reactor
▧ = Circulating current reactor (not necessarily required)

		Load displacement factor unity			
		Input displacement factor	Input, rms VA	Transformer primary, total rms VA	Transformer secondary, total rms VA
One-phase output	$r = 1.0$	0.843	$1.48 V_{W\max} I_o$	$1.48 V_{W\max} I_o$	$2.09 V_{W\max} I_o$
	$r = 0.1$	0.078	$1.48 V_{W\max} I_o$	$1.48 V_{W\max} I_o$	$2.09 V_{W\max} I_o$
Three-phase output	$r = 1.0$	0.843	$1.21(3 V_{W\max} I_o)$	$1.21(3 V_{W\max} I_o)$	$1.32(3 V_{W\max} I_o)$
	$r = 0.1$	0.078	$1.32(3 V_{W\max} I_o)$	$1.32(3 V_{W\max} I_o)$	$1.32(3 V_{W\max} I_o)$

212 CYCLOCONVERTER CIRCUITS

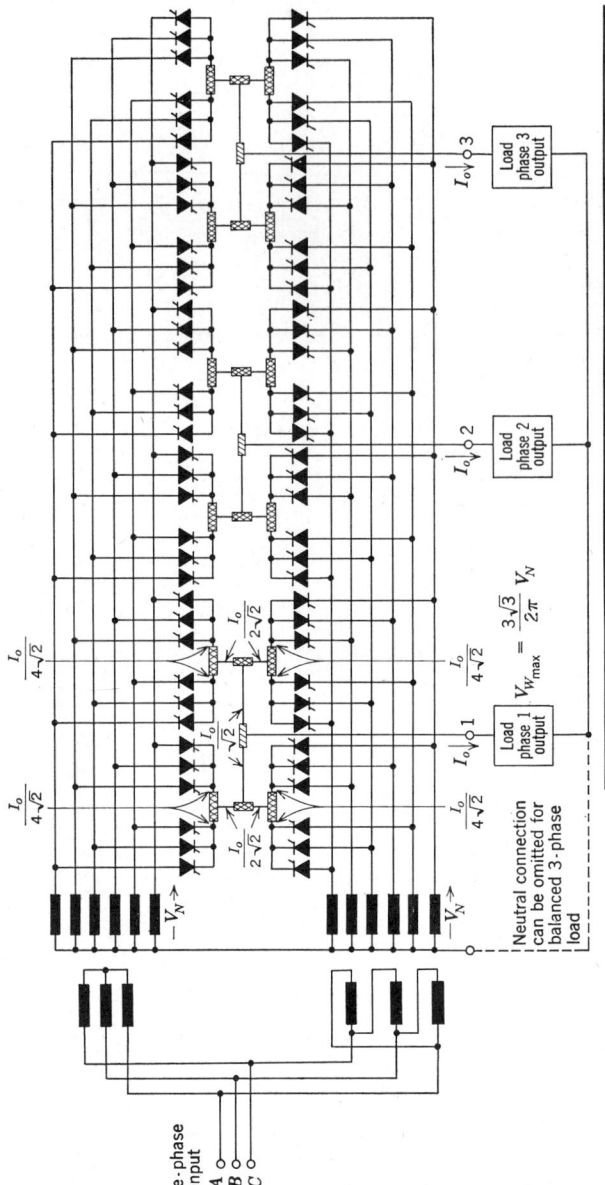

$V_{W\max}$ = Maximum wanted rms output voltage
I_o = rms output current
V_N = rms line-to-neutral voltage at transformer secondary
▨ = Interphase reactor
▱ = Circulating current reactor (not necessarily required)

Figure 8.3. Twelve-pulse midpoint cycloconverter circuit.

		Load displacement factor unity			
		Input displacement factor	Input, rms VA	Transformer primary, total rms VA	Transformer secondary, total rms VA
One-phase output	$r = 1.0$	0.843	$1.43 V_{W\max} I_o$	$1.48 V_{W\max} I_o$	$2.09 V_{W\max} I_o$
	$r = 0.1$	0.078	$1.43 V_{W\max} I_o$	$1.48 V_{W\max} I_o$	$2.09 V_{W\max} I_o$
Three-phase output	$r = 1.0$	0.843	$1.19(3 V_{W\max} I_o)$	$1.21(3 V_{W\max} I_o)$	$1.32(3 V_{W\max} I_o)$
	$r = 0.1$	0.078	$1.29(3 V_{W\max} I_o)$	$1.32(3 V_{W\max} I_o)$	$1.32(3 V_{W\max} I_o)$

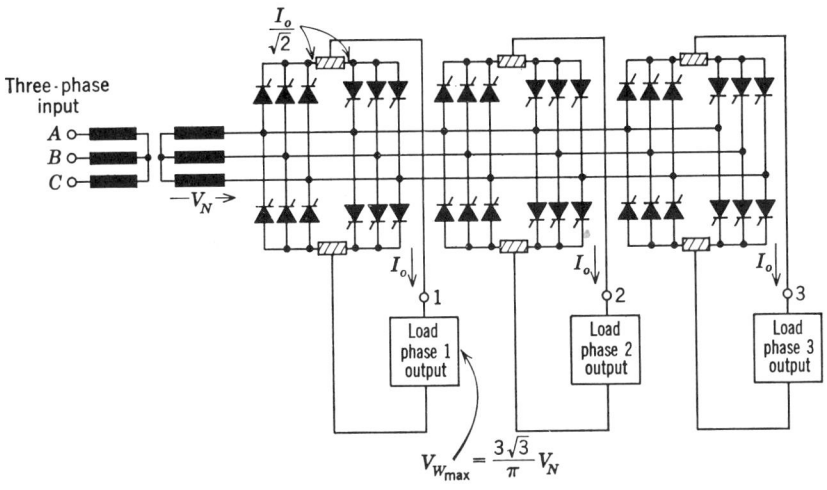

		Load displacement factor unity			
		Input displacement factor	Input, rms VA	Transformer primary, total rms VA	Transformer secondary, total rms VA
One-phase output	$r = 1.0$	0.843	$1.48 V_{W_{max}} I_o$	$1.48 V_{W_{max}} I_o$	$1.48 V_{W_{max}} I_o$
	$r = 0.1$	0.078	$1.48 V_{W_{max}} I_o$	$1.48 V_{W_{max}} I_o$	$1.48 V_{W_{max}} I_o$
Three-phase output	$r = 1.0$	0.843	$1.21(3 V_{W_{max}} I_o)$	$1.21(3 V_{W_{max}} I_o)$	$1.21(3 V_{W_{max}} I_o)$
	$r = 0.1$	0.078	$1.32(3 V_{W_{max}} I_o)$	$1.32(3 V_{W_{max}} I_o)$	$1.32(3 V_{W_{max}} I_o)$

$V_{W_{max}}$ = Maximum wanted rms output voltage
I_o = Rms output current
V_N = Rms line-to-neutral voltage at transformer secondary
▨ = Circulating current reactor (not necessarily required)

Figure 8.4. Six-pulse bridge cycloconverter circuit with isolated loads.

terminals of each of the three converters are fed from an isolated secondary winding on the input transformer. Thus there is no connection between the output terminals of the bridge circuits via their input connections, and it is permissible to make connections between the 3-phase loads.

With this circuit, the total volt-amps handled by the three isolated transformer secondary windings are approximately 22% larger than the primary volt-amps. This is because, in effect, each isolated transformer secondary "sees" a separate cycloconverter supplying a single phase load, and therefore

214 CYCLOCONVERTER CIRCUITS

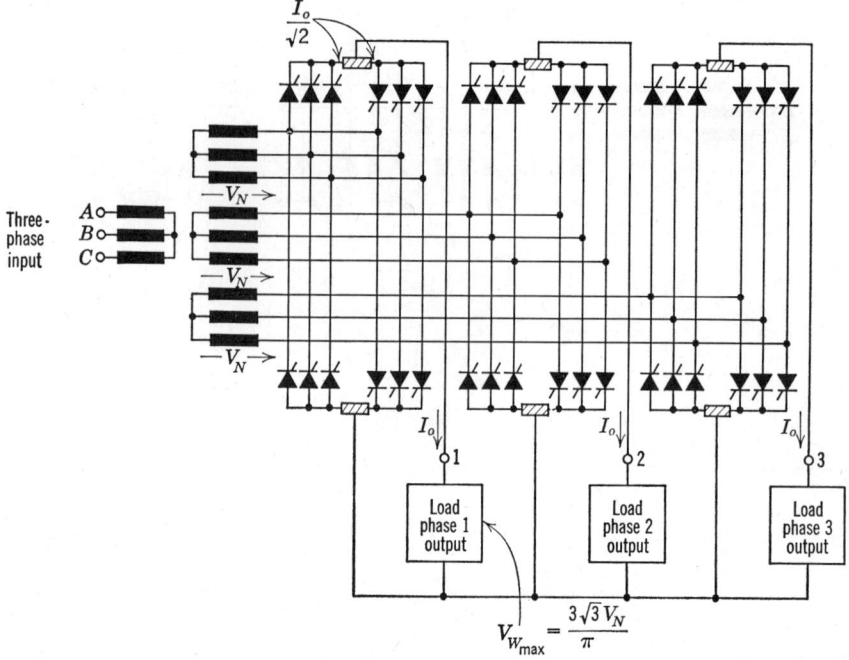

		Load displacement factor unity			
		Input displacement factor	Input, rms VA	Transformer primary, total rms VA	Transformer secondary, total rms VA
One-phase output	$r = 1.0$	0.843	$1.48 V_{W_{max}} I_o$	$1.48 V_{W_{max}} I_o$	$1.48 V_{W_{max}} I_o$
	$r = 0.1$	0.078	$1.48 V_{W_{max}} I_o$	$1.48 V_{W_{max}} I_o$	$1.48 V_{W_{max}} I_o$
Three-phase output	$r = 1.0$	0.843	$1.21(3 V_{W_{max}} I_o)$	$1.21(3 V_{W_{max}} I_o)$	$1.48(3 V_{W_{max}} I_o)$
	$r = 0.1$	0.078	$1.32(3 V_{W_{max}} I_o)$	$1.32(3 V_{W_{max}} I_o)$	$1.48(3 V_{W_{max}} I_o)$

$V_{W_{max}}$ = Maximum wanted rms output voltage
I_o = rms output current
V_N = rms line-to-neutral voltage at transformer secondary
▨ = Circulating current reactor (not necessarily required)

Figure 8.5. Six-pulse bridge cycloconverter circuit with nonisolated loads.

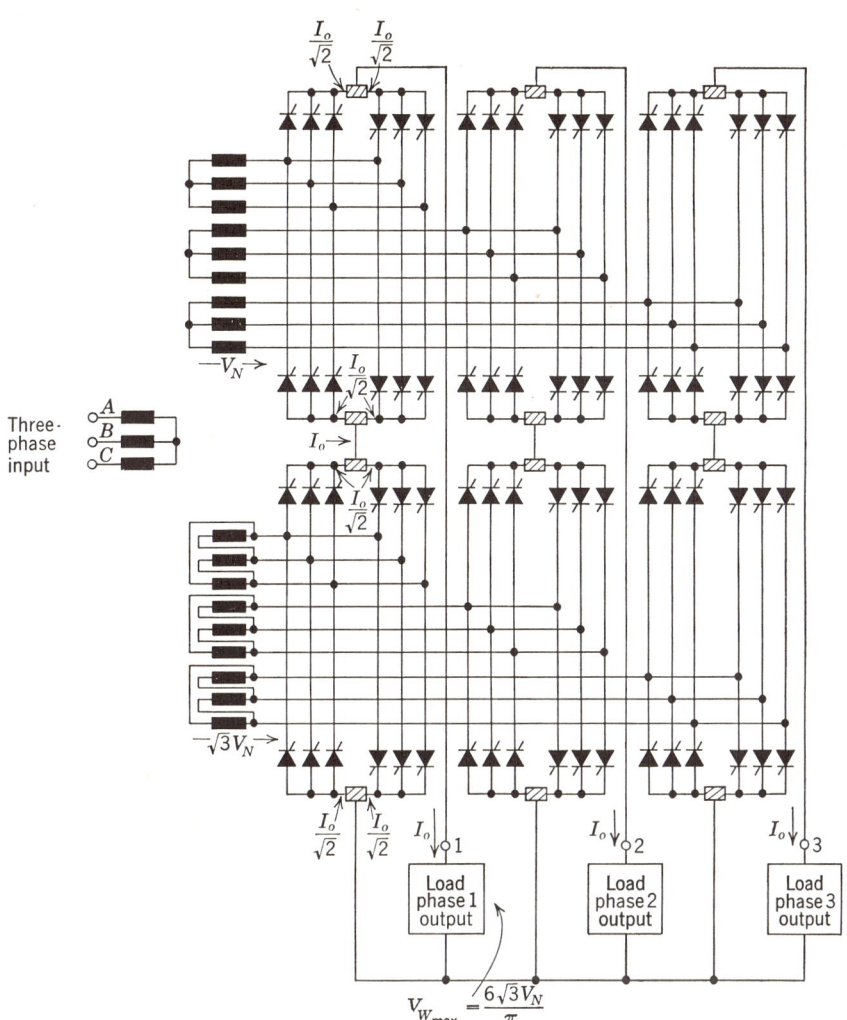

		Load displacement factor unity			
		Input displacement factor	Input, rms VA	Transformer primary, total rms VA	Transformer secondary, total rms VA
One-phase output	$r = 1.0$	0.843	$1.43 V_{W_{max}} I_o$	$1.43 V_{W_{max}} I_o$	$1.48 V_{W_{max}} I_o$
	$r = 0.1$	0.078	$1.43 V_{W_{max}} I_o$	$1.43 V_{W_{max}} I_o$	$1.48 V_{W_{max}} I_o$
Three-phase output	$r = 1.0$	0.843	$1.19(3 V_{W_{max}} I_o)$	$1.19(3 V_{W_{max}} I_o)$	$1.48(3 V_{W_{max}} I_o)$
	$r = 0.1$	0.078	$1.29(3 V_{W_{max}} I_o)$	$1.29(3 V_{W_{max}} I_o)$	$1.48(3 V_{W_{max}} I_o)$

$V_{W_{max}}$ = Maximum wanted rms output voltage
I_o = Rms output current
V_N = Rms line-to-neutral voltage at transformer secondary
▨ = Circulating current reactor (not necessarily required)

Figure 8.6. Twelve-pulse bridge cycloconverter circuit.

it carries the harmonic currents due to the pulsating power of the associated single phase output.

"Symmetrical" Twelve-Pulse Bridge Circuit

A diagram of a "symmetrical" 12-pulse bridge cycloconverter is shown in Fig. 8.6. This comprises three identical 4-quadrant 12-pulse bridge converters, one for each output phase. The input terminals of each of the 6 individual 6-pulse converters are fed from separate secondary windings on the input transformer. It should be noted that it is not permissible to use the same secondary winding for more than one converter. This is because each 12-pulse converter, by itself, requires two completely isolated transformer secondary windings.

OPEN-DELTA CYCLOCONVERTER CIRCUITS

If a 3-phase output is required, actually it is not essential to use three separate single phase cycloconverter circuits to generate independently three output voltages. Basically, this is because, for a set of balanced three-phase voltages, any one of the voltage vectors can be derived from the summation of the other two. Thus it is possible to use only two independent cycloconverter circuits, with their output terminals connected in an "open delta" configuration, to produce a 3-phase output.

With this type of connection, then, only two, instead of three, separate 4-quadrant converters are required, and thus, for a circuit of a given pulse number, the total number of thyristors required, together with the associated firing circuits and other auxiliary components, are reduced by a factor of 33%. However, as will be seen, the utilization factor of the converters and input transformer is less than that of the symmetrical circuit arrangement, and this therefore to some extent offsets the advantage of the reduced complexity of the power circuit. Also, it may be more difficult to control the converters to produce a properly balanced 3-phase output, because of the inherent asymmetrical nature of the power circuit. Nonetheless, the economic advantages to be gained from the use of the open-delta circuit may sometimes outweigh its technical limitations.

The basic principle of the open-delta connection is illustrated by the simplified equivalent circuit diagram, together with the associated voltage and current vector diagrams, of Fig. 8.7. In this simplified representation the ripple voltages generated at the output terminals of the converters are neglected, and it is assumed that each converter produces only the sinusoidal wanted voltage component.

OPEN DELTA CYCLOCONVERTER CIRCUITS 217

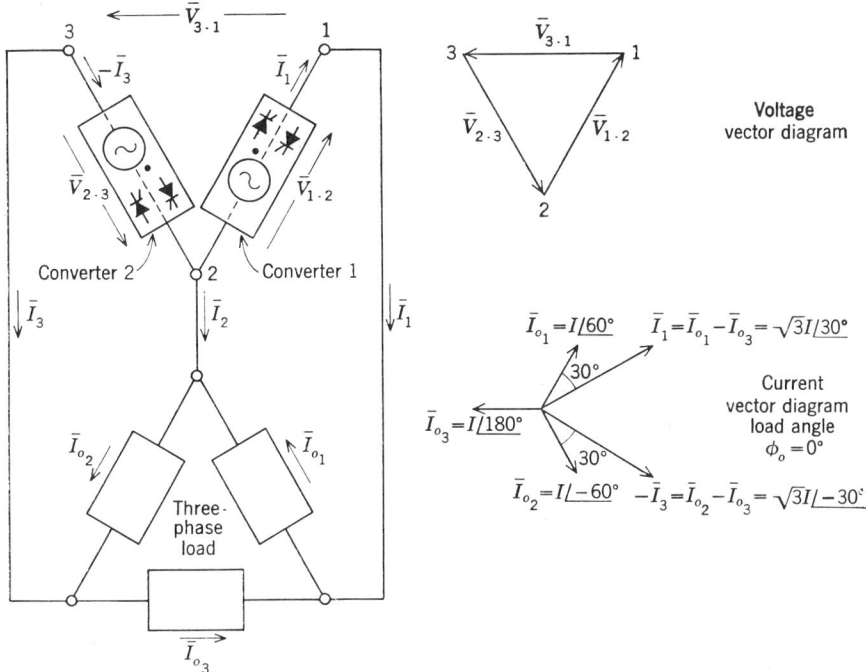

Figure 8.7. Schematic illustration of cycloconverter "open-delta" connection. In general, for load angle ϕ_o: Converter 1 current = $\sqrt{3}I$ at $/(\phi_o - 30°)$ with respect to $\bar{V}_{1-2}$; and Converter 2 current = $\sqrt{3}I$ at $/(\phi_o + 30°)$ with respect to $\bar{V}_{2-3}$.

It is clear that the voltage $\bar{V}_{3-1}$ which appears across the "open-delta"—between the points 3 and 1—which is the summation of the voltages $\bar{V}_{3-2}$ and $\bar{V}_{2-1}$, is the same as that which would be produced by a third converter connected between these points.

It is evident, however, that since no current is able to flow through the "open-delta," of necessity the currents carried by each of the two converters are not the same as would be obtained in a closed delta circuit. Thus, the current through converter 1, from 2 to 1, is the line current $\bar{I}_1$, and the current through converter 2, from 3 to 2, is the inverted line current, $-\bar{I}_3$. Thus, considering, for example, the specific case of a unity displacement factor load, the current carried by converter 1 is $\sqrt{3}$ × the load delta current, at an angle of 30° lagging the voltage, and the current through converter 2 is $\sqrt{3}$ × the load delta current, at an angle of 30° leading the voltage. Thus the total volt-amps handled by the two converters are 1.15 × the total load volt-amps. Also, since the displacement factor reflected to the input side of the

218 CYCLOCONVERTER CIRCUITS

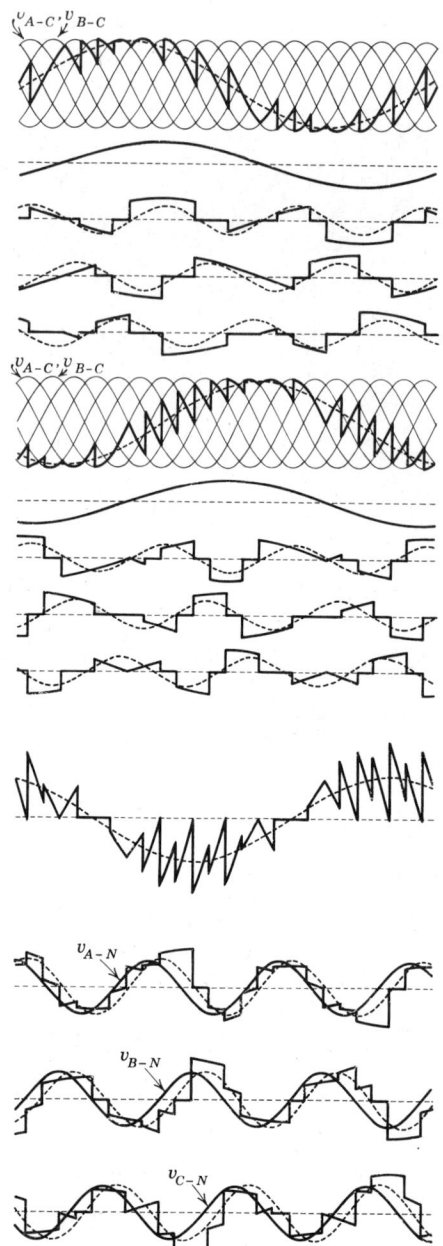

Output voltage 1-2,
generated by converter 1
----- Wanted component

Output current of converter 1
(30° lagging voltage)

Current in input line A, due to
converter 1 output current
----- Fundamental component

Current in input line B, due to
converter 1 output current
----- Fundamental component

Current in input line C, due to
converter 1 output current
----- Fundamental component

Output voltage 2-3,
generated by converter 2
----- Wanted component

Output current of converter 2
(30° leading voltage)

Current in input line A, due to
converter 2 output current
----- Fundamental component

Current in input line B, due to
converter 2 output current
----- Fundamental component

Current in input line C, due to
converter 2 output current
----- Fundamental component

Output voltage 3-1,
across open-delta
----- Wanted component

Voltage of line A, and total
current in input line A,
due to both converters
----- Fundamental component

Voltage of line B, and total
current in input line B,
due to both converters
----- Fundamental component

Voltage of line C, and total
current in input line C,
due to both converters
----- Fundamental component

Figure 8.8. Theoretical waveforms for a 6-pulse bridge "open-delta" cycloconverter. Load displacement angle $= 0°$; output frequency $= \frac{1}{3} \times$ input frequency; and maximum output voltage, that is $r = 1.0$.

cycloconverter is inherently lagging, regardless of whether the output displacement factor is lagging or leading, this means that the displacement factor at the input side of the system is appropriate to a load displacement angle of 30°, even though, in fact, the displacement angle of the load itself is 0°.

For the more general case of a nonunity displacement factor load, the input displacement factor of the open-delta circuit is invariably less than that of the symmetrical circuit. This is illustrated by the curves of Fig. 8.16.

A further factor is that although it is possible to generate a balanced set of wanted voltage components across the delta, it is inevitable that the harmonic voltages appearing across the 3 points of the delta are dissimilar to one another, and this asymmetry of the voltage harmonics may give rise to quite appreciable unbalance of the current harmonics.

Typical theoretical waveforms for a 6-pulse bridge open-delta connected cycloconverter are illustrated in Fig. 8.8. These waveforms are for a load displacement factor of unity, with maximum output voltage, and it is assumed that each of the converter currents are pure sinusoids. The dissimilarity between the harmonic distortion of the voltage waveforms across the 3 points of the delta is evident from these waveshapes. A comparison of the waveforms of the input line currents of Fig. 8.8. with those of Fig. 6.11, confirms the fact that, for the open-delta circuit, the displacement angle between the fundamental component of the input line current, and the associated input voltage, is greater than that for the symmetrical circuit. It can be seen, also, that the harmonic distortion of the input line current waveform is somewhat greater than for the symmetrical circuit.

Two alternative "open-delta" circuits are presented in the following sections.

Open-Delta Three-Pulse Midpoint Circuit

A diagram of a 3-pulse midpoint "open-delta" cycloconverter circuit is shown in Fig. 8.9. This comprises two identical 4-quadrant 3-pulse converters. Two of the load connections are taken from the output terminals of the two converters, and the third connection comprises the neutral point of the transformer secondary. It is inherent, then, that current flows in this connection, and thus a zig-zag transformer secondary winding is required.

Open-Delta Six-Pulse Bridge Circuit

A diagram of a 6-pulse bridge "open-delta" cycloconverter circuit is shown in Fig. 8.10. This comprises two identical 4-quadrant 6-pulse bridge converters. The input terminals of each of the converters are connected to separate,

220 CYCLOCONVERTER CIRCUITS

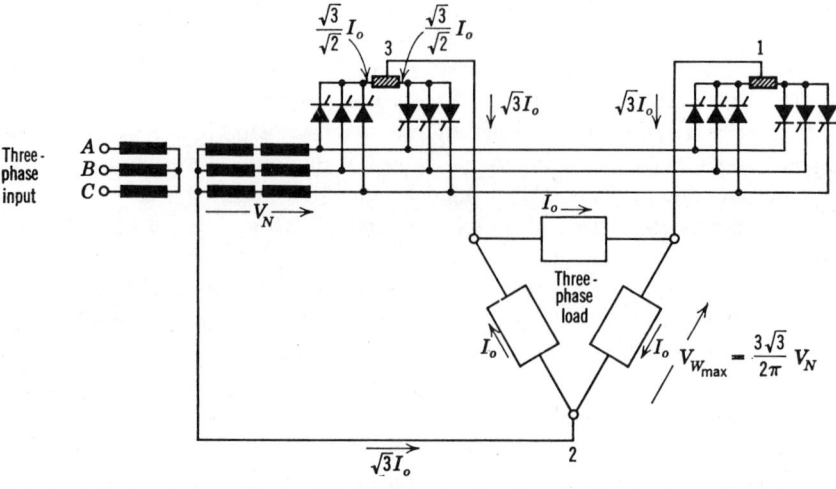

	Load displacement factor unity, $r = 1.0$			
Input displacement factor	Input, rms VA (approx.)	Transformer primary, total rms VA (approx.)	Transformer secondary, total rms VA (approx.)	Total rms output VA handled by converters
0.77	$2.2(3V_{W_{max}}I_o)$	$2.2(3V_{W_{max}}I_o)$	$2.6(3V_{W_{max}}I_o)$	$1.15(3V_{W_{max}}I_o)$

$V_{W_{max}}$ = Maximum wanted rms output voltage (across delta)
I_o = Rms output current (in delta)
V_N = Rms line-to-neutral voltage at transformer secondary
▨ = Circulating current reactor (not necessarily required)

Figure 8.9. Three-pulse midpoint "open-delta" cycloconverter circuit.

isolated, secondary windings on the input transformer, thus permitting the "open-delta" connection to be made between the output terminals of the two converters.

"RING-CONNECTED" CYCLOCONVERTER CIRCUITS

With the object in view of reducing the complexity of the cycloconverter power circuit, but without sacrificing the benefits of a given circuit pulse number, an alternative arrangement to the open-delta circuit, for a 3-phase output, is the "ring-connected" circuit. This arrangement uses three 2-quadrant converters, with their output terminals connected in a closed ring.

OPEN DELTA CYCLOCONVERTER CIRCUITS

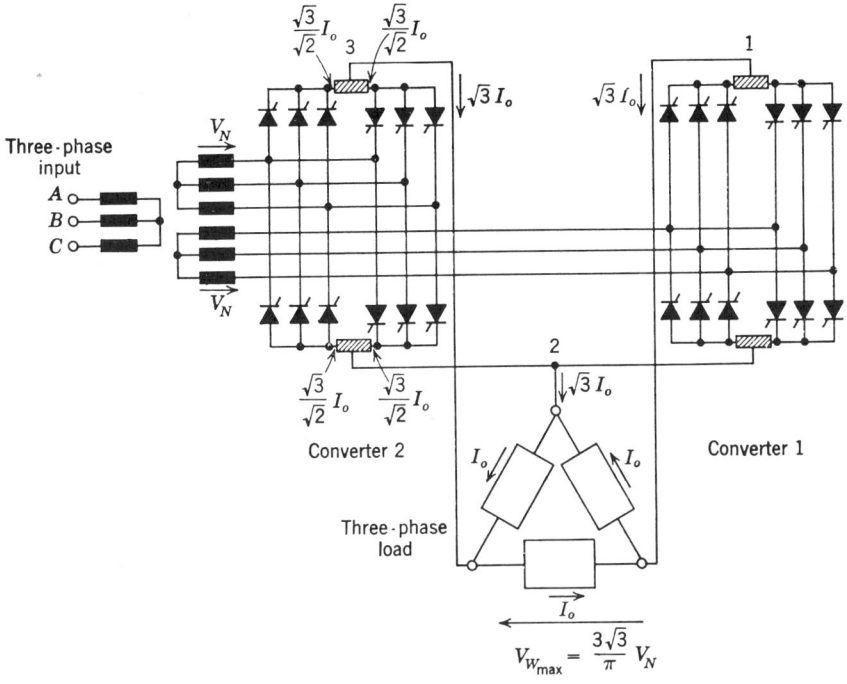

Load displacement factor unity, $r = 1.0$				
Input displacement factor	Input, rms VA (approx.)	Transformer primary, total rms VA (approx.)	Transformer secondary, total rms VA (approx.)	Total rms output VA handled by converters
0.77	$1.69(3V_{W_{\max}}I_o)$	$1.69(3V_{W_{\max}}I_o)$	$1.71(3V_{W_{\max}}I_o)$	$1.15(3V_{W_{\max}}I_o)$

$V_{W_{\max}}$ = Maximum wanted rms output voltage (across delta)
I_o = Rms output current (in delta)
V_N = Rms line-to-neutral voltage at transformer secondary
▨ = Circulating current reactor (not necessarily required)

Figure 8.10. Six-pulse bridge "open-delta" cycloconverter circuit.

The 3-phase load is connected at each point of connection of one converter to the next.

A simplified schematic representation of the "ring-connected" arrangement is illustrated in Fig. 8.11. Although the converter connected in each arm of the delta is capable of carrying current only in one direction, the output line current can, of course, flow in either direction. Thus, for example,

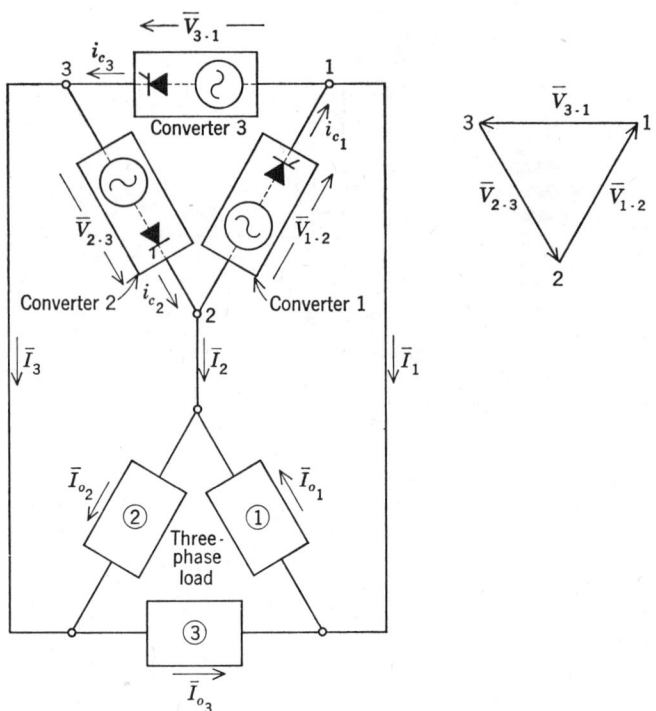

Figure 8.11. Schematic illustration of "ring-connected" cycloconverter circuit.

positive current in line 1 flows through converter 1, and negative current in line 1 flows through converter 3.

With this type of connection, then, three 2-quadrant converters are required, and therefore, for a given circuit pulse number, the total number of thyristors, together with the associated firing circuits and other auxiliary components, are reduced by a factor of 50%, as compared to the "symmetrical" circuit arrangement. However, as might be expected intuitively, and as is indeed the case in practice, a quite severe penalty is paid in terms of the utilization factor of the converters and the input transformer. Nonetheless, for certain applications, the relative simplicity of this type of connection may be a decisive factor.

Waveforms of the currents obtained in the idealized ring-connected circuit of Fig. 8.11 are illustrated in Fig. 8.12. It is assumed here that the external line current waveforms are pure sinusoids. The waveforms of the currents in the individual converters are obtained from the basic relationships between the instantaneous line and converter currents, in conjunction with the

"RING-CONNECTED" CYCLOCONVERTER CIRCUITS 223

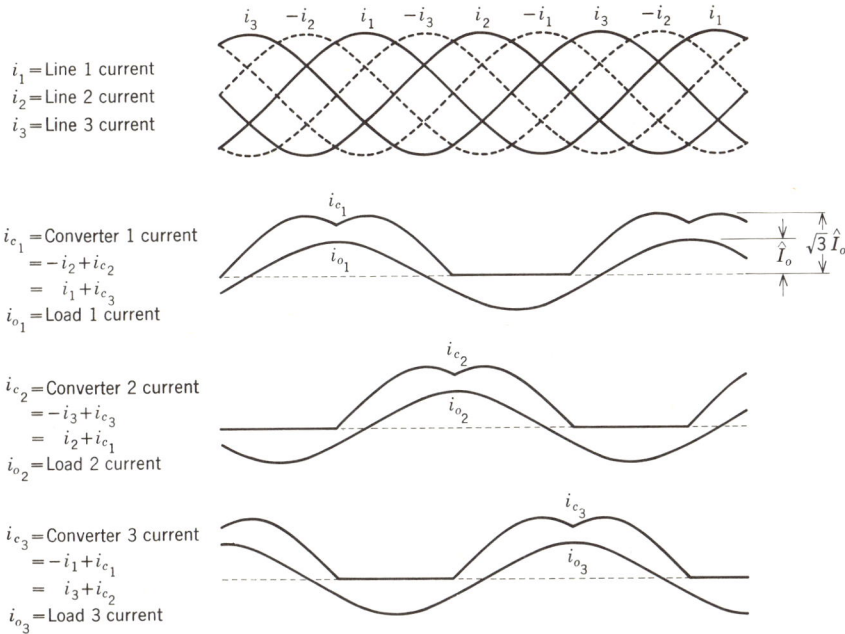

Figure 8.12. Current waveforms associated with the idealized "ring-connected" cycloconverter circuit of Fig. 8.11.

basic principle that the current in each converter can flow only in the "positive" direction. It can be seen that two of the three converters are always in conduction, with the third one idle. Thus each converter conducts for a period of 240° of each output cycle, and can be blocked, if desired, during the remaining 120° period, thus preventing the flow of circulating current around the ring.

It can be shown that the rms value of this "ideal" converter current waveform is 1.55 × the rms value of the sinusoidal load delta-current. Thus each converter must be rated to carry 1.55 × the rms load current. Since, moreover, the input terminals of each converter are connected to isolated transformer secondary windings, the total rms volt-amps carried by the transformer secondary are large, by comparison with the output volt-amps.

Typical theoretical waveforms for a 6-pulse bridge ring-connected cycloconverter circuit are illustrated in Fig. 8.13. These waveforms are for a load displacement factor of 1.0, with maximum output voltage, and it is assumed that the output line currents are pure sinusoids. A comparison of the waveforms of these input line currents with those of Fig. 8.8, for the open-delta

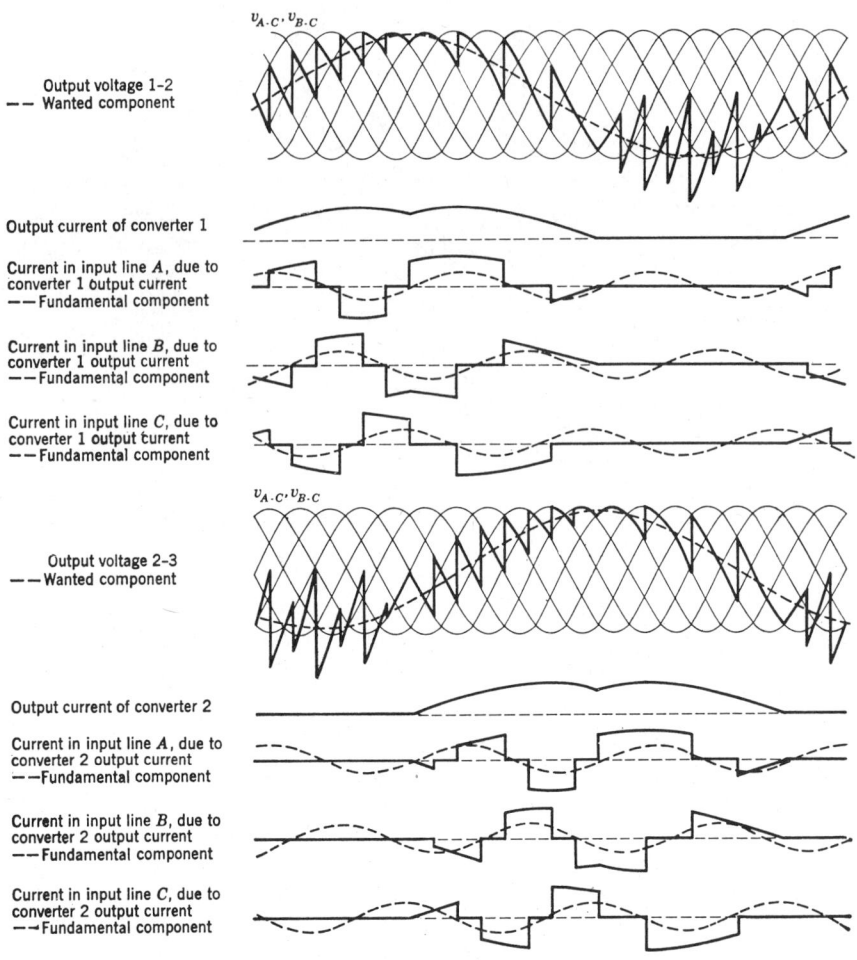

Figure 8.13. Theoretical waveforms for a 6-pulse bridge ring connected cycloconverter. Load displacement angle $= 0°$; output frequency $= \frac{1}{3} \times$ input frequency; and maximum output voltage, that is, $r = 1.0$.

circuit, and of Fig. 6.11, for the "symmetrical" circuit, illustrates the fact that the displacement angle between the fundamental component of the input line current and the associated input voltage, is greatest for the ring-connected circuit. This is also shown by the curves of Fig. 8.16. It is interesting to note, however, that the harmonic distortion of the input line current of the ring-connected circuit is less than that of the open-delta circuit, so that the input power factor of the ring-connected circuit is, in fact, slightly higher

"RING-CONNECTED" CYCLOCONVERTER CIRCUITS 225

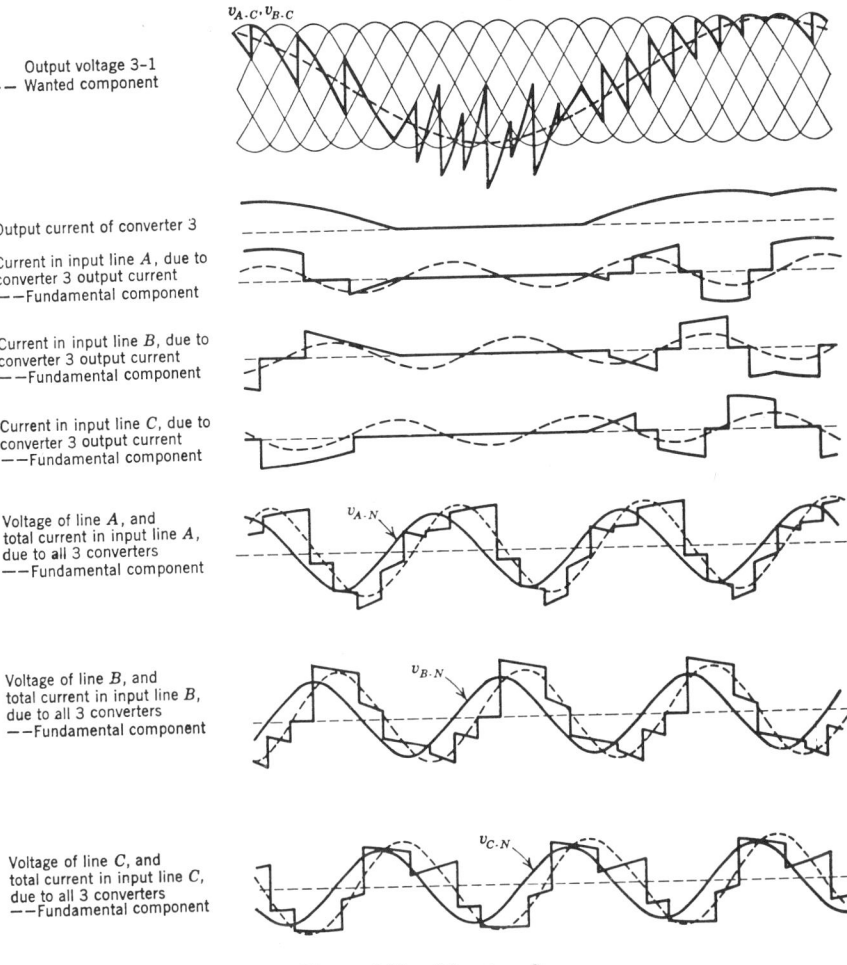

Figure 8.13 (*Continued*)

than that of the open delta. It may also be pointed out here that the ring-connected circuit has the "natural" advantage, over the open-delta circuit, of being a perfectly symmetrical arrangement.

A further point, which is evident from the waveforms of Fig. 8.13, is that the peak voltage impressed across the output terminals of the converter, when it is temporarily "idle," is in excess of the peak line-to-line input voltage. This is due to the ripple voltages of the two "active" converters adding together across the terminals of the idle converter. Because of this effect,

226   CYCLOCONVERTER CIRCUITS

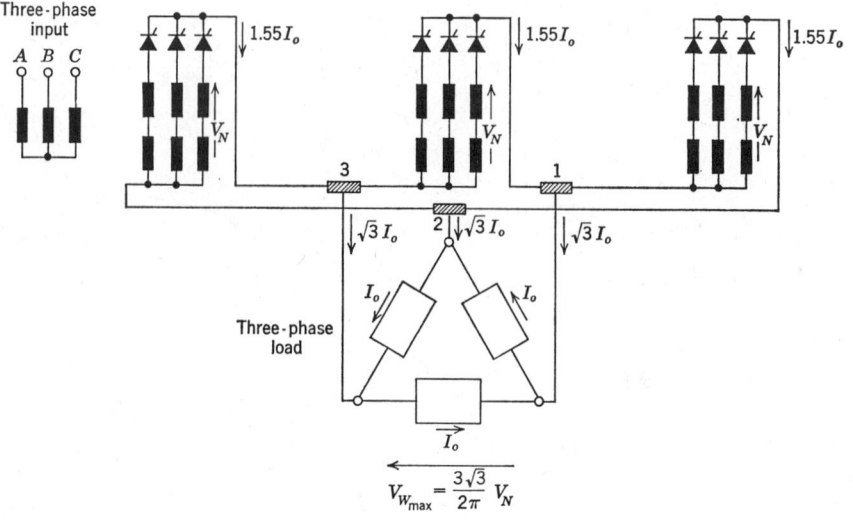

	Load displacement factor unity, $r = 1.0$			
Input displacement factor	Input, rms VA (approx.)	Transformer primary, total rms VA (approx.)	Transformer secondary, total rms VA	Total rms output VA handled by converters
0.688	$1.75(3V_{W_{max}}I_o)$	$1.75(3V_{W_{max}}I_o)$	$3.76(3V_{W_{max}}I_o)$	$1.55(3V_{W_{max}}I_o)$

$V_{W_{max}}$ = Maximum wanted rms output voltage (across delta)
I_o = Rms output current (in delta)
V_N = Rms line-to-neutral voltage at transformer secondary
▨ = Circulating current reactor (not necessarily required)

Figure 8.14. Three-pulse "ring-connected" cycloconverter circuit.

the peak voltages developed across the thyristors may be 1.5 or more, of the peak line-to-line voltage. This, of course, increases the required voltage rating of the thyristors of the converters, beyond that required for the symmetrical, or open-delta, circuit arrangements. Alternatively, if the ring-circuit is operated with a continuous circulating current, so that all three converters are kept in continuous conduction, then this effect does not manifest itself.

Two alternative ring-connected circuits are presented in the following sections.

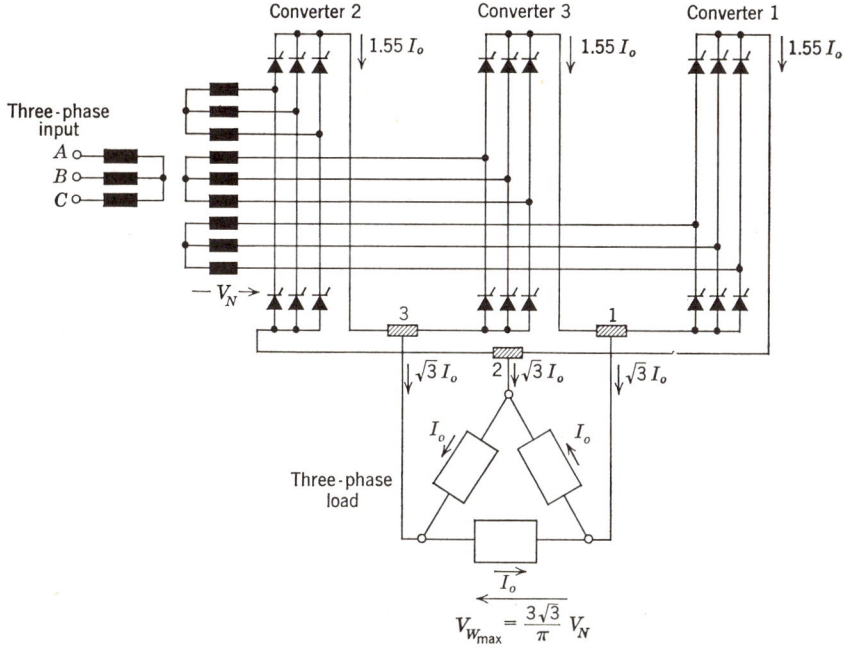

Load displacement factor unity, $r = 1.0$					
Input displacement factor	Input, rms VA (approx.)	Transformer primary, total rms VA (approx.)	Transformer secondary, total rms VA	Total rms output VA handled by converters	
0.688	$1.6(3V_{w_{max}}I_o)$	$1.6(3V_{w_{max}}I_o)$	$2.3(3V_{w_{max}}I_o)$	$1.55(3V_{w_{max}}I_o)$	

$V_{w_{max}}$ = Maximum wanted rms output voltage (across delta)
I_o = Rms output current (in delta)
V_N = Rms line-to-neutral voltage at transformer secondary
▨ = Circulating current reactor (not necessarily required)

Figure 8.15. Six-pulse bridge "ring-connected" cycloconverter circuit.

Ring-Connected Three-Pulse Midpoint Circuit

A diagram of a 3-pulse ring-connected cycloconverter circuit is shown in Fig. 8.14. This comprises 3 identical 3-pulse 2-quadrant converters, connected in a closed ring. Each converter is supplied from a separate secondary winding on the input transformer. Because of the flow of current in the neutral connection, zig-zag transformer secondary windings are required.

228 CYCLOCONVERTER CIRCUITS

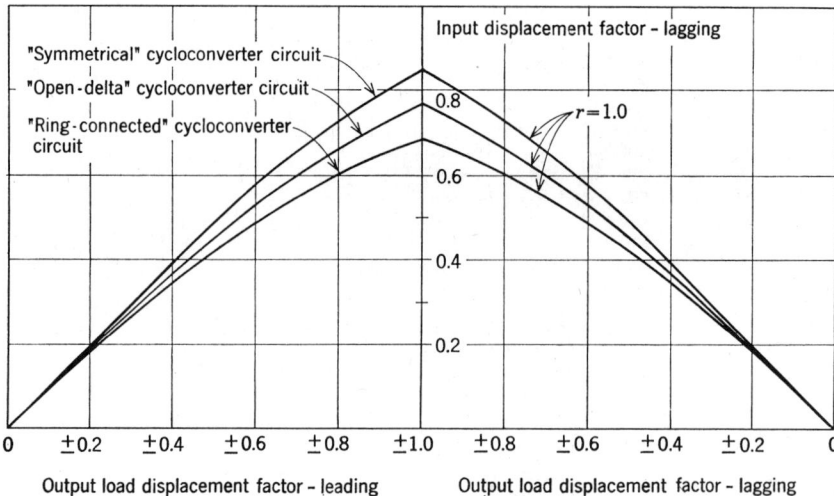

Figure 8.16. Curves showing the relationship between the output load displacement factor and the input displacement factor, for "symmetrical," "open–delta," and "ring-connected" cycloconverter circuits. Maximum output voltage; that is, $r = 1.0$.

Ring-Connected Six-Pulse Bridge Circuit

A diagram of a 6-pulse bridge ring-connected cycloconverter circuit is shown in Fig. 8.15. This comprises 3 identical 6-pulse 2-quadrant bridge converters, connected in a closed ring. Each converter is supplied from a separate secondary winding on the input transformer.

Chapter Nine

Theoretical Principles of Firing Pulse Timing Control

In order to control the output voltage of the phase-controlled converter or cycloconverter, it is necessary to control the phase of the thyristor firing pulses. Many alternative principles exist for achieving this end.

In this chapter, various alternative principles of pulse timing control are discussed. The treatment is not exhaustive; however, it is considered to be a reflection of the most modern and sophisticated pulse timing control principles.

A discussion is also presented on the question of the theoretically correct firing instants for the thyristors of the cycloconverter. It is shown that the "cosine wave crossing" control principle for determining the firing instants can be regarded as being the most natural one for the cycloconverter, inasmuch as it has the unique property that it produces the minimum possible total distortion of the output voltage waveform.

THE COSINE WAVE CROSSING PULSE TIMING METHOD

The Basic Principle

In Chapter 5, it was seen that if the firing angle is made to respond to an analog "control" or "reference" voltage, in such a way that the cosine of the firing angle is proportional to this reference voltage, then for a steady d-c output, with continuous conduction, the resulting relationship between the reference voltage and the mean voltage at the d-c terminals of the converter is linear. It would seem, therefore, that a cosine relationship between the firing angle and the reference voltage is the "naturally correct" one for a steady d-c output, because, with this relationship, the phase-controlled converter becomes essentially an amplifier with a linear voltage transfer characteristic.

230 PRINCIPLES OF PULSE TIMING CONTROL

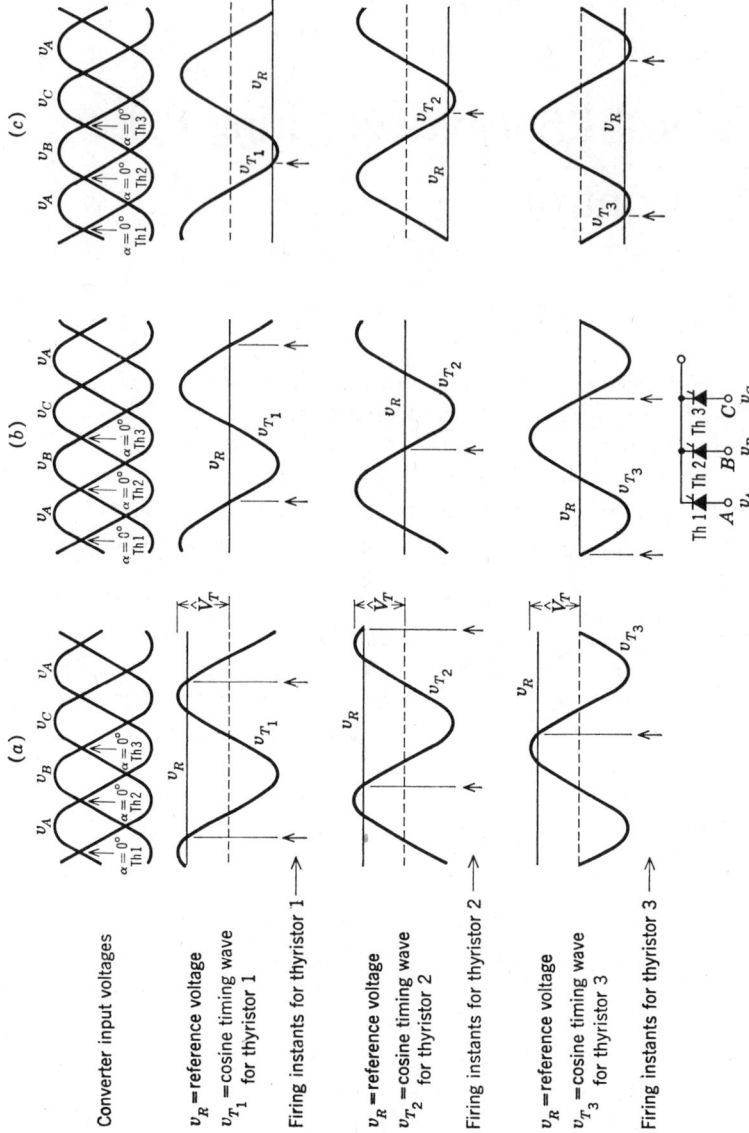

Figure 9.1. Waveforms illustrating the basic principle of the "cosine wave crossing" control method for determining the firing instants of the thyristors of a phase-controlled converter or cycloconverter. (a) $\alpha = 0°$; (b) $\alpha = 90°$; and (c) $\alpha = 150°$. Note: Waveforms are shown for 3-pulse converter, but principle is applicable to circuits of any pulse number.

COSINE WAVE CROSSING PULSE TIMING METHOD

Considering the cycloconverter, it might be expected, intuitively, that with a cosine relationship between the firing angle and the reference voltage, substitution of the d-c "reference" voltage with an alternating sinusoidal "reference" voltage would produce an output voltage waveform with a "mean envelope" corresponding exactly to the input reference voltage. Of course, this expectation would be based on the assumption that the firing angle is capable of responding to the continuously changing level of the reference voltage, as rapidly as is permitted by the natural limitations of the waveform fabrication process of the cycloconverter.

The desired relationship between the firing angle and the analog reference voltage can be realized by means of the "cosine wave crossing" control method. The basic principle, quite simply, is to determine the firing point for each thyristor from the "crossing" point of an associated "cosine timing wave" with the analog reference voltage. The cosine timing wave is derived from, and synchronized to, the converter a-c input voltage, and its phase is such that its peak occurs at the earliest possible commutation angle (i.e., $\alpha = 0°$) of the associated thyristor.

The cosine wave crossing control principle is illustrated by the waveforms of Fig. 9.1. Each firing pulse is initiated at the point at which the associated cosine timing wave becomes instantaneously equal to the reference voltage. That is, when

$$\hat{V}_T \cos \theta_i = v_R \tag{9.1}$$

where $\hat{V}_T$ = peak value of the timing wave

v_R = value of the reference voltage.

By definition, at this instant, θ_i is equal to α:

$$\therefore \quad \hat{V}_T \cos \alpha = v_R$$

$$\therefore \quad \cos \alpha = \frac{v_R}{\hat{V}_T} \tag{9.2}$$

"Self Regulating" Property of the Cosine Wave Crossing Control Method

A natural feature of the cosine wave crossing timing method is that if the amplitudes of the cosine timing waves are permitted to vary in correspondence with variations in the amplitudes of the a-c voltages feeding the converter—which, in practice, arises as a natural result of deriving the former directly from the latter—then, with a constant reference voltage, the mean d-c terminal voltage of the converter theoretically remains constant. The reason

232 PRINCIPLES OF PULSE TIMING CONTROL

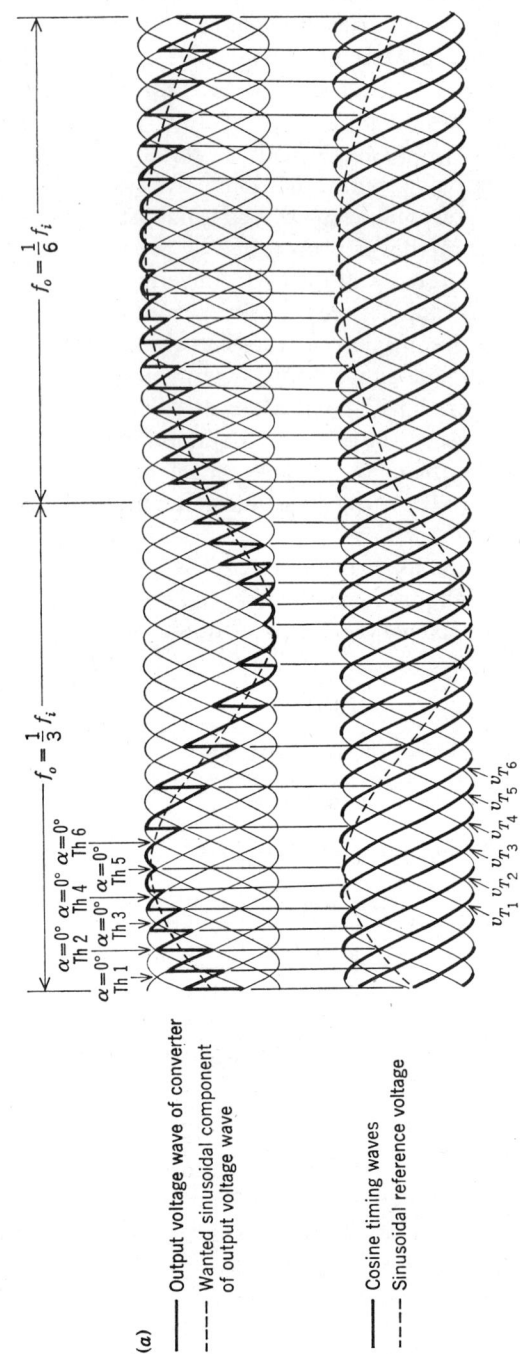

(a)
——— Output voltage wave of converter
- - - - Wanted sinusoidal component of output voltage wave
——— Cosine timing waves
- - - - Sinusoidal reference voltage

COSINE WAVE CROSSING PULSE TIMING METHOD

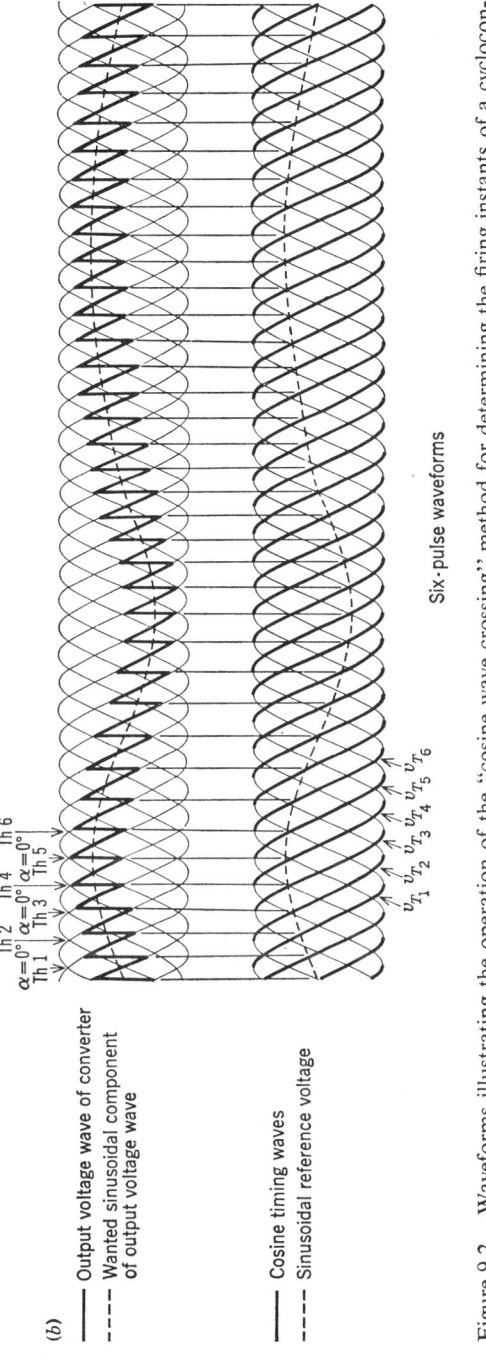

Figure 9.2. Waveforms illustrating the operation of the "cosine wave crossing" method for determining the firing instants of a cycloconverter. (a) 100% output voltage; (b) 50% output voltage.

for this is that any variation in the amplitude of the timing waves, caused by a corresponding amplitude variation in the converter a-c voltages, results in a shift of firing angle which is inherently such as to maintain a constant mean voltage at the d-c terminals. (This, of course, assumes that there is no tendency for the limits of the firing angle control range to be exceeded.)

The Cosine Wave Crossing Control Method Applied to a Cycloconverter

The cosine wave crossing method of controlling the pulse timing produces a firing angle in advance of 90° with a "positive" reference voltage, and a firing angle in retard of 90° with a "negative" reference voltage. Thus, if the reference is an alternating voltage, the result is to produce a "to and fro" phase modulation of the firing angle about the 90° "quiescent" point. The frequency of this modulation is the frequency of the reference voltage, and the depth of the modulation is determined by the amplitude of the reference voltage.

Figure 9.2 shows output voltage waveforms for a 6 pulse 2-quadrant converter, with continuous conduction, obtained with the cosine wave crossing control method. It is clear from these waveforms that the amplitude and frequency of the "mean envelope" of the output voltage waveform correspond with the amplitude and frequency of the reference wave, and thus the general effect can be seen to be that of a linear amplifier. However, although the output waveform has the desired general appearance, it is not easy to judge, from a superficial inspection, whether it is theoretically the "best" waveform which is possible to generate.

In order to examine this point, it is instructive, firstly, to consider, once more, the case of a steady d-c output. In this case, the "correct" firing instants are those which result in each and every "segment" of the ripple voltage wave being exactly the same as any other segment. In other words, the interval between any two firing instants should ideally be the same as between any other two. This is illustrated in Fig. 9.3a, which shows the output voltage waveform of a 6-pulse 2-quadrant converter with a fixed mean output voltage. However, this ideal firing pulse pattern is not unique in producing this mean output voltage. In fact, it is merely one of an infinite variety of possible timing patterns for the firing pulses, which would result in this same mean d-c terminal voltage. In order to illustrate this point, Fig. 9.3b shows that the same mean voltage would result by alternately advancing one pulse and retarding the next. In this case, however, it is evident that the distortion of the output waveform has been increased. Indeed, it can easily be shown that for any pulse timing pattern other than the ideal one of Fig. 9.3a, the distortion of the d-c terminal voltage waveform is increased.

COSINE WAVE CROSSING PULSE TIMING METHOD

(*a*) D-c terminal voltage waveform of 6-pulse converter, with evenly spaced firing pulses

Distortion is minimum possible

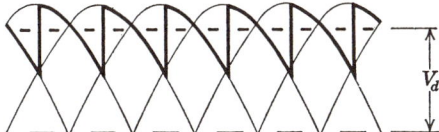

(*b*) D-c terminal voltage waveform of 6-pulse converter, with unevenly spaced firing pulses

Distortion is greater than the minimum possible

Figure 9.3. Waveforms illustrating that with a steady d-c output, the firing pulses of the phase-controlled converter should be evenly spaced for minimum distortion. (*a*) d-c terminal voltage waveform of 6-pulse converter, with evenly spaced firing pulses. Distortion is minimum possible. (*b*) d-c terminal voltage waveform of 6-pulse converter, with unevenly spaced firing pulses. Distortion is greater than the minimum possible.

This phenomenon suggest a criterion for establishing the "natural" firing instants for a cycloconverter. That is to say, the "natural" firing instants for a cycloconverter could be defined as being those which result in the minimum possible overall distortion of the output voltage wave.

It will now be shown that the cosine wave crossing method for determining the timing of the firing pulses has the unique property that it produces the theoretically minimum possible overall rms harmonic distortion of the output voltage wave. As a result of this deduction, it can be concluded that the firing instants determined by this control principle are the "natural" ones for a cycloconverter. (However, as is explained later, for practical reasons some small correction to these "natural" firing instants may often be desirable.)

In order to prove this result, imagine a "hypothetical" control method for a phase controlled cycloconverter, which causes an alternating output voltage waveform to be fabricated in such a way that the actual output waveform always has the nearest possible instantaneous value to a sinusoidal "reference" waveform of the desired output frequency; it being assumed, of course, that the correct firing sequence is always maintained. Such a result could, in fact, be realized by a control scheme which makes a continuous comparison between the sinusoidal reference voltage and the actual a-c input voltage, which is, for the time being, connected to the output terminals, and which makes a second continuous comparison between the reference voltage and the a-c input voltage which is "next due" to be switched to the output terminals. So long as the instantaneous difference between the reference and the actual output voltage is less than the difference which would result if the "next" a-c wave were to take its place at the output terminals, then the "existing" voltage wave is allowed to remain. At the instant at which the voltage wave "next due" at the output terminals starts to come closer to the reference than the "existing" voltage wave, a firing pulse is generated (assuming, of course, that a natural commutation is possible at this time).

The basic operating principle of this "hypothetical" control scheme is illustrated by the waveforms shown in Fig. 9.4. The operation can be described mathematically as follows:

Consider for example, the time period $T1$. (The conclusion reached concerning the operation during this time period can be applied to the operation during any other period of time occurring between two consecutive firing pulses.) The "existing" input voltage, v_1, is retained at the output terminals, so long as it remains "closer" to the reference voltage than the voltage "next due" at the output terminals (i.e., v_2). That is, so long as

$$(v_R - v_1) < (v_2 - v_R)$$

that is,

$$v_R < \frac{v_1 + v_2}{2}$$

A commutation to v_2 occurs when

$$(v_R - v_1) = (v_2 - v_R)$$

that is when

$$v_R = \frac{v_1 + v_2}{2} \tag{9.3}$$

Since v_1 and v_2 are both sinusoids, it is clear that $(v_1 + v_2)/2$, which is the mean of these two waves (shown dotted in Fig. 9.4), is also a sinusoid. Furthermore, the peak of this waveform occurs at the point at which a commutation would take place from v_1 to v_2, when $\alpha = 0°$. In other words, this waveform is exactly equivalent to a "cosine timing wave" associated with thyristor 2. Since, moreover, the commutation is produced at the point at which this wave becomes instantaneously equal to the "reference" wave, it is clear, for given relative levels of "timing" wave and "reference" wave, that this "hypothetical" control method produces exactly the same output voltage waveform as the cosine wave crossing method.

If it is assumed, for the moment, that the actual wanted sinusoidal component of the output waveform, fabricated by this "hypothetical" control method, corresponds exactly with the reference wave, then, since the instantaneous deviation of the actual output waveform from the reference waveform is, by definition, always the minimum possible, it follows that the rms

COSINE WAVE CROSSING PULSE TIMING METHOD

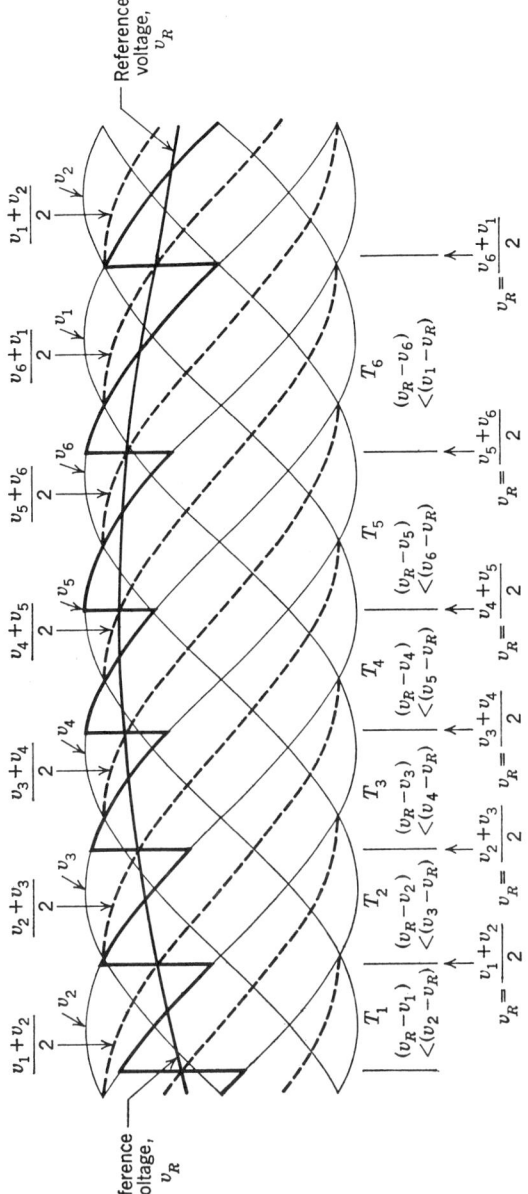

Figure 9.4. The diagram illustrates the basic operating principle of a "hypothetical" firing angle control scheme which produces minimum possible instantaneous deviation of the actual output voltage waveform from the reference waveform, and shows that this produces exactly the same results as the "cosine wave crossing" control principle.

deviation from this waveform must necessarily have the theoretically minimum possible value.

In point of fact, the wanted sinusoidal component of the output voltage waveform which is produced by this "hypothetical" control method has a somewhat larger amplitude than the "reference" voltage which produces it (by a certain constant factor, which depends upon the pulse number of the circuit). However, it will now be shown that the rms deviation from the actual "wanted" sinusoidal component still has the minimum possible theoretical value.

Assume that the actual sinusoidal component having the wanted output frequency is $(\hat{V}_R + \Delta \hat{V}_R)$, where $\hat{V}_R$ is the amplitude of the "reference" wave.

Thus, the instantaneous output voltage is given by

$$v_o = (V_R + \Delta V_R) \sin \theta_o + \sum_0^\infty V_n \sin \theta_n \quad (9.4)$$

↑ Wanted component of output voltage   ↑ Superimposed harmonic components

And the rms value of the deviation from the "reference" voltage wave is

$$V_{\text{dev}} = \frac{1}{\sqrt{2}} \left[(\Delta \hat{V}_R)^2 + \sum_0^\infty \hat{V}_n^2 \right]^{1/2} \quad (9.5)$$

Assume now that an adjustment is made to the timing of the firing pulses, but that the amplitude of the wanted component, that is $(\hat{V}_R + \Delta V_R)$, remains unaltered. The rms value of the deviation from the reference voltage must necessarily be increased, since, by definition, it originally had the minimum possible theoretical value. Since $\Delta \hat{V}_R$ has not changed, it follows that $\sum_0^\infty \hat{V}_n^2$ which, of course, is a measure of the rms distortion of the output voltage, must have increased. It can thus be concluded that the output waveform with the "original" pulse timing control, which corresponds precisely with that produced by a cosine wave crossing control, must have the minimum possible rms distortion.*

Limitations of Open-Loop Pulse Timing for a Cycloconverter

The cosine wave crossing control method is, in itself, an "open-loop" control, inasmuch as its function is to generate the firing instants for the

---

* It should be noted, however, that this conclusion is valid only so long as $\hat{V}_{W\max} \leqslant V_{d\max}$. ($\hat{V}_{W\max}$ could be made greater than $V_{d\max}$, at the expense of the introduction of direct integer multiple harmonic distortion at the output.)

thyristors of the cycloconverter in accordance with a preordained and purely "mechanical" procedure, which is without direct heed to the output voltage waveform actually produced as a result of this process. It has already been seen, in Chapter 7, that an open-loop firing angle control gives rise to unnecessary distortion of the output voltage, in the event that the load current becomes discontinuous. Quite apart from this, however, even if the load current is completely continuous, an open-loop control of the pulse timing in practice may produce certain highly objectionable distortion components at the output, which can give rise to unacceptable system operation. For this reason, some means of closed-loop feedback from the output waveform of the cycloconverter to the pulse timing circuits, the effect of which is to apply a relatively small, but, from a practical point of view, highly significant, correction to the firing instants, thereby suppressing the objectionable distortion terms, is often necessary.

One source of objectionable distortion terms has theoretical origins. As will be seen in Chapter 11, a detailed harmonic analysis of the output voltage waveform of the cycloconverter, produced by the basic open-loop cosine wave crossing control method, leads to the conclusion that distortion components are generated which have subharmonic, or even zero, frequency—except at certain discrete output to input frequency ratios.* Although the amplitudes of these voltage components are relatively small, nonetheless, if the load circuit has a low impedance at subharmonic frequency, then the result is that an excessive amount of subharmonic current flows at the output. This is objectionable, since it comprises an additional and unnecessary load on the system, as well as possibly giving rise to saturation of magnetic components. Thus, for example, with an induction motor load, the presence of small amounts of subharmonic voltage at the output of the cycloconverter could give rise to large subharmonic components of current, which would have a very adverse effect upon the system performance.

Another source of "objectionable" distortion, with an open-loop firing angle control, arises from practical imperfections in the pulse timing circuits, which give rise to errors in the timing of the firing pulses. These pulse timing errors are directly reflected in the output voltage waveform of the cycloconverter, as additional distortion terms. Timing errors commonly arise, for example, as a result of slight phase inaccuracies in the filtering circuits for the cosine timing waveforms. Under some circumstances, small timing errors of this type may be of little practical significance; under other conditions, however, particularly where it is required to operate the cycloconverter with a relatively low output voltage ratio, they may assume a major importance.

* More generally, this is true for any open-loop firing angle control method.

In order to illustrate this, consider an application in which a cycloconverter is required to control the speed of an induction motor over a range of 100:1, by means of frequency control, with a substantially constant voltage to frequency ratio. If the cycloconverter is designed to produce its maximum possible output voltage at maximum speed, then at minimum speed the required output voltage is 1% of maximum. In order to produce this output voltage, the required depth of modulation of the firing angle, about the 90° quiescent point, is approximately 0.60 degree. It is clear that if an "undistorted" output voltage waveform is to be obtained, any errors which exist in the timing of the firing instants, and hence also in the phase position of the filtered cosine timing waveforms, should be much less than 0.6 of a degree. In practice, this is not possible to achieve with any degree of reliability, with an open-loop pulse timing control system. Moreover, even if it was possible, there would still be a problem due to small phase unbalances which might exist between the 3-phase input voltages.

The Use of Feedback for Suppression of "Objectionable" Distortion

It has been seen that the cycloconverter can be regarded as being a linear voltage amplifier—with the realization, of course, that distortion components are inevitably produced, as a result of the basic waveform fabrication process. It has also been seen (with continuous conduction) that the basic open-loop cosine wave crossing control method produces the minimum possible total distortion of the output voltage waveform; albeit at the expense of the production of relatively low amplitude subharmonic distortion components—which, in practice, may have undesirable effects upon the system operation. In addition, small practical pulse timing errors can produce a relatively large "nontheoretical" distortion of the wanted output voltage component, particularly when the cycloconverter operates with a low output voltage ratio.

In either case, since the objectionable voltage distortion components have low amplitude in relation to the maximum possible wanted output voltage, it can be conjectured that only a small amount of judiciously applied correction to the "open-loop" firing instants would be sufficient to eliminate these components.

This leads to the concept of using a negative feedback from the output of the cycloconverter to the analog reference voltage input stage, the purpose of which is to eliminate the "objectionable" distortion terms appearing at the output. Clearly, since the major portion of the distortion present in the output voltage is inherent in the basic mechanism of the cycloconverter,

COSINE WAVE CROSSING PULSE TIMING METHOD

it is necessary for the feedback circuit to be able to make a clear distinction between the "necessary" and the "objectionable" distortion terms. This is because the former inherently cannot be eliminated, and transmission of these components back to the input can in fact lead to an unstable mode of operation, with a consequent deterioration of the output waveform. Thus the control must be such that the firing instants are determined basically by the cosine wave crossing points, with the feedback signal merely providing the necessary "fine" correction to the pulse timing, which eliminates the "objectionable" distortion terms.

Such a control system can be implemented in several ways, and it is possible, depending upon the application, to use either a voltage or a current feedback from the output of the cycloconverter. The latter is sometimes preferred, since the "necessary" distortion of this waveform is usually considerably smaller than that of the voltage, and, in addition, the effects of the "objectionable" voltage distortion terms are more readily measurable in the current waveform.

Figure 9.5 shows, in simplified diagrammatic form, the basic elements of the type of feedback control scheme under discussion. The objectionable distortion terms to be suppressed, through the action of the firing pulse timing circuits, are separated, by means of a suitable filter circuit, and are then added, in a negative sense, along with the sinusoidal reference voltage, as an input to the pulse timing circuits. By suitable design, this type of closed-loop control system can provide a very effective means of suppression of the objectionable distortion terms at the output of the cycloconverter.

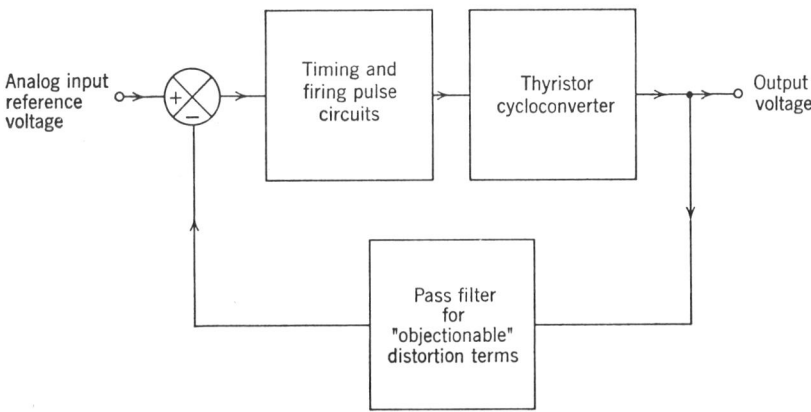

Figure 9.5. Simple diagrammatic representation of a negative feedback control scheme for suppression of "objectionable" harmonic distortion terms at the output of the cycloconverter.

OTHER PULSE TIMING CONTROL PRINCIPLES

The discussion so far has been concerned only with the cosine wave crossing pulse timing control principle, for the reason that this principle is the theoretically correct one on which to base the timing of the firing pulses of the cycloconverter. This is not to say, however, that the cosine wave crossing control method is invariably the best *practical* approach for determining the timing of the firing pulses, either for the cycloconverter, or for the phase-controlled converter producing a variable d-c output. Indeed, alternative pulse timing principles are commonly used, either because of their particular suitability to the application, or because of the relative simplicity of the associated circuitry, or often, simply as a matter of expedience in utilizing already existing "standard" control circuitry.

In the first category of alternative pulse timing control principles, various methods can be classified which use the same basic principle of determining the timing of the firing pulses from the intersection of a reference voltage wave with a set of timing waves synchronized to the input voltages. The essential differences between these various methods reside only in the shape of the timing waves, and in the manner of deriving these waves. Thus, for example, a linear sawtooth timing wave, synchronized to the zero-crossing of the converter input voltage wave, is often used in place of the cosine timing wave. These types of pulse timing control method represent a relatively minor conceptual departure from the cosine wave crossing control method, and will not be discussed further here.

In addition, various alternative pulse timing control methods exist, which are based upon more fundamentally different principles. Two such pulse timing methods are discussed in the following sections.

"Integral Control"

A practical difficulty with the cosine wave crossing control method arises from the presence of distortions or "spikes" appearing on the input voltage waves. These "spikes," which may originate from "external" disturbances on the supply system, or may be a direct result of the commutations of the converter itself, can cause spurious intersections of the timing waves with the reference voltage, thereby giving rise to incorrectly timed firing pulses. In order to overcome this difficulty, the cosine timing waves can be obtained from the converter input voltage waves through filter circuits which remove the voltage spikes and deliver smooth timing waves to the control system. Typically, the timing wave filters might comprise resistance-capacitance circuits, giving a phase shift of 60°. By providing this precise amount of phase shift, the timing waves can be given the correct phase position with respect to the converter input voltages.

For applications in which the input frequency is variable, however, it is difficult to provide adequate filtering and, at the same time, always maintain the desired phase relationship of the timing waves. In this case, a fundamentally different firing pulse timing principle, referred to here as the "integral control" principle, can provide a highly satisfactory basis for the pulse timing control. (This is not to imply, however, that the "integral control" might not also be preferred in some applications in which the supply frequency is fixed.)

The basic principle of the "integral control" method can be explained by considering a simple example in which a 6-pulse 2-quadrant converter operates at a steady firing angle, and produces a steady mean output voltage. A typical output waveform obtained with firing pulses which have perfect "tracking" accuracy is shown in Fig. 9.6a, and in b and c it is shown how this output voltage is comprised of a steady d-c component, with a superimposed a-c ripple component. Examining the a-c ripple voltage waveform, it is seen that during the interval between any two successive firing points, the net voltage-time integral of this wave is zero; in other words, the areas of waveform above and below the zero axis are exactly equal to one another. Thus, if this voltage waveform is applied to the input of an integrating circuit, the output of the integrator would be instantaneously zero at each firing point, as illustrated in Fig. 9.6d.

This phenomenon suggests that a simple timing principle is to generate a firing pulse each time the integral of the ripple voltage waveform becomes instantaneously equal to zero. Since, by definition, the mean output voltage is required to be proportional to the reference voltage, the ripple voltage waveform can be obtained simply by subtracting the reference from the (suitably scaled) actual output waveform. With such a scheme it is ensured that each and every segment of the ripple voltage has zero mean value, and therefore between every two firing points the mean value of the output voltage is equal to the reference voltage. Thus a very tight "pulse-by-pulse" control is exercised over the output voltage waveform, and, in fact, this principle automatically provides a closely regulated closed-loop control of the output voltage.

This control principle has two important features. Firstly, since the firing pulses are generated at the zero values of the integral of the ripple voltage, it is insensitive to changes in the supply frequency. In other words, although the amplitude of the waveform of the integral of the ripple voltage changes with changing supply frequency, its zero values always correspond with the desired firing instants. Secondly, any "spikes" which appear on the output voltage waveform of the converter do not have any immediate or drastic effect upon

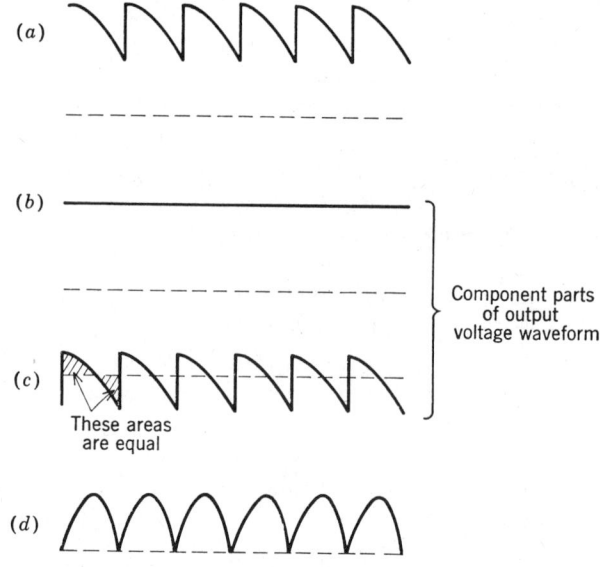

Figure 9.6. Waveforms illustrating the basic principle of the "integral control" pulse timing method. (a) Output voltage waveform of converter; (b) direct voltage component; (c) a-c ripple voltage component; (d) integrated ripple voltage waveform.

the timing of the firing pulses, since the integral value of the output ripple voltage is hardly influenced by these. This may not necessarily be the case with other types of pulse timing control, in which the timing of the firing pulses is determined from the instantaneous intersection of a timing waveform with the reference voltage. For this latter type of control, distortions appearing on the timing waveforms can conceivably give rise to grossly mistimed firing pulses. With the "integral control," on the other hand, the effect of spikes and "commutation notches" is simply to "compensate" the waveform of the integral of the ripple voltage, so that the timing of the firing pulses is adjusted just sufficiently to maintain the desired mean output voltage.

In practice, although the basic "integral control" principle described is theoretically feasible if it is required to produce a steady d-c output, it is not satisfactory, by itself, for producing an alternating output voltage. For an a-c output, it can be shown mathematically that although this basic pulse timing control may operate apparently satisfactorily for a number of successive firing points, eventually the time intervals between consecutive firing pulses become more and more irregular, until finally the control is completely lost. Moreover, even with a steady d-c output, this basic "integral control" shows a tendency, under some conditions, to settle into an asymmetrical operating mode, with the timing of the firing pulses occurring at irregular intervals.

Thus, in order to make the integral control principle function satisfactorily in practice, it is necessary to complement it with a stabilizing feedback which tends to create an "even" spacing between the firing pulses. This can be achieved, for example, by deriving an analog voltage signal which is a measure of the difference between the elapsed time since the last pulse, and the time between the last pulse and the last-but-one pulse. This "correction" signal is added, in the proper sense, to the integral of the ripple voltage, and the firing pulse is generated when the sum of these signals becomes instantaneously zero.

Phase-Locked Oscillator

Another technique for timing the firing pulses is based upon the simple fact that under steady state conditions (with a steady d-c output), the pulses for successive thyristors are produced at evenly spaced intervals of time. Thus, the time period between any two consecutive commutations is equal to the period of the input voltage wave, divided by the pulse number of the converter.

In theory, then, for any given steady converter firing angle, the firing instants could be timed from an independent "clock pulse" oscillator, so long as this precisely maintains the desired frequency and phase. This is illustrated by the waveforms of Fig. 9.7a. The waveforms of Fig. 9.7b and c demonstrate that an increase in the clock frequency above the synchronous value results in a steadily increasing level of output voltage, by virtue of the fact that each successive firing point is relatively more advanced than the previous one, whereas a decrease in the clock frequency results in a steadily decreasing level of output voltage, because, in this case, each successive firing point is relatively more retarded than the previous one.

Of course, in practice it is not possible to devise an independently free-running oscillator which precisely maintains the desired frequency; thus a basic open-loop pulse timing method of this type is not practicable, since it would inevitably produce a continuous drifting of the converter firing angle. Fortunately, however, the scheme can be transformed into an elegant practical control method, simply by using a negative feedback loop to "lock" the oscillator, so that its frequency and phase are forced to correspond with the desired conditions at the output of the converter. This can be achieved, for example, by designing the clock pulse oscillator so that its frequency is controlled in accordance with, say, a direct voltage. If, in turn, this voltage

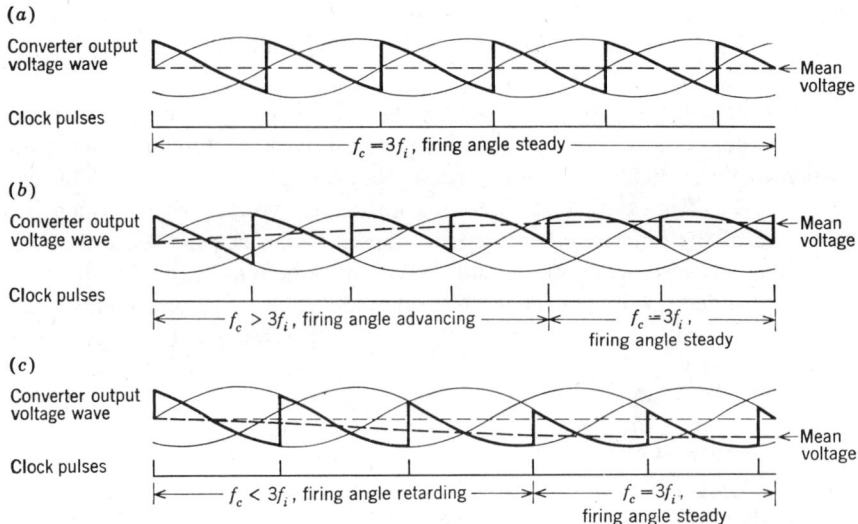

Figure 9.7. Waveforms for a 3-pulse converter demonstrating the basic principle of the phase-locked oscillator pulse timing method. (a) With $f_c = f_i/3$, the converter firing angle is steady; (b) with $f_c > f_i/3$, the converter firing angle progressively advances; (c) with $f_c < f_i/3$, the converter firing angle progressively retards. f_c = Clock frequency = frequency at which firing pulses are delivered to converter.

is obtained from the error between a reference value, and the desired converter output parameter to be controlled, then the action of the complete control loop is automatically such as to lock the oscillator to the desired firing angle.

For example, assume that it is required to provide a closed-loop control of the converter output voltage. In this case, the input voltage to the clock pulse oscillator would be the amplified error between a reference voltage and the converter output voltage. Let it be assumed that nominally zero voltage at the input of the voltage controlled oscillator gives the desired "synchronous" clock frequency, and that a positive input voltage gives an increased clock frequency, and a negative voltage gives a reduced clock frequency. Under steady state conditions, then, the error voltage is nominally zero, and the oscillator clock frequency is precisely such as to maintain the required steady converter firing angle. That this is so can be seen by considering, for example, the effect of a small increase in the clock frequency, due to, say, a small drift in the voltage to frequency transfer characteristic of the oscillator. The initial effect of an increasing clock frequency is to advance the converter firing angle, and hence to increase the output voltage. Thus a negative error signal is produced at the input of the oscillator, and this automatically corrects the clock frequency back to the synchronous value. By making the gain of

OTHER PULSE TIMING CONTROL PRINCIPLES 247

the error amplifier sufficiently high, virtually any desired accuracy of regulation of the output voltage can be obtained with such a scheme.

In virtually all practical applications a closed-loop control of one or other of the converter output parameters is required. Thus the fact that a closed-loop feedback is an inherent ingredient of the phase-locked oscillator pulse timing method is not a practical limitation; on the contrary, one of the basic features of this pulse timing principle is its elegance and relative simplicity.

Chapter Ten

Firing Pulse Generator Circuit Principles

The function of the firing pulse generator is to deliver correctly timed, properly shaped, firing pulses to the gates of the thyristors in the power converter. Almost invariably, the phase of the firing pulses, relative to the converter input voltage, is controlled in accordance with an analog "reference" signal, typically by means of one or other of the pulse timing principles discussed in Chapter 9.

Numerous approaches to the design of the firing pulse generator circuit are possible, and it is not the intention here to present an exhaustive treatment of this topic. Rather, a general view is presented, on the level of simplified functional schematic diagrams, of some of the circuit techniques most commonly used in the firing pulse generator. A discussion is also presented of the firing angle "end stop" control problem; and various circuit techniques for implementing the end-stop control are discussed.

ASSUMPTIONS AND SCOPE OF DISCUSSION

Generally, as a matter of convenience, the discussion in this chapter is related to firing pulse generators for the 3-pulse commutating group, shown in Fig. 10.1. The circuit techniques under discussion are, however, generally applicable to firing pulse generators of higher (or lower) pulse number.

It is also generally assumed that the design of the firing pulse generator is compatible with the production of "extended" firing pulses. As here defined, an "extended" firing pulse is one whose duration spans the complete period up to the point at which the next thyristor in succession, in the same commutating group, is fired. In practice, under most circumstances, only the leading edge of the firing pulse actually is used to fire the thyristor, and the remainder of the pulse is redundant. However, in order to ensure correct operation of the power converter under all circumstances, the use of an extended firing pulse is desirable.

SOME FUNCTIONAL SCHEMES 249

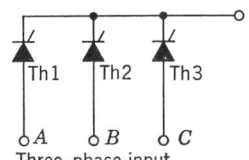

Figure 10.1. Three-pulse thyristor commutating group.   Three-phase input

SOME FUNCTIONAL SCHEMES

Schemes Using Cosine Wave Crossing Control

INDIVIDUAL PULSE TIMING COMPARATORS. A functional diagram of a 3-pulse firing pulse generator, using the cosine wave crossing control principle, along with associated waveforms, is illustrated in Fig. 10.2. This scheme basically consists of three identical channels, one for each firing pulse. In each channel, a cosine timing wave, obtained through a filter and a transformer from the converter input voltage, is applied to one input terminal of a comparator; the analog reference voltage is applied to the other input terminal of the comparator. The output voltage of the comparator changes level at the intersection point of the cosine timing wave with the reference voltage, and this produces a corresponding "clock pulse" at the output of the associated clock pulse generator. The clock pulse "sets" the associated pulse output flip-flop, thereby initiating a firing pulse in this output channel. The pulse output flip-flop is reset by the clock pulse from the following channel. By this means, the width of the firing pulse is automatically adjusted to cover the complete time period between consecutive firing instants, independently of the firing angle.

TIMING WAVE MULTIPLEXING. A practical simplification of the scheme just described results from the realization of the fact that it is not necessary to continuously compare each cosine timing voltage waveform with the reference voltage. In fact, there is no need to commence the sampling of any given timing waveform until the point at which the preceding timing waveform intersects the reference voltage. By the same token, there is no further need, for the time being, to sample a given timing waveform once its "crossing point" has been reached, and its associated pulse initiated. Thus, considering a 3-pulse circuit, once the "1" firing pulse has been initiated, further sampling of timing waveform v_{T_1} during the intervening period up to the point when the "3" pulse has been initiated, is quite unnecessary, since it is never required to reinitiate the "1" pulse before the "2" and "3" pulses have occurred. Thus, at any given time, it is necessary to sample only that timing waveform which will next initiate an output pulse.

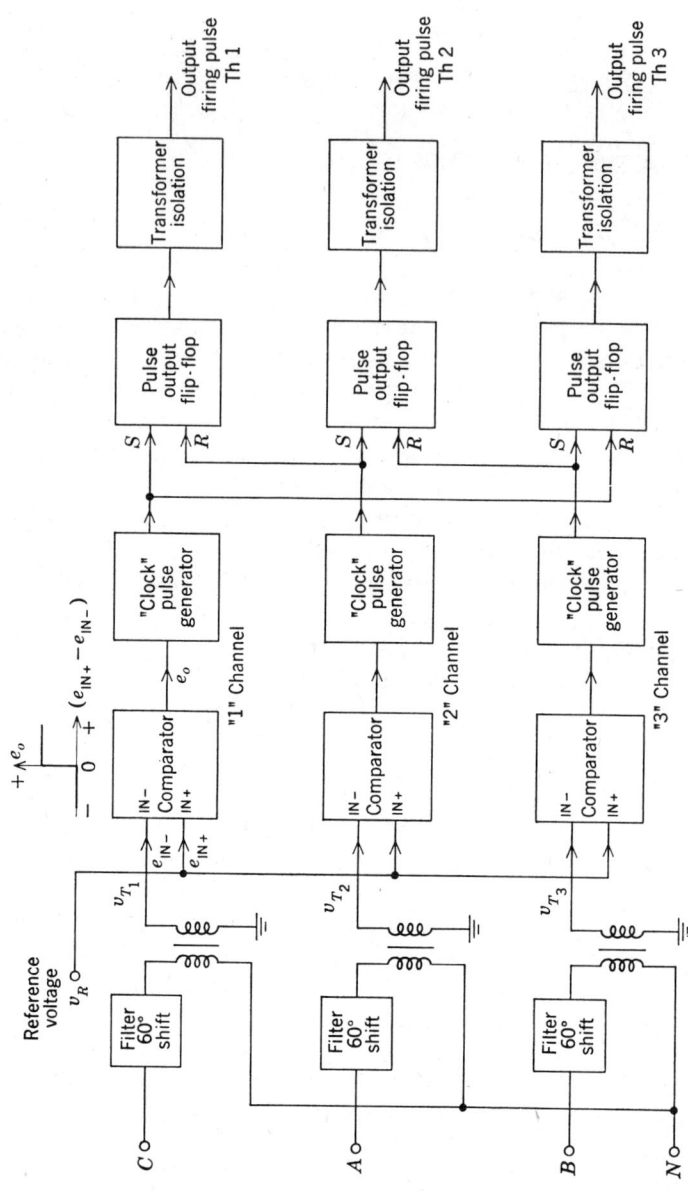

SOME FUNCTIONAL SCHEMES 251

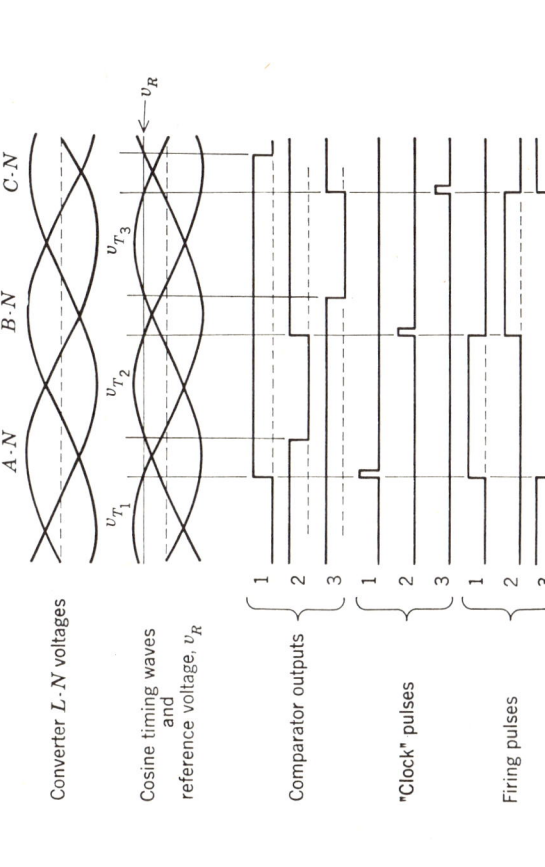

Figure 10.2. Simplified functional diagram of a firing pulse generator using the cosine wave crossing pulse timing control principle, and associated waveforms.

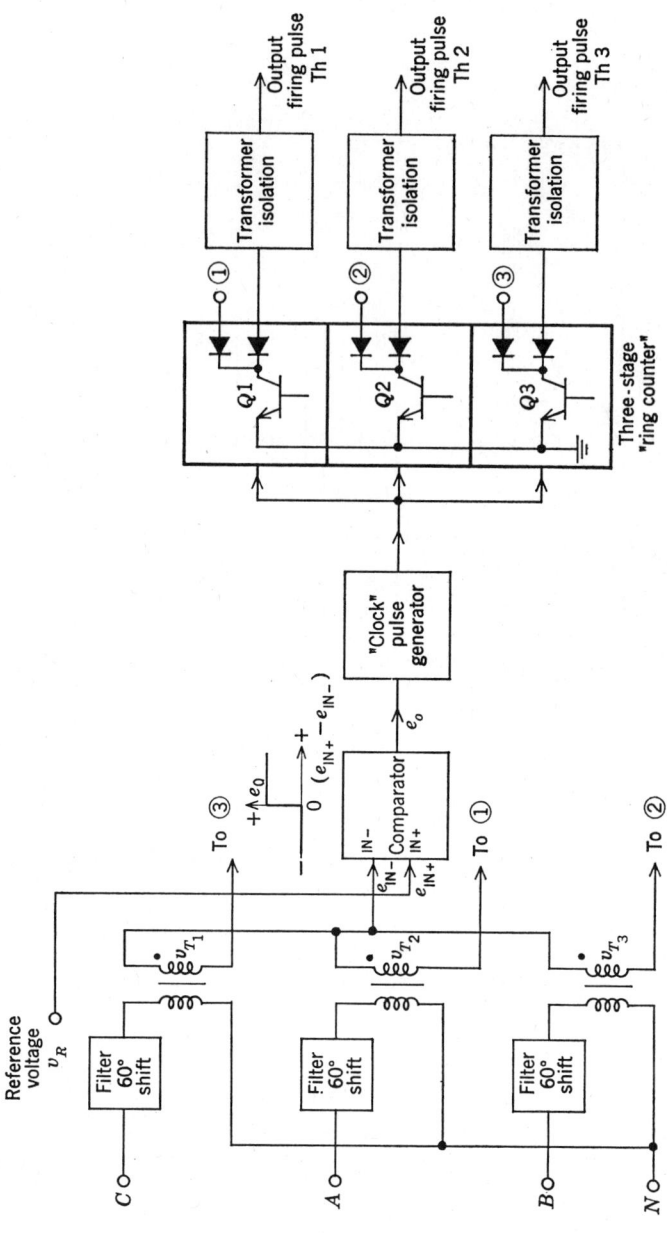

SOME FUNCTIONAL SCHEMES 253

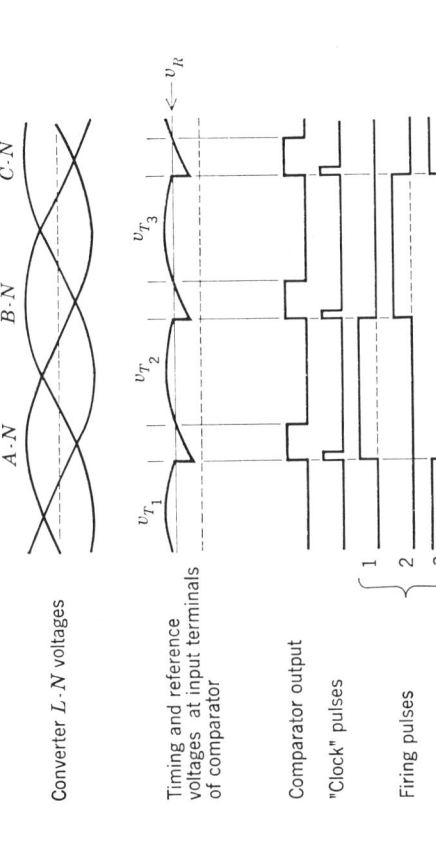

Figure 10.3. Simplified functional diagram of a firing pulse generator using the cosine wave crossing pulse timing control principle, with timing wave multiplexing, and associated waveforms.

It is evident that if each timing waveform is sampled only during the necessary period—which can be accomplished by means of a multiplexing arrangement which automatically selects the timing voltage waves in sequence, one after the other—then a single comparator can be used for the timing of all of the firing pulses. Assuming that this is done, so that the output signal of the comparator produces a "clock" pulse at each firing instant, then, for a 3-pulse circuit, these pulses occur at $3 \times$ the line frequency (assuming a steady firing angle), and it is necessary to translate them into firing pulses which occur in rotation in the 3 output channels. This can be achieved by means of feeding the clock pulses as the trigger input to a 3-stage "ring counter" circuit, the purpose of which is simultaneously to distribute and shape the firing pulses. The design of this "ring counter" circuit is such that only one of the 3 stages is ON at any given time. Each successive clock pulse changes the state of the circuit, so that the conduction states of the 3 stages occur in regular sequence, one after the other. Thus, for example, when the "1" stage is ON, the "2" and "3" stages are OFF, and a firing pulse is delivered to thyristor 1. The next clock pulse to be generated switches the "2" stage ON, and the "1" stage OFF, thus delivering a firing pulse to the thyristor 2; and so on.

Considering, now, the means for switching the timing waveforms to the comparator in the correct sequence: since the period of a given output pulse corresponds exactly with the period during which it is required to sample the next timing waveform, it is possible (but not essential) to use the *same* switching device in the ring counter circuit to perform the dual function of shaping the final drive pulse, and of connecting the next timing waveform to the input of the comparator. The resulting scheme, along with typical associated waveforms, is shown in diagrammatic form in Fig. 10.3.

Scheme Using "Integral Control"

A functional diagram of a 3-pulse firing pulse generator, using the "integral control" principle (shown, for simplicity, without the stabilizing circuit required to prevent irregular pulse timing—as discussed in Chapter 9), along with associated theoretical waveforms, is illustrated in Fig. 10.4.

The difference between the reference voltage, and a suitably scaled portion of the raw output voltage wave of the converter, is fed as the input to an integrator. The waveform of this difference voltage is a replica of the ripple voltage appearing at the output terminals of the converter, and therefore the output voltage of the integrator is the integral of the converter ripple voltage, and this has a zero-value at each firing point. This zero-value is translated into a timing clock pulse, through the action of the comparator and the clock pulse generator. The clock pulses are fed as the trigger input to a 3-stage "ring counter," the function of which is both to shape and distribute the output firing pulses. Each successive clock pulse changes the state of this circuit, so that the conduction states of the 3 stages occur in regular sequence, one after the other.

It will be appreciated that this firing angle control scheme provides a closed-loop control of the output voltage of the converter. Moreover, since the d-c gain of the integrator is theoretically infinite, the steady state error between the reference voltage and the mean output voltage is theoretically zero (within the normal range of operation), regardless of the amplitude of the a-c terminal voltage of the converter.

As an alternative to using the voltage waveform appearing at the output terminals of the power converter as a basis for comparison with the reference voltage, it is possible to construct, at a low level, within the firing pulse generator circuit itself, a replica of the output voltage wave of the converter, and to use this simulated output waveform as the "feedback" signal to the input of the integrator. The simulated output voltage wave can be conveniently constructed by connecting the suitably transformed a-c input voltages of the power converter to the switches of the ring-counter circuit, as shown by the dotted circuitry in Fig. 10.4. Since the conduction periods of these switches are in exact coincidence with the conduction periods of the thyristors of the power converter (assuming that this is in continuous conduction), therefore the voltage waveform appearing at point P is a replica of the converter output voltage waveform. This arrangement has the feature that the firing pulse generator circuit, by itself, comprises a self-contained functional unit, which is capable of operating quite independently of the power converter.

Scheme Using Phase-Locked Oscillator

A functional diagram of a 3-pulse firing pulse generator, using a phase-locked oscillator for determining the timing of the firing pulses, together with associated waveforms, is illustrated in Fig. 10.5.

The difference between the reference voltage and an analog feedback voltage, representing the converter output parameter under control (voltage, current, machine speed, etc.), is amplified, and applied as the frequency-controlling input signal to the voltage controlled oscillator. This oscillator comprises an integrator, a level comparator, and a reset circuit. The input voltage to the integrator is the sum of a fixed "bias" voltage, E, and the output voltage from the error amplifier. The bias voltage E is set so that with no output from the error amplifier, the time taken for the output voltage of the integrator

256 FIRING PULSE GENERATOR CIRCUIT PRINCIPLES

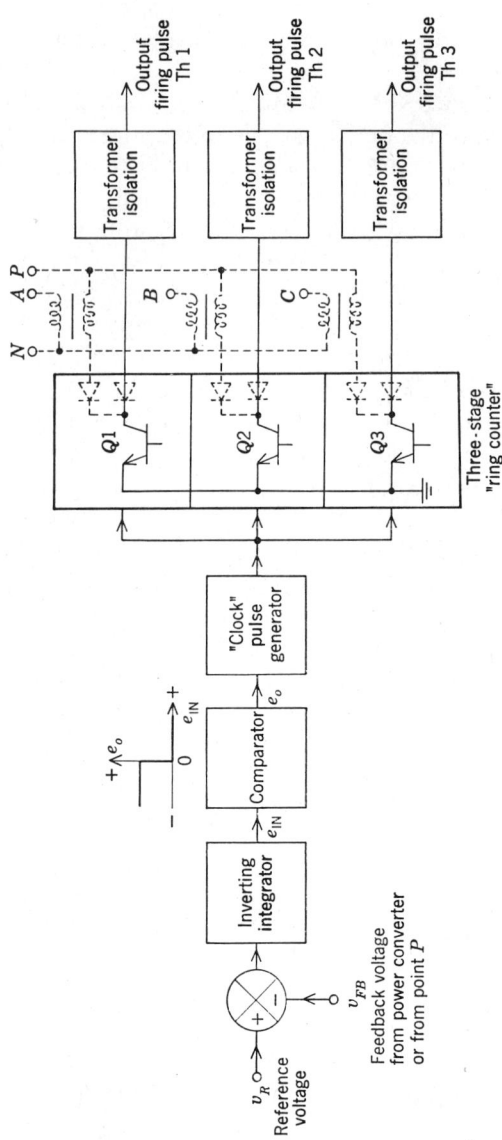

SOME FUNCTIONAL SCHEMES 257

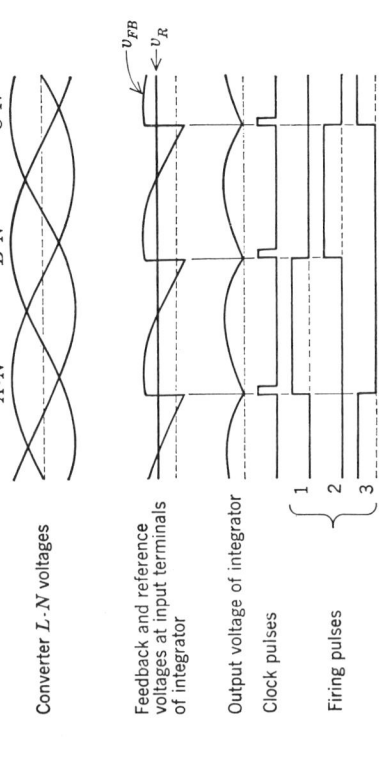

Figure 10.4. Simplified diagram of a firing pulse generator using the "integral control" pulse timing principle, and associated waveforms.

258 FIRING PULSE GENERATOR CIRCUIT PRINCIPLES

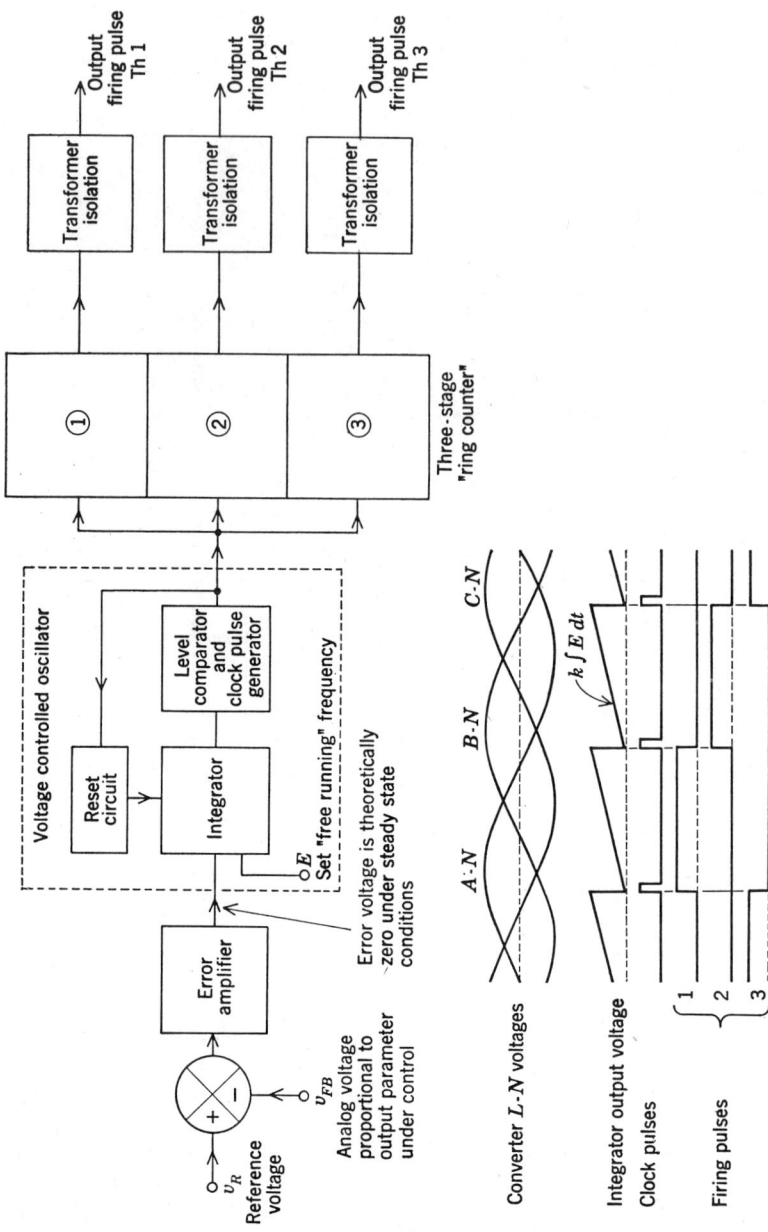

Figure 10.5. Simplified functional diagram of a firing pulse generator using the phase-locked oscillator pulse timing control principle, and associated waveforms.

to rise from zero, to the switching level of the comparator, is nominally $\frac{1}{3}$ of a cycle (for this 3-pulse circuit). At the point at which the output voltage of the integrator reaches the switching level of the comparator, a clock-pulse is produced. This clock-pulse is fed as the trigger input to the 3-stage "ring-counter" circuit, the function of which is both to shape and distribute the output firing pulses. Each successive clock pulse changes the state of this circuit, so that the conduction states of the 3 stages occurs in regular sequence, one after the other. The clock pulses are also fed to the integrator reset circuit, so that the output voltage of the integrator is reset to zero at each firing pulse.

It is evident that with zero output voltage from the error amplifier, the free-running frequency of the voltage controlled oscillator is nominally 3 × the frequency of the supply; further, a positive error voltage results in an increased frequency, and hence a progressively advancing converter firing angle, whereas a negative error voltage results in a reduced frequency, and hence a progressively retarding converter firing angle. Thus it can be seen that the action of this complete closed-loop scheme is such as to automatically lock the frequency and phase of the oscillator, so that it produces the precise firing angle necessary to maintain the desired value of the converter output parameter under control.

END-STOP CONTROL

So far, it has been tacitly assumed that the operation of the timing circuit is such that the firing pulses are initiated at some point within the desired firing angle control range of the converter; that is to say, that the firing angle lies between 0° and some limiting angle slightly less than 180°. In practice, there may be a tendency, under some conditions, for the firing angle to be "pushed" beyond the limits of the permissible control range. In the absence of an overriding "end-stop" control, this can give rise to missing or ineffective firing pulses, with consequent commutation failures in the power converter; generally, the result is to produce an intolerable, if not catastrophic, mode of system operation.

In practice, then, it is essential to incorporate some means within the firing pulse timing circuit for ensuring that the firing angle is not permitted to exceed the boundaries of the "safe" control range. Various methods for achieving this result are discussed in the following sections.

Simple Method Using Reference Voltage Clamp

For the cosine wave crossing, and similar, firing angle control circuits, a simple method for ensuring that the firing angle lies within the desired range is to set a limit on the possible range of variaiton of the analog reference voltage as "seen" by the timing comparator, so that this always lies inside the peak-to-peak levels of the timing waveform. Thus it is ensured that a crossing point between the reference and timing voltages is always obtained, and hence there can be no missing firing pulses. At the same time, an insurance against commutation failures can be achieved by appropriately setting the limiting level of the reference voltage in the inverting region of operation. A simple way of achieving this result is to clamp the positive and negative reference voltage levels, at the comparator input terminal, with, say, a pair of inversely connected zener diodes.

Although satisfactory for some applications, tihs simple method of providing a firing angle end-stop has its limitations, due to the fact that any drift in the clamping voltage level, as well as in the voltage offset of the comparator (due to temperature changes, etc.), as well as any variations in the amplitudes of the timing waves, are reflected as corresponding changes in the boundaries of the firing angle control range. Of course, it is feasible to devise quite sophisticated clamping circuits which are, at least partially, self-compensating against such effects. Nonetheless, due to the inherent tendency for drift in analog circuitry of this type, this end-stop technique is generally satisfactory only for applications in which it is permissible to set the nominal boundaries of the control range within a safe margin of the theoretical limits.

Methods Using Time-Dependent Information

A more sophisticated approach to the problem of providing the necessary end-stop control is to derive, from the prevailing waveshapes in the power converter, up-to-date time-dependent digital information, which defines the earliest and latest firing instants for each thyristor. This information is superimposed onto the normal pulse timing control mechanism, in such a way that the production of pulses outside of the permissible control range, as well as missing pulses, is automatically prevented. At the same time, the operation of the normal pulse timing control is unaffected, so long as the firing pulses are produced within the boundaries of the permissible range of firing angle control.

The following discussion of this general approach to the end-stop control is subdivided into various topics. Firstly, the definition of the permissible range of firing angle control, and how this is related to the prevailing waveshapes in the power converter, are discussed. Secondly, methods of deriving the required time-dependent digital information from the power converter are considered. And thirdly, a general functional method of applying this information to provide the desired end-stop control, is presented.

END-STOP CONTROL

THEORETICAL LIMITS OF THE FIRING ANGLE CONTROL RANGE. The basic function of the end-stop control is to generate a firing pulse at the boundary of the required firing angle control range, in the event that the normal pulse timing control mechanism tends to produce a firing pulse outside of this range.

In practice, the required range of firing angle control is determined by the requirements of the particular application, and often it is the case that the boundaries of the control range lie well within the theoretically permitted limits. It may also be the case that the range of firing angle control is restricted by practical limitations in the pulse timing circuit itself; for example, with the cosine wave crossing control, the minimum firing angle attainable is usually greater than $0°$, due to the practical difficulty of detecting a crossover point in the "flat topped" peak region of the cosine timing wave.

Such questions, however, depend upon the specific circumstances, and are not of a fundamental nature. The more basic question to be considered here is: what are the theoretical limits of the firing angle control range beyond which the proper "symmetrical" operation of the power converter cannot be obtained.

So far, it has been assumed, for a 2- (or 4-) quadrant converter, that the theoretical limits of the firing angle control range are, in the rectifying region of operation, $0°$, and, in the inverting region, almost $180°$. This assumption was based upon the premise that the commutation of current from one thyristor to the next takes place instantaneously. Whereas this simplifying assumption is useful, and valid, so far as a general understanding of the basic operating principles of the phase-controlled converter or cycloconverter is concerned, as will be seen, it is not valid when considering, in detail, the question of the theoretical limiting firing angle for the *inverting* region of operation.

In order to consider this question further, it is necessary to draw upon some of the subject matter of Chapter 13. As is explained there, with inductance connected between the a-c source voltage and the converter input terminals (which, to a lesser or greater extent, is always present in practice), the process of commutating the current from one thyristor to the next occupies a finite period of time. During this commutation "overlap" time, both the outgoing and incoming thyristors are in conduction. Thus, the point at which reverse voltage is applied across the outgoing thyristor does not correspond with the point at which the firing pulse is applied to the incoming thyristor, but it is delayed by the commutation overlap time. Furthermore, this overlap time subtracts directly from the time available for reverse biasing of the thyristor. Thus, for a given firing angle, the sum of the "commutation overlap" and "reverse bias" times is constant; and the longer is the overlap time, the shorter is the reverse bias time.

The overlap time is a function of the voltage available for commutating

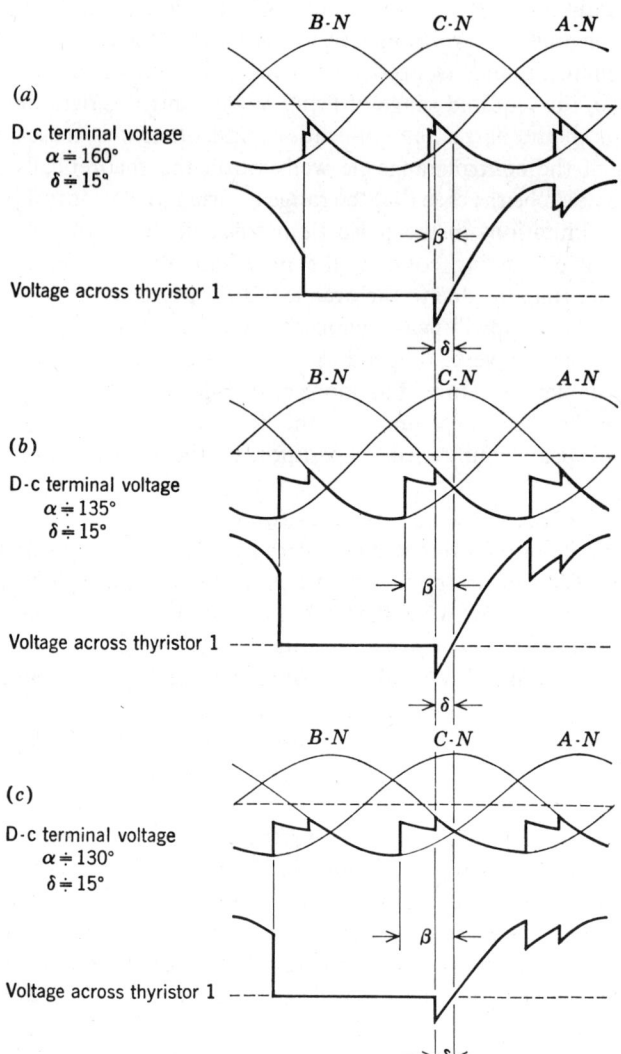

Figure 10.6. Waveforms illustrating the dependence of the limiting inversion end-stop firing angle upon the amplitudes of the current and the a-c voltages. (a) "Light" load—"normal" amplitude of a-c voltage; (b) "full" load—"normal" amplitude of a-c voltage; (c) "full" load—"low" amplitude of a-c voltage.

the current from one line inductance to the next, the amplitude of the current to be commutated, and the value of the source inductance itself. Thus, if the limiting inversion end-stop firing angle is defined as being such as to result in a fixed reverse bias time for the thyristors (in other words, it is the latest possible firing point at which the operation of the converter can be maintained, without running the risk of commutation failures), then it is clear that this firing angle is not rigidly fixed, but it is a function of the prevailing amplitude of the current to be commutated, and of the voltage at the a-c terminals of the converter.

The waveforms of Fig. 10.6 illustrate the dependence of the inversion end-stop firing point upon the amplitudes of the voltage and current. For each of the three cases illustrated, the thyristor recovery angle, δ, is the same. It is seen that in order to maintain this fixed recovery angle, it is necessary to advance the limiting inversion end-stop firing angle, both with increasing current, and with decreasing a-c input voltage.

In conclusion, then, whereas the limiting firing angle for the rectifying mode of operation has a fixed angular position with respect to the input voltage wave (i.e., $\alpha` = 0°$), irrespective of the amplitude of the voltage or current, the limiting firing angle for the inverting mode of operation is not fixed, but depends upon the value of the input source inductance, the amplitude of the converter a-c voltage, and the current to be commutated.

METHODS OF DETERMINING END-STOP FIRING POINTS FROM THE CONVERTER WAVEFORMS. For the rectifying region of operation, the limiting position of the end-stop firing pulse can be easily determined from the converter a-c voltage waveforms. The earliest possible point in the input cycle at which application of the firing pulse to the thyristor has a fruitful result, is the point at which the anode voltage first becomes positive. (Although, actually, a somewhat earlier firing pulse is permissible, so long as the duration of the pulse is sufficient that the thyristor turns on as soon as its anode does become positive.) Thus the limiting rectification end-stop firing angle can be determined simply by detecting the zero-crossing of the appropriate line-to-line voltage, as illustrated by the schematic diagram of Fig. 10.7.

Determination of the timing of the limiting inversion end-stop firing pulse is more complicated, especially if it is required to provide automatic adjustment of the pulse timing in correspondence with the prevailing amplitudes of the voltage and current, so as to result in a fixed recovery time, or a fixed recovery angle, for the thyristors. The basic requirement is to make a prediction of the firing instant which, once the ensuing commutation overlap period has expired, leaves the desired period of reverse bias for the outgoing thyristor. Since this involves a prediction of future events, it is inherently so that no method of inversion end-stop control can be completely infallible.

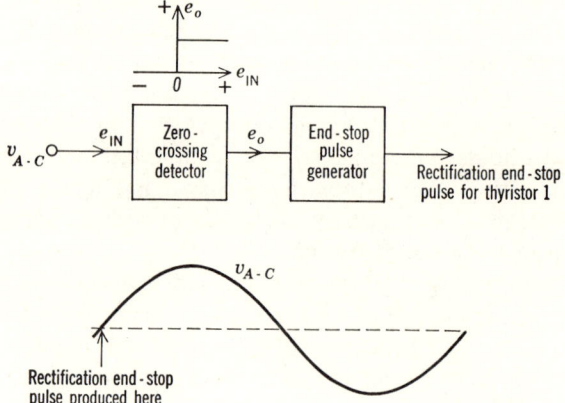

Figure 10.7. Simplified functional diagram of circuit for determining the position of the rectification end-stop firing pulse.

At best, the determination of the timing of the firing pulse can be based upon information obtained from immediately preceding voltage and current waveforms, on the assumption that the waveforms which ensue once the firing pulse has been delivered, will follow the normally predictable course of events. If, once the firing pulse has been generated, the waveforms in the power converter depart appreciably from the predicted ones—as might be caused, for example, by a sudden "dip" in the incoming voltage, or a sudden increase of load—then it is quite possible for a commutation failure to occur.

In practice, the design and complexity of the circuitry required to determine the timing of the inversion end-stop firing pulse depends upon a variety of considerations, such as how close to the theoretical limiting end-stop firing angle it is desired to operate the converter, whether it is required to operate close to this limit at all levels of load, or only under certain extreme conditions, how wide is the range of variation of the current, and of the supply voltage and frequency, the susceptibility of the supply voltage to sudden erratic distortions, the undesirability of occasional commutation failures, and so on.

Since many considerations are involved, many variations are possible in the circuit design techniques for determining the timing of the inversion end-stop firing pulses. The following discussion, although by no means exhaustive, should serve to provide an acquaintance with some of the teechniques commonly used.

Figure 10.8 shows, in schematic form, a circuit which determines the timing of the inversion end-stop firing pulse so as to result in a fixed thyristor recovery angle, regardless of the amplitude of the current or the a-c voltage

END-STOP CONTROL

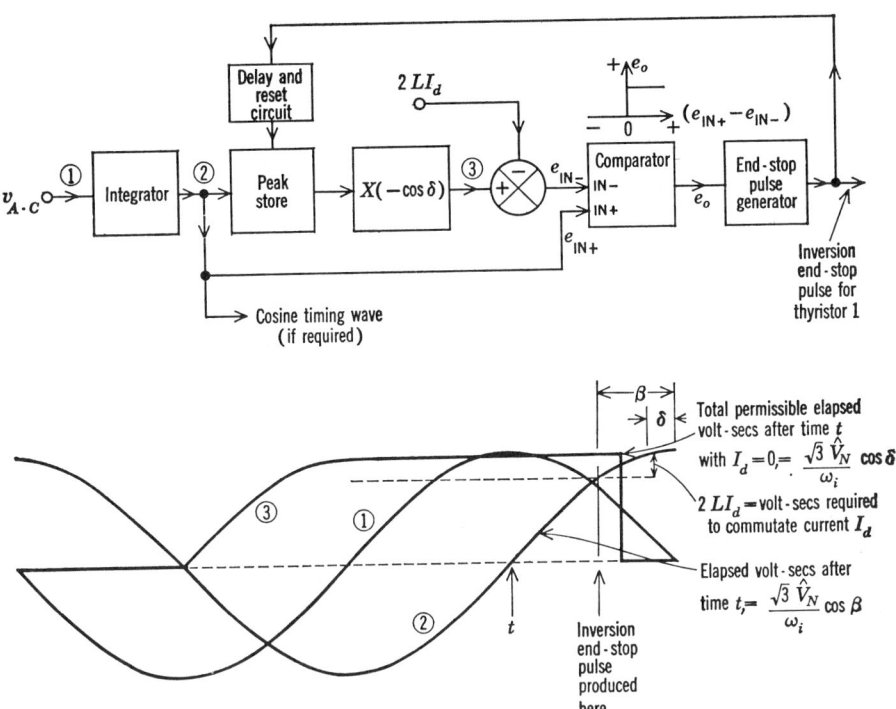

Figure 10.8. Simplified functional diagram of a circuit for determining the position of the inversion end-stop firing pulse so as to produce a constant recovery angle δ, and associated waveforms.

(within certain limits). The principle of this circuit is to compute the timing of the end-stop firing pulse so that the necessary volt-second integral of the line-to-line voltage, required to effect commutation of the current, remains between the point of delivery of the pulse, and the point at which commutation of the current is required to be completed.

The appropriate line-to-line commutating voltage, (1), is integrated, to yield the voltage wave (2). From time t onwards, this latter waveform is a measure of the elapsed volt-second integral of the commutating voltage. If this cmmoutating voltage can be assumed to be a sinusoid having a known peak value, then the amplitude of the elapsed volt-second integral of this wave at any given point in the $\frac{1}{4}$-cycle is predictable, and hence, by implication, the amplitude of the volt-second integral remaining for commutation of the current, is also predictable.

The output waveform of the peak store circuit, multiplied by $-\cos \delta$ [waveform (3)], represents the predicted level of the elapsed volt-second integral of the commutating voltage at angle δ in advance of the commutating voltage zero-crossing. (This information actually is based upon the previous peak value of the integral of the commutating voltage, rather than upon the actual peak value of the commutating voltage at time t. It would be possible, instead, to store this latter voltage, thus providing slightly more up-to-date information; however, the complexity of the resulting circuit is somewhat greater, especially if the input frequency is variable.) Thus the intersection point of waveform (3) with waveform (2) represents the desired position of the end-stop firing pulse, for the condition when zero volt-second integral is required for commutating the current; that is to say, when $I_d = 0$. A voltage representing the required volt-second integral, $2LI_d$, for commutation of the current, obtained from a feedback signal representing the converter current, is subtracted from waveform (3), and the intersection of the resulting wave, with waveform (2), represents the predicted position of the end-stop firing pulse, necessary to provide a recovery angle δ, after the completion of the commutation overlap period.

In mathematical terms, the function of this circuit is to determine the timing of the inversion end-stop firing pulse in accordance with the following equation:

$$\sqrt{3}\,\frac{\hat{V}_N}{\omega_i}\cos\delta - 2LI_d = \sqrt{3}\,\frac{\hat{V}_N}{\omega_i}\cos\beta \qquad (10.1)$$

- $\sqrt{3}\,\frac{\hat{V}_N}{\omega_i}\cos\delta$: Elapsed volt-second integral at angle δ in advance of voltage zero-crossing
- $2LI_d$: Volt-second integral required to commutate current I_d
- $\sqrt{3}\,\frac{\hat{V}_N}{\omega_i}\cos\beta$: Elapsed volt-second integral at angle β in advance of voltage zero-crossing

The derivation of this equation is presented in Chapter 13.

It should be appreciated, of course, that this functional scheme relies for its operation upon the validity of the assumption that the commutating voltage is sinusoidal; furthermore, it also relies upon the assumption that sudden drastic changes of voltage and current do not take place during subcycle periods (other than, of course, the "commutation notches" themselves).

It should be pointed out, also, that the commutating voltage connected as the input to this circuit is the a-c source voltage "behind" the "line" inductance. In practice, this voltage may not be immediately physically available; however, it can always be "reconstructed," at a low level, from the voltage waveform appearing at the a-c terminals of the converter, by means of circuitry which constructs and "reinstates" the "notches" removed from the line-to-line voltage by the commutations in the power converter.

In many applications it is not actually required to operate the converter at the limiting inversion end-stop firing angle, at least not over a wide range of current and voltage, and in this event, the complication and expense associated with the type of circuit of Fig. 10.8 is unjustified.

A less sophisticated end-stop pulse position detecting scheme, for a 3-pulse converter, is shown in Fig. 10.9. The basic principle is to determine both the rectification and inversion end-stop positions from the instantaneous level of the line-to-line commutating voltage itself. For each thyristor, a digital waveform is produced which represents the permissible range of firing angle control. The earliest permissible firing point is determined from the zero-crossing of the appropriate line-to-line commutating voltage, and the latest permissible firing point is determined from the point at which the commutating voltage falls to a given level. In the particular scheme shown, the level of the commutating voltage which determines the latest permissible firing point is the sum of a fixed level, plus a level proportional to the load current.

This timing principle for the inversion end-stop firing pulse does not, of course, result in a theoretically constant recovery angle, with varying voltage and current. However, at least the position of the inversion end-stop firing pulse moves in the correct direction as the amplitudes of the voltage and current change. This method of end-stop position sensing is quite satisfactory, so long as it is not required to operate within too close limits of the theoretical limiting angle, at any rate not over a wide range of conditions.

The approach can be further simplified—at the expense of a correspondingly coarser end-stop position control—by omitting the compensating signal proportional to the current, and simply determining the inversion end-stop position from the point at which the commutating voltage reaches a given fixed level.

APPLICATION OF THE END-STOP INFORMATION TO THE PULSE TIMING CONTROL. Once the logic information which defines the permissible firing angle control range has been derived—by whatever method—it is necessary to apply this information, in the pulse timing control circuit, in such a manner that the production of firing pulses outside of the permissible range is automatically prevented; at the same time, the normal operation of the pulse timing control must not be interfered with, so long as the firing pulses are produced within the boundaries of the permissible control range.

In practice, many variations are possible in the circuit techniques for applying the end-stop information, to achieve the desired result. The intention here is to provide an insight into the type of circuit approach which may be used, by considering just one basic functional circuit for providing the required end-stop control. The particular scheme considered is applicable to each of the firing pulse generator arrangements of Figs. 10.3 through 10.5, and is therefore of general interest.

The method is based upon the use of logic signals, for each thyristor, which define the permissible range of control of firing angle, and take the

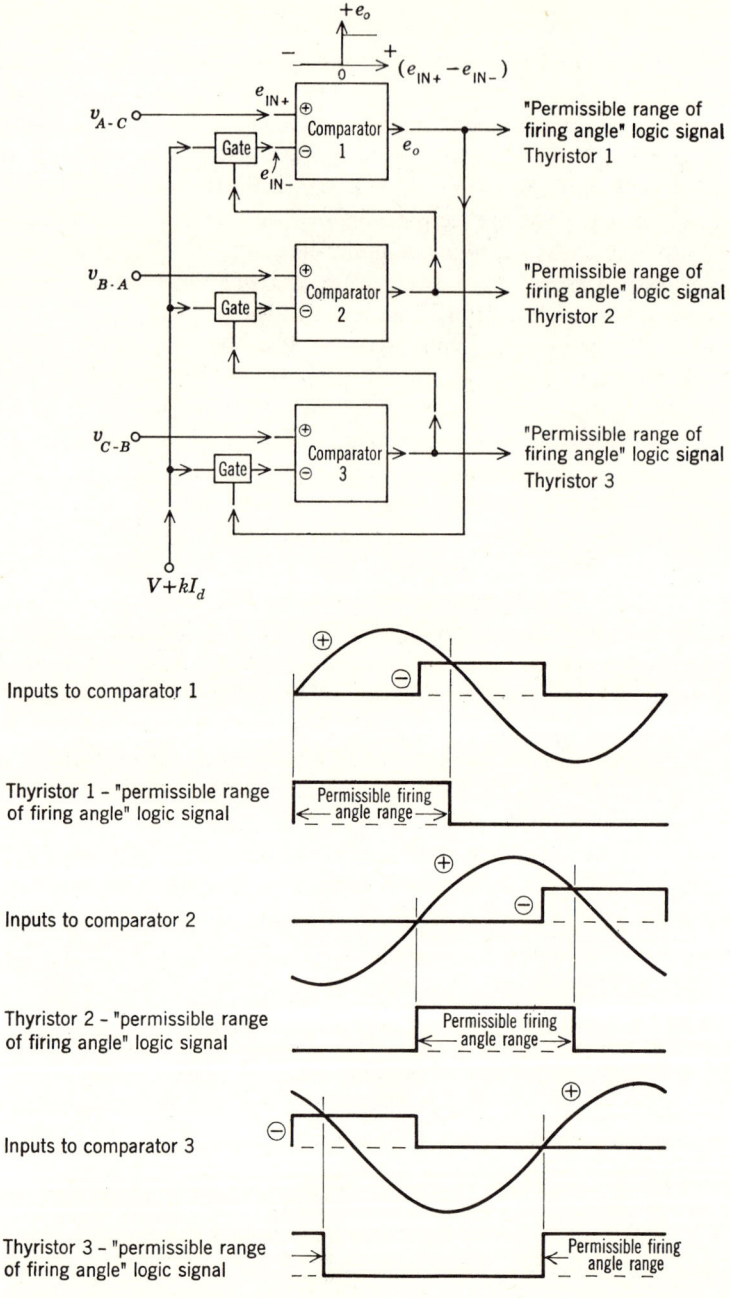

Figure 10.9. Simplified functional diagram of a circuit for producing logic signals representing the "permissible range of firing angle," in accordance with the instantaneous level of voltage available for commutation, and associated waveforms.

form shown in Fig. 10.9. These signals, in conjunction with other signals generated in the pulse timing circuits, are used to control the opening and closing of logic gates connected at the input to the clock pulse generator, which determines the timing of the firing pulses.

A complete functional end-stop control scheme, for a 3-pulse circuit, is shown in Fig. 10.10. The part of the scheme labeled "rectification end-stop" consists of three 4-input AND gates, one associated with each thyristor, which are interposed between the timing comparator and the clock pulse generator. The output signal of the timing comparator switches from a "0" to a "1" state at the desired instant of initiation of each firing pulse (this is the case in each of the functional schemes of Figs. 10.3 through 10.5), and this

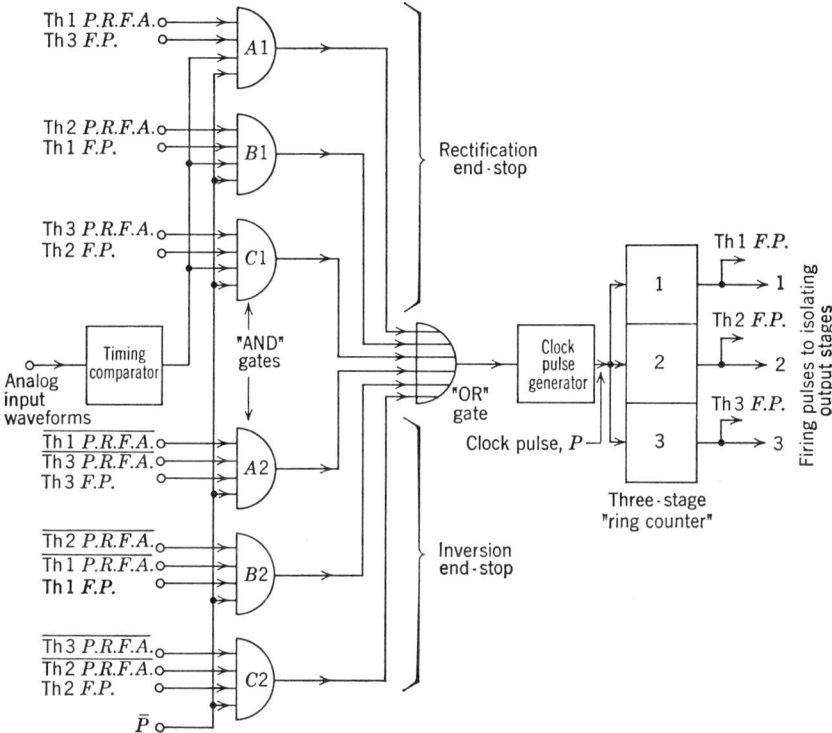

Figure 10.10. Simplified functional end-stop control scheme. Th 1.P.R.F.A. = Logic signal representing the permissible range of firing angle of thyristor 1; Th 2.P.R.F.A. = logic signal representing the permissible range of firing angle of thyristor 2; Th 3.P.R.F.A. = logic signal representing the permissible range of firing angle of thyristor 3; Th 1.F.P. = logic signal representing the firing pulse for thyristor 1; Th 2.F.P. = logic signal representing the firing pulse for thyristor 2; Th 3.F.P. = logic signal representing the firing pulse for thyristor 3.

signal is directed, through the 3 AND gates, to the input of the clock pulse generator.

Consider, for example, the conditions just prior to the initiation of a firing pulse for thyristor 1, somewhere within its permissible range of firing angle control. At this time, the Th 1.P.R.F.A. ("thyristor 1, permissible range of firing angle") logic signal has a "1" value; also, just prior to initiation of firing pulse 1, it is necessarily so that firing pulse 3 is in existence, and hence the Th 3.F.P. ("thyristor 3 firing pulse") logic signal has a "1" value; also, the $\bar{P}$ signal (which is the complement to the clock pulse) has a "1" value, since there is no clock pulse at this time. Hence, at the point at which the output signal of the comparator switches from "0" to "1", the output signal of AND gate A1 assumes a "1" value, a clock pulse is initiated, and the firing pulse is delivered to thyristor 1.

Depending upon the particular firing angle, and the timing mechanism of the particular firing pulse generator, the output signal of the comparator may or may not return immediately to a "0" value at this time. If it does not, then it is necessarily the case that it will do so before the Th 2.P.R.F.A. signal assumes a "1" value, and hence there is no further possibility for a new clock pulse to be generated until the earliest permissible firing instant for thyristor 2 is reached. (This can be confirmed from a detailed study of the operating waveforms of the circuits of Figs. 10.3 and 10.4—for the circuit of Fig. 10.5, this consideration does not apply.) If the firing angle is such that the output signal of the comparator does return immediately to zero as soon as the clock pulse has been initiated, then, in any case, there is no further possibility for another clock pulse to be initiated until, once again, the output of the comparator assumes a "1" value, which occurs at the point at which the firing instant for thyristor 2 is reached. (The presence of the $\bar{P}$ signal at the inputs of each of the AND gates positively prevents pulse "race" conditions, which might possibly otherwise exist at the firing instants.)

The foregoing discussion illustrates that the normal operation of the pulse timing control, within the permissible firing angle control range, is unaffected by the superimposed "end-stop" control circuit.

Now consider the operation of the system when there is a tendency for the normal pulse timing control mechanism to initiate an early firing pulse. Under this condition, the output signal of the timing comparator already has a "1" value at the instant at which the appropriate "permissible range of firing angle" logic signal changes from a "0" to a "1" value. (This can be confirmed from an examination of the operation of each of the pulse timing schemes of Figs. 10.3 through 10.5.) Thus, the firing pulse is now automatically initiated at the earliest permissible point.

This is further explained by considering, for example, the conditions during the period immediately preceding the permissible range of firing angle for thyristor 1. Assuming that there is already a tendency for the firing pulse to be applied to this thyristor, then the output signal from the timing comparator, as well as the Th 3.F.P. logic signal, both have a "1" value, and thus as soon as the Th 1.P.R.F.A. logic signal assumes a "1" value, the output signal of AND gate A1 also assumes a "1" value, and a firing pulse for thyristor 1 is initiated. The next clock pulse is produced as soon as the Th 2.P.R.F.A. signal assumes a "1" value, and so on—so long as the tendency continues for the normal pulse timing control mechanism to initiate early firing pulses.

The part of the scheme labeled "inversion end-stop" has no effect upon the operation of the system, so long as the firing pulses are initiated in advance of the latest permissible firing point. If, however, the firing pulse for any thyristor has not been initiated by the time this limiting point in the cycle is reached, then the output signal of the appropriate AND gate immediately assumes a "1" value, thereby initiating the inversion end-stop firing pulse. For example, assume that the Th 3.F.P. logic signal still has a "1" value at the point at which the $\overline{\text{Th 1.P.R.F.A.}}$ logic signal changes from 0 to 1—that is to say, at the latest permissible firing point for thyristor 1. At this point, the output signal of AND gate A2 changes from 0 to 1 (since the $\overline{\text{Th 3.P.R.F.A.}}$ signal already has a "1" value), and thus a firing pulse for thyristor 1 is initiated immediately.

The reason for the connection of the logic signal $\bar{P}$ to the inputs of the inversion end-stop AND gates is to prevent pulse "race" conditions which might otherwise exist at the leading edges of the P.R.F.A. logic signals.

COMPLEMENTARY FIRING PULSE GENERATORS FOR DUAL-CONVERTERS

As has been seen, the dual-converter or cycloconverter consists of a pair of complementary "positive" and "negative" converter banks. This arrangement basically requires 2 firing pulse generator circuits, one for each converter, and (with continuous conduction) it is necessary to control the two converter firing angles so that their sum is always substantially 180°.

In some applications, in which there is never a requirement for simultaneous application of firing pulses to both converters, it is feasible, for the sake of circuit economy, to use a single pulse timing circuit for both converters, and to switch the firing pulses between two separate sets of output circuits, at the same time reversing the polarity of the analog reference voltage which controls the firing angle, in accordance with the direction of flow of the load current.

It is more common, however, to use essentially separate firing pulse generator circuits for the positive and negative converters. The question arises, then, as to how to control the firing angles of the two circuits, in response to a common input reference voltage, so that their sum is always 180°.

One possible approach is to use two identical, but completely independent, firing pulse generator circuits, the "ground" rails of which are "floating" with respect to one another, and to connect the pair of "reference" voltage input terminals of one circuit in inverse parallel with those of the other. This approach, however, gives rise to certain inconveniences, inasmuch as it is generally desirable to use one common "ground" rail for the complete control system.

With two identical firing pulse generator circuits, with a common ground rail, the required firing angle control characteristics for the two converters can be obtained by applying equal and opposite input reference voltages to the two timing circuits. This can be achieved by inverting the input reference voltage to say, the "negative" firing circuit. At the same time, for a given 3-pulse dual converter group, it is necessary to displace the a-c synchronizing voltages for the two firing circuits by 180°, with respect to one another.

An alternative, and more elegant method, which permits the same analog reference voltage to be directly applied to both the positive and negative timing circuits, is simply to interchange the connections to the input terminals of the timing comparator of the "negative" firing pulse generator. This simple approach eliminates the requirement for inverting the reference voltage for the "negative" firing pulse generator; furthermore, the same a-c synchronizing voltages, with no-phase displacement, can be used for both circuits.

THE PULSE ISOLATING OUTPUT STAGE

In most power converter circuits, differences of potential exist between the gates of the various thyristors, as well as between the thyristors and the control circuit, and therefore it is usually necessary for the output channels of the firing pulse generator to be isolated from one another, as well as from the control circuit itself. This requires the use of pulse output transformers, connected as the isolating links between the thyristors and the firing pulse generator circuit. By providing several isolated secondaries on the pulse transformer, it is possible to provide simultaneous firing pulses for several thyristors. This feature is useful in applications where thyristors are connected in series, or in parallel, with one another, or, for example, where it is required to simultaneously fire pairs of thyristors located in different circuit positions.

The design of the pulse transformer itself presents some problems, because this transformer is usually required to exhibit properties which, from the design viewpoint, tend to conflict with one another. Thus, one important

THE PULSE ISOLATING OUTPUT STAGE 273

requirement is that the coupling between the primary and secondary windings should be as "tight" as possible, in order to keep the leakage inductance low, and thereby ensure that the rise-time of the output pulse is not significantly impaired by the output transformer itself. A second, conflicting, requirement is that the insulation level between the windings should be relatively high (typically up to several kilovolts), in order that the differences of potential between the windings, applied as a result of the normal converter circuit operation, can be safely withstood.

It is not the intention here, however, to examine the detailed aspects of the design of the pulse transformer itself. Rather, various different circuit arrangements of the pulse isolating output stage, which are commonly used, and their relative merits, will be briefly discussed.

Simple Output Stage

The simplest isolating output circuit arrangement comprises a pulse transformer, connected directly to the collector of an output drive transistor, as illustrated in Fig. 10.11. When the transistor is switched into saturation, the voltage V is applied across the primary of the pulse transformer; this causes a corresponding voltage pulse to appear across the secondary of the transformer, which is connected (usually in conjunction with certain circuit components) to the gate of the thyristor to be fired. When the transistor is switched OFF, the magnetizing current flowing in the transformer primary is diverted into the diode, D; as a result, the voltage across the winding reverses, and the flux in the transformer core is reset by this "backswing" voltage, which corresponds to the forward voltage drop of the diode. (Actually, D may not necessarily be a simple diode, and it is possible to clamp the "backswing" voltage to any desired level.)

This simple circuit, although quite commonly used, has certain drawbacks. First, the magnetizing current drawn by the transformer is inherently

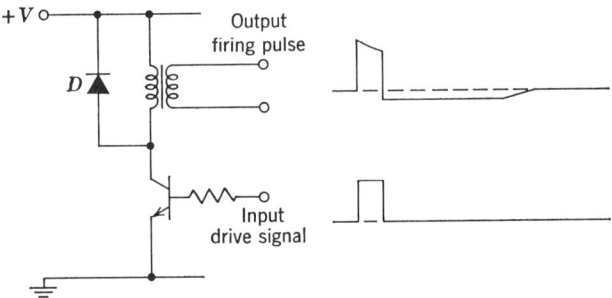

Figure 10.11. Simple firing pulse isolating output stage.

unidirectional; hence the flux excursion in the core of the transformer is also unidirectional, and the utilization of the core is poor. Second, as a result of the direct component of current, it is generally necessary to use a gapped core; this results in a relatively large magnetizing current, and this has to be carried by the switching transistor, in addition to the useful "output" component of current. Because of these considerations, in order to keep the physical size of the pulse transformer within reasonable limits, this simple type of pulse isolating output stage is generally suitable only for delivering relatively short-duration firing pulses (say 25–50 μsec). Thus, for example, this simple circuit would not be well suited to delivering extended 120° firing pulses to the thyristors of a converter operating from a 60 Hz supply.

Blocking Oscillator Output Stage

The performance of the simple pulse output circuit just considered can be improved by the addition of another winding to the pulse transformer, which is connected, through a suitable coupling circuit, so as to provide a regenerative feedback drive current to the base of the output transistor. This is illustrated by the simplified circuit shown in Fig. 10.12.

This is a so-called "blocking oscillator," and its operation is such that it free-runs indefinitely, producing a train of relatively short-duration output pulses, as illustrated, so long as the input drive signal is maintained. Typically, the duration of each pulse might be 30 μsec, with a time between pulses of 150 μsec.

The main feature of this type of pulse output stage is that it is capable of providing a "pseudo-extended" firing pulse, yet it uses relatively simple circuitry, with a small output transformer. Thus, whereas the size of the pulse

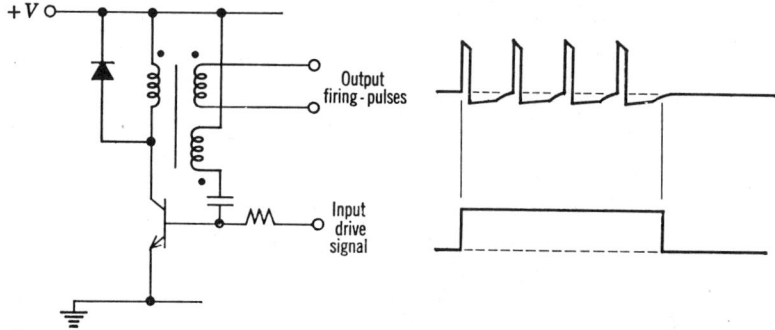

Figure 10.12. Simplified diagrammatic representation of a blocking oscillator pulse isolating output stage.

transformer is compatible only with the duration of each of the individual pulses within the train, nonetheless, for some purposes, the effective width of the output pulse corresponds to the duration of the complete train of pulses.

In order to appreciate this point, it should be remembered that the purpose of the extended firing pulse is to allow the thyristor to pick up anode current at any point within the period of the firing pulse. This may be necessary if the system operation is such that the anode voltage of the thyristor is not yet positive at the point of initiation of the firing pulse, or, if the voltage momentarily goes negative shortly after the firing pulse has been initiated (due, possibly, to the commutations of another converter connected to the same line); or, if the converter current is discontinuous, and it is required to fire the thyristor more than once within the period of the extended pulse.

For the former purpose, provided that each pulse within the train is capable, by itself, of properly firing the thyristor, then the application of an extended train of firing pulses may be virtually equivalent to the application of a continuous firing pulse. Thus, whereas with a continuous pulse, the thyristor switches on as soon as its anode voltage becomes positive, with a train of pulses, on the other hand, there is a maximum delay before firing takes place corresponding to the duration between two successive pulses within the train. Typically, this delay may correspond to about 3° of a 60 Hz wave; for most practical purposes, this is relatively negligible.

For operation with discontinuous current, on the other hand, this type of pseudo-extended pulse may not be satisfactory. Thus, for example, for a 6-pulse bridge converter with discontinuous conduction, it is necessary to fire each thyristor twice during each input cycle, at 60° intervals, and, at each firing instant, it is necessary for two thyristors to be simultaneously pulsed. Thus, the use of 6 separately timed free-running blocking oscillator pulse output stages would not be satisfactory, since there is no guarantee that the individual pulses generated by successive stages "align" with one another.

It is mentioned also that, because of the regenerative nature of the blocking oscillator, this type of pulse output stage often proves to be somewhat susceptible to "stray" noise pick-up phenomena. This can cause spurious triggering of the oscillator, and hence, also, the production of spurious firing pulses.

Output Stages Using Square-Wave Carrier Oscillators

The basic principle of the blocking oscillator pulse output stage just discussed is to deliver a pseudo-extended firing pulse output, through a relatively small output transformer, by means of exciting this transformer with a high "carrier frequency."

A more sophisticated approach to producing an extended firing pulse through a small pulse transformer, by means of a "carrier frequency" technique, is to excite the transformer with a high frequency square wave of voltage. This square voltage wave is rectified at the secondary of the pulse transformer, to provide a virtually perfectly continuous output firing pulse. Because the firing pulse is now virtually completely continuous, this approach does not suffer from the shortcomings of the pseudo-extended firing pulse of the blocking oscillator.

INDIVIDUAL OSCILLATORS. Figure 10.13 illustrates, in simplified schematic form, a firing pulse output stage using a square-wave "carrier" oscillator to deliver an extended firing pulse to the gate of the thyristor, through a high-frequency pulse transformer. Typically, the operating frequency of the "carrier" oscillator might be 20 kHz, or higher.

The circuit is arranged so that the oscillator is switched ON and OFF, in accordance with the state of the gating input signal. The final output firing pulse is, for all practical purposes, completely continuous, and its duration corresponds precisely with the duration of the gating signal applied to the oscillator.

MASTER OSCILLATOR AND INDIVIDUAL GATING TRANSISTORS. An alternative arrangement is to use a single continuously running "master" oscillator for all firing pulse output channels. The output of this oscillator is switched to the various pulse output transformers, in sequence, in accordance with the

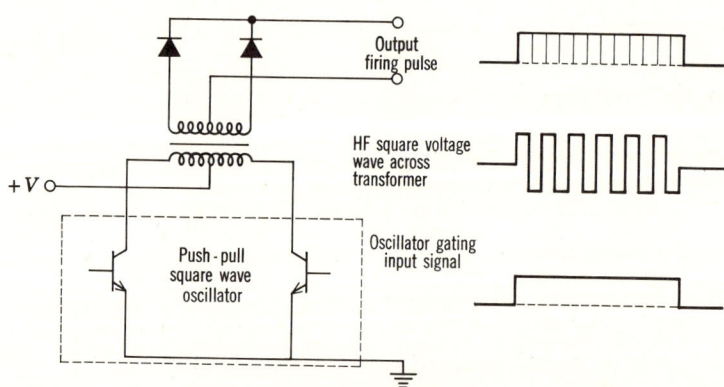

Figure 10.13. Simplified diagrammatic representation of an isolating output stage using a HF "carrier" oscillator to deliver an extended firing pulse.

THE PULSE ISOLATING OUTPUT STAGE

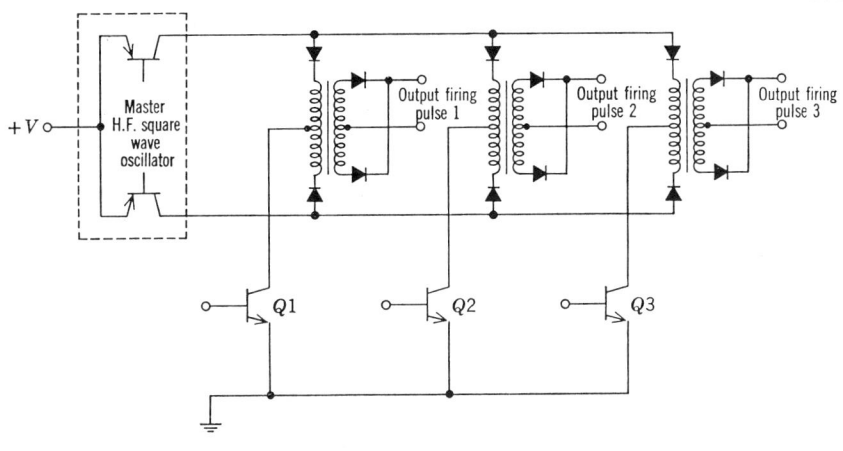

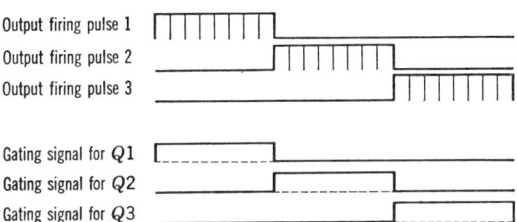

Figure 10.14. Simplified diagrammatic representation of a scheme for providing extended output firing pulses, using a single "master" carrier oscillator for all output channels.

required pattern and timing of firing pulses at the output. The arrangement is illustrated, in simplified diagrammatic form, in Fig. 10.14.

Because this scheme uses only one oscillator to supply all the output channels, the practical circuitry is generally simpler and less expensive than that required for the individual oscillator scheme of Fig. 10.13. Moreover, this scheme lends itself particularly well to the types of firing pulse generator circuit discussed in this chapter, which use ring-counter circuits for distributing the firing pulses in sequence around the output channels. Thus the gating transistors of Fig. 10.14 actually can be the transistors of the ring-counter distributor circuit itself. For dual converters and cycloconverters, in which normally the facility for inhibiting the firing pulses is required, this can be achieved simply by inhibiting the master oscillator. This avoids disturbing the sequential operation of the ring distributor circuits, which are left in normal operation at all times, regardless of whether or not the output firing pulses are being delivered.

Chapter Eleven

Harmonic Analysis of the Output Voltage of the Cycloconverter

A knowledge of the harmonic* content of the output voltage is a necessary factor in determining the filtering requirements for any particular application, as well as in assessing the limits of performance of the phase-controlled cycloconverter.

In this chapter, a theoretical harmonic analysis is presented of the output voltage of the cycloconverter. Comprehensive quantitative data is given which defines the harmonic distortion of the output voltage of 3-, 6-, and 12-pulse cycloconverter circuits, for a wide range of conditions.

SCOPE OF THE ANALYSIS

In Chapters 6 and 7, it has been seen that the cycloconverter may be operated either with no circulating current, or with a relatively small amount of circulating current, or possibly, under special circumstances, with a continuous circulating current. Since the waveshape of the output voltage depends upon whether one or both converters are in conduction at one time, the

---

* As will be seen, the frequencies of the distortion components present in the output voltage of the cycloconverter generally are not integer multiples of the wanted output frequency. Thus, according to the strict definition of the word "harmonic," the distortion components generally are not strictly "harmonics" of the wanted output frequency. In this book, however, the term "harmonic," when applied to the output voltage of the cycloconverter, is used in a broad sense to apply to all the distortion components present, regardless of their frequency relationship to the wanted component. By the same token, the term "subharmonic" frequency is taken to mean any frequency below the wanted output frequency; and the term "superharmonic" frequency is taken to mean any frequency above the wanted output frequency.

THE BASIC PROBLEM, AND THE APPROACH USED

harmonic content of the output voltage is dependent upon the mode of operation.

Under most practical conditions, the cycloconverter is operated with little or no circulating current, and therefore a harmonic analysis for the circulating current-free mode is of most general practical interest. Accordingly, the analysis presented in this chapter is concerned primarily with this mode of operation of the cycloconverter.

As a by-product of this main analysis, analytical data for the special case of operation with a continuous circulating current are also obtained. As has been explained already, the disadvantages of this latter mode of operation limit its use in practice to special circumstances only.

The analysis and quantitative data presented are applicable to an "open-loop" cosine wave crossing firing angle control, and a continuous load current waveform is assumed. As was mentioned in Chapter 9, and as is confirmed by the results of this analysis, a strictly open-loop firing angle control inevitably results in the production of small amounts of distortion terms which may be objectionable, inasmuch as they have subharmonic frequency. In practice, it is usually necessary to use a corrective feedback, the effect of which is to apply a small adjustment to the basic "open-loop" firing instants, and thereby, to suppress the objectionable distortion terms.

Thus the results of a harmonic analysis of the voltage waveform obtained with a purely "open-loop" firing angle control, are not strictly applicable in detail to a practical control scheme which uses a negative feedback to suppress the small amounts of objectionable distortion terms which would otherwise be present. These results are, nevertheless, very closely representative of the predominant harmonic components to be found in the output voltage, for the reason that these components inherently cannot be suppressed, and therefore remain virtually unmodified by feedback. In addition, of course, the quantitative data related to the objectionable subharmonic components, obtained with an open-loop firing angle control, is useful, inasmuch as a knowledge of the "open-loop" amplitudes of these components enables an assessment to be made of the feasibility, or otherwise, of suppressing these by means of corrective feedback.

THE BASIC ANALYTICAL PROBLEM, AND THE APPROACH USED

The output voltage of the cycloconverter consists of selected time-segments of the input voltage waves, which are pieced together so as to form an overall waveshape in which the predominant component is a sinusoid of the desired output frequency. The exact form and structure of the output voltage waveshape mainly depends upon the following factors:

(1) The pulse number of the converter
(2) The ratio between the output and input frequencies
(3) The relative level of the output voltage
(4) The displacement angle of the load
(5) The method of control of the firing instants

If an attempt is made to apply the conventional techniques of Fourier harmonic analysis, the conclusion is reached that this is not a satisfactory approach for a general analysis of the output voltage waveshape of the cycloconverter. Of course, the output voltage waveshape for any particular set of conditions can be constructed, and the coefficients of the Fourier harmonic series computed. This one computation, by itself, may be formidable, especially if the output to input frequency ratio is such that the output waveform does not repeat itself precisely every output cycle; in this event, of course, it is not sufficient merely to analyze the waveform over a single cycle of the wanted output frequency.

Even once such a computation has been carried out, the results have limited usefulness, because they are applicable only to one specific chosen set of conditions. Even if many calculations of this type are made, to cover a range of conditions, such a point-by-point analytical approach is not really satisfactory, since it does not reveal the natural laws governing the harmonic spectrum of the output voltage waveshape, nor does it directly indicate how the harmonic spectrum is related to, and influenced by, the several independent variables.

What is required, then, is a general analytical approach which enables the harmonic series for the output voltage to be expressed quite generally, in terms of each of the independent variables. Such an analytical technique will now be outlined.

In Chapter 4, the analysis of the d-c terminal voltage waveform of the phase-controlled converter was performed by expressing this wave as the mathematical sum of the voltage segments generated by each of the individual thyristors within the converter. Each individual voltage segment is expressed mathematically as the product of the appropriate sinusoidal input voltage, and a "switching function"; this switching function has unity amplitude whenever the associated thyristor is ON, and zero amplitude whenever it is OFF. By expressing each "switching function" as a harmonic series, so the harmonic series for the d-c terminal voltage waveform is obtained.

Actually, for a steady firing angle, this method of analysis has no particular advantage over a more conventional approach. For the cycloconverter voltage waveform, on the other hand, this analytical method is extremely useful, because it is a relatively simple matter to express the switching function as a "phase modulated harmonic series," and thence to arrive at a general harmonic series for the output voltage waveform, in terms of each of the independent variables. This, then, is the method of analysis which is used in this chapter.

SELECTION OF THE THREE-PULSE WAVEFORM FOR DETAILED ANALYSIS, AND THE REASONS FOR THIS

Before proceeding to show how the analysis is performed, firstly it is necessary to decide upon the pulse number of the waveform to be analyzed. At the outset, the waveform of a 2-pulse cycloconverter will be eliminated from consideration, because such a circuit has little practical application, and hence a detailed analysis of its waveform is of little interest.

It was seen in Chapter 4 that once the harmonic series for the waveform of the 3-pulse phase-controlled converter has been obtained, then, in order to arrive at the corresponding result for any other "multipulse" circuit, containing any number of combinations of the 3-pulse group, it is simply a question of eliminating certain harmonic terms from the basic series for the 3-pulse waveform. This simple procedure is also valid for the cycloconverter. Thus, since virtually all practical cycloconverter circuits consist of combinations of the basic 3-pulse group, the analysis of the waveforms of a complete range of practical circuits is essentially reduced to an analysis of the basic 3-pulse waveform. On this basis, it is clear that the 3-pulse waveform is the one which should be chosen for detailed analysis.

DERIVATION OF THE GENERAL EXPRESSION FOR THE THREE-PULSE WAVEFORM, FOR AN ARBITRARY FIRING ANGLE CONTROL METHOD

The output voltage waveform of the cycloconverter is constructed from the voltage waveforms of two oppositively poled 2-quadrant converters. If the cycloconverter operates with a continuous circulating current, then each of the two converters continuously generates a voltage waveform at its output terminals, and the external output voltage waveform of the cycloconverter is the average of these two voltage waveforms. If the cycloconverter operates with no circulating current, then each 2-quadrant converter generates a voltage waveform at its output terminals for a half-period of each output cycle, and, in this case, the external output voltage waveform of the cycloconverter is comprised of alternate segments produced by the positive and negative converters.

CYCLOCONVERTER OUTPUT VOLTAGE ANALYSIS

As the first step in the analysis, it is necessary to obtain general expressions for the voltage waveforms generated by each of the individual converters, assuming that each is in continuous conduction. Once these basic expressions have been obtained, they can then be used to derive the general expression for the external output voltage waveform of the cycloconverter, for both the circulating current, and circulating current-free modes of operation.

For each 2-quadrant converter of the cycloconverter, the quiescent firing angle is 90°. In other words, if both converters are permanently operated at a 90° firing angle, then each delivers zero mean voltage, and the external

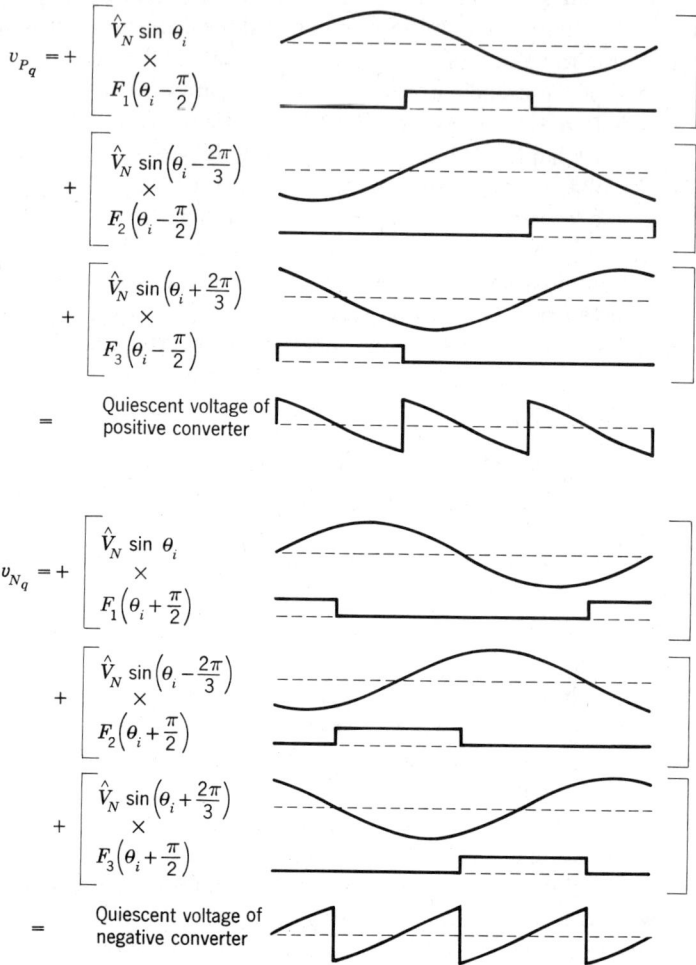

Figure 11.1. Waveforms illustrating the synthesis of the general mathematical expressions for the quiescent voltages of the positive and negative converters.

DERIVATION OF GENERAL EXPRESSION

output voltage of the cycloconverter is also zero. In order to generate an output voltage, the firing angles of each of the converters are oscillated to and fro, in opposite directions to one another, about the 90° quiescent "midpoint."

The logical starting point for the analysis, then, is to obtain general expressions, in terms of the thyristor switching functions, for the output voltage waveforms of each of the converters, with a steady 90° "quiescent" firing angle.

The waveforms of Fig. 11.1 illustrate how these general expressions are obtained. The quiescent output voltage of the positive converter, v_{P_q} is given by

$$v_{P_q} = \hat{V}_N \sin \theta_i \cdot F_1\left(\theta_i - \frac{\pi}{2}\right) + \hat{V}_N \sin\left(\theta_i - \frac{2\pi}{3}\right) \cdot F_2\left(\theta_i - \frac{\pi}{2}\right)$$
$$+ \hat{V}_N \sin\left(\theta_i + \frac{2\pi}{3}\right) \cdot F_3\left(\theta_i - \frac{\pi}{2}\right) \quad (11.1)$$

The quiescent output voltage of the negative converter, v_{N_q} is given by

$$v_{N_q} = \hat{V}_N \sin \theta_i \cdot F_1\left(\theta_i + \frac{\pi}{2}\right) + \hat{V}_N \sin\left(\theta_i - \frac{2\pi}{3}\right) \cdot F_2\left(\theta_i + \frac{\pi}{2}\right)$$
$$+ \hat{V}_N \sin\left(\theta_i + \frac{2\pi}{3}\right) \cdot F_3\left(\theta_i + \frac{\pi}{2}\right) \quad (11.2)$$

$F_1(\theta_i)$, $F_2(\theta_i)$ and $F_3(\theta_i)$ are the same "switching functions" used in Chapter 4, and these are defined mathematically by expressions (4.6) through (4.8) respectively.

It is now necessary to determine how the above general expressions, for the "quiescent" voltages of each of the converters, must be modified, in order to obtain the corresponding expressions applicable to a continuously time-varying "to and fro" phase modulation of the firing angles.

Consider firstly the positive converter; in order to generate a voltage waveform having frequency f_o, the firing angle is oscillated "to and fro" about the quiescent point, in accordance with some function of the wanted output frequency, which, for the present, will be expressed, in quite general terms, as $f(\theta_o)$.

The value of $f(\theta_o)$ oscillates symmetrically to and fro about zero, at a repetition frequency equal to the wanted output frequency. Since the theoretical limits of control of the firing angle, either side of the quiescent point, are $\pm\pi/2$, the maximum possible absolute value for $f(\theta_o)$ is $\pi/2$; but, of course, for output voltage ratios less than unity, the peak value of $f(\theta_o)$ will be less than this absolute maximum.

Since the phase modulation of the firing angle of the negative converter is equal, but in the opposite sense, to that of the positive converter, the corresponding phase-modulating function for the negative converter is $-f(\theta_o)$.

284 CYCLOCONVERTER OUTPUT VOLTAGE ANALYSIS

Thus, for a continuously time-varying "to and fro" phase modulation of the converter firing angles, in accordance with an arbitrary modulating function, defined as $f(\theta_o)$, the general expressions for the voltages generated by the positive and negative converters become:

$$v_P = \hat{V}_N \sin \theta_i \cdot F_1\left(\theta_i - \frac{\pi}{2} + f(\theta_o)\right)$$

$$+ \hat{V}_N \sin\left(\theta_i - \frac{2\pi}{3}\right) \cdot F_2\left(\theta_i - \frac{\pi}{2} + f(\theta_o)\right)$$

$$+ \hat{V}_N \sin\left(\theta_i + \frac{2\pi}{3}\right) \cdot F_3\left(\theta_i - \frac{\pi}{2} + f(\theta_o)\right) \quad (11.3)$$

$$v_N = \hat{V}_N \sin \theta_i \cdot F_1\left(\theta_i + \frac{\pi}{2} - f(\theta_o)\right)$$

$$+ \hat{V}_N \sin\left(\theta_i - \frac{2\pi}{3}\right) \cdot F_2\left(\theta_i + \frac{\pi}{2} - f(\theta_o)\right)$$

$$+ \hat{V}_N \sin\left(\theta_i + \frac{2\pi}{3}\right) \cdot F_3\left(\theta_i + \frac{\pi}{2} - f(\theta_o)\right) \quad (11.4)$$

From expressions (4.6) through (4.8), which define $F_1(\theta_i)$ through $F_3(\theta_i)$ in terms of their associated harmonic series, the voltage of the positive converter, v_P, can now be written as

$$v_P = \hat{V}_N \sin \theta_i \left\{ \frac{1}{3} + \frac{\sqrt{3}}{\pi}\left[\sin\left(\theta_i - \frac{\pi}{2} + f(\theta_o)\right) - \tfrac{1}{2}\cos 2\left(\theta_i - \frac{\pi}{2} + f(\theta_o)\right) \right.\right.$$

$$\left.\left. - \tfrac{1}{4}\cos 4\left(\theta_i - \frac{\pi}{2} + f(\theta_o)\right) \cdots \right]\right\}$$

$$+ \hat{V}_N \sin\left(\theta_i - \frac{2\pi}{3}\right)\left\{\frac{1}{3} + \frac{\sqrt{3}}{\pi}\left[\sin\left(\theta_i - \frac{\pi}{2} + f(\theta_o) - \frac{2\pi}{3}\right)\right.\right.$$

$$- \tfrac{1}{2}\cos 2\left(\theta_i - \frac{\pi}{2} + f(\theta_o) - \frac{2\pi}{3}\right)$$

$$\left.\left. - \tfrac{1}{4}\cos 4\left(\theta_i - \frac{\pi}{2} + f(\theta_o) - \frac{2\pi}{3}\right) \cdots \right]\right\}$$

$$+ \hat{V}_N \sin\left(\theta_i + \frac{2\pi}{3}\right)\left\{\frac{1}{3} + \frac{\sqrt{3}}{\pi}\left[\sin\left(\theta_i - \frac{\pi}{2} + f(\theta_o) + \frac{2\pi}{3}\right)\right.\right.$$

$$- \tfrac{1}{2}\cos 2\left(\theta_i - \frac{\pi}{2} + f(\theta_o) + \frac{2\pi}{3}\right)$$

$$\left.\left. - \tfrac{1}{4}\cos 4\left(\theta_i - \frac{\pi}{2} + f(\theta_o) + \frac{2\pi}{3}\right) \cdots \right]\right\}$$

$$(11.5)$$

By trigonometric manipulation, this reduces to:

$$v_P = \frac{3\sqrt{3}\,\hat{V}_N}{2\pi} \{\sin f(\theta_o) + \tfrac{1}{2}[\sin 3\theta_i \cos 2f(\theta_o) + \cos 3\theta_i \sin 2f(\theta_o)]$$
$$+ \tfrac{1}{4}[\sin 3\theta_i \cos 4f(\theta_o) + \cos 3\theta_i \sin 4f(\theta_o)]$$
$$+ \tfrac{1}{5}[\sin 6\theta_i \cos 5f(\theta_o) + \cos 6\theta_i \sin 5f(\theta_o)]$$
$$+ \tfrac{1}{7}[\sin 6\theta_i \cos 7f(\theta_o) + \cos 6\theta_i \sin 7f(\theta_o)] \cdots \}$$

(11.6)

In a similar manner, the voltage of the negative converter, v_N, can be shown to be given by

$$v_N = \frac{3\sqrt{3}\,\hat{V}_N}{2\pi} \{\sin f(\theta_o) + \tfrac{1}{2}[\sin 3\theta_i \cos 2f(\theta_o) - \cos 3\theta_i \sin 2f(\theta_o)]$$
$$+ \tfrac{1}{4}[\sin 3\theta_i \cos 4f(\theta_o) - \cos 3\theta_i \sin 4f(\theta_o)]$$
$$+ \tfrac{1}{5}[-\sin 6\theta_i \cos 5f(\theta_o) + \cos 6\theta_i \sin 5f(\theta_o)]$$
$$+ \tfrac{1}{7}[-\sin 6\theta_i \cos 7f(\theta_o) + \cos 6\theta_i \sin 7f(\theta_o)] \cdots \}$$

(11.7)

From expressions (11.6) and (11.7), the general expressions for the external voltage waveform of the cycloconverter, for both the circulating current, and circulating current-free modes of operation, can be obtained.

The Circulating Current Mode

As illustrated by the waveforms of Fig. 11.2, with both converters continuously in conduction, the external voltage waveform is given by

$$v_o = \frac{v_P + v_N}{2} \qquad (11.8)$$

Hence, by taking the average of expressions (11.6) and (11.7), the general expression for the output voltage of the cycloconverter, operating with a continuous circulating current, becomes

$$v_o = \frac{3\sqrt{3}\,\hat{V}_N}{2\pi} [\sin f(\theta_o) + \tfrac{1}{2}\sin 3\theta_i \cos 2f(\theta_o) + \tfrac{1}{4}\sin 3\theta_i \cos 4f(\theta_o)$$
$$+ \tfrac{1}{5}\cos 6\theta_i \sin 5f(\theta_o) + \tfrac{1}{7}\cos 6\theta_i \sin 7f(\theta_o) \cdots] \quad (11.9)$$

286 CYCLOCONVERTER OUTPUT VOLTAGE ANALYSIS

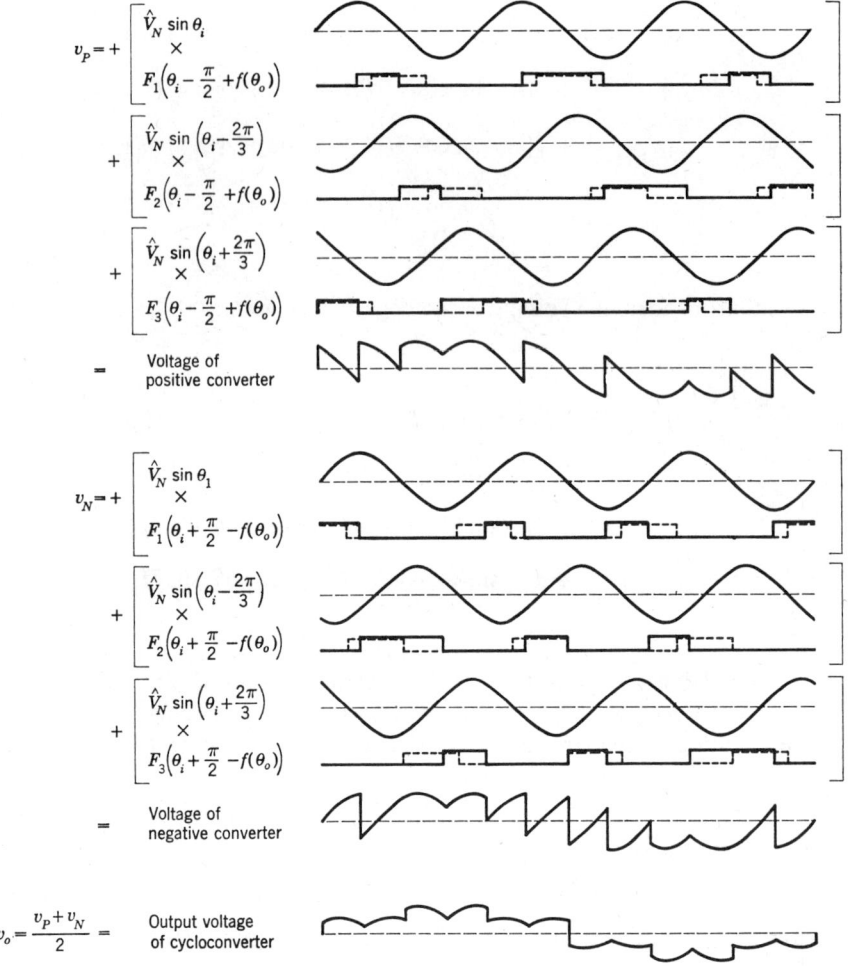

Figure 11.2. Waveforms illustrating the method of derivation of the general mathematical expression for the output voltage of the cycloconverter operating with a continuous circulating current. Quiescent positions of thyristor switching functions shown dashed.

The Circulating Current-Free Mode

In the circulating current-free mode of operation, each converter conducts in turn for half the total output cycle time, and the conduction period of each converter, relative to the output voltage waveform, is determined by the displacement angle of the load. For the purposes of analysis, it will be assumed that the waveform of the load current is a sinusoid, which is displaced from the wanted component of output voltage by the load displacement angle. Thus, the positive converter is in conduction over the period of the output voltage wave from ϕ_o, to $(\phi_o + \pi)$, and the negative converter from $(\phi_o + \pi)$ to $(\phi_o + 2\pi)$, and so on.

In order to obtain the general expression for the external voltage waveform of the cycloconverter, it is necessary to introduce a complementary pair of "converter switching functions," designated $F_P(\theta_o)$ and $F_N(\theta_o)$, for the positive and negative converters respectively. $F_P(\theta_o)$ has unity amplitude whenever the positive converter is in conduction, and zero amplitude whenever the positive converter is blocked. $F_N(\theta_o)$ has unity amplitude whenever the negative converter is in conduction, and zero amplitude whenever the negative converter is blocked. These two functions are expressed mathematically as follows:

$$F_P(\theta_o) = \frac{1}{2} + \frac{2}{\pi} [\sin(\theta_o + \phi_o) + \tfrac{1}{3}\sin 3(\theta_o + \phi_o)$$
$$+ \tfrac{1}{5}\sin 5(\theta_o + \phi_o) + \tfrac{1}{7}\sin 7(\theta_o + \phi_o) \cdots] \quad (11.10)$$

$$F_N(\theta_o) = \frac{1}{2} - \frac{2}{\pi} [\sin(\theta_o + \phi_o) + \tfrac{1}{3}\sin 3(\theta_o + \phi_o)$$
$$+ \tfrac{1}{5}\sin 5(\theta_o + \phi_o) + \tfrac{1}{7}\sin 7(\theta_o + \phi_o) \cdots] \quad (11.11)$$

As illustrated by the waveforms of Figure 11.3, the voltages v_P' and v_N' generated by the positive and negative converters respectively, are given by

$$v_P' = v_P \cdot F_P(\theta_o) \quad (11.12)$$

$$v_N' = v_N \cdot F_N(\theta_0) \quad (11.13)$$

288 CYCLOCONVERTER OUTPUT VOLTAGE ANALYSIS

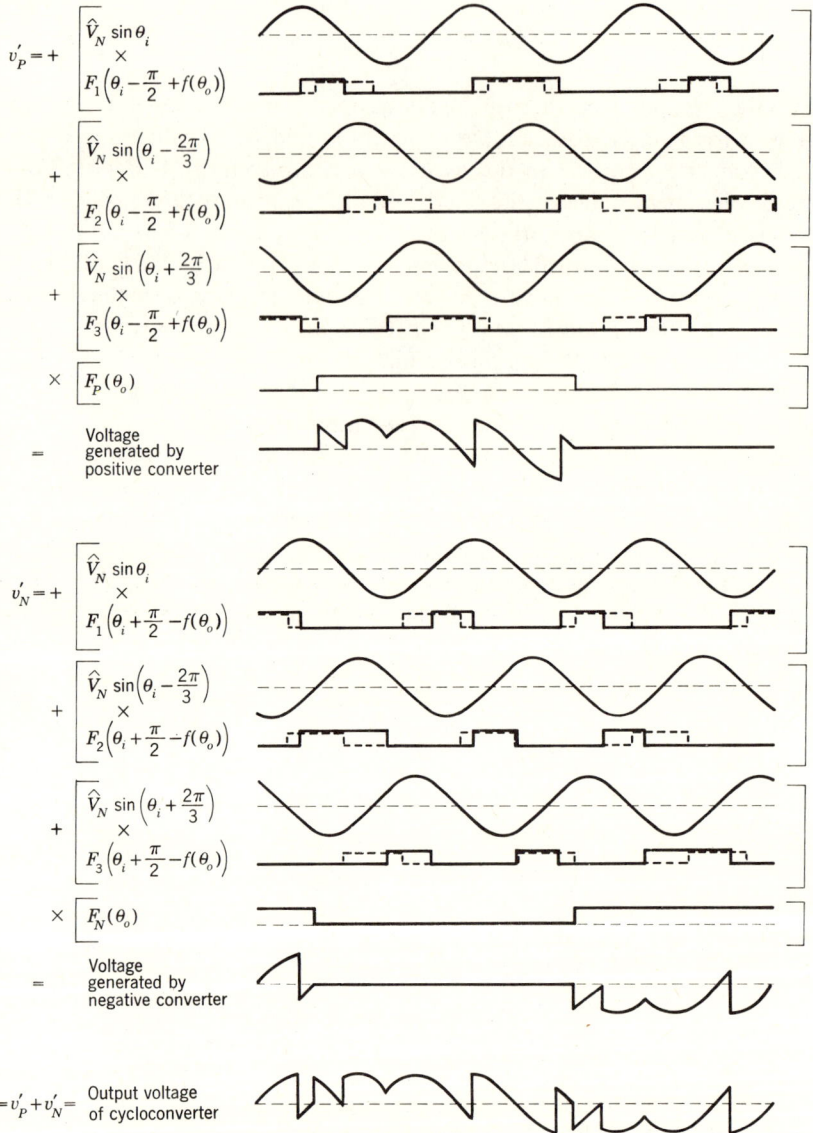

Figure 11.3. Waveforms illustrating the method of derivation of the general mathematical expression for the output voltage of the cycloconverter operating in the circulating current-free mode. Quiescent positions of thyristor switching functions shown dashed.

CHOICE OF COSINE CONTROL FOR ANALYSIS

And the external voltage waveform of the cycloconverter is given by

$$v_o = v'_P + v'_N = v_P F_P(\theta_o) + v_N F_N(\theta_o) \tag{11.14}$$

By substituting into expression (11.14) for v_P and v_N, as given by expressions (11.6) and (11.7) respectively, and for $F_P(\theta_o)$ and $F_N(\theta_o)$, as given by expressions (11.10) and (11.11) respectively, the general expression for the output voltage of the 3-pulse cycloconverter, operating with no circulating current, becomes

$$\begin{aligned}
v_o = {} & \frac{3\sqrt{3}\,\hat{V}_N}{2\pi} \Big\{ \sin f(\theta_o) + \tfrac{1}{2} \sin 3\theta_i \cos 2f(\theta_o) + \tfrac{1}{4} \sin 3\theta_i \cos 4f(\theta_o) \\
& + \tfrac{1}{5} \cos 6\theta_i \sin 5f(\theta_o) + \tfrac{1}{7} \cos 6\theta_i \sin 7f(\theta_o) \cdots \Big\} \\
& + \frac{3\sqrt{3}\,\hat{V}_N}{2\pi} \Big\{ [\tfrac{1}{2} \cos 3\theta_i \sin 2f(\theta_o) + \tfrac{1}{4} \cos 3\theta_i \sin 4f(\theta_o) \\
& + \tfrac{1}{5} \sin 6\theta_i \cos 5f(\theta_o) + \tfrac{1}{7} \sin 6\theta_i \cos 7f(\theta_o) \cdots] \\
& \times \frac{4}{\pi} [\sin(\theta_o + \phi_o) + \tfrac{1}{3} \sin 3(\theta_o + \phi_o) \\
& + \tfrac{1}{5} \sin 5(\theta_o + \phi_o) + \tfrac{1}{7} \sin 7(\theta_o + \phi_o) \cdots] \Big\} \tag{11.15}
\end{aligned}$$

It is to be noted that the first series contained in this expression is identical with expression (11.9), for the output voltage of the cycloconverter when operating with a continuous circulating current.

SELECTION OF THE COSINE WAVE CROSSING CONTROL METHOD FOR DETAILED ANALYSIS, AND THE REASONS FOR THIS

Expressions (11.9) and (11.15), for the output voltage of the cycloconverter for the circulating-current, and circulating current-free modes of operation respectively, are presented in terms of an arbitrary function, $f(\theta_o)$, which represents the continuous "to and fro" oscillation of the firing angle about the quiescent point. The exact mathematical form of this function is directly determined by the method of control of the timing of the thyristor firing pulses.

Thus, in order to carry the analysis to its conclusion, it is necessary, firstly, to decide upon the method of firing pulse timing for which the analysis is to be performed.

In Chapter 9 it is argued that the "cosine wave crossing" control method for determining the timing of the firing pulses is the "naturally correct" one for the cycloconverter, and this hypothesis is strengthened by the mathematical proof of the fact that this control method produces the theoretically minimum possible total r.m.s. distortion of the output voltage waveform.

It can be concluded, then, that the cosine wave crossing control method is the most natural one on which to base a detailed harmonic analysis of the output voltage waveform. As will be seen, this conclusion is confirmed by the analytical results, to be obtained later in this chapter, which show that this control method has the unique property that it does not produce "unnecessary" direct integer multiple harmonics of the wanted output frequency.

MATHEMATICAL DEFINITION OF THE MODULATING FUNCTION FOR THE COSINE WAVE CROSSING CONTROL METHOD

Having decided, then, that the output voltage waveform, as fabricated by the cosine wave crossing control, is to be subjected to detailed analysis, the next step is to obtain the precise mathematical definition of the firing angle modulating function for this control method.

Figure 11.4 shows the thyristor switching functions resulting from a "to and fro" phase modulation of the firing angle about the quiescent point, as produced by the cosine wave crossing control method.

The reference voltage waveform is a sinusoid, given by:

$$v_R = r\hat{V}_T \sin \theta_o \qquad (11.16)$$

where $\hat{V}_T$ = peak value of the cosine timing waves.

From these waveforms it is evident that at any given time, the phase of each thyristor switching function is shifted by an angle $\sin^{-1} r \sin \theta_o$, with respect to the quiescent position. In other words, the firing angle phase modulating function is defined by

$$f(\theta_o) = \sin^{-1} r \sin \theta_o \qquad (11.17)$$

Having established this, it is now possible to substitute the exact mathematical expression for $f(\theta_o)$ into the general expressions, (11.9) and (11.15), for the output voltage of the cycloconverter, with and without circulating current, and, thereby, to derive the specific harmonic series for the output voltage.

DEFINITION OF THE MODULATING FUNCTION

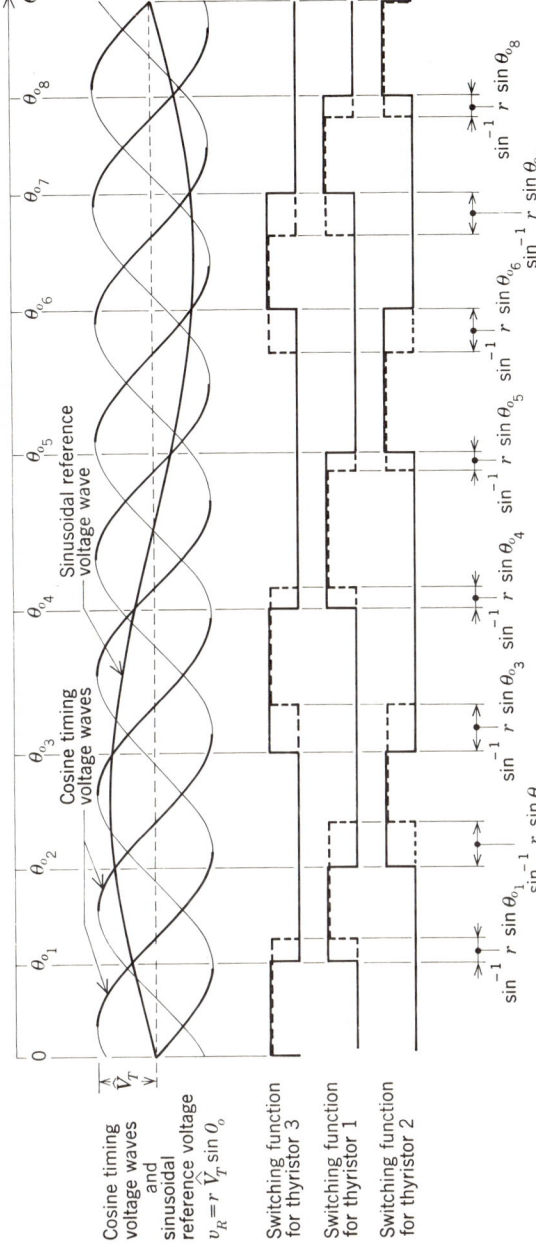

Figure 11.4. Waveforms illustrating that with the cosine wave crossing control method for determining the firing instants, the phase-modulating function is given by: $f(\theta_o) = \sin^{-1} r \cdot \sin \theta_o$. Quiescent positions of thyristor switching functions are shown dashed.

DERIVATION OF THE HARMONIC SERIES FOR THE THREE-PULSE VOLTAGE

The Circulating Current Mode

Expression (11.9) defines the output voltage waveform in terms of the arbitrary phase-modulating function, $f(\theta_o)$. By substituting into expression (11.9) for $f(\theta_o)$, as given by expression (11.17), the following expression is obtained:

$$v_o = \frac{3\sqrt{3}\,\hat{V}_N}{2\pi} [\sin(\sin^{-1} r \sin \theta_o)$$
$$+ \tfrac{1}{2} \sin 3\theta_i \cos(2 \sin^{-1} r \sin \theta_o) + \tfrac{1}{4} \sin 3\theta_i \cos(4 \sin^{-1} r \sin \theta_o)$$
$$+ \tfrac{1}{5} \cos 6\theta_i \sin(5 \sin^{-1} r \sin \theta_o) + \tfrac{1}{7} \cos 6\theta_i \sin(7 \sin^{-1} r \sin \theta_o) \cdots] \quad (11.18)$$

If each of the terms in this series can now be expressed in terms of simple trigonometric functions, then the desired expression for the output voltage, in terms of a series of simple harmonic components, can be obtained.

Clearly, the first term in the series can be written as

$$\sin(\sin^{-1} r \sin \theta_o) = r \sin \theta_o \quad (11.19)$$

This term, of course, represents the wanted component of output voltage.

For the remaining terms, the following general identities can be shown to exist:

$$\sin([6p - 1] \sin^{-1} r \sin \theta_o) = a_{(6p-1)_1} \sin \theta_o + a_{(6p-1)_3} \sin 3\theta_o$$
$$+ \cdots a_{(6p-1)(6p-1)} \sin(6p-1)\theta_o \quad (11.20)$$

$$\sin([6p + 1] \sin^{-1} r \sin \theta_o) = a_{(6p+1)_1} \sin \theta_o + a_{(6p+1)_3} \sin 3\theta_o$$
$$+ \cdots a_{(6p+1)(6p+1)} \sin(6p+1)\theta_o \quad (11.21)$$

$$\cos((3[2p - 1] - 1) \sin^{-1} r \sin \theta_o)$$
$$= a_{(3[2p-1]-1)_0} + a_{(3[2p-1]-1)_2} \cos 2\theta_o$$
$$+ \cdots a_{(3[2p-1]-1)(3[2p-1]-1)} \cos(3[2p-1]-1)\theta_o \quad (11.22)$$

$$\cos((3[2p - 1] + 1) \sin^{-1} r \sin \theta_o)$$
$$= a_{(3[2p-1]+1)_0} + a_{(3[2p-1]+1)_2} \cos 2\theta_o$$
$$+ \cdots a_{(3[2p-1]+1)(3[2p-1]+1)} \cos(3[2p-1]+1)\theta_o \quad (11.23)$$

(Note: These are *not* infinite series.)

DERIVATION OF THE THREE-PULSE EXPRESSION

where:

$$a_{(6p\pm 1)(2n-1)} = \frac{1}{\pi}\int_0^{2\pi} \sin[(6p\pm 1)(\sin^{-1} r\sin\theta_o)]\cdot \sin(2n-1)\theta_o \cdot d\theta_o$$

$$a_{(3[2p-1]\pm 1)_0} = \frac{1}{2\pi}\int_0^{2\pi} \cos[(3[2p-1]\pm 1)(\sin^{-1} r\sin\theta_o)]\cdot d\theta_o$$

$$a_{(3[2p-1]\pm 1)_{2n}} = \frac{1}{\pi}\int_0^{2\pi} \cos[(3[2p-1]\pm 1)(\sin^{-1} r\sin\theta_o)]\cdot \cos 2n\theta_o \cdot d\theta_o$$

The computed values of the above coefficients, for values of r from 1.0 to 0.1 (in increments of 0.1), for $p = 1$ and $p = 2$, are shown in Tables 11.1.

It is now necessary to substitute expressions (11.20) through (11.23) into expression (11.18). By performing this substitution, the following general harmonic series, representing the output voltage of the 3-pulse cycloconverter, when operating with a continuous circulating current, is obtained:

$$v_o = \frac{3\sqrt{3}\,\hat{V}_N}{2\pi}\Bigg\{ r\sin\theta_o \leftarrow \text{Wanted component of output voltage}$$

Superimposed harmonic components
$$+ \tfrac{1}{2}\sum_{p=1}^{p=\infty}\Bigg(\sum_{n=0}^{2n=3[2p-1]+1}\left[\frac{a_{(3[2p-1]-1)_{2n}}}{3[2p-1]-1} + \frac{a_{(3[2p-1]+1)_{2n}}}{3[2p-1]+1}\right]$$

$$\times [\sin(3[2p-1]\theta_i + 2n\theta_o) + \sin(3[2p-1]\theta_i - 2n\theta_o)]$$

$$+ \sum_{n=0}^{2n+1=6p+1}\left[\frac{a_{(6p-1)(2n+1)}}{6p-1} + \frac{a_{(6p+1)(2n+1)}}{6p+1}\right]$$

$$\times [\sin(6p\theta_i + [2n+1]\theta_o) - \sin(6p\theta_i - [2n+1]\theta_o)]\Bigg)\Bigg\}$$

(11.24)

It is of interest to note, for the particular case of $r = 1.0$, that the above expression reduces to

$$v_o = \frac{3\sqrt{3}\,\hat{V}_N}{2\pi}\{\sin\theta_o \leftarrow \text{Wanted component of output voltage}$$

Superimposed harmonic components
$$+ \tfrac{1}{2}(\tfrac{1}{2}[\sin(3\theta_i + 2\theta_o) + \sin(3\theta_i - 2\theta_o)]$$

$$+ \tfrac{1}{4}[\sin(3\theta_i + 4\theta_o) + \sin(3\theta_i - 4\theta_o)]$$

$$+ \tfrac{1}{5}[\sin(6\theta_i + 5\theta_o) - \sin(6\theta_i - 5\theta_o)]$$

$$+ \tfrac{1}{7}[\sin(6\theta_i + 7\theta_o) - \sin(6\theta_i - 7\theta_o)]\cdots)\} \quad (11.25)$$

Tables 11.1 Showing Values of the Coefficients of the Harmonic Series Described by Expressions (11.20) through (11.23)

r	a_{20}	a_{22}	a_{40}	a_{42}	a_{44}	a_{80}	a_{82}	a_{84}	a_{86}	a_{88}	a_{10_0}	a_{10_2}	a_{10_4}	a_{10_6}	a_{10_8}	$a_{10_{10}}$
1.0	.000	1.000	.000	.000	1.000	.000	.000	.000	.000	1.000	.000	.000	.000	.000	.000	1.000
0.9	.190	.810	−.272	.616	.656	−.043	.139	−.334	.808	.430	−.109	.197	−.108	−.146	.818	.349
0.8	.360	.640	−.331	.922	.410	.237	−.466	.307	.755	.168	−.119	.302	−.479	.585	.604	.107
0.7	.510	.490	−.240	1.000	.240	.172	−.478	.769	.480	.058	.205	−.355	.050	.777	.294	.028
0.6	.640	.360	−.051	.922	.130	−.129	.050	.823	.239	.017	.227	−.562	.660	.561	.107	.006
0.5	.750	.250	.188	.750	.063	−.363	.656	.609	.094	.004	−.100	−.029	.820	.278	.029	.001
0.4	.840	.160	.437	.538	.026	−.329	.967	.334	.028	.001	−.377	.685	.591	.096	.006	.000
0.3	.910	.090	.664	.328	.008	−.010	.876	.129	.005	.000	−.268	.977	.270	.021	.001	.000
0.2	.960	.040	.845	.154	.002	.451	.520	.029	.000	.000	.218	.712	.067	.002	.000	.000
0.1	.990	.010	.960	.040	.000	.846	.152	.002	.000	.000	.765	.231	.005	.000	.000	.000

r	a_{51}	a_{53}	a_{55}	a_{71}	a_{73}	a_{75}	a_{77}	a_{11_1}	a_{11_3}	a_{11_5}	a_{11_7}	a_{11_9}	$a_{11_{11}}$	a_{13_1}	a_{13_3}	a_{13_5}	a_{13_7}	a_{13_9}	$a_{13_{11}}$	$a_{13_{13}}$
1.0	.000	.000	1.000	.000	.000	.000	1.000	.000	.000	.000	.000	.000	1.00	.000	.000	.000	.000	.000	.000	1.000
0.9	−.530	.693	.591	.276	−.417	.785	.478	−.300	.284	−.208	−.050	.810	.314	.285	−.313	.353	−.349	.134	.775	.254
0.8	−.403	.922	.328	−.306	.103	.826	.210	.012	.110	−.384	.664	.532	.086	−.374	.356	−.248	−.107	.729	.402	.055
0.7	.036	.875	.168	−.624	.649	.600	.082	.488	−.508	.294	.716	.226	.020	.058	.089	−.401	.607	.551	.131	.010
0.6	.538	.691	.078	−.409	.890	.348	.028	.017	−.306	.758	.434	.071	.004	.498	−.540	.317	.692	.238	.030	.002
0.5	.938	.469	.031	.164	.820	.164	.008	−.612	.419	.693	.177	.016	.000	−.133	−.195	.766	.390	.067	.005	.000
0.4	1.142	.269	.010	.771	.572	.060	.002	−.454	.863	.381	.048	.002	.000	−.674	.613	.603	.130	.012	.000	.000
0.3	1.119	.123	.002	1.128	.298	.015	.000	.406	.734	.127	.008	.000	.000	−.092	.862	.246	.024	.001	.000	.000
0.2	.883	.038	.000	1.086	.101	.002	.000	1.107	.328	.021	.001	.000	.000	.933	.476	.045	.002	.000	.000	.000
0.1	.485	.005	.000	.659	.014	.000	.000	.943	.051	.001	.000	.000	.000	1.045	.082	.002	.000	.000	.000	.000

The Circulating Current-Free Mode

Expression (11.15) defines the output voltage waveform in terms of the arbitrary phase-modulating function, $f(\theta_o)$. By substituting into expression (11.15) for $f(\theta_o)$, as given by expression (11.17), the following expression is obtained:

(A)
$$v_o = \frac{3\sqrt{3}\,\hat{V}_N}{2\pi} \{\sin(\sin^{-1} r \sin \theta_o) \\
+ \tfrac{1}{2} \sin 3\theta_i \cos(2 \sin^{-1} r \sin \theta_o) \\
+ \tfrac{1}{4} \sin 3\theta_i \cos(4 \sin^{-1} r \sin \theta_o) \\
+ \tfrac{1}{5} \cos 6\theta_i \sin(5 \sin^{-1} r \sin \theta_o) \\
+ \tfrac{1}{7} \cos 6\theta_i \sin(7 \sin^{-1} r \sin \theta_o) \cdots \}$$

(B)
$$+ \frac{3\sqrt{3}\,\hat{V}_N}{2\pi} \{[\tfrac{1}{2} \cos 3\theta_i \sin(2 \sin^{-1} r \sin \theta_o) \\
+ \tfrac{1}{4} \cos 3\theta_i \sin(4 \sin^{-1} r \sin \theta_o) \\
+ \tfrac{1}{5} \sin 6\theta_i \cos(5 \sin^{-1} r \sin \theta_o) \\
+ \tfrac{1}{7} \sin 6\theta_i \cos(7 \sin^{-1} r \sin \theta_o) + \cdots] \\
\times \frac{4}{\pi} [\sin(\theta_o + \phi_o) + \tfrac{1}{3} \sin 3(\theta_o + \phi_o) \\
+ \tfrac{1}{5} \sin 5(\theta_o + \phi_o) + \tfrac{1}{7} \sin 7(\theta_o + \phi_o) \cdots]\} \quad (11.26)$$

The first term, (A), in this expression is identical with expression (11.18), and thus the general harmonic series to which this term reduces is given by expression (11.24). [And, for the particular case of $r = 1$, this reduces to expression (11.25).]

In order to reduce the second term, (B), of expression (11.26) into a series of harmonic components, it is necessary, firstly, to express the individual terms as simple trigonometric functions.

Tables 11.2a Showing values of the Coefficients of the Harmonic Series Described by Expressions (11.27) and (11.28)

Note: For $r = 1.0$ only, the following relationships can be used to calculate additional coefficients;

$$a_{(2n+1)_0} = \frac{2}{\pi} \frac{1}{(2n+1)} (-1)^n$$

$$a_{(2n+1)_{2m}} = \frac{2}{\pi} \left[\frac{1}{2m + (2n+1)} - \frac{1}{2m - (2n+1)} \right] (-1)^{m+n}$$

r	a_{1_0}	a_{1_2}	a_{1_4}	a_{1_6}	a_{1_8}	$a_{1_{10}}$	$a_{1_{12}}$	$a_{1_{14}}$	$a_{1_{16}}$
1.0	.637	.424	−.085	.036	−.020	.013	−.009	.007	−.005
0.9	.746	.277	−.027	.005	−.001	.000	.000	.000	.000
0.8	.813	.198	−.012	.002	.000	.000	.000	.000	.000
0.7	.863	.142	−.006	.000	.000	.000	.000	.000	.000
0.6	.903	.100	−.003	.000	.000	.000	.000	.000	.000
0.5	.934	.067	−.001	.000	.000	.000	.000	.000	.000
0.4	.959	.042	.000	.000	.000	.000	.000	.000	.000
0.3	.977	.023	.000	.000	.000	.000	.000	.000	.000
0.2	.990	.010	.000	.000	.000	.000	.000	.000	.000
0.1	.997	.003	.000	.000	.000	.000	.000	.000	.000

r	a_{5_0}	a_{5_2}	a_{5_4}	a_{5_6}	a_{5_8}	$a_{5_{10}}$	$a_{5_{12}}$	$a_{5_{14}}$	$a_{5_{16}}$	a_{7_0}	a_{7_2}	a_{7_4}	a_{7_6}	a_{7_8}	$a_{7_{10}}$	$a_{7_{12}}$	$a_{7_{14}}$	$a_{7_{16}}$
1.0	.127	−.303	.707	.579	−.163	.085	−.053	.037	−.028	−.091	.198	−.270	.686	.594	−.175	.094	−.061	.043
0.9	−.014	−.079	.921	.187	−.019	.004	−.001	.000	.000	.169	−.322	.193	.828	.142	−.013	.002	−.001	.000
0.8	−.260	.460	.727	.077	−.004	.001	.000	.000	.000	.208	−.513	.736	.526	.045	−.002	.000	.000	.000
0.7	−.360	.840	.490	.031	−.001	.000	.000	.000	.000	−.043	−.102	.857	.274	.014	.000	.000	.000	.000
0.6	−.290	.987	.291	.011	.000	.000	.000	.000	.000	−.304	.483	.697	.120	.004	.000	.000	.000	.000
0.5	−.083	.929	.150	.004	.000	.000	.000	.000	.000	−.377	.890	.443	.043	.001	.000	.000	.000	.000
0.4	.201	.733	.065	.001	.000	.000	.000	.000	.000	−.203	.971	.220	.012	.000	.000	.000	.000	.000
0.3	.502	.476	.021	.000	.000	.000	.000	.000	.000	.154	.764	.080	.002	.000	.000	.000	.000	.000
0.2	.763	.233	.004	.000	.000	.000	.000	.000	.000	.563	.420	.017	.000	.000	.000	.000	.000	.000
0.1	.938	.061	.000	.000	.000	.000	.000	.000	.000	.881	.118	.001	.000	.000	.000	.000	.000	.000

Table 11.2a (Continued)

r	a_{11_0}	a_{11_2}	a_{11_4}	a_{11_6}	a_{11_8}	$a_{11_{10}}$	$a_{11_{12}}$	$a_{11_{14}}$	$a_{11_{16}}$	a_{13_0}	a_{13_2}	a_{13_4}	a_{13_6}	a_{13_8}	$a_{13_{10}}$	$a_{13_{12}}$	$a_{13_{14}}$	$a_{13_{16}}$
1.0	−.058	.120	−.133	.165	−.246	.667	.609	−.187	.104	.049	−.100	.108	−.124	.158	−.240	.662	.613	−.190
0.9	.075	−.181	.277	−.437	.547	.641	.083	−.007	.001	.064	−.110	.042	.103	−.360	.646	.555	.064	−.005
0.8	−.209	.430	−.418	.169	.753	.259	.016	.001	.000	.044	−.132	.271	−.465	.444	.652	.178	.010	.000
0.7	.038	.016	−.325	.717	.472	.080	.003	.000	.000	−.221	.459	−.441	.128	.723	.308	.042	.001	.000
0.6	.278	−.566	.371	.697	.200	.019	.000	.000	.000	.078	−.054	−.296	.703	.466	.095	.007	.000	.000
0.5	.070	−.341	.803	.405	.060	.003	.000	.000	.000	.279	−.604	.473	.651	.181	.020	.001	.000	.000
0.4	−.308	.438	.704	.154	.012	.000	.000	.000	.000	−.061	−.118	.815	.318	.043	.003	.000	.000	.000
0.3	−.348	.954	.357	.035	.001	.000	.000	.000	.000	−.394	.756	.547	.085	.005	.000	.000	.000	.000
0.2	.105	.797	.095	.004	.000	.000	.000	.000	.000	−.102	.923	.169	.010	.000	.000	.000	.000	.000
0.1	.719	.274	.007	.000	.000	.000	.000	.000	.000	.619	.367	.013	.000	.000	.000	.000	.000	.000

Tables 11.2b Showing Values of the Coefficients of the Harmonic Series Described by Expressions (11.29) and (11.30)

r	a_{21}	a_{23}	a_{25}	a_{27}	a_{29}	a_{211}	a_{213}	a_{215}	a_{217}	a_{41}	a_{43}	a_{45}	a_{47}	a_{49}	a_{411}	a_{413}	a_{415}	a_{417}
1.0	.849	.509	−.121	.057	−.033	.022	−.015	.012	−.009	−.340	.728	.566	−.154	.078	−.049	.033	−.024	.019
0.9	1.094	.273	−.029	.006	.001	.000	.000	.000	.000	−.249	.966	.215	−.022	.004	−.001	.000	.000	.000
0.8	1.141	.169	−.011	.001	.000	.000	.000	.000	.000	.199	.845	.101	−.006	.001	.000	.000	.000	.000
0.7	1.109	.104	−.004	.000	.000	.000	.000	.000	.000	.638	.647	.046	−.002	.000	.000	.000	.000	.000
0.6	1.023	.062	−.002	.000	.000	.000	.000	.000	.000	.964	.447	.020	−.001	.000	.000	.000	.000	.000
0.5	.901	.034	−.001	.000	.000	.000	.000	.000	.000	1.134	.276	.008	.000	.000	.000	.000	.000	.000
0.4	.750	.017	.000	.000	.000	.000	.000	.000	.000	1.143	.148	.002	.000	.000	.000	.000	.000	.000
0.3	.579	.007	.000	.000	.000	.000	.000	.000	.000	1.003	.065	.001	.000	.000	.000	.000	.000	.000
0.2	.394	.002	.000	.000	.000	.000	.000	.000	.000	.741	.020	.000	.000	.000	.000	.000	.000	.000
0.1	.199	.000	.000	.000	.000	.000	.000	.000	.000	.393	.002	.000	.000	.000	.000	.000	.000	.000

r	a_{81}	a_{83}	a_{85}	a_{87}	a_{89}	a_{811}	a_{813}	a_{815}	a_{817}	a_{101}	a_{103}	a_{105}	a_{107}	a_{109}	a_{1011}	a_{1013}	a_{1015}	a_{1017}
1.0	−.162	.185	−.261	.679	.599	−.179	.097	−.063	.045	.129	−.140	.170	−.250	.670	.606	−.185	.102	−.067
0.9	.391	−.397	.304	.780	.124	−.011	−.002	.000	.000	−.286	.345	−.449	.480	.686	.095	−.008	.001	.000
0.8	.158	−.367	.788	.444	.035	−.002	.000	.000	.000	.362	−.290	−.004	.787	.311	.021	.001	.000	.000
0.7	−.442	.221	.776	.203	.009	.000	.000	.000	.000	.327	−.503	.631	.571	.109	.004	.000	.000	.000
0.6	−.631	.736	.534	.076	.002	.000	.000	.000	.000	−.339	.043	.788	.284	.030	.001	.000	.000	.000
0.5	−.235	.889	.281	.023	.000	.000	.000	.000	.000	−.658	.680	.568	.103	.006	.000	.000	.000	.000
0.4	.460	.716	.111	.005	.000	.000	.000	.000	.000	−.197	.879	.274	.026	.001	.000	.000	.000	.000
0.3	1.025	.407	.030	.001	.000	.000	.000	.000	.000	.648	.634	.085	.004	.000	.000	.000	.000	.000
0.2	1.141	.146	.004	.000	.000	.000	.000	.000	.000	1.151	.260	.013	.000	.000	.000	.000	.000	.000
0.1	.738	.020	.000	.000	.000	.000	.000	.000	.000	.881	.039	.000	.000	.000	.000	.000	.000	.000

NOTE: For $r = 1.0$ only, the following relationship can be used to calculate additional coefficients:

$$a_{2n_{(2m+1)}} = \frac{2}{\pi}\left[\frac{1}{(2m+1)-2n} - \frac{1}{(2m+1)+2n}\right](-1)^{m+n}$$

DERIVATION OF THE THREE-PULSE EXPRESSION

The following general identities can be shown to exist:

$$\cos[(6p-1)\sin^{-1} r\sin\theta_o]$$
$$= a_{(6p-1)_0} + a_{(6p-1)_2}\cos 2\theta_o + \cdots a_{(6p-1)_{2n}}\cos 2n\theta_o + \cdots \quad (11.27)$$

$$\cos[(6p+1)\sin^{-1} r\sin\theta_o]$$
$$= a_{(6p+1)_0} + a_{(6p+1)_2}\cos 2\theta_o + \cdots a_{(6p+1)_{2n}}\cos 2n\theta_o + \cdots \quad (11.28)$$

$$\sin[(3[2p-1]-1)\sin^{-1} r\sin\theta_o]$$
$$= a_{(3[2p-1]-1)_1}\sin\theta_o + a_{(3[2p-1]-1)_3}\sin 3\theta_o$$
$$+ \cdots a_{(3[2p-1]-1)_{(2n-1)}}\sin(2n-1)\theta_o + \cdots \quad (11.29)$$

$$\sin[(3[2p-1]+1)\sin^{-1} r\sin\theta_o]$$
$$= a_{(3[2p-1]+1)_1}\sin\theta_o + a_{(3[2p-1]+1)_3}\sin 3\theta_o$$
$$+ \cdots a_{(3[2p-1]+1)_{(2n-1)}}\sin(2n-1)\theta_o + \cdots \quad (11.30)$$

(Note: Unlike (11.20) through (11.23), expressions (11.27) through (11.30) are infinite series.)
where

$$a_{(6p\pm1)_0} = \frac{1}{2\pi}\int_0^{2\pi}\{\cos[(6p\pm1)\sin^{-1} r\sin\theta_o]\}\,d\theta_o$$

$$a_{(6p\pm1)_{2n}} = \frac{1}{\pi}\int_0^{2\pi}\{\cos[(6p\pm1)\sin^{-1} r\sin\theta_o]\}\{\cos 2n\theta_o\}\cdot d\theta_o$$

$$a_{(3[2p-1]\pm1)_{(2n-1)}} = \frac{1}{\pi}\int_0^{2\pi}\{\sin[(3[2p-1]\pm1)\sin^{-1} r\sin\theta_o]\}$$
$$\times\{\sin(2n-1)\theta_o\}\cdot d\theta_o$$

The computed values of the coefficients of expressions (11.27) and (11.28), for values of p up to 2, and n up to 8, for values of r from 1.0 to 0.1, are shown in Tables 11.2a; and the computed values of the coefficients of expressions (11.29) and (11.30), for values of p up to 2, and n up to 9, for values of r from 1.0 to 0.1, are shown in Tables 11.2b.

It is now necessary to substitute expressions (11.27) through (11.30) into part (B) of expression (11.26). By performing this substitution, the following general harmonic series, representing the output voltage of the 3-pulse cycloconverter, in the circulating current-free mode of operation, is obtained:

$$v_o = \frac{3\sqrt{3}\,\hat{V}_N}{2\pi}\Bigg[r\sin\theta_o \;\longleftarrow\; \text{Wanted component of output voltage}$$

Superimposed harmonic components

$$+\frac{1}{2}\sum_{p=1}^{p=\infty}\Bigg\{\sum_{n=0}^{2n=3[2p-1]+1}\Bigg[\frac{a_{(3[2p-1]-1)2n}}{3[2p-1]-1}+\frac{a_{(3[2p-1]+1)2n}}{3[2p-1]+1}\Bigg]$$
$$\times\,[\sin(3[2p-1]\theta_i+2n\theta_o)+\sin(3[2p-1]\theta_i-2n\theta_o)]$$
$$+\sum_{n=0}^{2n+1=6p+1}\Bigg[\frac{a_{(6p-1)(2n+1)}}{6p-1}+\frac{a_{(6p+1)(2n+1)}}{6p+1}\Bigg]$$
$$\times\,[\sin(6p\theta_i+[2n+1]\theta_o)-\sin(6p\theta_i-[2n+1]\theta_o)]$$

$$+\frac{2}{\pi}\sum_{n=0}^{n=\infty}\Bigg[\frac{a_{(3[2p-1]-1)(2n+1)}}{3[2p-1]-1}+\frac{a_{(3[2p-1]+1)(2n+1)}}{3[2p-1]+1}\Bigg]$$
$$\times\,\Bigg[\frac{2\cos(2n+1)\phi_o}{2n+1}\cdot\cos(3[2p-1]\theta_i)\Bigg]$$

$$+\frac{2}{\pi}\sum_{m=1}^{m=\infty}\sum_{n=0}^{n=\infty}\Bigg[\frac{1}{2n+1-2m}\Bigg]$$
$$\times\Bigg[\frac{a_{(3[2p-1]-1)(2n+1)}}{3[2p-1]-1}+\frac{a_{(3[2p-1]+1)(2n+1)}}{3[2p-1]+1}\Bigg]$$
$$\times\,[\cos\{3(2p-1)\theta_i-(2m\theta_o-[2n+1-2m]\phi_o)\}$$
$$+\cos\{3(2p-1)\theta_i+(2m\theta_o-[2n+1-2m]\phi_o)\}]$$

$$+\Bigg[\frac{1}{2n+1+2m}\Bigg]$$
$$\times\Bigg[\frac{a_{(3[2p-1]-1)(2n+1)}}{3[2p-1]-1}+\frac{a_{(3[2p-1]+1)(2n+1)}}{3[2p-1]+1}\Bigg]$$
$$\times\,[\cos\{3(2p-1)\theta_i-(2m\theta_o+[2n+1+2m]\phi_o)\}$$
$$+\cos\{3(2p-1)\theta_i+(2m\theta_o+[2n+1+2m]\phi_o)\}]$$

$$+\frac{2}{\pi}\sum_{m=0}^{m=\infty}\sum_{n=0}^{n=\infty}\Bigg[\frac{1}{2n-(2m+1)}\Bigg]\Bigg[\frac{a_{(6p-1)2n}}{6p-1}+\frac{a_{(6p+1)2n}}{6p+1}\Bigg]$$
$$\times\,[-\cos\{6p\theta_i-([2m+1]\theta_o-[2n-(2m+1)]\phi_o)\}$$
$$+\cos\{6p\theta_i+([2m+1]\theta_o-[2n-(2m+1)]\phi_o)\}]$$

$$+\Bigg[\frac{1}{2n+2m+1}\Bigg]\Bigg[\frac{a_{(6p-1)2n}}{6p-1}+\frac{a_{(6p+1)2n}}{6p+1}\Bigg]$$
$$\times\,[\cos\{6p\theta_i-([2m+1]\theta_o+[2n+2m+1]\phi_o)\}$$
$$-\cos\{6p\theta_i+([2m+1]\theta_o+[2n+2m+1]\phi_o)\}]\Bigg\}\Bigg]$$

(11.31)

DERIVATION OF THE THREE-PULSE EXPRESSION

It is of interest to note, for the special case of $r = 1.0$, and $\phi_o = 0$, that expression (11.31) reduces to

$$v_o = \frac{3\sqrt{3}\,\hat{V}_N}{2\pi} \bigg\{ \sin\theta_o \;\leftarrow\; \text{Wanted component of output voltage}$$

Superimposed harmonic components
$$+ \tfrac{1}{2}(\tfrac{1}{2}[\sin(3\theta_i + 2\theta_o) + \sin(3\theta_i - 2\theta_o)]$$
$$+ \tfrac{1}{4}[\sin(3\theta_i + 4\theta_o) + \sin(3\theta_i - 4\theta_o)]$$
$$+ \tfrac{1}{5}[\sin(6\theta_i + 5\theta_o) - \sin(6\theta_i - 5\theta_o)]$$
$$+ \tfrac{1}{7}[\sin(6\theta_i + 7\theta_o) - \sin(6\theta_i - 7\theta_o)] \cdots)$$

$$+ \frac{1}{\pi}([\tfrac{1}{2} + \tfrac{1}{2}]\cos 3\theta_i$$

$$+ \tfrac{1}{4}[\tfrac{1}{1} + \tfrac{1}{3}][\cos(3\theta_i + 2\theta_o) + \cos(3\theta_i - 2\theta_o)]$$
$$+ \tfrac{1}{2}[-\tfrac{1}{1} + \tfrac{1}{3}][\cos(3\theta_i + 4\theta_o) + \cos(3\theta_i - 4\theta_o)]$$
$$+ \tfrac{1}{4}[-\tfrac{1}{1} + \tfrac{1}{5}][\cos(3\theta_i + 6\theta_o) + \cos(3\theta_i - 6\theta_o)]$$
$$+ \tfrac{1}{2}[-\tfrac{1}{3} + \tfrac{1}{5}][\cos(3\theta_i + 8\theta_o) + \cos(3\theta_i - 8\theta_o)] + \cdots$$
$$+ [-\tfrac{1}{3}\cdot\tfrac{1}{5} + \tfrac{1}{3}\cdot\tfrac{1}{7}][\cos(6\theta_i + \theta_o) - \cos(6\theta_i - \theta_o)]$$
$$+ [\tfrac{1}{1}\cdot\tfrac{1}{5} - \tfrac{1}{5}\cdot\tfrac{1}{7}]\;[\cos(6\theta_i + 3\theta_o) - \cos(6\theta_i - 3\theta_o)]$$
$$+ [-\tfrac{1}{5}\cdot\tfrac{1}{5} + \tfrac{1}{1}\cdot\tfrac{1}{7}][\cos(6\theta_i + 5\theta_o) - \cos(6\theta_i - 5\theta_o)]$$
$$+ [-\tfrac{1}{1}\cdot\tfrac{1}{5} - \tfrac{1}{7}\cdot\tfrac{1}{7}][\cos(6\theta_i + 7\theta_o) - \cos(6\theta_i - 7\theta_o)]$$

$$+ \cdots)\bigg\} \quad (11.32)$$

Furthermore, for $r = 1$ and $\phi_o = 90°$, expression (11.31) becomes

$$v_o = \frac{3\sqrt{3}\,\hat{V}_N}{2\pi} [\sin\theta_o \;\leftarrow\; \text{Wanted component of output voltage}$$

Superimposed harmonic components
$$+ \tfrac{1}{2}\sin(3\theta_i + 2\theta_o) + \tfrac{1}{4}\sin(3\theta_i + 4\theta_o)$$
$$+ \tfrac{1}{5}\sin(6\theta_i + 5\theta_o) + \tfrac{1}{7}\sin(6\theta_i + 7\theta_o)\cdots] \quad (11.33)$$

And, for $r = 1.0$, and $\phi_o = -90°$, expression (11.31) becomes:

$$v_o = \frac{3\sqrt{3}\,\hat{V}_N}{2\pi} [\sin\theta_o \;\leftarrow\; \text{Wanted component of output voltage}$$

Superimposed harmonic components
$$+ \tfrac{1}{2}\sin(3\theta_i - 2\theta_o) + \tfrac{1}{4}\sin(3\theta_i - 4\theta_o)$$
$$- \tfrac{1}{5}\sin(6\theta_i - 5\theta_o) - \tfrac{1}{7}\sin(6\theta_i - 7\theta_o)\cdots] \quad (11.34)$$

HARMONIC SERIES FOR THE SIX-PULSE VOLTAGE WAVEFORM

The 6-pulse voltage waveform is comprised of 2 3-pulse waveforms, the ripple voltages of which are mutually displaced with respect to one another. It was seen in Chapter 4, for the case of a steady d-c output, that in order to derive the harmonic series for the 6-pulse voltage waveform, it is necessary simply to add together the harmonic series for 2 3-pulse waveforms, the first series containing terms in θ_i, and the second series containing terms in $(\theta_i + \pi)$. This simple procedure is equally applicable to the voltage waveform of the cycloconverter.

The Circulating Current Mode

By applying the above procedure to expression (11.24), the following general harmonic series, representing the output voltage of the 6-pulse cycloconverter, when operating with a continuous circulating current, is obtained:

$$v_o = k \cdot \frac{3\sqrt{3}\,\hat{V}_N}{2\pi} \Bigg\{ r \sin \theta_o \;\leftarrow\; \text{Wanted component of output voltage}$$

Superimposed harmonic components
$$+ \frac{1}{2} \sum_{p=1}^{p=\infty} \sum_{n=0}^{2n+1=6p+1} \left[\frac{a_{(6p-1)(2n+1)}}{6p-1} + \frac{a_{(6p+1)(2n+1)}}{6p+1} \right]$$

$$\times \left[\sin(6p\theta_i + [2n+1]\theta_o) - \sin(6p\theta_i - [2n+1]\theta_o) \right] \Bigg\} \quad (11.35)$$

where

$$k = 1.0 \text{ for the 6-pulse midpoint circuit}$$

$$k = 2.0 \text{ for the 6-pulse bridge circuit} \quad (11.36)$$

The Circulating Current-Free Mode

By application of the same procedure to expression (11.31), the following general harmonic series, representing the output voltage of the 6-pulse

EXPRESSION FOR THE TWELVE-PULSE VOLTAGE

cycloconverter, in the circulating current-free mode of operation, is obtained:

$$v_o = k \cdot \frac{3\sqrt{3}\,\hat{V}_N}{2\pi} \Bigg[r \sin \theta_o \longleftarrow \text{Wanted component of output voltage}$$

Superimposed harmonic components

$$+ \frac{1}{2} \sum_{p=1}^{p=\infty} \left\{ \sum_{n=0}^{2n+1=6p+1} \left[\frac{a_{(6p-1)(2n+1)}}{6p-1} + \frac{a_{(6p+1)(2n+1)}}{6p+1} \right] \right.$$

$$\times [\sin(6p\theta_i + [2n+1]\theta_o) - \sin(6p\theta_i - [2n+1]\theta_o)]$$

$$+ \frac{2}{\pi} \sum_{m=0}^{m=\infty} \sum_{n=0}^{n=\infty} \left[\frac{1}{2n-(2m+1)} \right] \left[\frac{a_{(6p-1)2n}}{6p-1} + \frac{a_{(6p+1)2n}}{6p+1} \right]$$

$$\times [-\cos\{6p\theta_i - ([2m+1]\theta_o - [2n-(2m+1)]\phi_o)\}$$

$$+ \cos\{6p\theta_i + ([2m+1]\theta_o - [2n-(2m+1)]\phi_o)\}]$$

$$+ \left[\frac{1}{2n+2m+1} \right] \left[\frac{a_{(6p-1)2n}}{6p-1} + \frac{a_{(6p+1)2n}}{6p+1} \right]$$

$$\times [\cos\{6p\theta_i - ([2m+1]\theta_o + [2n+2m+1]\phi_o)\}$$

$$\left. - \cos\{6p\theta_i + ([2m+1]\theta_o + [2n+2m+1]\phi_o)\}] \right\} \Bigg]$$

(11.37)

where k is defined by (11.36).

HARMONIC SERIES FOR THE TWELVE-PULSE VOLTAGE WAVEFORM

As was seen in Chapter 4, in order to obtain the harmonic series for the 12-pulse voltage waveform, it is necessary simply to add together the harmonic series for 2 6-pulse waveforms, the first containing terms in θ_i, and the second containing terms in $(\theta_i - \pi/6)$.

The Circulating Current Mode

By application of the above procedure to expression (11.35) the following general harmonic series, representing the output voltage of the 12-pulse

304   CYCLOCONVERTER OUTPUT VOLTAGE ANALYSIS

cycloconverter, when operating with a continuous circulating current, is obtained:

$$v_o = C \cdot \frac{3\sqrt{3}\,\hat{V}_N}{2\pi} \Bigg\{ r \sin \theta_o \;\leftarrow\; \text{Wanted component of output voltage}$$

Superimposed harmonic components
$$+ \frac{1}{2} \sum_{p=1}^{p=\infty} \sum_{n=0}^{2n+1=12p+1} \left[\frac{a_{(12p-1)(2n+1)}}{12p - 1} + \frac{a_{(12p+1)(2n+1)}}{12p + 1} \right]$$
$$\times \left[\sin(12p\theta_i + [2n+1]\theta_o) - \sin(12p\theta_i - [2n+1]\theta_o) \right] \Bigg\}$$

(11.38)

where

$c = 1.0$ for the 12-pulse midpoint circuit
$c = 2.0$ for 2 6-pulse bridge circuits connected in parallel
$c = 4.0$ for 2 6-pulse bridge circuits connected in series   (11.39)

The Circulating Current-Free Mode

By application of the same procedure to expression (11.37), the following general harmonic series, representing the output voltage of the 12-pulse cycloconverter, in the circulating current-free mode of operation, is obtained:

$$v_o = C \cdot \frac{3\sqrt{3}\,\hat{V}_N}{2\pi} \bigg[r \sin \theta_o \;\leftarrow\; \text{Wanted component of output voltage}$$

Superimposed harmonic components
$$+ \frac{1}{2} \sum_{p=1}^{p=\infty} \Bigg\{ \sum_{n=0}^{2n+1=12p+1} \left[\frac{a_{(12p-1)(2n+1)}}{12p - 1} + \frac{a_{(12p+1)(2n+1)}}{12p + 1} \right]$$
$$\times \left[\sin(12p\theta_i + [2n+1]\theta_o) - \sin(12p\theta_i - [2n+1]\theta_o) \right]$$
$$+ \frac{2}{\pi} \sum_{m=0}^{m=\infty} \sum_{n=0}^{n=\infty} \left[\frac{1}{2n - (2m+1)} \right] \left[\frac{a_{(12p-1)2n}}{12p - 1} + \frac{a_{(12p+1)2n}}{12p + 1} \right]$$
$$\times \left[-\cos\{12p\theta_i - ([2m+1]\theta_o - [2n - (2m+1)]\phi_o)\} \right.$$
$$\left. + \cos\{12p\theta_i + ([2m+1]\theta_o - [2n - (2m+1)]\phi_o)\} \right]$$
$$+ \left[\frac{1}{2n + 2m + 1} \right] \left[\frac{a_{(12p-1)2n}}{12p - 1} + \frac{a_{(12p+1)2n}}{12p + 1} \right]$$
$$\times \left[\cos\{12p\theta_i - ([2m+1]\theta_o + [2n + 2m + 1]\phi_o)\} \right.$$
$$\left. - \cos\{12p\theta_i + ([2m+1]\theta_o + [2n + 2m + 1]\phi_o)\} \right] \Bigg\} \bigg]$$

(11.40)

where C is defined by (11.39).

HARMONIC SERIES FOR VOLTAGE WAVEFORMS WITH OTHER PULSE NUMBERS

The harmonic series for voltage waveforms with other pulse numbers, which are integer multiples of 3, can be deduced directly from the basic series for the 3-pulse waveform. Invariably it is found that those harmonic components are present which contain terms in integer multiples of the input frequency and the circuit pulse number, with the same relative amplitudes as in the 3-pulse voltage, whereas all other harmonic terms are absent.

Thus, for example, the harmonic terms present in the output voltage of the 24-pulse cycloconverter have frequencies given by:

$$f_H = 24pf_i \pm (2n+1)f_o$$

The relative amplitudes of these components are the same as in the 3-pulse waveform.

THE WANTED COMPONENT OF THE OUTPUT VOLTAGE

An inspection of the general harmonic series derived for the output voltage of cycloconverters of various pulse numbers, shows that the wanted component of the output voltage invariably can be represented by the following expression:

$$v_W = \hat{V}_{W_{max}} r \sin \theta_o \qquad (11.41)$$

Furthermore, as might be expected, the value of $\hat{V}_{W_{max}}$ is invariably identical with the maximum d-c voltage attainable from the circuit—that is, the mean component of voltage obtained with a steady firing angle of 0°. Thus

$$\hat{V}_{W_{max}} = \frac{s \cdot 3\sqrt{3}\, \hat{V}_N}{2\pi} = V_{d_{max}} \qquad (11.42)$$

where s is the number of 3-pulse groups connected in series with one another·

According to expression (11.16), the analog reference voltage which controls the timing of the firing pulses is given by

$$v_R = \hat{V}_T r \sin \theta_o$$

Thus it is clear that the wanted component of the output voltage is directly proportional to, and is in phase with, the analog reference voltage. Thus, so far as the wanted component of output voltage is concerned, the cycloconverter, taken in conjunction with a firing circuit using the cosine wave crossing control principle, can be regarded as being an amplifier with a linear voltage transfer characteristic, with no phase shift between the input and output voltages. This conclusion is generally valid, irrespective of the output to input frequency ratio, or the load displacement angle.

A further point; since $\hat{V}_{W_{max}}$ is directly proportional to the cycloconverter input voltage, this means to say that if the amplitude of the cosine timing voltage, $\hat{V}_T$, is also directly proportional to the cycloconverter input voltage, then the amplitude of the wanted component of the output is dependent only upon the amplitude of the reference voltage, and is independent of the amplitude of the input voltage of the cycloconverter. (This, of course, assumes that there is no tendency for the firing angle to be "pushed" beyond the boundaries of the permissible control range.)

THE HARMONIC FREQUENCIES

General Comments

From an inspection of the general formulae for the output voltage, it is possible to perceive the harmonic frequencies present in the output of the cycloconverter. Before proceeding to an examination of the quantitative results of a detailed computation of the coefficients of the harmonic series, it is of interest, firstly, simply to make an examination of the spectrum of harmonic frequencies to be found in the output voltage. The following discussion, then, is specifically concerned with the frequencies of the distortion components, and, for the time being, their amplitudes will be ignored.

It can be seen from the general formulae that the harmonic distortion components have frequencies which are sums or differences between multiples of the output and input frequencies. Thus each harmonic frequency is a function of both the input and output frequency, and variations in either of these frequencies result in corresponding alterations to the spectrum of harmonic frequencies at the output.

It can also be seen, from the general formulae for the output voltage, that harmonic frequencies which are direct integer multiples of the output frequency are absent, except at certain discrete output to input frequency ratios, at which the "necessary" "beat" frequencies happen to be integer multiples of the output frequency, simply because, at these discrete frequency ratios, the output and input waves are synchronous with one another. (It should be noted, however, that the conclusion that "integer multiple" harmonics are generally absent, applies only to the "theoretically ideal" waveform, as fabricated by the cosine wave crossing control method. As mentioned in Chapters 9 and 13, due to practical effects, small amounts of "integer multiple" harmonics are usually to be found in the output voltage of the cycloconverter.)

From an inspection of the general expressions for the output voltage of the (3-pulse) cycloconverter, in the circulating current, and circulating current-free modes, (expressions (11.24) and (11.31) respectively), it is evident that a definite difference exists between the harmonic frequency spectra for the two cases. For the circulating current mode, the series for each "family" of harmonic components terminates at a specified term. For the circulating current-free mode, the general expression for the output voltage consists of the sum of two expressions. The first of these is identical with the expression for the output voltage in the circulating current mode; the second expression contains the same harmonic families as the first, but each of these families has an infinite number of terms. Thus the spectrum of frequencies for the circulating current-free mode is theoretically considerably more "diversified" than with the circulating current mode. This is illustrated by the harmonic frequency charts discussed in the following sections.

Frequency Spectrum for the Circulating Current-Free Mode

THE THREE-PULSE VOLTAGE. An inspection of expression (11.31) shows that for the 3-pulse cycloconverter, operating in the circulating current-free mode, the frequencies of the harmonic components present in the output voltage are given by the following general relationships

$$f_H = |3(2p - 1)f_i \pm 2nf_o| \tag{11.43}$$

and

$$f_H = |6pf_i \pm (2n + 1)f_o| \tag{11.44}$$

where p is any integer from 1 to infinity
n is any integer from 0 to infinity

The chart of Fig. 11.5 is a graphical representation of the relationship between the spectrum of harmonic frequencies in the output voltage of the 3-pulse cycloconverter, operating in the circulating current-free mode, and the output to input frequency ratio. For convenience, the harmonic frequencies are expressed as multiples of the input frequency. By expressing them thus, a series of linear relationships are obtained. This is evident from (11.43) and (11.44), which can be rearranged as follows:

$$\frac{f_H}{f_i} = \left| 3(2p - 1) \pm 2n\frac{f_o}{f_i} \right| \tag{11.45}$$

$$\frac{f_H}{f_i} = \left| 6p \pm (2n + 1)\frac{f_o}{f_i} \right| \tag{11.46}$$

For clarity, only the first few terms in each "family" of harmonic frequencies are shown in Fig. 11.5. As will be seen later, these are the terms which have

308   CYCLOCONVERTER OUTPUT VOLTAGE ANALYSIS

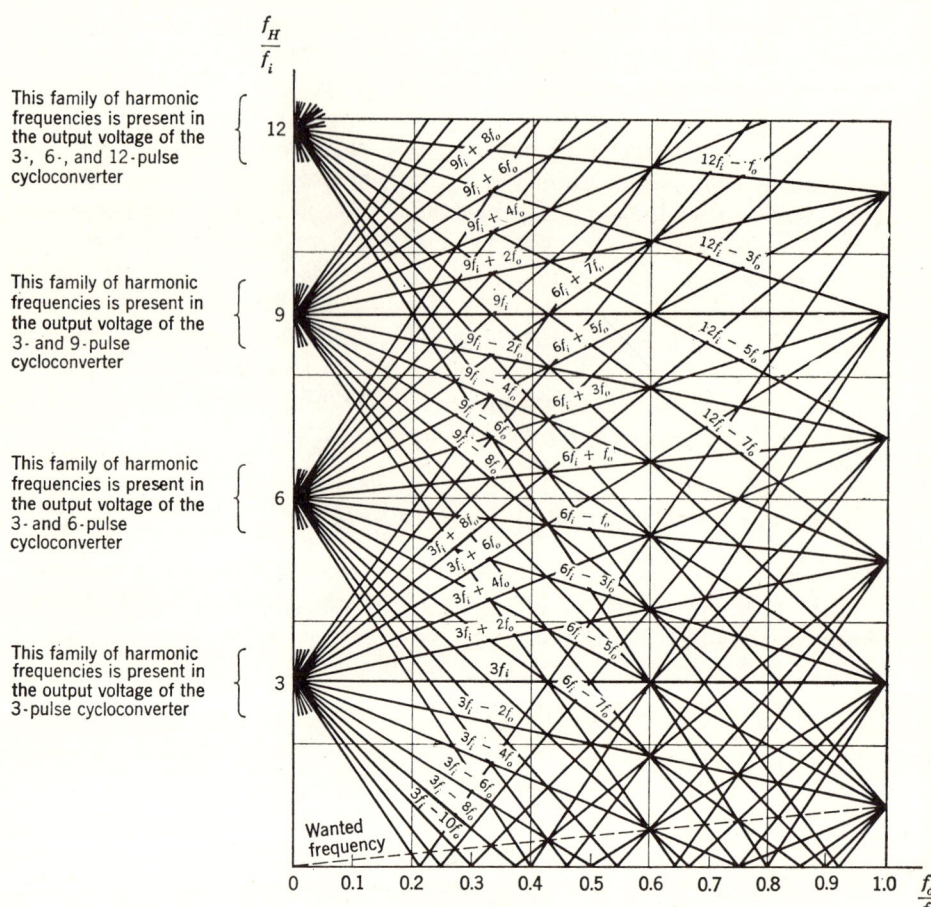

Figure 11.5. Chart showing the relationships between the "predominant" harmonic frequencies present in the output voltage of the 3-pulse cycloconverter, operating without circulating current, and the output to input frequency ratio. For cycloconverters with higher pulse numbers, certain harmonic families are eliminated as indicated.

the largest amplitudes, and therefore generally have the most practical significance. The following discussion should clarify the meaning and physical interpretation of this "harmonic frequency chart."

Consider, firstly, the particular case of $f_o/f_i = 0$—in other words, a steady d-c output. It is seen that the harmonic frequencies present in the output voltage have values of $3 \times f_i$, $6 \times f_i$, $9 \times f_i$, and so on. Of course, this result, for a 3-pulse converter with a steady d-c output, is already well known.

Consider, now, the situation with a very low output to input frequency ratio, say 0.01. It is seen that each discrete "parent" harmonic frequency associated with the condition of zero output frequency, has now been replaced by a "family" of frequencies, each of which are very close to one another, and to the "parent" frequency. Thus, at this very low output frequency ratio, the spectrum of harmonic frequencies is, for all practical purposes, virtually the same as with zero output frequency. This is not surprising, and it is no more than would be expected intuitively. Nonetheless, a process of "migration" of the harmonic frequencies away from their "parent" harmonics has already commenced.

As the output to input frequency ratio is further increased, so the separation of the individual members of each harmonic family from one another, and from the parent harmonic, becomes progressively greater, and it is evident that an increasing ratio between output and input frequency results in an increasing departure of the spectrum of harmonic frequencies from that obtained at zero output frequency.

Consider, for example, the point at which the output to input frequency ratio is exactly 0.3. At this point, the lowest harmonic frequency is actually zero—this is the $(3f_i - 10f_o)$ term, which clearly now has digressed a long way from the $3 \times f_i$ "parent" harmonic. The next harmonic frequency, due to both the $(3f_i - 8f_o)$, and $(3f_i - 12f_o)$ components, is $0.6 \times$ the input frequency, and so on.

The dashed line, having unity slope, represents the wanted output frequency itself. Thus all harmonic frequencies below this line are "subharmonic" frequencies, whereas all frequencies above it are "superharmonic" frequencies.

In general, harmonic components having frequencies close to the wanted output frequency, and, in particular, subharmonic and zero-frequency components, are potentially objectionable in practice, because they cannot be readily filtered off, and they can cause excessive amounts of harmonic current to flow.

The fact that the harmonic frequency chart shows an increasing preponderance of "objectionable" harmonic frequencies, with increasing output to input frequency ratio, would seem, intuitively, to be indicative of a progressive deterioration in the quality of the output voltage waveform. This is indeed the case, although actually the harmonic frequency chart, by itself, does not constitute sufficient information to make this deduction, because it is necessary to know also the relative amplitudes of the components to which the objectionable frequencies belong. This information is not furnished by the frequency chart, which is concerned solely with the frequencies of the harmonic components, and not with their amplitudes.

So long as the amplitudes of the objectionable harmonic components are relatively small, they do not constitute a serious practical threat to the successful application of the cycloconverter, especially since it is possible, by means of closed loop negative feedback techniques, to virtually completely suppress small amounts of objectionable harmonics which would otherwise be present with an "open-loop" firing angle control.

A point, illustrated by this harmonic frequency chart, is that at certain discrete ratios of output to input frequency, several harmonic frequencies take on the same values as one another. Moreover, several of the harmonic frequencies may assume the same value as the wanted output frequency. Thus, at these discrete points, several "harmonic" components become, as it were, "mated" to the wanted component. The result of this is that the net amplitude of the component at the wanted output frequency, theoretically, is no longer exactly proportional to the reference voltage. Since, however, the amplitudes of the "mated" harmonic components generally are relatively small, this effect is mainly of theoretical interest, and, for most practical purposes it does not detract from the basic concept of the cycloconverter as a linear amplifier, which invariably produces a wanted component of output voltage which is directly proportional to, and in phase with, the reference voltage, regardless of the output to input frequency ratio.

A further related point, illustrated by the harmonic frequency chart, is that at certain discrete frequency ratios, the spectrum of the harmonic frequencies assumes the form of a "conventional" Fourier series, in which each harmonic frequency is a direct integer multiple of the wanted output frequency. Thus, for example, at output frequencies given by $f_o = 3f_i/(2n + 1)$, only odd harmonics of the output frequency are present. And, at output frequencies given by $f_o = 3f_i/2n$, both odd and even harmonic frequencies—as well as a "zero frequency" component—are present.

In fact, this result is to be expected from a consideration of the waveforms of the 3-pulse cycloconverter at these discrete frequencies. Thus, for $f_o = 3f_i/(2n + 1)$, all half cycles of the output voltage wave are exactly similar to one another, which indicates the presence of odd harmonics only. On the other hand, for $f_o = 3f_i/2n$, all complete cycles of the output voltage wave—but not half cycles—are the same as one another, indicating the presence of both odd and even harmonics.

THE SIX-PULSE VOLTAGE. An inspection of expression (11.37) shows that for the 6-pulse cycloconverter, operating in the circulating current-free mode, the frequencies of the harmonic components present in the output voltage are given by:

$$f_H = |6pf_i \pm (2n + 1)f_o| \qquad (11.47)$$

Thus, the $3(2p - 1)f_i \pm 2nf_o$ harmonic families, associated with the 3-pulse voltage, are absent from the 6-pulse voltage, whereas the remaining harmonic families are exactly the same.

Thus, the frequency chart of Fig. 11.5, for the 3-pulse cycloconverter, is also applicable to the 6-pulse cycloconverter, with the provision that the $3(2p - 1)f_i \pm 2nf_o$ harmonic families are absent.

THE TWELVE-PULSE VOLTAGE. An inspection of expression (11.40) shows that for the 12-pulse cycloconverter, operating in the circulating current-free mode, the frequencies of the harmonic components present in the output voltage are given by

$$f_H = |12pf_i \pm (2n + 1)f_o| \qquad (11.48)$$

Thus, the frequency chart of Fig. 11.5, for the 3-pulse cycloconverter, is also applicable to the 12-pulse circuit, with the provision that all but the $12f_i \pm (2n + 1)f_o$ harmonic families are absent.

Frequency Spectrum for the Circulating Current Mode

THE THREE-PULSE VOLTAGE. An inspection of expression (11.24) shows that for a 3-pulse cycloconverter, operating with a continuous circulating current, the frequencies of the harmonic components present in the output voltage are given by

$$f_H = |3(2p - 1)f_i \pm 2nf_o| \qquad (11.49)$$

where $n \leqslant 3(2p - 1) + 1$ and

$$f_H = |6pf_i \pm (2n + 1)f_o| \qquad (11.50)$$

where

$$(2n + 1) \leqslant (6p + 1)$$

Expressions (11.49) and (11.50) are similar to (11.43) and (11.44), for the circulating current-free mode, but with the difference that each harmonic family now terminates at a definite term, rather than having an infinite number of components.

The chart of Fig. 11.6 shows the relationship between the harmonic frequencies, and the output to input frequency ratio, for the 3-pulse cycloconverter operating with a continuous circulating current. This chart shows all the harmonic frequencies which theoretically are present, up to and including the $12f_i \pm (2n + 1)f_o$ harmonic family. A comparison of this chart, with that of Fig. 11.5 for the circulating current-free mode, in which, for clarity, only the harmonic frequencies which are "predominant" among those theoretically present, are actually shown, demonstrates the relative "compactness" of the harmonic frequency spectrum for the circulating current mode.

THE SIX-PULSE VOLTAGE. An inspection of expression (11.35) shows that for the 6-pulse cycloconverter operating in the circulating current mode, the frequencies of the harmonic components present in the output voltage are given by

$$f_H = |6pf_i \pm (2n+1)f_o| \qquad (11.51)$$

where

$$(2n+1) \leqslant (6p+1)$$

Thus the frequency chart of Fig. 11.6, for the 3-pulse cycloconverter, is also applicable to the 6-pulse cycloconverter, with the provision that the $3(2p-1)f_i \pm 2nf_o$ harmonic families are absent.

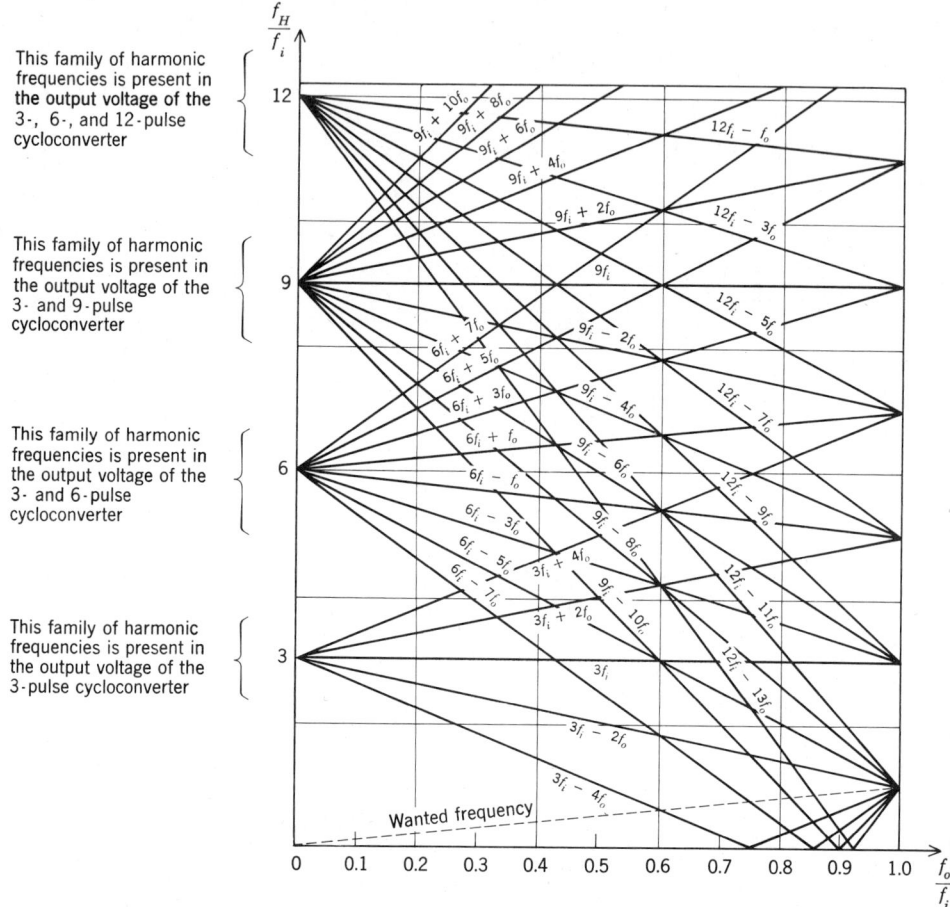

Figure 11.6. Chart shows the relationships between the harmonic frequencies present in the output voltage of the cycloconverter, operating with a continuous circulating current, and the output to input frequency ratio.

AMPLITUDES OF THE HARMONIC COMPONENTS

THE TWELVE-PULSE VOLTAGE. An inspection of expression (11.38) shows that for the 12-pulse cycloconverter operating in the circulating current mode, the frequencies of the harmonic components present in the output voltage are given by

$$f_H = |12pf_i \pm (2n + 1)f_o| \tag{11.52}$$

where

$$(2n + 1) \leqslant (12p + 1).$$

Thus, the frequency chart of Fig. 11.6, for the 3-pulse cycloconverter, is also applicable to the 12-pulse circuit, with the provision that all but the $12f_i \pm (2n + 1)f_o$ harmonic families are absent.

THE AMPLITUDES OF THE HARMONIC COMPONENTS

General Comments

The discussion so far has been concerned only with the frequencies of the harmonic components present in the output voltage, and their amplitudes have been ignored. An inspection of the general expressions obtained for the output voltage shows that the amplitude of each harmonic component is a function of the output voltage ratio and the load displacement angle, (or, for the circulating current mode, just of the output voltage ratio), but is *independent* of the frequency of the component. This means to say, for a given output voltage ratio and load displacement angle, that each line representing a given harmonic "beat" frequency on the chart of Fig. 11.5 (or Fig. 11.6) has associated with it a fixed amplitude, which is invariably the same regardless of the frequency ratio.

It has been seen already that the presence or absence of given families of harmonic frequencies at the output is determined by the pulse number of the converter. An inspection of the general formulae obtained for the output voltage of cycloconverters of various pulse numbers reveals the fact, for a given output voltage ratio and load displacement angle, that those harmonic components which are present always have the same relative amplitudes, independent of the pulse number of the converter. This means to say that a single set of quantitative data, related to the amplitudes of the harmonic components of the output voltage of the 3-pulse cycloconverter, is equally applicable to the output voltage of any cycloconverter having a pulse number which is a multiple of 3 (in other words, to virtually all practical cycloconverter circuits), it being necessary simply to ignore the data related to those harmonic components which are known to be absent from a given circuit.

Harmonic Amplitudes for the Circulating Current-Free Mode

From the general expression (11.31), for the output voltage of the 3-pulse cycloconverter, certain general identities can be shown to exist between the amplitudes of given harmonic components, with various given load displacement angles. These identities are as follows:

For any given output voltage ratio:

Amplitude of component having any given beat frequency, with load displacement angle ϕ_o = Amplitude of component having the same beat frequency at load displacement angle $(\pi - \phi_o)$ (11.53)

and

Amplitude of component having frequency $nf_i + mf_o$ with load displacement angle ϕ_o = Amplitude of component having frequency $nf_i - mf_o$ with load displacement angle $-\phi_o$ (11.54)

Conversely

Amplitude of component having frequency $nf_i - mf_o$ with load displacement angle ϕ_o = Amplitude of component having frequency $nf_i + mf_o$ with load displacement angle $-\phi_o$ (11.55)

The physical explanation for these results is illustrated, in general terms, by the voltage waveforms of Figs. 11.7 and 11.8, for a 6-pulse cycloconverter.

In order for the positive converter to generate a positive-going level of output voltage, it is necessary for its firing pulses to be applied at a "switching frequency" which is greater than 6 × the input frequency. By so doing, each firing point is relatively farther advanced than the previous one, and the level of the output voltage moves progressively in a positive direction. Conversely, for the positive converter to produce a negative-going level of output voltage, the firing pulses must be applied at a switching frequency which is less than 6 × the input frequency; in this case, each firing point is relatively farther retarded than the previous one, and the level of the output voltage moves progressively in a negative direction.

Thus, for the positive converter, a "fast" switching frequency (i.e., $>6f_i$) produces an output voltage waveform with a positive slope, whereas a "slow" switching frequency (i.e., $<6f_i$) produces an output voltage waveform with a negative slope. Conversely, for the negative converter, a "fast" switching frequency produces an output voltage waveform with a negative slope, whereas a slow switching frequency produces an output voltage waveform with a positive slope.

AMPLITUDES OF THE HARMONIC COMPONENTS 315

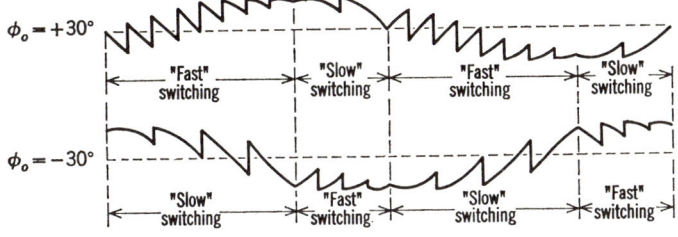

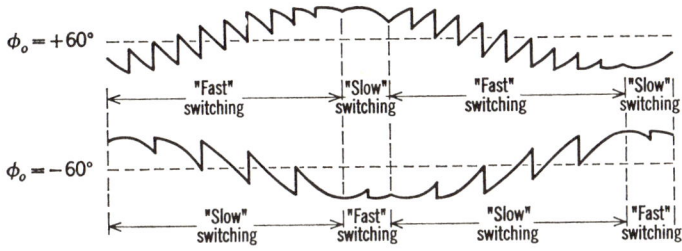

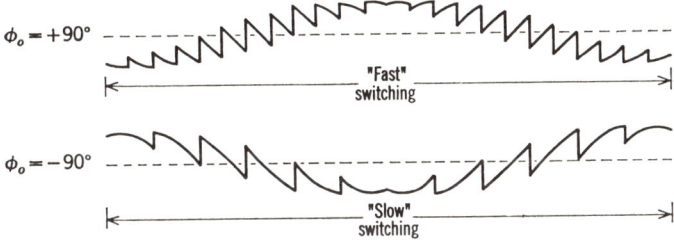

Figure 11.7. Six-pulse voltage waveforms illustrating that interchange of the sign of the load displacement angle results in a corresponding interchange between the durations of the "fast" and "slow" switching periods.

From the waveforms of Figs. 11.7 and 11.8, it is clear that the duration of the periods of fast switching and slow switching for each converter depend upon the point in the output cycle at which the bank selection takes place; this, of course, is determined directly by the displacement angle of the load.

In general terms, it can be said that the "fast switching" portions of the voltage waveform are responsible for the presence of beat frequencies consisting of sums of multiples of f_i and f_o, whereas the "slow switching" portions

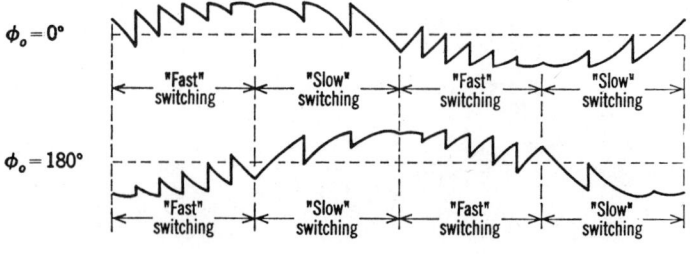

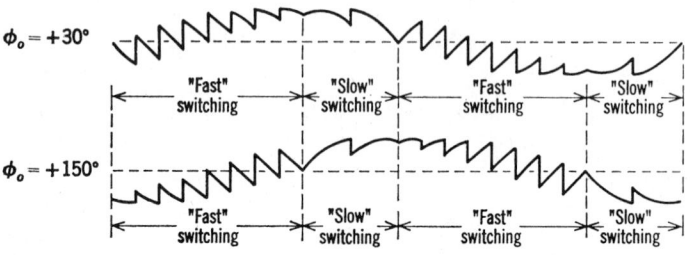

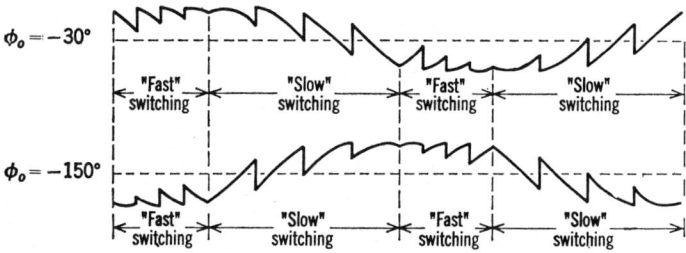

Figure 11.8. Six-pulse voltage waveforms illustrating that for load displacement angles of ϕ_o and $(\pi - \phi_o)$, the durations of the "fast" and "slow" switching periods are the same.

of the voltage waveform are responsible for the presence of beat frequencies consisting of differences of multiples of f_i and f_o. Moreover, the relative amplitudes of the complementary "sum" and "difference" harmonic components are related to the relative durations of the "fast switching" and "slow switching" periods, and interchanging the duration of these periods, which corresponds to interchanging the sign of the load displacement angle, as illustrated by Fig. 11.7, results in a corresponding interchange between the amplitudes of the complementary "sum" and "difference" frequency components. Conversely, for voltage waveforms for load displacement angles of ϕ_o and $(\pi - \phi_o)$, which have the same relative durations of "fast" and "slow" switching periods, as illustrated by Fig. 11.8, the amplitudes of the harmonic components are also the same.

For the particular load displacement angles of 0° and 180°, the "fast switching" and "slow switching" periods are equal to one another; and as might be expected, the amplitudes of the complementary sum and difference frequency components are also equal to one another, and the same for either case.

For loads having positive displacement angle, the "fast" switching periods have relatively longer duration than the "slow" switching periods, and the amplitudes of the "sum frequency" harmonic components are generally larger than those of the complementary "difference frequency" harmonic components. In the extreme case, for $\phi_o = +90°$ the "fast" switching mode of operation persists for the whole of the output cycle; and (for $r = 1$), the "sum frequency" harmonic components only are present.

Conversely, for loads having negative displacement angle, the "slow switching" periods have relatively longer duration than the "fast switching" periods, and the amplitudes of the "difference frequency" harmonic components are now generally larger than those of the complementary "sum frequency" harmonic components. In the extreme case, for $\phi_o = -90°$ the "slow" switching mode of operation persists for the whole of the output cycle; and (for $r = 1$), the "difference frequency" harmonic components only are present.

Bearing in mind the relationships of (11.53) through (11.55), it is evident that the results of a computation of the amplitudes of the harmonic components for a range of load displacement angles in one quadrant only, say 0° to +90°, can also be applied to load displacement angles in the other three quadrants. This, of course, avoids considerable repetition of data.

QUANTITATIVE DATA. The amplitudes of the predominant members of each of the $3f_i \pm 2nf_o$, $6f_i \pm (2n + 1)f_o$, and $12f_i \pm (2n + 1)f_o$ harmonic families, for the circulating current-free mode of operation, for a range of output voltage ratios from 1.0 to 0.1, and for a range of load displacement angles, which [by virtue of the identities of expressions (11.53) through (11.55)] embrace the complete spectrum from 0 to 360°, are shown in Tables 11.3 through 11.5. Figure 11.9 gives a graphical portrayal of the amplitudes of these harmonic components, for the particular case of $r = 1.0$.

These "amplitude tables," when used in conjunction with the "frequency chart" of Fig. 11.5, provide a comprehensive set of quantitative data for 3-, 6-, and 12-pulse cycloconverters.

Tables 11.3 *Showing the Amplitudes of the Distortion Components having Frequencies of $3f_i \pm 2nf_o$, in the Output Voltage of the Cycloconverter, for Various Output Voltage Ratios and Output Load Displacement Angles*

NOTES:
1. Tables show peak amplitude of harmonic, as per unit value of $\hat{V}_{w\,max}$
2. These harmonic components are present in the output voltage of the 3-pulse cycloconverter, with no circulating current
3. Where there is only one column for a given value of $2n$, the values given apply to both the $3f_i + 2nf_o$ and $3f_i - 2nf_o$ harmonic components.
4. Where there are two columns for a given value of $2n$: The values in the "black" column apply to the $3f_i + 2nf_o$ harmonic components for negative ϕ_o, and to the $3f_i - 2nf_o$ harmonic components for positive ϕ_o. The values given in the "white" column apply to the $3f_i + 2nf_o$ harmonic components for negative ϕ_o, and to the $3f_i - 2nf_o$ harmonic components for positive ϕ_o.

$3f_i \pm 2nf_o$

	$\phi_o = 0°, 180°$								$\phi_o = \pm 30°, \pm 150°$								
$2n \rightarrow$	0	2	4	6	8	10	12	14	0	2 (b)	2 (w)	4 (b)	4 (w)	6	8	10	12	14
$r \downarrow$																		
1.0	0.318	.272	.164	.064	.021	.015	.009	.007	.179	.167	.333	.279	.134	.095	.062	.044	.037	.031
0.9	.394	.283	.137	.048	.020	.013	.008	.006	.264	.207	.362	.237	.138	.086	.058	.044	.036	.030
0.8	.470	.276	.112	.038	.018	.011	.007	.005	.354	.229	.364	.197	.133	.079	.054	.042	.034	.029
0.7	.538	.257	.090	.031	.015	.009	.006	.005	.438	.235	.348	.162	.122	.071	.050	.038	.032	.027
0.6	.596	.230	.071	.025	.013	.008	.005	.004	.516	.229	.317	.131	.108	.063	.044	.035	.028	.024
0.5	.644	.199	.055	.020	.011	.007	.004	.003	.584	.212	.028	.103	.092	.054	.038	.030	.025	.021
0.4	.682	.163	.040	.015	.008	.005	.004	.003	.642	.184	.227	.079	.074	.044	.031	.025	.020	.017
0.3	.712	.125	.028	.011	.006	.004	.003	.002	.689	.148	.173	.057	.056	.033	.024	.019	.016	.013
0.2	.734	.084	.018	.007	.004	.003	.002	.001	.722	.104	.116	.037	.037	.023	.016	.013	.011	.009
0.1	.746	.042	.009	.004	.002	.001	.001	.001	.743	.054	.057	.019	.019	.011	.008	.007	.005	.005

$3f_i \pm 2nf_o$

	$\phi_o = \pm 60°, \pm 120°$									$\phi_o = \pm 90°$								
$2n \rightarrow$	0	2	4 (b)	4 (w)	6	8	10	12	14	0	2 (b)	2 (w)	4 (b)	4 (w)	6	8	10	12	14
$r \downarrow$																			
1.0	0.020	.486	.020	.269	.017	.016	.014	.012	.010	.000	.000	.500	.000	.250	.000	.000	.000	.000	.000
0.9	.081	.540	.063	.210	.038	.030	.024	.020	.017	.027	.020	.579	.149	.015	.012	.010	.008	.007	.006
0.8	.167	.554	.110	.167	.054	.041	.033	.028	.024	.097	.062	.613	.059	.043	.032	.026	.021	.018	.015
0.7	.267	.536	.151	.141	.065	.049	.039	.033	.028	.195	.109	.604	.011	.071	.051	.039	.032	.027	.023
0.6	.372	.491	.180	.126	.070	.052	.042	.035	.030	.307	.150	.561	.059	.091	.063	.048	.039	.032	.028
0.5	.474	.426	.193	.113	.069	.051	.041	.034	.029	.422	.177	.490	.085	.101	.068	.052	.041	.034	.030
0.4	.566	.347	.189	.099	.062	.046	.037	.031	.026	.529	.185	.399	.092	.099	.066	.049	.039	.033	.029
0.3	.642	.260	.167	.080	.051	.038	.030	.205	.021	.621	.171	.298	.084	.086	.056	.042	.033	.028	.024
0.2	.701	.169	.127	.057	.036	.027	.021	.018	.015	.691	.135	.193	.063	.063	.041	.030	.024	.020	.017

Voltage of the Cycloconverter, for Various Output Voltage Ratios and Output Load Displacement Angles

NOTES:

1. Tables show peak amplitude of harmonic, as per unit value of $\hat{V}_{W\max}$
2. These harmonic components are present in the output voltage of 3- and 6-pulse cycloconverters, with no circulating current
3. Where there is only one column for a given value of $(2n+1)$, the values given apply to both the $6f_i + (2n+1)f_o$ and $6f_i - (2n+1)f_o$ harmonic components
4. Where there are two columns for a given value of $(2n+1)$:
   The values given in the "black" column apply to the $6f_i + (2n+1)f_o$ harmonic components for positive ϕ_o, and to the $6f_i - (2n+1)f_o$ harmonic components for negative ϕ_o.
   The values given in the "white" column apply to the $6f_i + (2n+1)f_o$ harmonic components for negative ϕ_o, and to the $6f_i - (2n+1)f_o$ harmonic components for positive ϕ_o

$6f_i \pm (2n+1)f_o$

$\phi_o = 0°, 180°$

$(2n+1) \rightarrow$ $r \downarrow$	1	3	5	7	9	11	13	15
1.0	.006	.055	.105	.100	.055	.026	.022	.017
0.9	.036	.097	.117	.081	.041	.026	.020	.017
0.8	.069	.123	.111	.065	.035	.023	.021	.017
0.7	.101	.136	.098	.053	.032	.023	.019	.016
0.6	.130	.138	.084	.045	.029	.022	.018	.015
0.5	.156	.132	.071	.040	.027	.021	.018	.015
0.4	.178	.120	.060	.036	.026	.021	.017	.015
0.3	.196	.105	.052	.034	.025	.020	.017	.015
0.2	.208	.090	.047	.032	.025	.020	.017	.015
0.1	.216	.078	.044	.031	.024	.020	.017	.015

$\phi_o = \pm 30°, \pm 150°$

$(2n+1) \rightarrow$ $r \downarrow$	1	3	5 (black)	5 (white)	7 (black)	7 (white)	9	11	13	15
1.0	.018	.045	.150	.058		.136	.055	.042	.027	.018	.015
0.9	.024	.057	.163	.071		.096	.051	.033	.022	.017	.014
0.8	.042	.100	.145	.067		.064	.041	.026	.019	.012	.011
0.7	.097	.130	.114	.055		.041	.030	.018	.013	.010	.008
0.6	.162	.142	.081	.040	.024		.020	.011	.008	.006	.004
0.5	.221	.134	.053	.027	.014		.013	.008	.005	.004	.003
0.4	.266	.112	.036	.025	.015		.015	.011	.008	.007	.006
0.3	.288	.098	.033	.031	.021		.021	.016	.013	.011	.009
0.2	.286	.084	.038	.038	.026		.026	.020	.017	.014	.012
0.1	.262	.074	.042	.042	.030		.030	.023	.019	.016	.014

$6f_i \pm (2n+1)f_o$

$\phi_o = \pm 60°, \pm 120°$

$(2n+1) \rightarrow$ $r \downarrow$	1	3	5	7 (black)	7 (white)	9	11	13	15
1.0	.023	.023	.022	.022		.181	.021	.019		
0.9	.072	.052	.097	.037		.209	.025	.017		
0.8	.087	.045	.183	.026		.188	.016	.011		
0.7	.065	.045	.238	.036		.147	.027	.017		
0.6	.088	.083	.247	.058		.103	.039	.025		
0.5	.174	.112	.213	.063		.066	.038	.025		
0.4	.260	.109	.152	.047		.032	.025	.016		
0.3	.316	.071	.084	.020		.007	.008	.004		
0.2	.327	.047	.046	.032		.021	.022	.016		
0.1	.291	.126	.062	.061		.038	.038	.027		

$\phi_o = \pm 90°$

$(2n+1) \rightarrow$ $r \downarrow$	1	3	5	7	9	11	13	15
1.0	.000	.000	.000		.000	.000	.000	.000
0.9	.039	.027	.020	.143	.010	.009	.007	.006
0.8	.081	.043	.026	.056	.009	.006	.008	.006
0.7	.067	.015	.002	.018	.003	.003	.002	.002
0.6	.010	.039	.032	.015	.015	.012	.010	.010
0.5	.123	.088	.054	.023	.021	.017	.015	.013
0.4	.235	.104	.051	.028	.016	.014	.011	.010
0.3	.313	.072	.023	.022	.005	.004	.003	.003
0.2	.337	.005	.019	.007	.009	.007	.006	.005
0.1	.301	.110	.057	.026	.020	.016	.014	.012

Tables 11.5 Showing the Amplitudes of the Distortion Components Having Frequencies of $12f_i \pm (2n+1)f_o$, in the Output Voltage of the Cycloconverter, for Various Output Voltage Ratios and Output Load Displacement Angles

NOTES:
1. Tables show peak amplitude of harmonic, as per unit value of $\hat{V}_{W\max}$.
2. These harmonic components are present in the output voltage of 3-, 6-, and 12-pulse cycloconverters, with no circulating current
3. Where there is only one column for a given value of $(2n+1)$, the values given apply to both the $12f_i + (2n+1)f_o$ and $12f_i - (2n+1)f_o$ harmonic components
4. Where there are two columns for a given value of $(2n+1)$: The values given in the "black" column apply to the $12f_i + (2n+1)f_o$ harmonic components for positive ϕ_o, and to the $12f_i - (2n+1)f_o$ harmonic components for negative ϕ_o. The values given in the "white" column apply to the $12f_i + (2n+1)f_o$ harmonic components for negative ϕ_o, and to the $12f_i - (2n+1)f_o$ harmonic components for positive ϕ_o

$12f_i \pm (2n+1)f_o$

$(2n+1) \rightarrow$	\multicolumn{10}{c|}{$\phi_o = 0°, 180°$}										
$r \downarrow$	1	3	5	7	9	11	13	15	17
1.0	.002	.001	.007	.005	.027	.050	.049	.027	.011
0.9	.011	.016	.010	.031	.053	.049	.030	.016	.011
0.8	.022	.021	.029	.052	.053	.036	.020	.013	.009
0.7	.029	.050	.057	.043	.025	.015	.011	.008
0.6	.035	.045	.061	.051	.032	.018	.012	.009	.008
0.5	.043	.061	.061	.041	.023	.014	.011	.009	.007
0.4	.057	.071	.056	.030	.017	.012	.010	.008	.007
0.3	.075	.070	.041	.022	.014	.011	.009	.008	.007
0.2	.091	.059	.029	.018	.013	.010	.009	.007	.006
0.1	.103	.044	.023	.016	.012	.010	.008	.007	.006

$(2n+1) \rightarrow$	\multicolumn{14}{c|}{$\phi_o = \pm 30°, \pm 150°$}														
$r \downarrow$	1	3	5	7	7	9	9	11	11	13	13	15	15	17	17
1.0	.002	.002	.003	.002	.010	.010	.020	.020	.027	.069	.026	.020	.012
0.9	.017	.015	.006	.007	.014	.033	.071	.026	.034	.018	.012	.008
0.8	.025	.011	.030	.031	.061	.036	.079	.026	.045	.017	.015	.010	.007	.005
0.7	.017	.035	.042	.044	.038	.085	.028	.062	.019	.024	.012	.008	.006	.004
0.6	.039	.044	.030	.043	.074	.035	.080	.026	.048	.015	.022	.012	.009	.008	.006
0.5	.044	.049	.055	.050	.038	.074	.026	.048	.017	.012	.010	.008	.007
0.4	.045	.070	.088	.057	.038	.036	.024	.017	.013	.012	.010	.007	.006	.005
0.3	.090	.079	.092	.031	.036	.020	.019	.013	.012	.009	.007	.003	.003	.002
0.2	.134	.060	.069	.025	.018	.012	.006	.006	.004	.004	.004	.003	.003	.002
0.1	.142	.058	.041	.031	.012	.007	.005	.007	.005	.005	.004	.003	.003	.002

$12f_i \pm (2n+1)f_o$

$(2n+1) \rightarrow$	\multicolumn{11}{c|}{$\phi_o = \pm 60°, \pm 120°$}											
$r \downarrow$	1	3	5	7	9	11	13	15	17
1.0	.001	.001	.003	.005	.007	.006	.008	.008	.007
0.9	.019	.018	.017	.016	.008	.013	.008	.010	.006	.011
0.8	.020	.024	.017	.023	.036	.019	.046	.015	.012	.008
0.7	.043	.032	.044	.022	.014	.013	.014	.103	.009	.069	.006
0.6	.031	.029	.049	.025	.074	.019	.102	.030	.011	.007	.009
0.5	.062	.047	.041	.030	.108	.019	.058	.102	.012	.058	.015
0.4	.064	.030	.109	.017	.094	.008	.014	.014	.008	.012
0.3	.051	.050	.126	.033	.051	.022	.013	.016	.012	.010	.006
0.2	.130	.058	.076	.025	.017	.013	.008	.008	.006	.006	.005

$(2n+1) \rightarrow$	\multicolumn{9}{c|}{$\phi_o = \pm 90°$}								
$r \downarrow$	1	3	5	7	9	11	13	15	17
1.0	.000	.000	.000	.000	.000	.000	.000	.000	.000
0.9	.000	.002	.000	.004	.003	.091	.077	.002	.002
0.8	.000	.003	.000	.003	.002	.086	.022	.002	.002
0.7	.000	.001	.000	.002	.010	.038	.005	.001	.000
0.6	.001	.000	.001	.013	.055	.019	.005	.006	.004
0.5	.003	.001	.000	.122	.026	.005	.007	.005	.001
0.4	.002	.001	.017	.092	.017	.002	.001	.001	.001
0.3	.002	.001	.002	.033	.013	.002	.001	.001	.001
0.2	.003	.003	.017	.011	.004	.008	.007	.006	.005
	.066	.114	.002	.004	.008	.002	.006	.001
	.028	.012	.011	.015	.004	.010	.003	.010	.005
	.003	.022	.016	.015	.012	.003	.010	.006	.005
	.056	.027	.016	.012	.011	.009	.007	.006	.005
	.013	.093	.008	.012	.011	.009	.007	.006	.005
	.117								

AMPLITUDES OF THE HARMONIC COMPONENTS 321

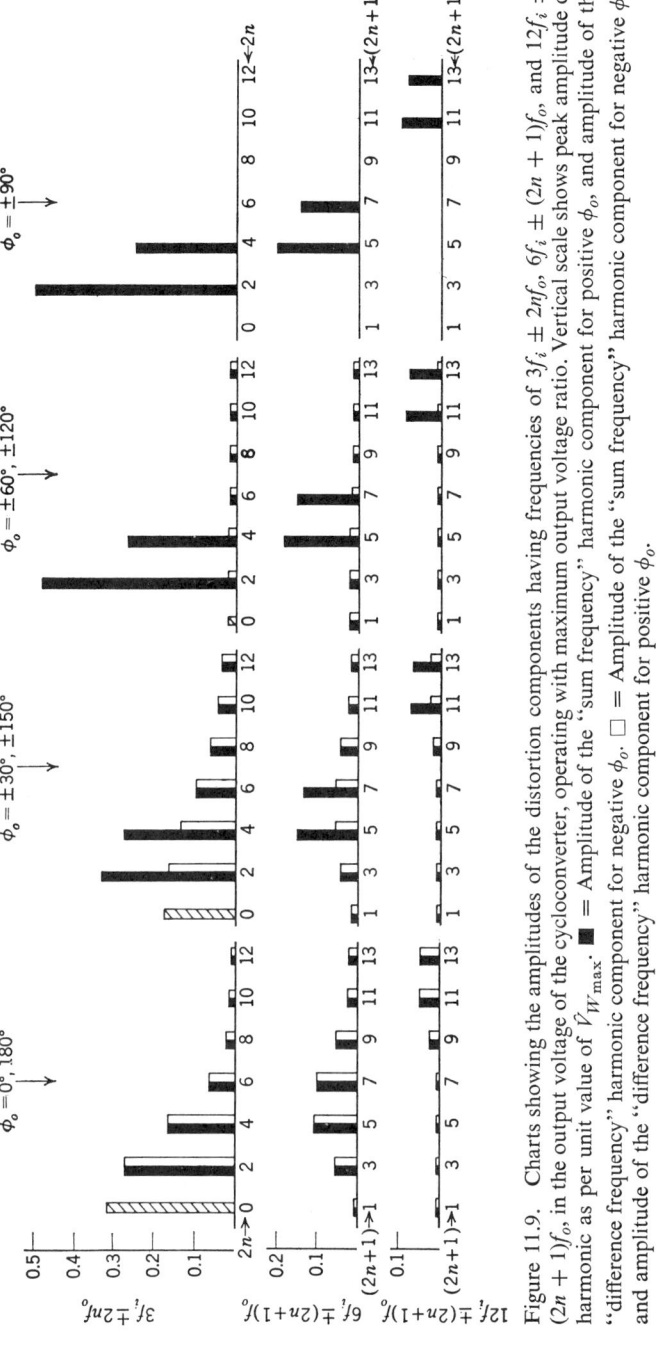

Figure 11.9. Charts showing the amplitudes of the distortion components having frequencies of $3f_i \pm 2nf_o$, $6f_i \pm (2n+1)f_o$, and $12f_i \pm (2n+1)f_o$, in the output voltage of the cycloconverter, operating with maximum output voltage ratio. Vertical scale shows peak amplitude of harmonic as per unit value of $\hat{V}_{W\max}$. ■ = Amplitude of the "sum frequency" harmonic component for positive ϕ_o, and amplitude of the "difference frequency" harmonic component for negative ϕ_o. ☐ = Amplitude of the "sum frequency" harmonic component for negative ϕ_o, and amplitude of the "difference frequency" harmonic component for positive ϕ_o.

Table 11.6 Showing the Amplitudes of the Distortion Components Having Frequencies of $3f_i \pm 2nf_o$, $6f_i \pm (2n + 1)f_o$, and $12f_i \pm (2n + 1)f_o$, in the Output Voltage of the Cycloconverter Operating with a Continuous Circulating Current, for Various Output Ratios

$2n \rightarrow$	$3f_i \pm 2nf_o$			$6f_i \pm (2n+1)f_o$				$12f_i \pm (2n+1)f_o$							$\leftarrow (2n+1)$
$r \downarrow$	0	2	4	1	3	5	7	1	3	5	7	9	11	13	$r \rightarrow$
1.0	.000	.250	.125	.000	.000	.100	.071	.000	.000	.000	.000	.000	.045	.038	1.0
0.9	.027	.279	.082	.033	.039	.115	.034	.003	.001	.004	.016	.042	.044	.010	0.9
0.8	.097	.275	.051	.062	.100	.092	.015	.014	.019	.027	.026	.052	.019	.002	0.8
0.7	.195	.247	.030	.041	.134	.060	.006	.024	.020	.002	.056	.031	.006	.000	0.7
0.6	.307	.205	.016	.025	.133	.033	.002	.020	.035	.047	.046	.012	.001	.000	0.6
0.5	.422	.156	.008	.105	.105	.015	.001	.033	.012	.061	.023	.003	.000	.000	0.5
0.4	.529	.107	.003	.169	.068	.005	.000	.047	.063	.040	.007	.001	.000	.000	0.4
0.3	.621	.063	.001	.193	.034	.001	.000	.015	.067	.015	.001	.000	.000	.000	0.3
0.2	.691	.029	.000	.166	.011	.000	.000	.086	.033	.003	.000	.000	.000	.000	0.2
0.1	.735	.007	.000	.096	.001	.000	.000	.083	.005	.000	.000	.000	.000	.000	0.1

NOTES:
1. Tables show peak amplitude of harmonic, as per unit value of $\hat{V}_{W\,max}$
2. The values given apply to both the complementary "sum" and "difference" frequency harmonic components
3. Harmonic amplitudes are independent of the output load displacement factor, for this mode of operation

Harmonic Amplitudes for the Circulating Current Mode

With the cycloconverter operating with a continuous circulating current, both converters are kept in continuous conduction, and the shape of the output voltage wave, which is the mean of the voltage waveforms of the positive and negative converters, is not dependent upon the displacement angle of the load. This is confirmed from an inspection of the general expression, (11.24), for the output voltage of the 3-pulse cycloconverter, in which it is seen that the amplitudes of the harmonic components are a function only of the output voltage ratio.

QUANTITATIVE DATA. The amplitudes of the $3f_i \pm 2nf_o$, $6f_i \pm (2n+1)f_o$ and $12f_i \pm (2n+1)f_o$ harmonic families, for the circulating current mode of operation, for a range of output voltage ratios from 1.0 to 0.1, are shown in Table 11.6.

ASSESSMENT OF THE LIMITS OF PERFORMANCE OF CIRCUITS OF DIFFERENT PULSE NUMBERS, AS DICTATED BY THE DISTORTION OF THE OUTPUT VOLTAGE

The question inevitably arises as to what is the maximum useful attainable output to input frequency ratio of the cycloconverter, as dictated by the distortion of the output voltage, and how is this related to the pulse number of the circuit.

In fact, a precise answer to this question cannot be given, for the reason that the deterioration in the quality of the output voltage waveshape, which takes place with increasing output to input frequency ratio, is a continuous process, and thus it is not possible to define clearly any one specific point beyond which the frequency ratio cannot be increased. Furthermore, as has been seen, the harmonic distortion of the output voltage is considerably effected by the displacement angle of the load, as well as by the output voltage ratio, and hence any assessment of the limits of performance of the cycloconverter must be made in terms of these factors. Thus, in practice, the maximum attainable frequency ratio of the cycloconverter is determined largely by the conditions and requirements of the particular application.

The following brief consideration of the distortion of the output voltages of 3-, 6-, and 12-pulse circuits, for the particular condition of maximum output voltage, and a load displacement angle of 0°, should nonetheless provide an approximate indication of the typical limits of performance of the various circuits.

324 CYCLOCONVERTER OUTPUT VOLTAGE ANALYSIS

Figures 11.10 through 11.12 show the frequency charts for the predominant components of the lowest order harmonic families for 3-, 6-, and 12-pulse circuits respectively. Also shown graphically alongside are the corresponding amplitudes of each of the harmonic components, for maximum output voltage, and a load displacement angle of 0°.

In comparing these charts, the fact is revealed that certain of the predominant harmonic components invariably assume subharmonic frequency, as the output to input frequency ratio is increased. The higher is the pulse number of the circuit, the higher is the output to input frequency ratio at which the predominant harmonic components take on subharmonic frequency, but this is always less than unity. Thus these charts demonstrate a basic characteristic of the phase controlled cycloconverter, namely, that it has associated with it, a natural frequency ratio "barrier," dependent upon the pulse number of

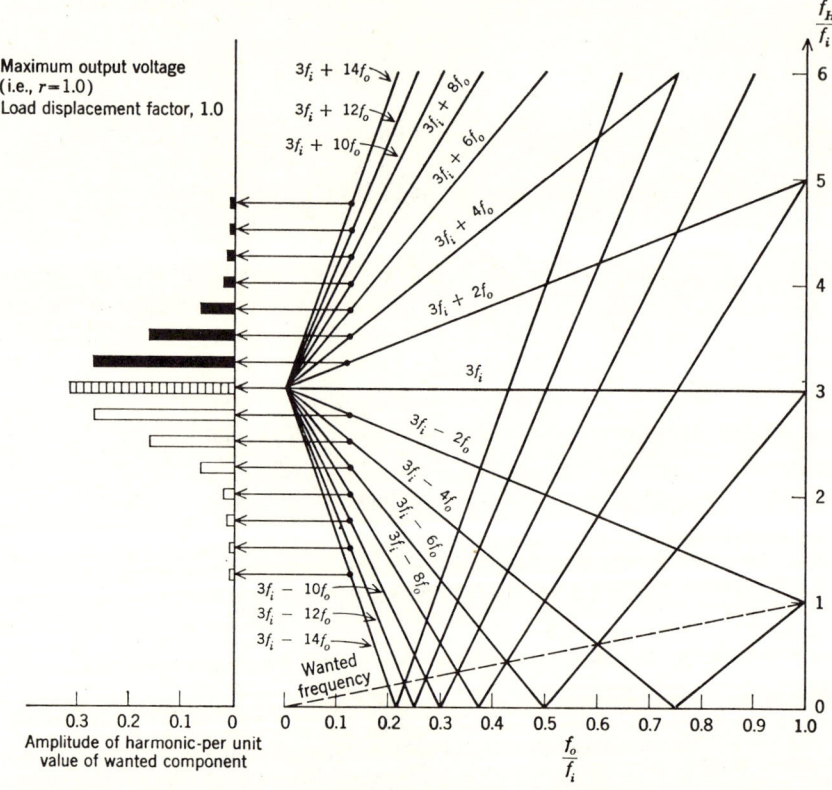

Figure 11.10. Charts showing the amplitudes of the predominant harmonic components in the output voltage of the 3-pulse cycloconverter, with maximum output voltage and unity load displacement factor, and the relationships between the harmonic frequencies and the output to input frequency ratio.

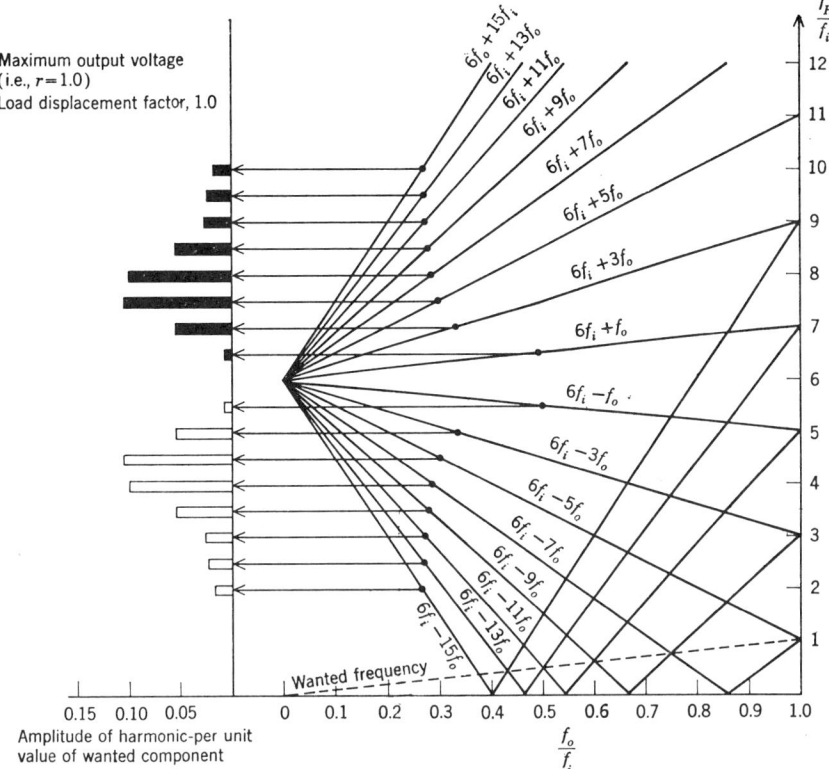

Figure 11.11. Charts showing the amplitudes of the predominant harmonic components in the output voltage of the 6-pulse cycloconverter, with maximum output voltage and unity load displacement factor, and the relationships between the harmonic frequencies and the output to input frequency ratio.

the circuit, but invariably less than unity, beyond which the predominant harmonic components assume subharmonic frequency.

The physical explanation for this, in very general terms, is that whereas during the "fast switching" periods (see Figs. 11.7 and 11.8, and the discussion related thereto), a relatively good resolution is possible in the fabrication of the output voltage waveform, during the "slow switching" periods, on the other hand, because of the need for always producing a natural commutation, the waveform resolution becomes progressively worse as the output to input frequency ratio is increased. For example, at an output frequency exactly equal to the input frequency, no commutations at all take place during the "slow switching" period, regardless of the circuit pulse number, since the output voltage wave must, of necessity, always "ride back down" on one or other of the input voltage waves.

It should be noted, however, that this characteristic does not imply that output frequencies greater than the input frequency are physically unattainable. This is not the case, and, in theory, the cycloconverter is perfectly functional at output to input frequency ratios greater than unity—although, in practice, with a circuit pulse number less than 12, this is generally unlikely to be a practical proposition.

In order to complete this brief discussion, it is of interest to make some— however arbitrary—comparison between the limiting output to input frequency ratios of circuits of different pulse numbers. In order to do this, it is necessary, firstly, to define some arbitrary criterion upon which to base the comparison. Let it be assumed, then, that the maximum useful output

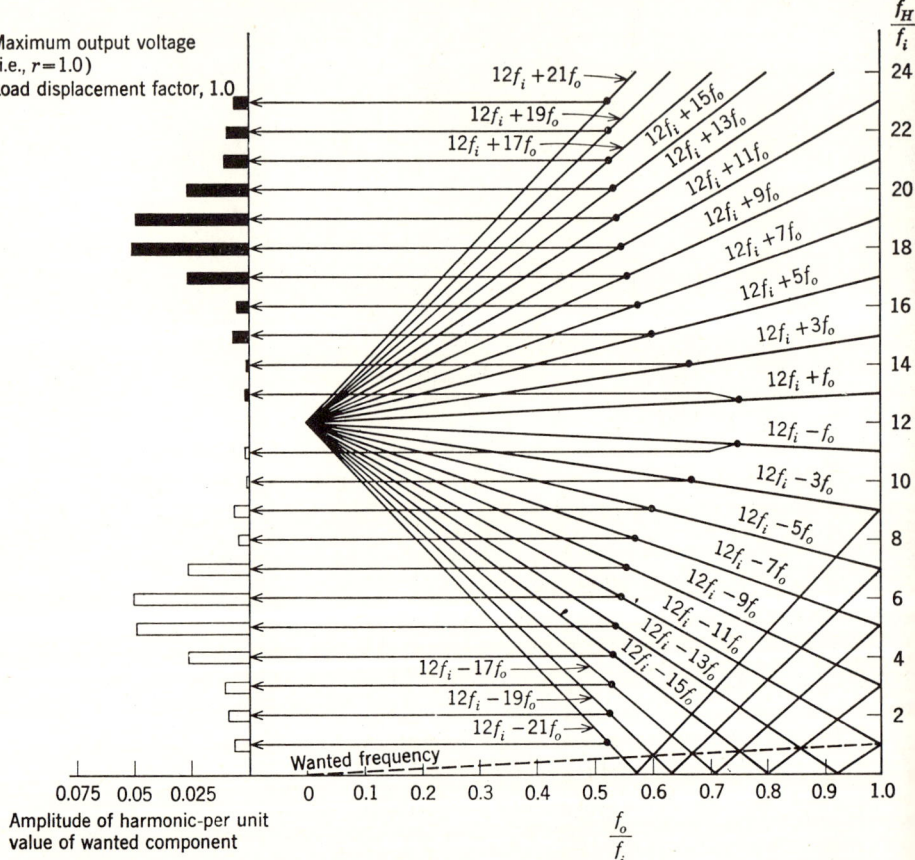

Figure 11.12. Charts showing the amplitudes of the predominant harmonic components in the output voltage of the 12-pulse cycloconverter, with maximum output voltage and unity load displacement factor, and the relationships between the harmonic frequencies and the output to input frequency ratio.

OTHER CONTROL METHODS   327

to input frequency ratio will be defined as that at which a harmonic component having an amplitude of approximately 0.025 per unit of the wanted component, with maximum output voltage, and a load displacement angle of 0°, first assumes a frequency which is less than the wanted output frequency. By reference to Figs. 11.10 through 11.12, it can be seen that the harmonic components concerned, for the 3-, 6-, and 12-pulse circuits, have frequencies of $(3f_i - 8f_o)$, $(6f_i - 11f_o)$, and $(12f_i - 15f_o)$ respectively. This leads to the following:

"Maximum" output to input frequency ratio
for the 3-pulse cycloconverter = 0.33
"Maximum" output to input frequency ratio
for the 6-pulse cycloconverter = 0.5
"Maximum" output to input frequency ratio
for the 12-pulse cycloconverter = 0.75

It is emphasized that these conclusions are based upon purely arbitrarily defined criteria, and, as such, they should not be regarded as being any more than generally indicative of the relative differences between the limits of performance of circuits of different pulse numbers.

UNNECESSARY DISTORTION GENERATED BY OTHER CONTROL METHODS

The cosine wave crossing control method has the unique property that it does not produce "unnecessary" direct integer multiple harmonics of the output frequency. This is explained as follows.

From the general expression, (11.15), for the output voltage of the (3-pulse) cycloconverter, it is seen that the first term in the series is the one responsible for the wanted component of output voltage. This first term is

$$v = \sin f(\theta_o) \tag{11.56}$$

It can easily be deduced that the only form that $f(\theta_o)$ can take, which, when substituted into (11.56) will yield but a single sinusoidal term having the wanted output frequency, is that due to the cosine wave crossing control method. That is,

$$f(\theta_o) = \sin^{-1} r \sin \theta_o$$

For any other control method, expression (11.56) inevitably takes on a more complex form, which contains a series of odd integer multiple harmonics, in addition to the wanted sinusoidal term.

Chapter Twelve

Analysis of the Input Current of the Cycloconverter

The waveform of the current drawn by the cycloconverter from the input supply is quite complex. As discussed in Chapter 6, it consists of a lagging fundamental component, with superimposed harmonic* terms.

In order to make an accurate determination of the required ratings and characteristics of the alternating power source, and of the input transformer, to which the cycloconverter is connected, it is essential to know the content of the input current wave, in terms of the constituent in-phase, quadrature, and harmonic components.

In this chapter, a harmonic analysis is made of the input current waveform of the cycloconverter, and comprehensive quantitative data, which precisely defines the constituent components of the input current waveforms of 3-, 6- and 12-pulse circuits, with both 1- and balanced 3-phase outputs, is presented.

This analysis clearly demonstrates the basic similarities which exist between the harmonic content of the input current waveforms of cycloconverters having different pulse numbers, and having different numbers of output phases. It is shown that the differences which exist between the input current waveforms of different cycloconverter circuit configurations can be resolved quite simply to a matter of whether given harmonic components are present or absent; moreover, for given conditions at the output, those harmonic

---

* As will be seen, the frequencies of the distortion components present in the input line current of the cycloconverter generally are not integer multiples of the input line frequency. Therefore, strictly speaking, these generally are not "harmonics" of the input line frequency. In this book, however, the term "harmonic," when applied to the input current of the cycloconverter, is used in a broad sense to apply to all the distortion components present, regardless of their frequency relationship to the "fundamental" line frequency. Likewise, the term "subharmonic" frequency is taken to mean any frequency below the input line frequency; and the term "superharmonic" frequency is taken to mean any frequency above the input line frequency.

components which are present in the input current invariably have the same relative amplitudes.

A transformer (or transformers) is often connected at the input side of the cycloconverter, either for the simple purpose of providing voltage transformation, or as a necessary part of the input circuit, or to serve both purposes. The exact waveform of the current at a given point in the input circuit of a transformer-fed cycloconverter, and hence also the exact mathematical expression applicable thereto, depends upon the particular transformer winding connections. However, as will be seen, the mathematical differences which exist between the harmonic series for current waveforms of a given pulse number, associated with alternative transformer connections, are simply differences of phase relationships of the harmonic components.

ASSUMPTIONS

The following assumptions are made in this chapter:

(1) The output current waveform of the cycloconverter is assumed to be a pure sinusoid, which is displaced from the wanted component of the output voltage by the displacement angle of the load. As has been seen, the output current waveform of the cycloconverter inevitably contains superimposed ripple components, so that this assumption is not strictly valid. It is, nonetheless, sufficiently accurate for most practical purposes.

(2) The internal impedance of the power source, as well as the winding resistance and leakage reactance of the input transformer (if used), are assumed to be negligible.

(3) The analysis is made on the assumption that the firing instants of the thyristors are determined by the cosine wave crossing control method. The reasons for choosing this control method, as a basis for a detailed analysis of the cycloconverter waveforms, have already been discussed in Chapter 11.

(4) The analysis is made for the circulating current-free mode of operation, and it is assumed that the "bank switching" takes place smoothly at the zero crossing of the sinusoidal output current wave. As discussed in Chapter 7, under some conditions the cycloconverter may be operated with a circulating current, in addition to the load current. However, in most practical applications, this circulating current is kept as small as possible, particularly under full load conditions, which, of course, are of most interest, so far as the present analysis is concerned.

ANALYTICAL APPROACH

The method of analysis of the input current waveform of the cycloconverter presented in this chapter uses the same basic technique as that employed for the analysis of the output voltage wave. Thus, the current in a given input line, due to a given thyristor, is expressed as the product of the thyristor "switching function," the converter "switching function," and the output current of the particular group; and the total current in a given input line is expressed as the sum of each of the currents of the thyristors connected to that line. By this method, it is possible to obtain a general series for the input line current of the cycloconverter, in terms of the in-phase, quadrature, and harmonic components, for any output voltage ratio and output load displacement angle. Once this series has been obtained, the input displacement, distortion, and power factors can be deduced.

NOTATION

For convenience, the following notation is used for describing the various input current waveforms associated with the cycloconverter:

A current waveform having a pulse number P, associated with a cycloconverter supplying a n-phase output, is denoted as waveform $(P.n.A, B,$ or $C)$. The letter A, B, or C refers to the waveform "type," which is determined by the particular input transformer connection.

For example, waveform $(6.1.A)$ is the primary input current waveform of a 6-pulse cycloconverter supplying a 1-phase output, with a wye-wye, or delta-delta, input transformer connection; waveform $(3.3.B)$ is the primary input current waveform of a 3-pulse cycloconverter supplying a (balanced) 3-phase output, with a wye-delta, or delta-wye, input transformer connection; and so on, as will become apparent.

DERIVATION OF THE HARMONIC SERIES FOR 3-PULSE CURRENT WAVEFORMS

Current at Converter Input Terminals—1-Phase Output

As illustrated by the waveforms of Fig. 12.1, the current waveform at the input terminals of the 3-pulse cycloconverter supplying a single phase load at its output, can be expressed in terms of the output current, and the thyristor and converter switching functions, as follows:

$$i_A = \hat{I}_o \sin(\theta_o + \phi_o) \cdot F_1\left(\theta_i - \frac{\pi}{2} + f(\theta_o)\right) \cdot F_P(\theta_o)$$

$$+ \hat{I}_o \sin(\theta_o + \phi_o) \cdot F_1\left(\theta_i + \frac{\pi}{2} - f(\theta_o)\right) \cdot F_N(\theta_o) \quad (12.1)$$

DERIVATION OF THREE-PULSE EXPRESSIONS 331

where $F_1\left(\theta_i \mp \frac{\pi}{2} \pm f(\theta_0)\right)$ takes the form of expression (4.6)

$F_P(\theta_i)$ is defined by expression (11.10)

$F_N(\theta_i)$ is defined by expression (11.11)

Substituting for $F_1\left(\theta_i \mp \frac{\pi}{2} \pm f(\theta_0)\right)$, $F_P(\theta_i)$ and $F_N(\theta_i)$ into (12.1) gives

$$i_A = \hat{I}_o \sin(\theta_o + \phi_o)\left\{\frac{1}{3} + \frac{\sqrt{3}}{\pi}\left[\sin\left(\theta_i - \frac{\pi}{2} + f(\theta_0)\right)\right.\right.$$
$$\left.\left. - \tfrac{1}{2}\cos 2\left(\theta_i - \frac{\pi}{2} + f(\theta_0)\right)\right.\right.$$
$$\left.\left. - \tfrac{1}{4}\cos 4\left(\theta_i - \frac{\pi}{2} + f(\theta_0)\right)\cdots\right]\right\}$$
$$\times \left\{\frac{1}{2} + \frac{2}{\pi}[\sin(\theta_o + \phi_o) + \tfrac{1}{3}\sin 3(\theta_o + \phi_o) + \tfrac{1}{5}\sin 5(\theta_o + \phi_o)\cdots]\right\}$$
$$+ \hat{I}_o \sin(\theta_o + \phi_o)\left\{\frac{1}{3} + \frac{\sqrt{3}}{\pi}\left[\sin\left(\theta_i + \frac{\pi}{2} - f(\theta_0)\right)\right.\right.$$
$$\left.\left. - \tfrac{1}{2}\cos 2\left(\theta_i + \frac{\pi}{2} - f(\theta_0)\right)\right.\right.$$
$$\left.\left. - \tfrac{1}{4}\cos 4\left(\theta_i + \frac{\pi}{2} - f(\theta_0)\right)\cdots\right]\right\}$$
$$\times \left\{\frac{1}{2} - \frac{2}{\pi}[\sin(\theta_o + \phi_o) + \tfrac{1}{3}\sin 3(\theta_o + \phi_o) + \tfrac{1}{5}\sin 5(\theta_o + \phi_o)\cdots]\right\} \quad (12.2)$$

By trigonometric manipulation, this reduces to

$$i_A = \hat{I}_o \sin(\theta_o + \phi_o)\left\{\frac{1}{3} + \frac{\sqrt{3}}{\pi}[\sin\theta_i \sin f(\theta_0)\right.$$
$$+ \tfrac{1}{2}\cos 2\theta_i \cos 2f(\theta_0) - \tfrac{1}{4}\cos 4\theta_i \cos 4f(\theta_0)$$
$$- \tfrac{1}{5}\sin 5\theta_i \sin 5f(\theta_0) + \tfrac{1}{7}\sin 7\theta_i \sin 7f(\theta_0)\cdots]$$
$$+ \frac{4\sqrt{3}}{\pi^2}[-\cos\theta_i \cos f(\theta_0)$$
$$- \tfrac{1}{2}\sin 2\theta_i \sin 2f(\theta_0) + \tfrac{1}{4}\sin 4\theta_i \sin 4f(\theta_0)$$
$$+ \tfrac{1}{5}\cos 5\theta_i \cos 5f(\theta_0) - \tfrac{1}{7}\cos 7\theta_i \cos 7f(\theta_0)\cdots]$$
$$\times [\sin(\theta_o + \phi_o) + \tfrac{1}{3}\sin 3(\theta_o + \phi_o)$$
$$\left. + \tfrac{1}{5}\sin 5(\theta_o + \phi_o) + \tfrac{1}{7}\sin 7(\theta_o + \phi_o)\cdots]\right\} \quad (12.3)$$

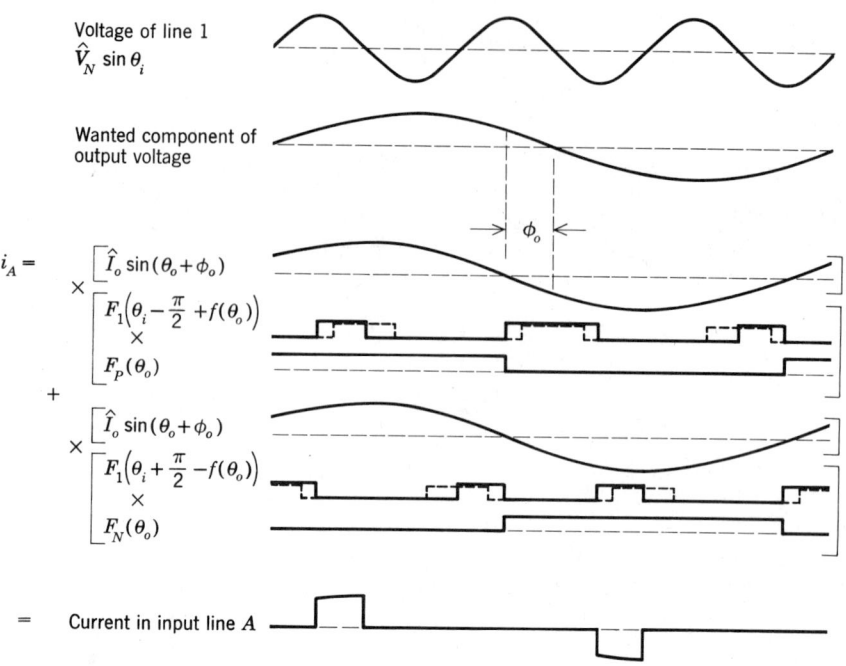

Figure 12.1. Waveforms illustrating the method of derivation of the general mathematical expression for the input line current of the cycloconverter supplying a single-phase load. Quiescent positions of thyristor switching functions shown dashed.

This is the general expression for the current, in terms of an arbitrary firing angle phase modulating function, $f(\theta_o)$.

According to expression (11.17), for the cosine wave crossing control method, the firing angle modulating function is represented mathematically by

$$f(\theta_o) = \sin^{-1} r \sin \theta_o \qquad (12.4)$$

Thus, by substituting for $f(\theta_o)$, as given by (12.4), into expression (12.3), and by use of the identities of expressions (11.20) through (11.23), and (11.27) through (11.30), it is possible to derive the general harmonic series for the input current waveform, in terms of any output voltage ratio and load displacement angle.

The algebraic manipulations are tedious, and only the result of this

DERIVATION OF THREE-PULSE EXPRESSIONS

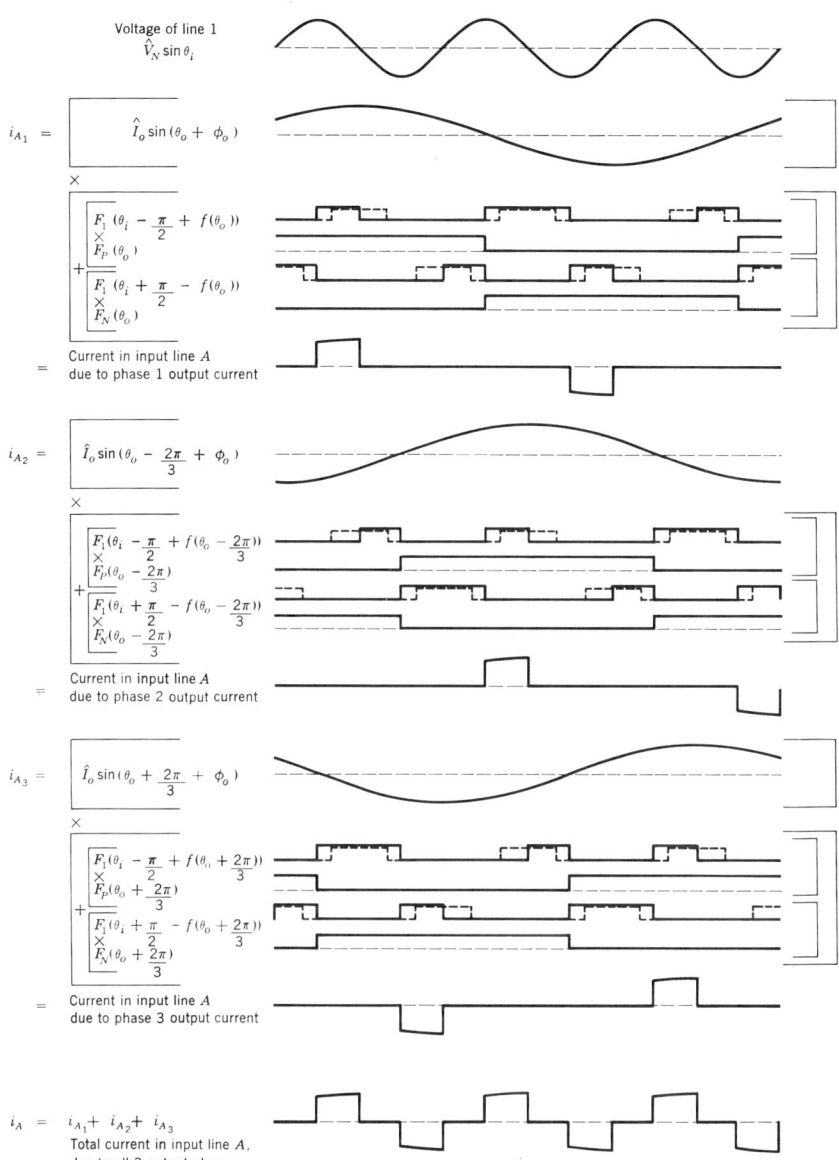

Figure 12.2. Waveforms illustrating the method of derivation of the general mathematical expression for the input line current of the cycloconverter supplying a balanced 3-phase load. Quiescent positions of thyristor switching functions shown dashed.

334   CYCLOCONVERTER INPUT CURRENT ANALYSIS

procedure is presented here:

(3.1.A) $\underline{I_o}$ In-phase component   Lagging quadrature component

$$i_A = \frac{\sqrt{3}\, r\hat{I}_o \cos\phi_o}{2\pi} \sin\theta_i - \left(\frac{2\sqrt{3}\,\hat{I}_o}{\pi^2} \sum_{n=0}^{n=\infty} \frac{-a_{12n} \cos 2n\phi_o}{(2n-1)(2n+1)}\right) \cos\theta_i$$

$$+ \frac{\hat{I}_o}{3} \sin(\theta_o + \phi_o) - \frac{\sqrt{3}\, r\hat{I}_o}{4\pi} \{\sin[\theta_i + (2\theta_o + \phi_o)] + \sin[\theta_i - (2\theta_o + \phi_o)]\}$$

$$- \frac{\sqrt{3}\,\hat{I}_o}{\pi^2} \sum_{m=1}^{m=\infty} \sum_{n=0}^{n=\infty} \frac{a_{12n}}{(2n+2m-1)(2n+2m+1)}$$

$$\times \{-\cos[\theta_i + (2m\theta_o + \{2n+2m\}\phi_o)]$$
$$\quad - \cos[\theta_i - (2m\theta_o + \{2n+2m\}\phi_o)]\}$$

$$+ \frac{a_{12n}}{(2n-2m+1)(2n-2m-1)}$$

$$\times \{-\cos[\theta_i + (2m\theta_o - \{2n-2m\}\phi_o)]$$
$$\quad - \cos[\theta_i - (2m\theta_o - \{2n-2m\}\phi_o)]\}$$

$$+ \frac{\sqrt{3}\,\hat{I}_o}{4\pi} \sum_{p=1}^{p=\infty} \sum_{n=0}^{2n=3[2p-1]\pm 1} \frac{\pm a_{(3[2p-1]\pm 1)\,2n}}{3[2p-1]\pm 1}$$

$$\times \begin{cases} \sin[(3[2p-1]\pm 1)\theta_i - ([2n+1]\theta_o + \phi_o)] \\ \quad - \sin[(3[2p-1]\pm 1)\theta_i + ([2n+1]\theta_o + \phi_o)] \\ -\sin[(3[2p-1]\pm 1)\theta_i - ([2n-1]\theta_o - \phi_o)] \\ \quad + \sin[(3[2p-1]\pm 1)\theta_i + ([2n-1]\theta_o - \phi_o)] \end{cases}$$

$$+ \frac{\sqrt{3}\,\hat{I}_o}{4\pi} \sum_{p=1}^{p=\infty} \sum_{n=0}^{2n+1=6p\pm 1} \frac{\pm a_{[6p\pm 1](2n+1)}}{6p\pm 1}$$

$$\times \begin{cases} -\sin[(6p\pm 1)\theta_i - ([2n+2]\theta_o + \phi_o)] \\ \quad - \sin[(6p\pm 1)\theta_i + ([2n+2]\theta_o + \phi_o)] \\ + \sin[(6p\pm 1)\theta_i - (2n\theta_o - \phi_o)] \\ \quad + \sin[(6p\pm 1)\theta_i + (2n\theta_o - \phi_o)] \end{cases}$$

$$+ \frac{\sqrt{3}\,\hat{I}_o}{\pi^2} \sum_{p=1}^{p=\infty} \sum_{m=0}^{m=\infty} \sum_{n=0}^{n=\infty} \frac{\pm a_{(3[2p-1]\pm 1)(2n+1)}}{(3[2p-1]\pm 1)([2n+1]+2m)([2n+1]+[2m+2])}$$

$$\times \left\{ \begin{array}{l} -\cos[(3[2p-1]\pm 1)\theta_i + ([2m+1]\theta_o \\ \qquad\qquad + [(2n+1)+(2m+1)]\phi_o)] \\ +\cos[(3[2p-1]\pm 1)\theta_i - ([2m+1]\theta_o \\ \qquad\qquad + [(2n+1)+(2m+1)]\phi_o)] \end{array} \right.$$

$$+ \frac{\pm a_{(3[2p-1]\pm 1)(2n+1)}}{(3[2p-1]\pm 1)([2n+1]-2m)([2n+1]-[2m+2])}$$

$$\times \left\{ \begin{array}{l} \cos[(3[2p-1]\pm 1)\theta_i + ([2m+1]\theta_o \\ \qquad\qquad - [(2n+1)-(2m+1)]\phi_o)] \\ -\cos[(3[2p-1]\pm 1)\theta_i - ([2m+1]\theta_o \\ \qquad\qquad - [(2n+1)-(2m+1)]\phi_o)] \end{array} \right.$$

$$+ \frac{\sqrt{3}\,\hat{I}_o}{\pi^2} \sum_{p=1}^{p=\infty} \sum_{m=1}^{m=\infty} \sum_{n=0}^{n=\infty} \frac{\pm a_{[6p\pm 1]2n}}{(6p\pm 1)(2n+[2m-1])(2n+[2m+1])}$$

$$\times \left\{ \begin{array}{l} \cos[(6p\pm 1)\theta_i + (2m\theta_o + [2n+2m]\phi_o)] \\ +\cos[(6p\pm 1)\theta_i - (2m\theta_o + [2n+2m]\phi_o)] \end{array} \right.$$

$$+ \frac{\pm a_{[6p\pm 1]2n}}{(6p\pm 1)(2n-[2m-1])(2n-[2m+1])}$$

$$\times \left\{ \begin{array}{l} \cos[(6p\pm 1)\theta_i + (2m\theta_o - [2n-2m]\phi_o)] \\ +\cos[(6p\pm 1)\theta_i - (2m\theta_o - [2n-2m]\phi_o)] \end{array} \right.$$

$$+ \frac{2\sqrt{3}\,\hat{I}_o}{\pi^2} \sum_{p=1}^{p=\infty} \sum_{n=0}^{n=\infty} \frac{\pm a_{[6p\pm 1]2n} \cos 2n\phi_o}{(6p\pm 1)(2n-1)(2n+1)} \cdot \cos[(6p\pm 1)\theta_i] \quad (12.5)$$

It is of interest to note, for the special case of $r = 1.0$, and $\phi_o = 0°$, that expression (12.5) reduces to

In-phase component ↓   Lagging quadrature component ↓

$$i_A = \frac{\sqrt{3}\,\hat{I}_o}{2\pi} \sin\theta_i - \frac{\sqrt{3}\,\hat{I}_o}{\pi^2} \cos\theta_i + \frac{\hat{I}_o}{3} \sin\theta_o \qquad \text{(continued)}$$

$$+ \frac{\sqrt{3}\,I_o}{4\pi}\{-\sin(\theta_i + 2\theta_o) - \sin(\theta_i - 2\theta_o)$$
$$+ \tfrac{1}{2}[\sin(2\theta_i + \theta_o) - \sin(2\theta_i - \theta_o)$$
$$+ \sin(2\theta_i + 3\theta_o) - \sin(2\theta_i - 3\theta_o)]$$
$$+ \tfrac{1}{4}[-\sin(4\theta_i + 3\theta_o) + \sin(4\theta_i - 3\theta_o)$$
$$- \sin(4\theta_i + 5\theta_o) + \sin(4\theta_i - 5\theta_o)]$$
$$+ \tfrac{1}{5}[-\sin(5\theta_i + 4\theta_o) - \sin(5\theta_i - 4\theta_o)$$
$$+ \sin(5\theta_i + 6\theta_o) + \sin(5\theta_i - 6\theta_o)]$$
$$+ \tfrac{1}{7}[\sin(7\theta_i + 6\theta_o) + \sin(7\theta_i - 6\theta_o)$$
$$- \sin(7\theta_i + 8\theta_o) - \sin(7\theta_i - 8\theta_o)]\cdots\}$$

$$+ \frac{\sqrt{3}\,I_o}{2\pi^2}\Big\{\frac{2}{5.3}\cos 5\theta_i + \frac{2}{7.3}\cos 7\theta_i - \frac{2}{11.5}\cos 11\theta_i - \frac{2}{13.7}\cos 13\theta_i \cdots$$
$$- (-\tfrac{1}{1} + \tfrac{1}{3})[\cos(\theta_i + 4\theta_o) + \cos(\theta_i - 4\theta_o)]$$
$$- (-\tfrac{1}{3} + \tfrac{1}{5})[\cos(\theta_i + 8\theta_o) + \cos(\theta_i - 8\theta_o)]\cdots$$
$$+ \tfrac{1}{2}(\tfrac{1}{1} + \tfrac{1}{1})[\cos(2\theta_i + \theta_o) - \cos(2\theta_i - \theta_o)]$$
$$+ \tfrac{1}{2}(\tfrac{1}{1} - \tfrac{1}{1})[\cos(2\theta_i + 3\theta_o) - \cos(2\theta_i - 3\theta_o)]\cdots$$
$$- \tfrac{1}{4}(-\tfrac{1}{1} - \tfrac{1}{3})[\cos(4\theta_i + \theta_o) - \cos(4\theta_i - \theta_o)]$$
$$- \tfrac{1}{4}(\tfrac{1}{1} + \tfrac{1}{3})[\cos(4\theta_i + 3\theta_o) - \cos(4\theta_i - 3\theta_o)]\cdots$$
$$+ \tfrac{1}{5}(-\tfrac{1}{1} - \tfrac{1}{3})[\cos(5\theta_i + 2\theta_o) + \cos(5\theta_i - 2\theta_o)]$$
$$+ \tfrac{1}{5}(\tfrac{1}{1} + \tfrac{1}{5})[\cos(5\theta_i + 4\theta_o) + \cos(5\theta_i - 4\theta_o)]\cdots$$
$$- \tfrac{1}{7}(\tfrac{1}{3} + \tfrac{1}{5})[\cos(7\theta_i + 2\theta_o) + \cos(7\theta_i - 2\theta_o)]$$
$$- \tfrac{1}{7}(-\tfrac{1}{1} - \tfrac{1}{5})[\cos(7\theta_i + 4\theta_o) + \cos(7\theta_i - 4\theta_o)]\cdots\Big\} \quad (12.6)$$

Furthermore, for $r = 1$ and $\phi_o = +90°$, expression (12.5) becomes

$$i_A = \frac{-\sqrt{3}\,I_o}{2\pi}\underbrace{\cos\theta_i}_{\text{Lagging quadrature component}} + \tfrac{1}{3}I_o \cos\theta_o$$

$$+ \frac{\sqrt{3}\, \hat{I}_o}{2\pi} \{-\cos(\theta_i + 2\theta_o)$$
$$+ \tfrac{1}{2}[\cos(2\theta_i + \theta_o) + \cos(2\theta_i + 3\theta_o)]$$
$$- \tfrac{1}{4}[\cos(4\theta_i + 3\theta_o) + \cos(4\theta_i + 5\theta_o)]$$
$$+ \tfrac{1}{5}[\cos(5\theta_i + 4\theta_o) + \cos(5\theta_i + 6\theta_o)]$$
$$- \tfrac{1}{7}[\cos(7\theta_i + 6\theta_o) + \cos(7\theta_i + 8\theta_o)] \cdots \} \quad (12.7)$$

And, for $r = 1.0$, and $\phi_o = -90°$, expression (12.5) becomes

$$i_A = -\frac{\sqrt{3}\, \hat{I}_o}{2\pi} \cos\theta_i \underset{\underset{\text{Lagging quadrature component}}{\uparrow}}{-} \tfrac{1}{3}\hat{I}_o \cos\theta_o$$

$$+ \frac{\sqrt{3}\, \hat{I}_o}{2\pi} \{-\cos(\theta_i - 2\theta_o)$$
$$- \tfrac{1}{2}[\cos(2\theta_i - \theta_o) + \cos(2\theta_i - 3\theta_o)]$$
$$+ \tfrac{1}{4}[\cos(4\theta_i - 3\theta_o) + \cos(4\theta_i - 5\theta_o)]$$
$$+ \tfrac{1}{5}[\cos(5\theta_i - 4\theta_o) + \cos(5\theta_i - 6\theta_o)]$$
$$- \tfrac{1}{7}[\cos(7\theta_i - 6\theta_o) + \cos(7\theta_i - 8\theta_o)] \cdots \} \quad (12.8)$$

Transformer Primary Currents—One-Phase Output

The schematic circuit diagram of Fig. 12.3 (circuit No. 1) shows the alternative wye and delta primary winding connections. Depending upon the connection, one or other of two alternative primary current waveforms may

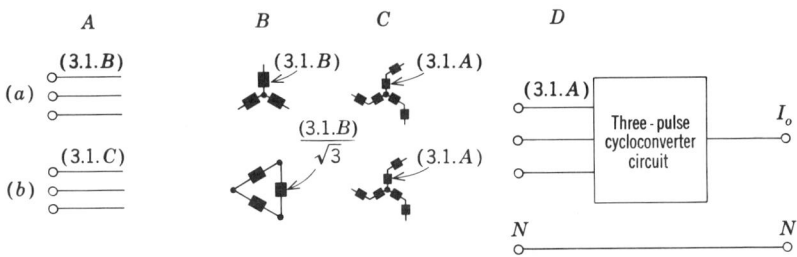

Circuit No. 1 Three-pulse circuit with 1-phase output

Figure 12.3. Schematic circuit diagram of 3-pulse cycloconverter with 1-phase output, with alternative input transformer connections. Primary and secondary L-N voltages are assumed to be the same. Quantitative data related to the input current to be found in Figs. 12.9 through 12.14, 12.16, 12.19, and Tables 12.1 through 12.4.

be obtained as follows:

$$(3.1.B)\underline{/0} = \frac{1}{\sqrt{3}}\left[(3.1.A)\underline{/0} - (3.1.A)\underline{/-\frac{2\pi}{3}}\right] \qquad (12.9)$$

$$(3.1.C)\phantom{\underline{/0}} = \frac{1}{\sqrt{3}}\left[(3.1.B)\underline{/0} - (3.1.B)\underline{/\frac{2\pi}{3}}\right] \qquad (12.10)$$

where $(3.1.A)\underline{/0}$ is given by (12.5).

In order to obtain the harmonic series for $(3.1.A)\underline{/-2\pi/3}$, it is necessary to substitute $(\theta_i - 2\pi/3)$ for θ_i into (12.5). Thus, by inspection of (12.5), it can be seen that the following modifications to this expression give the corresponding harmonic series for $(3.1.B)$:

(a) The $(\hat{I}_o/3) \times \sin(\theta_o + \phi_o)$ term is eliminated.
(b) All other terms are retained, with the same amplitudes, but with the following changes in phase (with respect to $(3.1.A)\underline{/0}$):

All terms containing θ_i change to terms containing $(\theta_i + \pi/6)$
All terms containing $[3(2p - 1) - 1]\theta_i$ change to terms containing

$$[3(2p - 1) - 1]\left[\theta_i - \frac{\pi}{6}\right]$$

All terms containing $[3(2p - 1) + 1]\theta_i$ change to terms containing

$$[3(2p - 1) + 1]\left[\theta_i + \frac{\pi}{6}\right]$$

All terms containing $[6p - 1]\theta_i$ change to terms containing

$$[6p - 1]\left[\theta_i - \frac{\pi}{6}\right]$$

All terms containing $[6p + 1]\theta_i$ change to terms containing

$$[6p + 1]\left[\theta_i + \frac{\pi}{6}\right]$$

Similarly, the following modification to expression (12.5) gives the corresponding expression for $(3.1.C)$:

The $(\hat{I}_o/3) \times \sin(\theta_o + \phi_o)$ term is eliminated, and all other terms are retained, with the same amplitudes and phase.

Rules for Determining Harmonic Series for Input Current with Three-Phase Output

The waveforms of Fig. 12.2 demonstrate that the mathematical expression, in terms of the thyristor and converter switching functions, for the input line current of the cycloconverter supplying a balanced 3-phase output, consists simply of the summation of the expressions for the currents due to each output phase separately:

$$i_A = i_{A_1} + i_{A_2} + i_{A_3} \tag{12.11}$$

The basic expressions for the individual components of input current due to the three output phases have exactly the same form as one another, the only difference between them being that whereas the expression for i_{A_1} contains terms in θ_o, the expressions for i_{A_2} and i_{A_3} contain terms in $(\theta_o - 2\pi/3)$, and $(\theta_o + 2\pi/3)$, respectively. Thus, it is evident that the harmonic series for the input line current $(i_{A_1} + i_{A_2} + i_{A_3})$ of the cycloconverter supplying a balanced 3-phase output, can be derived directly from the harmonic series for the input line current, i_{A_1}, of the cycloconverter supplying a single phase output, simply by substituting $(\theta_o - 2\pi/3)$ for θ_o into the expression for i_{A_1}, in order to obtain the corresponding expression for i_{A_2}, and by substituting $(\theta_o + 2\pi/3)$ for θ_o into the expression for i_{A_1}, to obtain the corresponding expression for i_{A_3}, and then adding together all three expressions. If this is done, it is found that certain harmonic terms cancel one another, whereas those which do not cancel add directly together.

Thus, the following simple rules can be formulated for deriving the expression for the input line current of the cycloconverter with a balanced three-phase output, from the corresponding expression for the input line current with a single phase output:

1. The amplitudes of all terms in which θ_o does not appear are multiplied by 3.
2. The amplitudes of all harmonic terms containing integer multiples of $3\theta_o$ are multiplied by 3.
3. All remaining harmonic terms disappear.

These rules, actually, are applicable to cycloconverter circuits of any pulse number.

Converter Input Current—Three-Phase Output

The schematic circuit diagram of Fig. 12.4 (circuit No. 2) shows various alternative transformer connections for a 3-pulse cycloconverter with a balanced 3-phase output. Depending upon the circuit connection, one or other of two alternative current waveforms may be obtained, as follows:

(3.3.A) $\underline{/0}$—The expression for this current waveform is obtained by application of the rules enumerated above to expression (12.5), for waveform (3.1.A) $\underline{/0}$

$$(3.3.B) = \frac{1}{\sqrt{3}}\left[(3.3.A)\underline{/0} - (3.3.A)\left/-\frac{2\pi}{3}\right.\right] \tag{12.12}$$

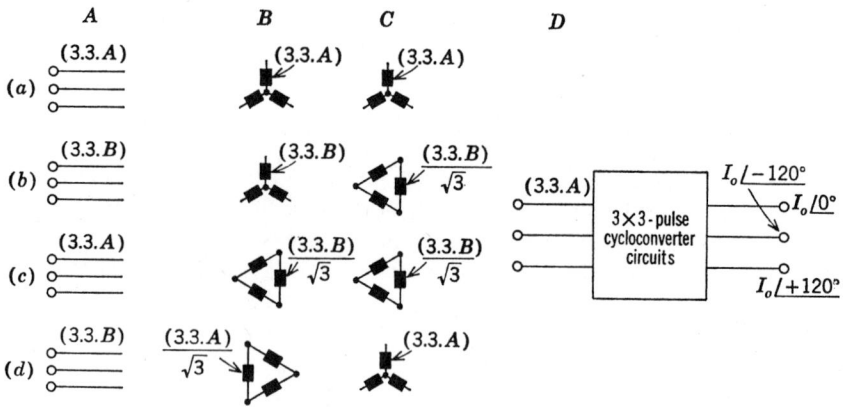

Circuit No. 2 Three-pulse circuit with 3-phase output

Figure 12.4. Schematic circuit diagram of 3-pulse cycloconverter with 3-phase output, with alternative input transformer connections. Primary and secondary L-N voltages are assumed to be the same. Quantitative data related to the input current to be found in Figs. 12.9 through 12.13, 12.15, 12.16, 12.19, and Tables 12.1 through 12.4.

The harmonic series for (3.3.B) is obtained by making the following modifications to the series for (3.3.A) $/0$:

All terms are retained, with the same amplitudes, but with the following changes in phase:

All terms containing θ_i change to terms containing $[\theta_i + \pi/6]$

All terms containing $[3(2p - 1) - 1]\theta_i$ change to terms containing

$$[3(2p - 1) - 1]\left[\theta_i - \frac{\pi}{6}\right]$$

All terms containing $[3(2p - 1) + 1]\theta_i$ change to terms containing

$$[3(2p - 1) + 1]\left[\theta_i + \frac{\pi}{6}\right]$$

All terms containing $[6p - 1]\theta_i$ change to terms containing

$$[6p - 1]\left[\theta_i - \frac{\pi}{6}\right]$$

All terms containing $[6p + 1]\theta_i$ change to terms containing

$$[6p + 1]\left[\theta_i + \frac{\pi}{6}\right]$$

HARMONIC SERIES FOR SIX-PULSE CURRENT WAVEFORMS

One-Phase Output

The schematic diagrams of Fig. 12.5 (circuits No. 3 and 4) show the 6-pulse midpoint and bridge cycloconverter circuits, with single phase outputs, with various input transformer connections. Depending upon the circuit connection, one or other of two alternative 6-pulse current waveforms may be obtained, as follows:

$$(6.1.A)\underline{/0} = \tfrac{1}{2}[(3.1.A)\underline{/0} - (3.1.A)\underline{/\pi}] \quad (12.13)$$

$$(6.1.B) = \frac{1}{\sqrt{3}}\left[(6.1.A)\underline{/0} - (6.1.A)\underline{/-\frac{2\pi}{3}}\right] \quad (12.14)$$

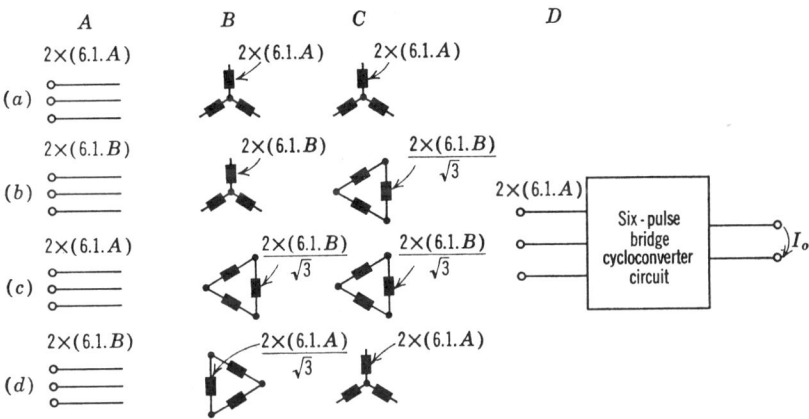

Circuit No. 3. Six-pulse midpoint circuit with 1-phase output

Circuit No. 4. Six-pulse bridge circuit with 1-phase output

Figure 12.5. Schematic circuit diagrams of 6-pulse cycloconverters with 1-phase output, with alternative input transformer connections. Primary and secondary L-N voltages are assumed to be the same. Quantitative data related to the input current to be found in Figs. 12.9 through 12.14, 12.17, 12.20, and Tables 12.1, 12.3, and 12.4.

342 CYCLOCONVERTER INPUT CURRENT ANALYSIS

Waveform (6.1.A) consists of half of the difference between two 3-pulse (3.1.A) waveforms, the first containing terms in (θ_i), and the second containing terms in $(\theta_i + \pi)$. Thus, by substituting $(\theta_i + \pi)$ for θ_i into expression (12.5), and by subtracting the resulting expression from the original one, the harmonic series for waveform (6.1.A) is obtained. An inspection of expression (12.5) shows that the procedure actually reduces to the simple matter of eliminating those harmonic components containing terms in $[3(2p-1) \pm 1]\theta_i$, as well as the component having the output frequency, whereas the remaining terms are unaltered:

$$(6.1.A)\underline{/0} = \overbrace{\frac{\sqrt{3}\, r\hat{I}_o \cos\phi_o}{2\pi}}^{\substack{\text{In-phase}\\\text{component}}} \sin\theta_i - \left(\frac{2\sqrt{3}\,\hat{I}_o}{\pi^2}\sum_{n=0}^{n=\infty}\overbrace{-\frac{a_{12n}\cos 2n\phi_o}{(2n-1)(2n+1)}}^{\substack{\text{Lagging quadrature}\\\text{component}}}\right)\cos\theta_i$$

$$-\frac{\sqrt{3}\,r\hat{I}_o}{4\pi}\{\sin[\theta_i + (2\theta_o + \phi_o)] + \sin[\theta_i - (2\theta_o + \phi_o)]\}$$

$$-\frac{\sqrt{3}\,\hat{I}_o}{\pi^2}\sum_{m=1}^{m=\infty}\sum_{n=0}^{n=\infty}\frac{a_{12n}}{(2n+2m-1)(2n+2m+1)}$$
$$\times\{-\cos[\theta_i + (2m\theta_o + \{2n+2m\}\phi_o)]$$
$$-\cos[\theta_i - (2m\theta_o + \{2n+2m\}\phi_o)]\}$$
$$+\frac{a_{12n}}{(2n-2m+1)(2n-2m-1)}$$
$$\times\{-\cos[\theta_i + (2m\theta_o - \{2n-2m\}\phi_o)]$$
$$-\cos[\theta_i - (2m\theta_o - \{2n-2m\}\phi_o)]\}$$

$$+\frac{\sqrt{3}\,\hat{I}_o}{4\pi}\sum_{p=1}^{p=\infty}\sum_{n=0}^{2n+1=6p\pm 1}\frac{\pm a_{[6p\pm 1](2n+1)}}{6p\pm 1}$$
$$\times\begin{cases}-\sin[(6p\pm 1)\theta_i - ([2n+2]\theta_o + \phi_o)]\\ \quad -\sin[(6p\pm 1)\theta_i + ([2n+2]\theta_o + \phi_o)]\\ +\sin[(6p\pm 1)\theta_i - (2n\theta_o - \phi_o)]\\ \quad +\sin[(6p\pm 1)\theta_i + (2n\theta_o - \phi_o)]\end{cases}$$

$$+ \frac{\sqrt{3}\,\hat{I}_o}{\pi^2} \sum_{p=1}^{p=\infty} \sum_{m=1}^{m=\infty} \sum_{n=0}^{n=\infty} \frac{\pm a_{[6p\pm 1]2n}}{(6p \pm 1)(2n + [2m - 1])(2n + [2m + 1])}$$

$$\times \left\{ \begin{array}{l} \cos\left[(6p \pm 1)\theta_i + (2m\theta_o + [2n + 2m]\phi_o)\right] \\ +\cos\left[(6p \pm 1)\theta_i - (2m\theta_o + [2n + 2m]\phi_o)\right] \end{array} \right\}$$

$$+ \frac{\pm a_{[6p\pm 1]2n}}{(6p \pm 1)(2n - [2m - 1])(2n - [2m + 1])}$$

$$\times \left\{ \begin{array}{l} \cos\left[(6p \pm 1)\theta_i + (2m\theta_o - [2n - 2m]\phi_o)\right] \\ +\cos\left[(6p \pm 1)\theta_i - (2m\theta_o - [2n - 2m]\phi_o)\right] \end{array} \right\}$$

$$+ \frac{2\sqrt{3}\,\hat{I}_o}{\pi^2} \sum_{p=1}^{p=\infty} \sum_{n=0}^{n=\infty} \frac{\pm a_{[6p\pm 1]2n} \cos 2n\phi_o}{(6p \pm 1)(2n - 1)(2n + 1)} \cos\left[(6p \pm 1)\theta_i\right]$$

(12.15)

From expressions (12.14) and (12.15), it can be deduced that the harmonic series for (6.1.B) is obtained by making the following modifications to the series for (6.1.A) $\underline{/0}$:

All terms are retained, with the same amplitudes, but with the following changes in phase:

All terms containing θ_i change to terms containing $[\theta_i + \pi/6]$
All terms containing $[6p - 1]\theta_i$ change to terms containing

$$[6p - 1]\left[\theta_i - \frac{\pi}{6}\right]$$

All terms containing $[6p + 1]\theta_i$ change to terms containing

$$[6p + 1]\left[\theta_i + \frac{\pi}{6}\right]$$

Three-Phase Output

The schematic diagrams of Fig. 12.6, (circuit Nos. 5, 6, and 7) show 6-pulse midpoint and bridge cycloconverter circuits, with three-phase outputs, with various input transformer connections. Depending upon the input circuit connection, one or other of two alternative 6-pulse current waveforms may be obtained, as follows:

(6.3.A) $\underline{/0}$—The expression for this current waveform is obtained by application of the rules enumerated on p. 339, to expression (12.15), for waveform (6.1.A) $\underline{/0}$

$$(6.3.B) = \frac{1}{\sqrt{3}}\left[(6.3.A)\underline{/0} - (6.3.A)\underline{/-\frac{2\pi}{3}}\right] \qquad (12.16)$$

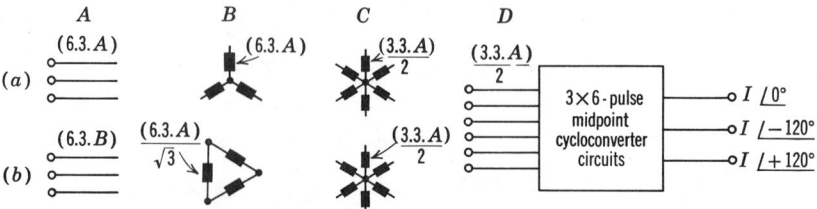

Circuit No. 5. Six-pulse midpoint circuit with 3-phase output

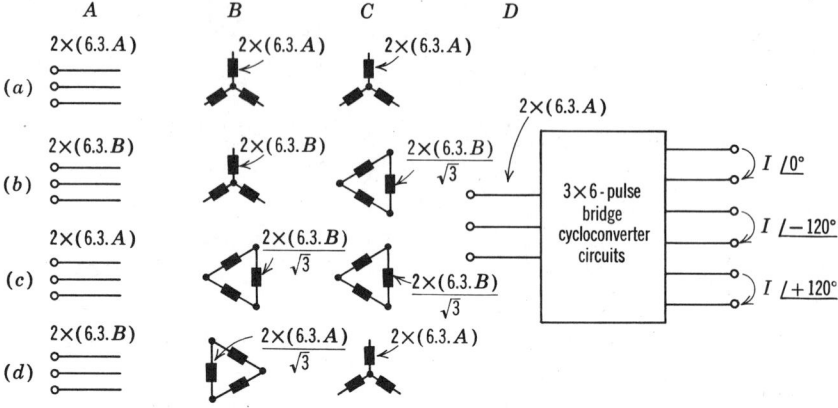

Circuit No. 6. Six-pulse bridge circuit with 3-phase output - isolated loads

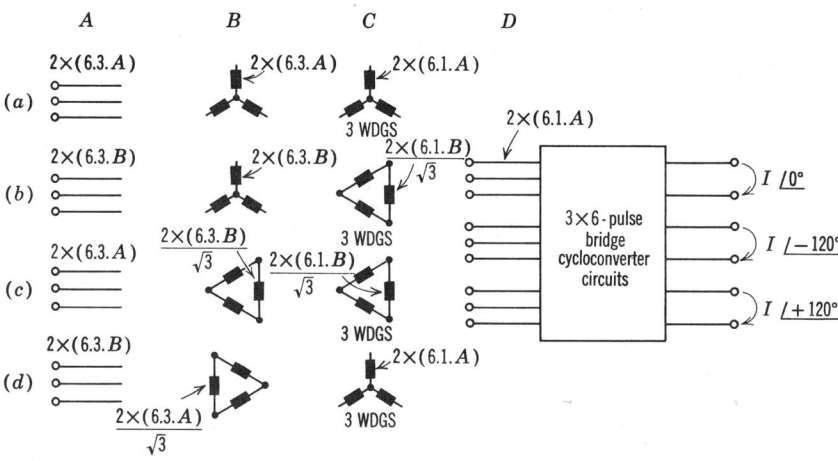

Circuit No. 7. Six-pulse bridge circuit with 3-phase output - nonisolated loads

Figure 12.6. Schematic circuit diagrams of 6-pulse cycloconverters with 3-phase output, with alternative input transformer connections. Primary and secondary L-N voltages are assumed to be the same. Quantitative data related to the input current to be found in Figs. 12.9 through 12.13, 12.15, 12.17, 12.20, and Tables 12.1, 12.3, and 12.4.

EXPRESSIONS FOR TWELVE-PULSE WAVES

The harmonic series for (6.3.B) is obtained by making the following modifications to the series for (6.3.A)$/0$:

All terms are retained, with the same amplitudes, but with the following changes in phase:
All terms containing θ_i change to terms containing $[\theta_i + \pi/6]$
All terms containing $[6p - 1]\theta_i$ change to terms containing

$$[6p - 1]\left[\theta_i - \frac{\pi}{6}\right]$$

All terms containing $[6p + 1]\theta_i$ change to terms containing

$$[6p + 1]\left[\theta_i + \frac{\pi}{6}\right]$$

HARMONIC SERIES FOR TWELVE-PULSE INPUT CURRENT WAVEFORMS

One-Phase Output

The schematic circuit diagrams of Fig. 12.7 (Circuit Nos. 8 and 9) show 12-pulse midpoint and bridge cycloconverter circuits, with single phase outputs, with various input transformer connections. Depending upon the circuit connection, one or other of two alternative 12-pulse current waveforms may be obtained, as follows:

$$(12.1.A) = \frac{1}{2}\left[(6.1.A)/\underline{0} + \frac{(6.1.A)}{\sqrt{3}}\bigg/+\frac{\pi}{6} + \frac{(6.1.A)}{\sqrt{3}}\bigg/-\frac{\pi}{6}\right] \quad (12.17)$$

$$(12.1.B) = \frac{1}{2}\left[\sqrt{\frac{2}{3}}\left((6.1.A)/\underline{0} + (6.1.A)\bigg/-\frac{\pi}{6}\right)\right.$$
$$\left. + \frac{\sqrt{3}-1}{\sqrt{6}}\left((6.1.A)\bigg/-\frac{\pi}{3} + (6.1.A)\bigg/+\frac{\pi}{6}\right)\right] \quad (12.18)$$

From the general expression, (12.15), for waveform (6.1.A), the corresponding expression for waveform (12.1.A)$/\underline{0}$ can be readily deduced:

$$(12.1.A)/\underline{0} = \underbrace{\frac{\sqrt{3}\, r\hat{I}_o \cos\phi_o}{2\pi}\sin\theta_i}_{\text{In-phase component}} - \underbrace{\left(\frac{2\sqrt{3}\,\hat{I}_o}{\pi^2}\sum_{n=0}^{n=\infty} - \frac{a_{12n}\cos 2n\phi_o}{(2n-1)(2n+1)}\right)\cos\theta}_{\text{Lagging quadrature component}}$$

$$-\frac{\sqrt{3}\, r\hat{I}_o}{4\pi}\{\sin[\theta_i + (2\theta_o + \phi_o)] + \sin[\theta_i - (2\theta_o + \phi_o)]\}$$

(continued)

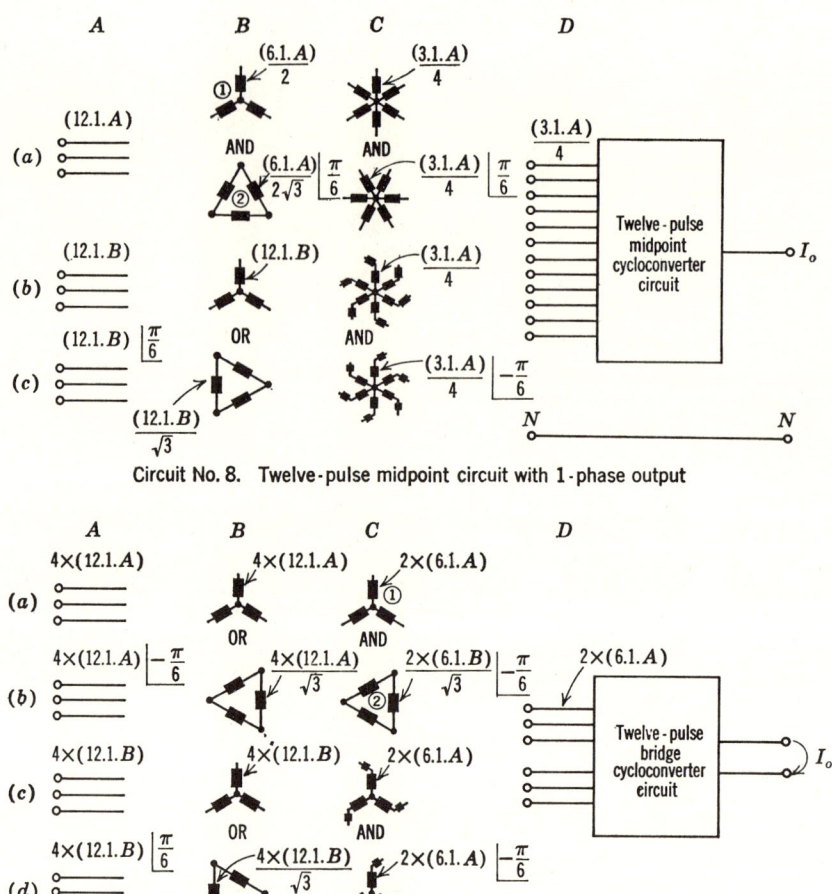

Circuit No. 8. Twelve-pulse midpoint circuit with 1-phase output

Circuit No. 9. Twelve-pulse bridge circuit with 1-phase output

Figure 12.7. Schematic circuit diagrams of 12-pulse cycloconverters with 1-phase output, with alternative input transformer connections. Primary and secondary L-N voltages are assumed to be the same. Quantitative data related to the input current to be found in Figs. 12.9 through 12.14, 12.18, 12.21, and Tables 12.1, and 12.4.

EXPRESSIONS FOR TWELVE-PULSE WAVES 347

$$-\frac{\sqrt{3}\,\hat{I}_o}{\pi^2}\sum_{m=1}^{m=\infty}\sum_{n=0}^{n=\infty}\frac{a_{12n}}{(2n+2m-1)(2n+2m+1)}$$

$$\times\{-\cos[\theta_i+(2m\theta_o+\{2n+2m\}\phi_o)]$$
$$\qquad-\cos[\theta_i-(2m\theta_o+\{2n+2m\}\phi_o)]\}$$

$$+\frac{a_{12n}}{(2n-2m+1)(2n-2m-1)}$$

$$\times\{-\cos[\theta_i+(2m\theta_o-\{2n-2m\}\phi_o)]$$
$$\qquad-\cos[\theta_i-(2m\theta_o-\{2n-2m\}\phi_o)]\}$$

$$+\frac{\sqrt{3}\,\hat{I}_o}{4\pi}\sum_{p=1}^{p=\infty}\sum_{n=0}^{2n+1=12p\pm1}\frac{\pm a_{[12p\pm1](2n+1)}}{12p\pm1}$$

$$\times\begin{pmatrix}-\sin[(12p\pm1)\theta_i-([2n+2]\theta_o+\phi_o)]\\\quad-\sin[(12p\pm1)\theta_i+([2n+2]\theta_o+\phi_o)]\\+\sin[(12p\pm1)\theta_i-(2n\theta_o-\phi_o)]\\\quad+\sin[(12p\pm1)\theta_i+(2n\theta_o-\phi_o)]\end{pmatrix}$$

$$+\frac{\sqrt{3}\,\hat{I}_o}{\pi^2}\sum_{p=1}^{p=\infty}\sum_{m=1}^{m=\infty}\sum_{n=0}^{n=\infty}\frac{\pm a_{[12p\pm1]2n}}{(12p\pm1)(2n+[2m-1])(2n+[2m+1])}$$

$$\times\left\{\begin{matrix}\cos[(12p\pm1)\theta_i+(2m\theta_o+[2n+2m]\phi_o)]\\+\cos[(12p\pm1)\theta_i-(2m\theta_o+[2n+2m]\phi_o)]\end{matrix}\right.$$

$$+\frac{\pm a_{[12p\pm1]2n}}{(12p\pm1)(2n-[2m-1])(2n-[2m+1])}$$

$$\times\left\{\begin{matrix}\cos[(12p\pm1)\theta_i+(2m\theta_o-[2n-2m]\phi_o)]\\+\cos[(12p\pm1)\theta_i-(2m\theta_o-[2n-2m]\phi_o)]\end{matrix}\right.$$

$$+\frac{2\sqrt{3}\,\hat{I}_o}{\pi^2}\sum_{p=1}^{p=\infty}\sum_{n=0}^{n=\infty}\frac{\pm a_{[12p\pm1]2n}\cos 2n\phi_o}{(12p\pm1)(2n-1)(2n+1)}\cos[(12p\pm1)\theta_i]$$

(12.19)

From expressions (12.17), (12.18) and (12.19), it can be deduced that the harmonic series for (12.1.*B*) is obtained by making the following modifications to the series for (12.1.*A*)/0:

All terms are retained, with the same amplitudes, but with the following changes in phase:

All terms containing θ_i change to terms containing $[\theta_i - \pi/12]$
All terms containing $[12p - 1]\theta_i$ change to terms containing

$$[12p - 1]\left[\theta_i + \frac{\pi}{12}\right]$$

All terms containing $[12p + 1]\theta_i$ change to terms containing

$$[12p + 1]\left[\theta_i - \frac{\pi}{12}\right]$$

Three-Phase Output

The schematic circuit diagrams of Fig. 12.8 (circuit Nos. 10 and 11) show 12-pulse midpoint and bridge cycloconverter circuits, with 3-phase outputs, with various input transformer connections. Depending upon the input circuit connection, one or other of two alternative 12-pulse current waveforms may be obtained, as follows:

(12.3.A)/0—The expression for this current waveform is obtained by application of the rules enumerated on p. 339, to expression (12.19), for waveform (12.1.A)/0

$$(12.3.B) = \frac{1}{2}\left[\sqrt{\frac{2}{3}}\left((6.3.A)\underline{/0} + (6.3.A)\underline{/-\frac{\pi}{6}}\right)\right.$$

$$\left. + \frac{\sqrt{3}-1}{\sqrt{6}}\left((6.3.A)\underline{/-\frac{\pi}{3}} + (6.3.A)\underline{/+\frac{\pi}{6}}\right)\right] \quad (12.20)$$

The harmonic series for (12.3.B) is obtained by making the following modifications to the series for (12.3.A)/0:

All terms are retained, with the same amplitudes but with the following changes in phase:

All terms containing θ_i change to terms containing $[\theta_i - \pi/12]$
All terms containing $[12p - 1]\theta_i$ change to terms containing

$$[12p - 1]\left[\theta_i + \frac{\pi}{12}\right]$$

All terms containing $[12p + 1]\theta_i$ change to terms containing

$$[12p + 1]\left[\theta_i - \frac{\pi}{12}\right]$$

FUNDAMENTAL COMPONENT OF INPUT CURRENT

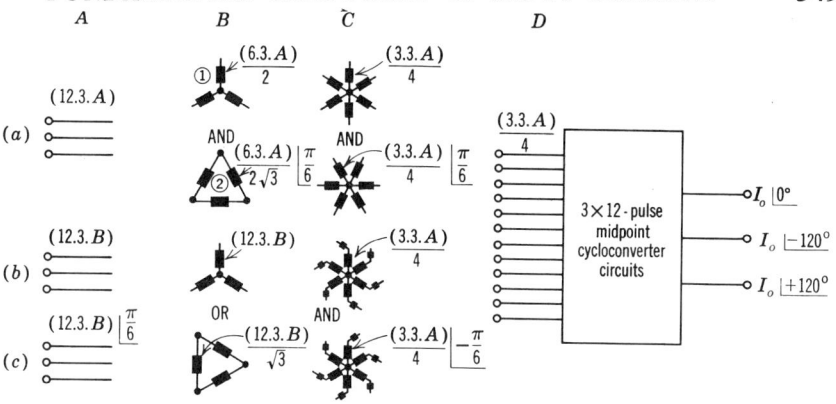

Circuit No. 10. Twelve-pulse midpoint circuit with 3-phase output

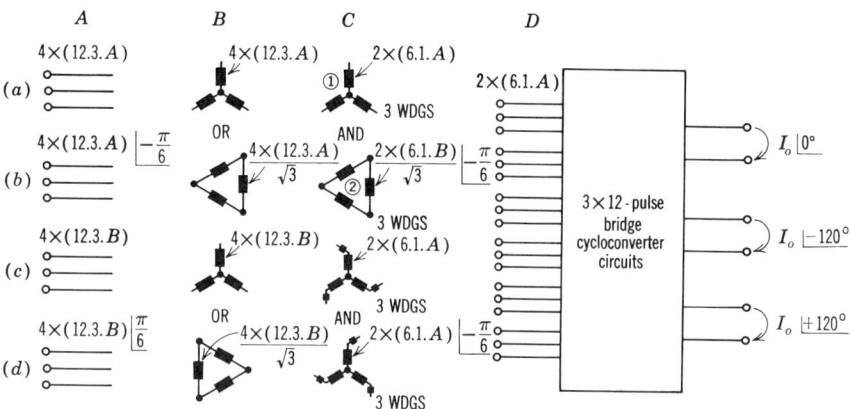

Circuit No. 11. Twelve-pulse bridge circuit with 3-phase output

Figure 12.8. Schematic circuit diagrams of 12-pulse cycloconverters with 3-phase output, with alternative input transformer connections. Primary and secondary L-N voltages are assumed to be the same. Quantitative data related to the input current to be found in Figs. 12.9 through 12.13, 12.15, 12.18, 12.21, and Tables 12.1 and 12.4.

THE FUNDAMENTAL COMPONENT OF INPUT CURRENT

General Comments

From the preceding analysis, it is seen that invariably two components of current having the input frequency are obtained—that is to say, an in-phase component, and a lagging quadrature component. It is seen, moreover, that the relative proportions of in-phase and quadrature current are independent of the pulse number of the converter, the number of output phases, and the output to input frequency ratio, and are a function only of the displacement factor of the load, and the output voltage ratio.

350 CYCLOCONVERTER INPUT CURRENT ANALYSIS

This result is to be expected, and it simply indicates, as in the case of the 2-quadrant phase-controlled converter with a steady d-c output, that the composition of the fundamental component of load consumed by the cycloconverter is a basic property, unrelated to the circuit configuration. Thus, for the cycloconverter, as for the phase-controlled converter, whereas the converter circuit configuration determines the presence or absence of given harmonic components of current, it does not determine the composition of the fundamental component of load at the input side.

The In-Phase Component

As discussed in Chapter 6, the in-phase component is the only component of current at the input which is capable of contributing to the mean power at the output of the cycloconverter. Thus the amplitude of this component necessarily assumes a value appropriate to the mean output power. This can be confirmed as follows:

From the preceding analysis, it is seen that the rms amplitude of the in-phase component of current at the input of the cycloconverter is given by the following general expression:

$$I_P = q \cdot s \cdot \frac{r\sqrt{3}}{2\pi} I_o \cos \phi_o \qquad (12.21)$$

where q = number of output phases
s = number of 3-pulse groups connected in series with one another
I_o = rms current of each output phase

Thus, the total power at the input is given by

$$P_i = 3V_N I_P$$

(3-Phase input)

$$= q \cdot s \frac{r 3\sqrt{3} \, V_N}{2\pi} I_o \cos \phi_o \qquad (12.22)$$

The total output power is given by

$$P_o = q \cdot \left(\frac{sr 3\sqrt{3} \, V_N}{2\pi} \right) \cdot I_o \cos \phi_o \qquad (12.23)$$

(Rms value of wanted output voltage component—see (11.41) and (11.42))

Hence, from (12.22) and (12.23)

$$P_i = P_o \qquad (12.24)$$

Figure 12.9 illustrates the linear relationships between the amplitude of the in-phase component of current at the input of the cycloconverter, the land displacement factor, and the output voltage ratio.

The Lagging Quadrature Component

The presence of the lagging quadrature component of current at the input side is attributable to two separate factors. First, any reactive component of load which may exist at the output of the cycloconverter can be visualized as producing a corresponding reactive component of load at the input; and, due to the phase control process, both leading and lagging loads at the output are invariably reflected as a lagging load at the input. Second, the phase-control mechanism of the cycloconverter itself gives rise to a lagging quadrature component of current at the input. Thus, even if the displacement factor of the load, and the output voltage ratio, are both unity, a quite substantial lagging quadrature component of current nonetheless exists at the input; and, in general, the displacement factor at the input is always less than that at the output.

From the preceding mathematical analysis, it is seen that the rms amplitude of the quadrature component of current, I_Q, at the input of the cycloconverter is given by the following general expression:

$$I_Q = q \cdot s \cdot \frac{2\sqrt{3}\,I_o}{\pi^2} \sum_{n=0}^{n=\infty} \frac{-a_{1_{2n}} \cos 2n\phi_o}{(2n-1)(2n+1)} \qquad (12.25)$$

where q = number of output phases
s = number of 3-pulse groups connected in series with one another
I_o = rms current of each output phase
$a_{1_{2n}}$ is defined by expression (11.28). Values for these coefficients are given in Tables 11.2a.

An inspection of expression (12.25) indicates that the amplitude of the quadrature component of input current is a function both of the load displacement factor and of the output voltage ratio. The curves of Fig. 12.10 are a graphical illustration of the relationships between I_Q, $\cos \phi_o$, and r.

By comparison of these curves with those of Fig. 12.9, it is seen that the amplitude of the quadrature component of current is generally quite appreciable in relation to the in-phase component.

Thus, considering, for example, the particular condition of unity load displacement factor and unity output voltage ratio, the amplitude of the quadrature component is $0.638 \times$ the in-phase component; and this factor further increases as the output voltage ratio is reduced. For a load displacement factor of unity, the quadrature component of current at the input is entirely attributable to the phase-control process and thus, for this load displacement factor, it is evident that the phase-control mechanism results in a quite substantial reactive loading of the supply.

More generally, for load displacement factors less than unity, in order to assess the relative contributions of the reactive component of load at the

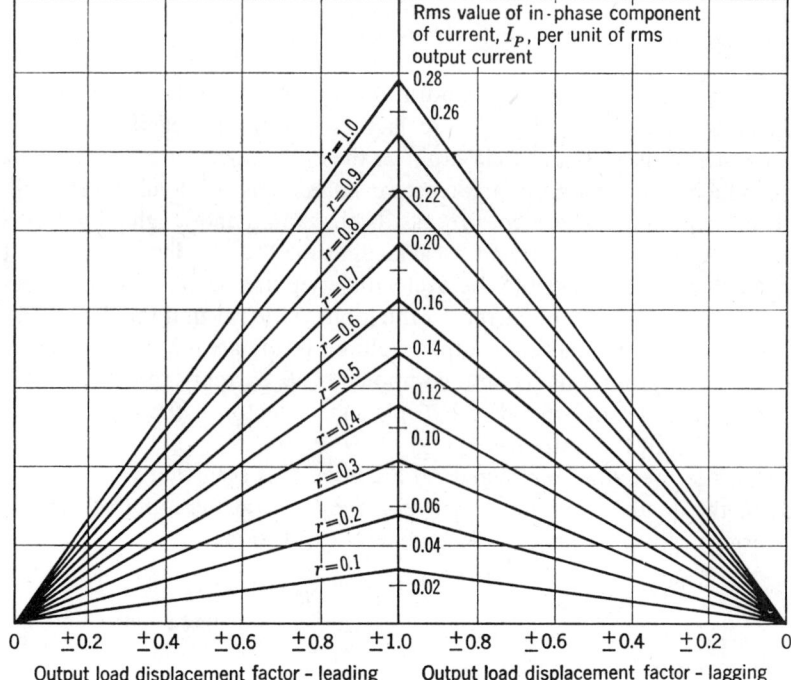

Figure 12.9. Curves showing the relationships between the per unit rms value of the fundamental in-phase component of current at the input of the cycloconverter, the output load displacement factor, and the output voltage ratio.

output, and of the phase-control process itself, to the total quadrature load at the input, it is convenient to visualize the quadrature input current as being comprised of two separate components. The first of these is a component which just balances the corresponding reactive load at the output; as such, it may be termed the "load" component of quadrature current. The second is a component whose presence can be visualized as being entirely brought about by the phase-control process. In other words, this component of reactive current at the input does not act as a "counterweight" to the reactive load at the output, and it may be termed the "phase-control" component.

The "load" component of quadrature current can be computed on the basis that if this was the sole component of quadrature current at the input, then the input displacement factor would be equal to the output displacement factor:

According to (12.21), the in-phase component of current is given by

$$I_P = q \cdot s \cdot \frac{r\sqrt{3}}{2\pi} I_o \cos \phi_o$$

FUNDAMENTAL COMPONENT OF INPUT CURRENT

Circuit No.	Curve multiplication factor			
	Point in circuit			
	A	B	C	D
1(a)	1.0	1.0	1.0	1.0
1(b)	1.0	0.577	1.0	1.0
2(a)	3.0	3.0	3.0	3.0
2(b)	3.0	3.0	1.732	3.0
2(c)	3.0	1.732	1.732	3.0
2(d)	3.0	1.732	3.0	3.0
3(a)	1.0	1.0	0.5	0.5
3(b)	1.0	0.577	0.5	0.5
4(a)	2.0	2.0	2.0	2.0
4(b)	2.0	2.0	1.154	2.0
4(c)	2.0	1.154	1.154	2.0
4(d)	2.0	1.154	2.0	2.0
5(a)	3.0	3.0	1.5	1.5
5(b)	3.0	1.732	1.5	1.5
6(a)	6.0	6.0	6.0	6.0
6(b)	6.0	6.0	3.464	6.0
6(c)	6.0	3.464	3.464	6.0
6(d)	6.0	3.464	6.0	6.0

Circuit No.	Curve multiplication factor			
	Point in circuit			
	A	B	C	D
7(a)	6.0	6.0	2.0	2.0
7(b)	6.0	6.0	1.154	2.0
7(c)	6.0	3.464	1.154	2.0
7(d)	6.0	3.464	2.0	2.0
8(a)	1.0	① 0.5 ② 0.288	0.25	0.25
8(b)	1.0	1.0	0.25	0.25
8(c)	1.0	0.577	0.25	0.25
9(a)	4.0	4.0	① 2.0 ② 1.154	2.0
9(b)	4.0	2.308	① 2.0 ② 1.154	2.0
9(c)	4.0	4.0	2.0	2.0
9(d)	4.0	2.308	2.0	2.0
10(a)	3.0	① 1.5 ② 0.866	0.75	0.75
10(b)	3.0	3.0	0.75	0.75
10(c)	3.0	1.732	0.75	0.75
11(a)	12.0	12.0	① 2.0 ② 1.154	2.0
11(b)	12.0	6.924	① 2.0 ② 1.154	2.0
11(c)	12.0	12.0	2.0	2.0
11(d)	12.0	6.924	2.0	2.0

Figure 12.9 (*Continued*).

Hence the "load" component of quadrature current at the input is given by

$$I_{Q_l} = q \cdot s \cdot \frac{r\sqrt{3}}{2\pi} I_o \sin \phi_o \qquad (12.26)$$

The "phase-control" component of quadrature current at the input is the difference between the total quadrature component and the "load" component:

$$I_{Q_P} = I_Q - I_{Q_l} \qquad (12.27)$$

where I_Q is given by (12.25), and I_{Q_l} by (12.26).

The curves of Fig. 12.11 show the relationships between the "phase-control" component of quadrature current at the input, expressed as a per unit value of the total quadrature component, the load displacement factor, and the output voltage ratio.

These curves illustrate that with an output voltage ratio at, or close to unity, the component of quadrature current at the input which is directly attributable to the phase-control process decreases quite rapidly as the

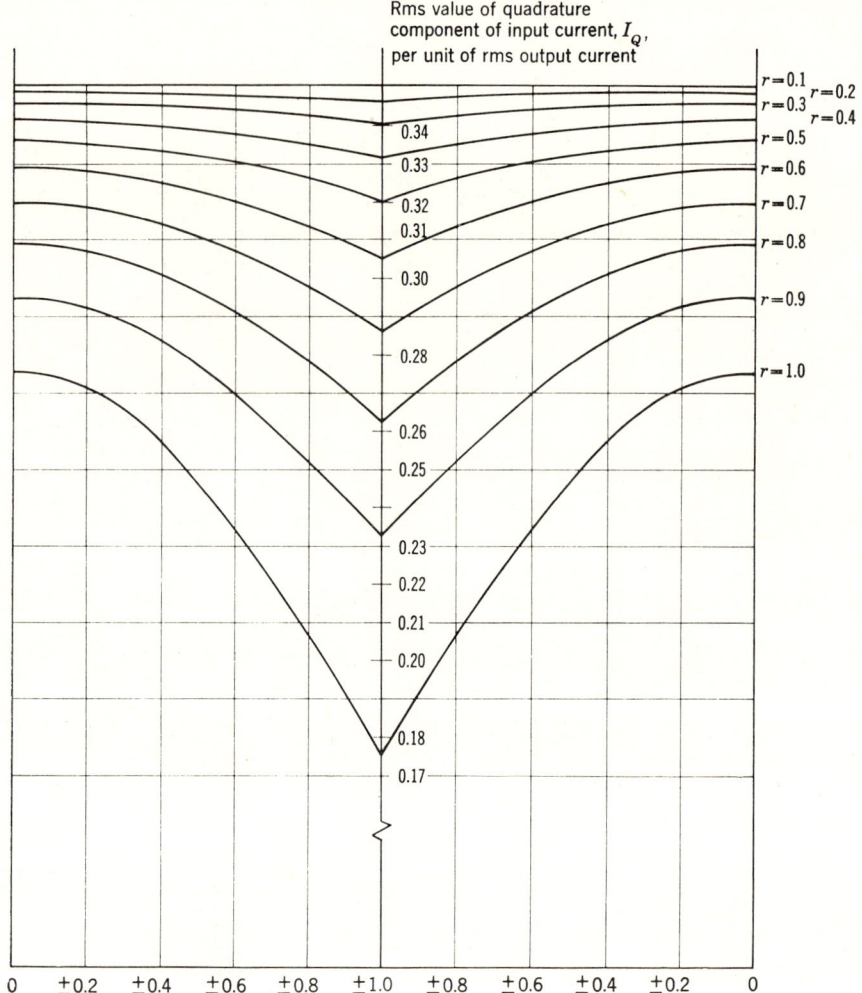

Figure 12.10. Curves showing the relationships between the per unit rms value of the fundamental quadrature component of current at the input of the cycloconverter, the output load displacement factor, and the output voltage ratio.

displacement factor of the load decreases. In other words, with unity output voltage ratio, with a load displacement factor less than unity, the "phase-control" component of quadrature current may be actually quite a small portion of the quadrature component of current attributable to the reactive load at the output. Thus, under these circumstances, the process of power conversion through the cycloconverter is accomplished with a relatively slight penalty in terms of the "phase-control" reactive VA produced at the

FUNDAMENTAL COMPONENT OF INPUT CURRENT

Circuit No.	Curve multiplication factor				Circuit No.	Curve multiplication factor			
	Point in circuit					Point in circuit			
	A	B	C	D		A	B	C	D
1(a)	1.0	1.0	1.0	1.0	7(a)	6.0	6.0	2.0	2.0
1(b)	1.0	0.577	1.0	1.0	7(b)	6.0	6.0	1.154	2.0
2(a)	3.0	3.0	3.0	3.0	7(c)	6.0	3.464	1.154	2.0
2(b)	3.0	3.0	1.732	3.0	7(d)	6.0	3.464	2.0	2.0
2(c)	3.0	1.732	1.732	3.0	8(a)	1.0	① 0.5 ② 0.288	0.25	0.25
2(d)	3.0	1.732	3.0	3.0	8(b)	1.0	1.0	0.25	0.25
3(a)	1.0	1.0	0.5	0.5	8(c)	1.0	0.577	0.25	0.25
3(b)	1.0	0.577	0.5	0.5	9(a)	4.0	4.0	① 2.0 ② 1.154	2.0
4(a)	2.0	2.0	2.0	2.0	9(b)	4.0	2.308	① 2.0 ② 1.154	2.0
4(b)	2.0	2.0	1.154	2.0	9(c)	4.0	4.0	2.0	2.0
4(c)	2.0	1.154	1.154	2.0	9(d)	4.0	2.308	2.0	2.0
4(d)	2.0	1.154	2.0	2.0	10(a)	3.0	① 1.5 ② 0.866	0.75	0.75
5(a)	3.0	3.0	1.5	1.5	10(b)	3.0	3.0	0.75	0.75
5(b)	3.0	1.732	1.5	1.5	10(c)	3.0	1.732	0.75	0.75
6(a)	6.0	6.0	6.0	6.0	11(a)	12.0	12.0	① 2.0 ② 1.154	2.0
6(b)	6.0	6.0	3.464	6.0	11(b)	12.0	6.924	① 2.0 ② 1.154	2.0
6(c)	6.0	3.464	3.464	6.0	11(c)	12.0	12.0	2.0	2.0
6(d)	6.0	3.464	6.0	6.0	11(d)	12.0	6.924	2.0	2.0

Figure 12.10 (*Continued*).

input side. For example, if the load displacement factor is 0.8, then, with unity output voltage ratio, the "phase-control" component of quadrature current is 0.2 per unit of the total quadrature component of current at the input. Thus the total fundamental component of input current is about 0.1 per unit greater than it would be under the hypothetical condition that the "phase control" component of quadrature current did not exist.

The Total Fundamental Component

The rms value of the total fundamental component of current at the input of the cycloconverter is given by

$$I_1 = \sqrt{I_P^2 + I_Q^2} \tag{12.28}$$

where I_P and I_Q are given by expressions (12.21) and (12.25) respectively.

The curves of Fig. 12.12 are a graphical illustration of the relationships between the rms value of the total fundamental component of current at the

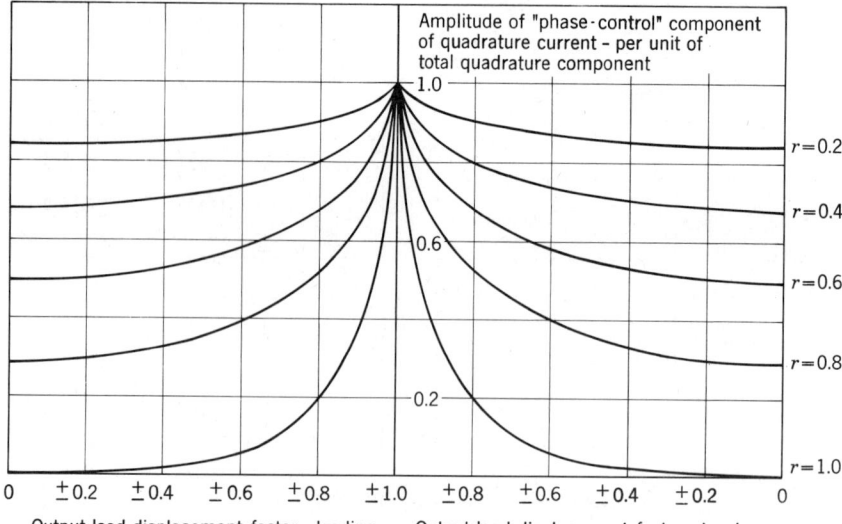

Figure 12.11. Curves showing the relationships between the "phase control" component of quadrature current at the input of the cycloconverter, the output load displacement factor, and the output voltage ratio.

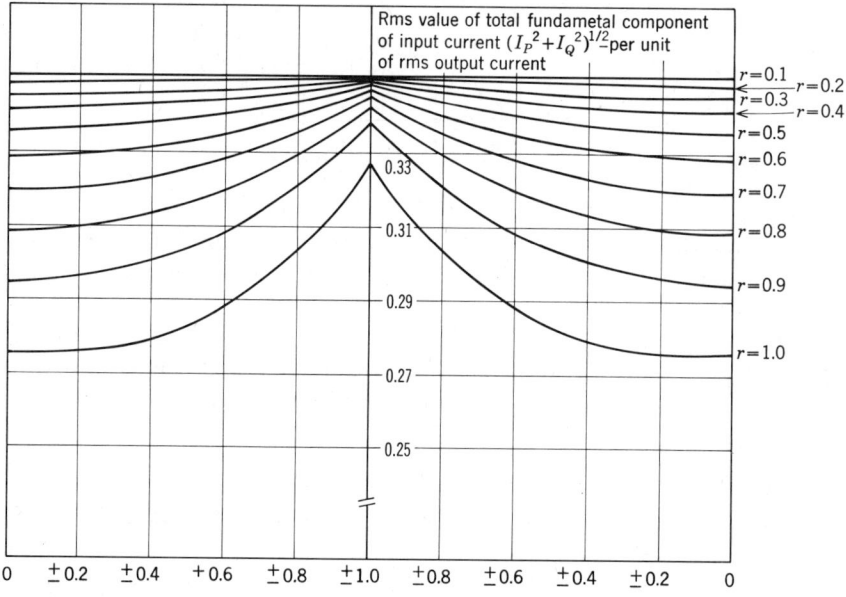

Figure 12.12. Curves showing the relationships between the per unit rms value of the total fundamental component of current at the input of the cycloconverter, the output load displacement factor, and the output voltage ratio.

Circuit No.	Curve multiplication factor Point in circuit			
	A	B	C	D
1(a)	1.0	1.0	1.0	1.0
1(b)	1.0	0.577	1.0	1.0
2(a)	3.0	3.0	3.0	3.0
2(b)	3.0	3.0	1.732	3.0
2(c)	3.0	1.732	1.732	3.0
2(d)	3.0	1.732	3.0	3.0
3(a)	1.0	1.0	0.5	0.5
3(b)	1.0	0.577	0.5	0.5
4(a)	2.0	2.0	2.0	2.0
4(b)	2.0	2.0	1.154	2.0
4(c)	2.0	1.154	1.154	2.0
4(d)	2.0	1.154	2.0	2.0
5(a)	3.0	3.0	1.5	1.5
5(b)	3.0	1.732	1.5	1.5
6(a)	6.0	6.0	6.0	6.0
6(b)	6.0	6.0	3.464	6.0
6(c)	6.0	3.464	3.464	6.0
6(d)	6.0	3.464	6.0	6.0

Circuit No.	Curve multiplication factor Point in curcuit			
	A	B	C	D
7(a)	6.0	6.0	2.0	2.0
7(b)	6.0	6.0	1.154	2.0
7(c)	6.0	3.464	1.154	2.0
7(d)	6.0	3.464	2.0	2.0
8(a)	1.0	① 0.5 ② 0.288	0.25	0.25
8(b)	1.0	1.0	0.25	0.25
8(c)	1.0	0.577	0.25	0.25
9(a)	4.0	4.0	① 2.0 ② 1.154	2.0
9(b)	4.0	2.308	① 2.0 ② 1.154	2.0
9(c)	4.0	4.0	2.0	2.0
9(d)	4.0	2.308	2.0	2.0
10(a)	3.0	① 1.5 ② 0.866	0.75	0.75
10(b)	3.0	3.0	0.75	0.75
10(c)	3.0	1.732	0.75	0.75
11(a)	12.0	12.0	① 2.0 ② 1.154	2.0
11(b)	12.0	6.924	① 2.0 ② 1.154	2.0
11(c)	12.0	12.0	2.0	2.0
11(d)	12.0	6.924	2.0	2.0

Figure 12.12. (*Continued*).

input of the cycloconverter, the load displacement factor, and the output voltage ratio.

A point illustrated by these curves is that, for any given load displacement factor, the relative amplitude of the fundamental component of current at the input of the cycloconverter increases somewhat with decreasing output voltage ratio. This characteristic is different to that obtained for the phase-controlled converter with a steady d-c output, for which the relative amplitude of the fundamental component is the same at all output voltage ratios. Depending upon the particular application, this characteristic of the cycloconverter could be quite objectionable, in as much as it may be necessary to size the windings of the input transformer for a current in excess of that obtained at full output power. This limitation can, however, be avoided, by means of the special control techniques, discussed later in this chapter, whereby the quadrature component of current can be reduced at reduced output voltage ratio.

The Universal Input/Output Displacement Factor Characteristics

The input displacement factor of the phase-controlled cycloconverter is given by

$$\cos \phi_i = \frac{I_P}{\sqrt{I_P^2 + I_Q^2}} \qquad (12.29)$$

where I_P and I_Q are given by (12.21) and (12.25) respectively.

The curves of Fig. 12.13 show the relationships between the input displacement factor, the output displacement factor, and the output voltage ratio. These characteristic curves embrace the complete spectrum of output

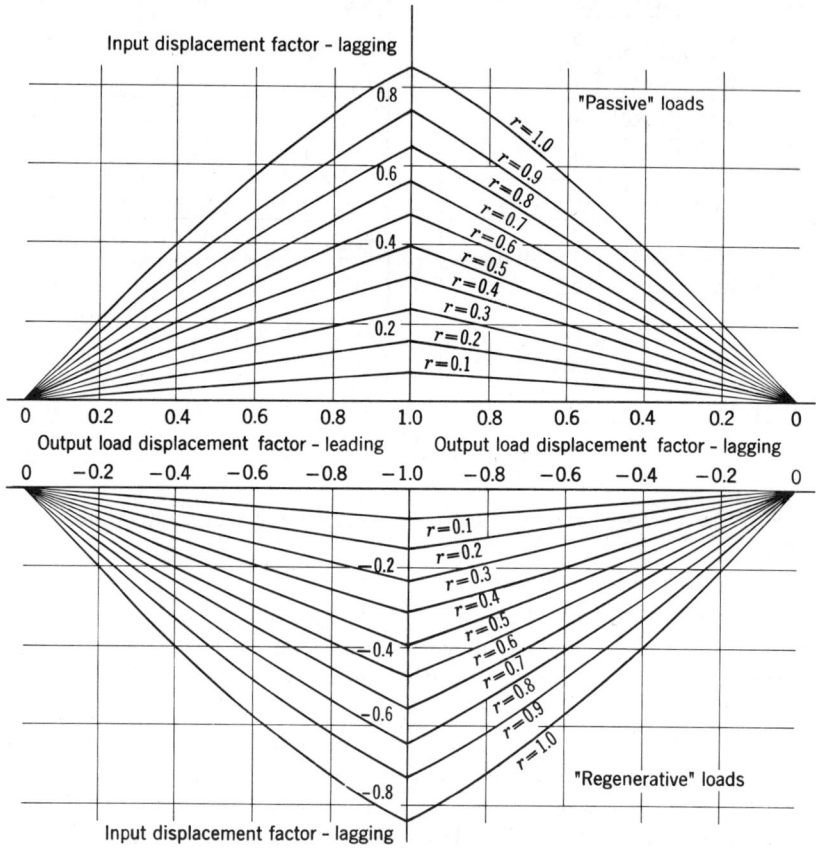

Figure 12.13. Universal relationships between the input displacement factor, the output load displacement factor, and the output voltage ratio, for all "symmetrical" phase-controlled cycloconverter circuits, with single- or balanced poly-phase outputs.

load displacement factors, both positive and negative (that is to say both "passive" and "regenerative" loads), and they are universally applicable to all "symmetrical" cycloconverter circuits, with either single or balanced polyphase outputs, regardless of the circuit configuration or pulse number. (It is to be noted, however, that these curves are not directly applicable to open-delta and ring connected circuits. These latter circuits generally have a somewhat lower input displacement factor than the "symmetrical" circuit, as discussed in Chapter 8.)

From the curves of Fig. 12.13, it is seen that the maximum theoretical input displacement factor of the phase-controlled cycloconverter is 0.843, and this occurs with unity output displacement factor and unity output voltage ratio. As the output voltage ratio decreases, so the input displacement factor decreases—but not quite in direct proportion. Conversely, for a given output voltage ratio, as the output displacement factor decreases, so the input displacement factor decreases—but, again, not quite in direct proportion.

The curves for "regenerative" loads are the "mirror image" of those for passive loads. Thus, for example, with a "fully regenerative" load at the output, and unity output voltage ratio, (a theoretical condition, actually not quite realizable in practice, because of the finite commutation "recovery" angle required), the input displacement factor is -0.843 lagging. This is exactly equal and opposite to that obtained with a "passive" load, of unity displacement factor, at the output.

It is pointed out, once more, that the input displacement factor is invariably lagging, regardless of the output displacement factor, or of whether the load is passive or regenerative. This means to say that the phase-controlled cycloconverter invariably consumes a lagging quadrature component of current from the input system to which it is connected, regardless of the direction of the power flow through the converter.

THE HARMONIC FREQUENCIES

General Comments

From an inspection of the general harmonic series for the various input current waveforms, it is possible to perceive the frequencies of the harmonic components present in the input line current of the cycloconverter. Before proceeding to a review of the quantitative results of a detailed computation of the amplitudes of the harmonic components, it is of interest, firstly, merely to make an examination of the spectrum of harmonic frequencies to be found in the input line current.

It can be seen from the general formulae that the harmonic distortion components in the input line current have frequencies which are generally sums or differences between multiples of the input and output frequencies; and, in addition, certain direct odd multiple harmonics of the input frequency are present. Thus, for the most part, the harmonic frequencies are functions of both the input and output frequencies, and variations in either of these frequencies result in corresponding alterations to the spectrum of harmonic frequencies in the input line current.

It is also evident, from the rules enumerated on p. 339 for a balanced 3-phase output, that a fairly high degree of cancellation of harmonic terms associated with the individual "single phase" converter groups takes place in the waveform of the net input current. Thus, the spectrum of harmonic frequencies in the input current of the cycloconverter supplying a balanced 3-phase output is considerably less "dense" than for a single-phase output. This, of course, would be expected from the relatively undistorted appearance of the input current waveforms associated with a 3-phase output, as compared to the waveforms associated with a single-phase output.

In the following sections, attention is drawn to the presence of two different types of family of harmonic frequencies in the input current waveform of the cycloconverter. The first type is a "characteristic" family of harmonic frequencies, which is invariably present, regardless of the circuit pulse number. The second type of harmonic family is termed "circuit-dependent," insofar as the presence or absence of this type is dictated by the pulse number of the cycloconverter.

The Characteristic Cycloconverter Harmonic Frequencies

As discussed in Chapter 6, for a cycloconverter supplying a single-phase output, harmonic components of current appear in the input lines, whose frequencies are given by $f_i \pm 2f_o$. The presence of these components was deduced, without regard to the circuit configuration or pulse number, purely on the grounds that the instantaneous input and output powers must be equal to one another.

Thus, for a cycloconverter with a single-phase output, the presence was predicted of two harmonic components of input current, which are independent of the circuit configuration or pulse number; as such, these components could be termed "characteristic cycloconverter harmonics."

The detailed harmonic analysis of the input line current waveforms, presented in this chapter, verifies the presence of the $f_i \pm 2f_o$ harmonic components. (It also verifies the prediction made in Chapter 6 that, for a balanced 3-phase output, these components are absent.)

Further than this, however, the detailed analysis also reveals a whole family of "characteristic cycloconverter harmonics," whose presence is independent of the circuit configuration, yet is not predictable on the simple basis of equality of the instantaneous input and output powers. Thus an inspection of the analytical results shows that the $f_i \pm 2f_o$ harmonic components are, in fact, but the "parent" terms of an infinite (but converging) series of "characteristic cycloconverter harmonics," the presence of which is independent of the circuit configuration or pulse number. The frequencies of these components are defined as follows:

For a single-phase output

$$f_{ch} = f_i \pm 2nf_o \qquad (12.30)$$

For a balanced 3-phase output

$$f_{ch} = f_i \pm 6nf_o \qquad (12.31)$$

where n is any integer from 1 to infinity.

The question naturally arises as to how the characteristic harmonic frequencies, other than the $f_i \pm 2f_o$ components, come to be present in the input line current of the cycloconverter, since their presence is not required on the basis of the equality of the instantaneous input and output powers; nor, on this basis, can they be capable of making any contribution to the instantaneous power at the input side. Indeed a computation of the net instantaneous power at the input due to the characteristic harmonic components of input current (other than the $f_i \pm 2f_o$ terms) shows that this is zero. The explanation for the presence of these characteristic harmonic components is that they are, in effect, purely "parasites," which are entirely attributable to the basic nature of the load presented by the cycloconverter to the input system, in exactly the same way that the phase-control mechanism also gives rise to a parasite "phase-control" quadrature component of current at the input, as already discussed.

A further related point is that the analytical expressions obtained in this chapter reveal the fact that for any condition other than unity load displacement factor, and unity output voltage ratio, the amplitudes of the $(f_i \pm 2f_o)$ harmonic components are larger than would be predicted on the basis of equality of instantaneous input and output power. The reason for this is that the phase relationships of the conjugate pair of harmonics is such that a partial cancellation of harmonic power due to the pair takes place, with the result that the net harmonic power at the input does (as must be the case) exactly balance the oscillatory component of power at the output.

Circuit-Dependent Harmonic Frequencies

In addition to the characteristic harmonic frequencies which appear in the input current waveform, which, as has been seen, are independent of the circuit pulse number, harmonic frequencies also exist whose presence or absence does depend upon the circuit pulse number.

An inspection of the general expressions obtained for cycloconverter input current waveforms of various pulse numbers, yields the following information concerning the circuit-dependent harmonic frequencies.

THE THREE-PULSE INPUT CURRENT WAVEFORM. The circuit dependent harmonic frequencies, appearing in the 3-pulse input current waveform, are as follows:

For a single-phase output
$f_H = f_o$ (appearing only in the transformer secondary current; assuming a zig-zag winding)

$$f_H = |[3(2p-1) \pm 1]f_i \pm (2n+1)f_o|$$

and

$$f_H = |[6p \pm 1]f_i \pm 2nf_o| \tag{12.32}$$

For a balanced 3-phase output:

$$f_H = |[3(2p-1) \pm 1]f_i \pm 3(2n+1)f_o|$$
$$f_H = |[6p \pm 1]f_i \pm 6nf_o| \tag{12.33}$$

where p is any integer from 1 to infinity
n is any integer from zero to infinity

THE SIX-PULSE INPUT CURRENT WAVEFORM. The circuit dependent harmonic frequencies, appearing in the 6-pulse input current waveform, are as follows:

For a single-phase output

$$f_H = |[6p \pm 1]f_i \pm 2nf_o| \tag{12.34}$$

For a balanced 3-phase output

$$f_H = |[6p \pm 1]f_i \pm 6nf_o| \tag{12.35}$$

where p is any integer from 1 to infinity
n is any integer from zero to infinity

THE TWELVE-PULSE INPUT CURRENT WAVEFORM. The circuit dependent harmonic frequencies, appearing in the 12-pulse input current waveform, are as follows:

For a single-phase output

$$f_H = |[12p \pm 1]f_i \pm 2nf_o| \qquad (12.36)$$

For a balanced 3-phase output

$$f_H = |[12p \pm 1]f_i \pm 6nf_o| \qquad (12.37)$$

where p is any integer from 1 to infinity
n is any integer from zero to infinity

Harmonic Frequency Charts

Figures 12.14 and 12.15 illustrate the relationships between the "predominant" harmonic frequencies present in the input current of the 3-pulse cycloconverter, for single- and balanced three-phase loads respectively. From expressions (12.34) through (12.37), it is evident that these charts are applicable also to 6- and 12- (as well as 9-) pulse cycloconverter input current waveforms, with the provision that the appropriate harmonic families, as indicated, are ignored.

Consider, for example, the harmonic frequency chart of Fig. 12.14, for a cycloconverter with single-phase output. This shows that at zero output to input frequency ratio, the spectrum of harmonic frequencies (for a 3-pulse circuit) are $2f_i$, $4f_i$, $5f_i$, $7f_i$ and so on. This, of course, is in agreement with the already known frequency spectrum for a 3-pulse converter with a steady d-c output.

As soon as the output frequency is raised above zero, each "parent" harmonic frequency "splits" into a family of harmonic frequencies, the members of which progressively diverge away from the parent frequency, with increasing output to input frequency ratio.

A point of some importance, illustrated by this harmonic frequency chart, is that harmonic frequencies less than the input frequency are present in the input current waveform. Moreover, at certain discrete output to input frequency ratios, these subharmonic components actually have zero frequency. Thus, to consider a particularly pertinent example, when the output frequency is exactly one-half of the input frequency, the "characteristic" ($f_i - 2f_o$) component has zero frequency, and this therefore appears as a direct component of current at the input.

The presence of low frequency subharmonic and direct components of current in the input lines is generally undesirable, since, depending upon their relative amplitudes, such components could give rise to appreciable

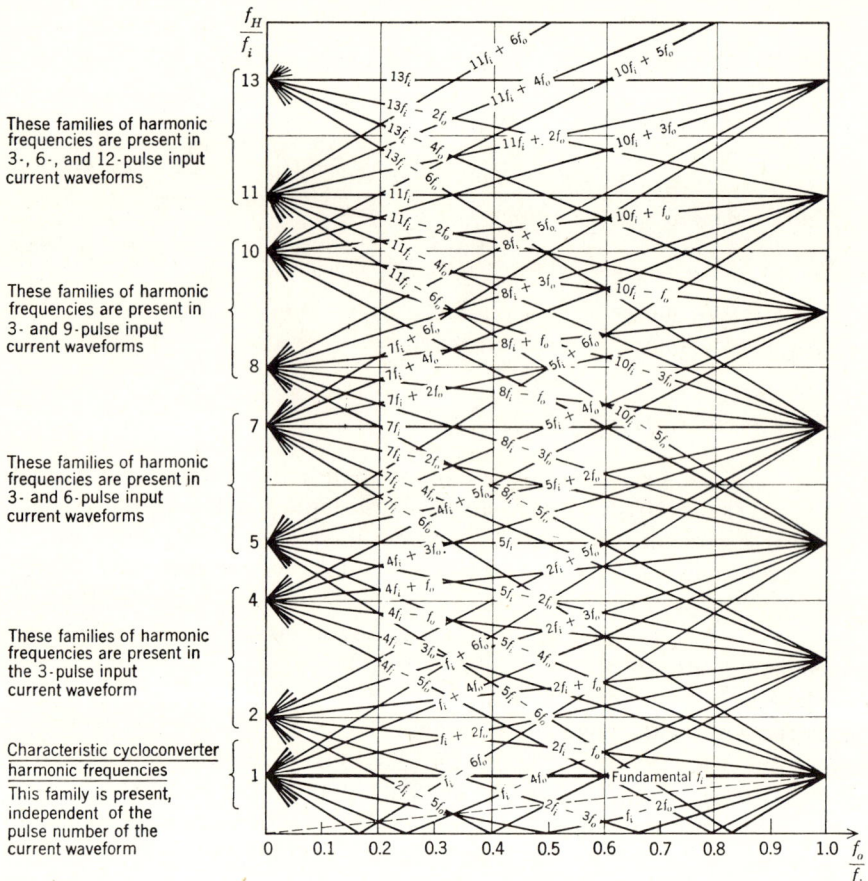

Figure 12.14. Chart showing the relationships between the "predominant" harmonic frequencies present in the 3-pulse input current waveform of the cycloconverter with a single-phase output, and the output to input frequency ratio. For input current waveforms with higher pulse numbers, certain harmonic families are eliminated, as indicated. *Note:* The f_o component (shown dashed) is present only in the 3-pulse input current; moreover, with a zig-zag transformer connection, it is confined to the secondary current.

subharmonic magnetization of the core of the input transformer, and this, of course, should be capable of handling such magnetization, without saturation.

An inspection of the chart of Fig. 12.15 shows that for a balanced 3-phase output, the subharmonic frequency region is much "less populated" than for a single-phase output. For this reason, the subharmonic components which do exist in the input current of a cycloconverter supplying a three-phase output are not generally of particular practical concern.

THE HARMONIC FREQUENCIES

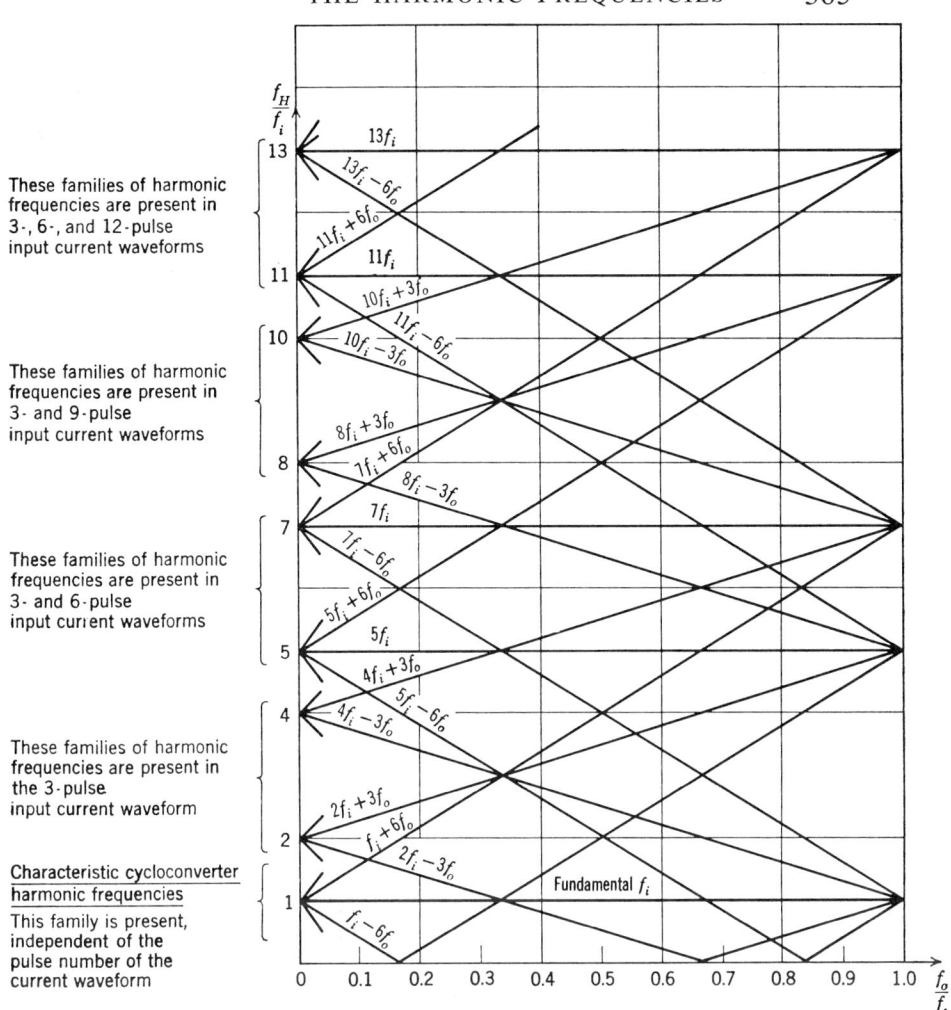

Figure 12.15. Chart showing the relationships between the "predominant" harmonic frequencies present in the 3-pulse input current waveform of the cycloconverter with a balanced 3-phase output, and the output to input frequency ratio. For input current waveforms with higher pulse numbers, certain harmonic families are eliminated as indicated.

THE UNIVERSAL RELATIONSHIPS BETWEEN THE OUTPUT VOLTAGE AND INPUT CURRENT HARMONICS OF THE CYCLOCONVERTER

From the analytical expressions obtained for the output voltage, and the input current, of the cycloconverter, it can be seen that a simple relationship exists between the converter pulse number, the harmonic frequencies present in the output voltage, and the harmonic frequencies present in the input current. These relationships can be stated as follows:

The harmonic frequencies present in the output voltage of the cycloconverter consist of integer multiples of the circuit pulse number and the input frequency, $(n \times P)f_i$, to which are both added, and subtracted, integer multiples of the output frequency; these integer multiples of the output frequency are ODD for $(n \times P)$EVEN, and EVEN for $(n \times P)$ODD.

For a cycloconverter with a single-phase output, the harmonic frequencies present in the input current are related to those in the output voltage as follows:

For each family of output voltage harmonics, $(n \times P)f_i \pm mf_o$, there are two corresponding families of input current harmonics:

$$|[(n \times P) - 1]f_i \pm (m - 1)f_o|$$
$$|[(n \times P) + 1]f_i \pm (m - 1)f_o|$$

where m is ODD for $(n \times P)$EVEN
m is EVEN for $(n \times P)$ODD

In addition, the characteristic family of harmonic frequencies is given by

$$|f_i \pm 2mf_o| \quad (m \geqslant 1)$$

For a cycloconverter with a balanced 3-phase output, for each family of output voltage harmonics, $(n \times P)f_i \pm mf_o$, the two corresponding families of input current harmonics are:

$$|[(n \times P) - 1]f_i \pm 3(m - 1)f_o|$$
$$|[(n \times P) + 1]f_i \pm 3(m - 1)f_o|$$

where m is ODD for $(n \times P)$EVEN
m is EVEN for $(n \times P)$ODD

In addition, the characteristic family of harmonic frequencies is given by

$$|f_i \pm 6mf_o| \quad (m \geqslant 1)$$

THE AMPLITUDES OF THE HARMONIC COMPONENTS

General Comments

In the previous sections, the discussion was mainly concerned with the frequencies of the harmonic components, and their amplitudes were ignored. An inspection of the general expressions obtained for the cycloconverter input current waveforms shows that the amplitude of each harmonic component is a function of the output voltage ratio and the load displacement angle, but is *independent* of the frequency of the component. This means to say that for a given output voltage ratio and load displacement angle, each line representing a given harmonic frequency on the charts of Figs. 12.14 and 12.15 has associated with it a fixed amplitude, which is invariably the same, regardless of the output to input frequency ratio.

It has been seen already that the presence or absence of given harmonic frequencies in the input current is determined by the pulse number of the waveform, and the number of output phases of the cycloconverter. An inspection of the general formulae obtained for current waveforms of various pulse numbers, reveals that for a given output voltage ratio and load displacement angle, those harmonic components which are present always have the same relative amplitudes, independently of the pulse number of the wave or the number of output phases. This means to say that a single set of quantitative data, related to the amplitudes of the harmonic components of the input current waveform of the 3-pulse cycloconverter with single-phase output, is equally applicable to all current waveforms having pulse numbers which are integer multiples of 3, for both single- and balanced 3-phase outputs, it being necessary simply to ignore the data related to those harmonic components which are known to be absent from a given waveform.

From the general expression, (12.5), for the input current waveform of the 3-pulse cycloconverter with single-phase output, certain general identities can be shown to exist between the amplitudes of given harmonic components, with various given load displacement angles. These identities are as follows:

For any given output voltage ratio:

Amplitude of component having any given frequency, with load displacement angle ϕ_o. = Amplitude of component having the same frequency, with load displacement angle $(\pi - \phi_o)$ (12.38)

And

Amplitude of component having frequency $nf_i + mf_o$ with load displacement angle ϕ_o. = Amplitude of component having frequency $nf_i - mf_o$ with load displacement angle $-\phi_o$. (12.39)

Conversely

Amplitude of component having frequency $nf_i - mf_o$ with load displacement angle ϕ_o. = Amplitude of component having frequency $nf_i + mf_o$ with load displacement angle $-\phi_o$. (12.40)

In Chapter 11, it was seen that relationships of exactly the same form [(11.53) through (11.55)] are applicable to the voltage harmonics at the output of the cycloconverter. These relationships were explained, in qualitative terms, by reference to the interrelationships between the "fast" and "slow" switching portions of the output voltage waveform. This same explanation also applies to the above relationships between the amplitudes of the input current harmonics.

Bearing in mind the relationships of (12.38) through (12.40), it is evident that the results of a computation of the amplitudes of the harmonic components of input current for a range of load displacement angles in one quadrant only, say 0° to +90°, can also be applied to load displacement angles in the other three quadrants.

Quantitative Data

The relative amplitudes of the predominant members of each of the $[f_i \pm 2nf_o]$, $[2f_i \pm (2n + 1)f_o]$, $[4f_i \pm (2n + 1)f_o]$, $[5f_i \pm 2nf_o]$, $[7f_i \pm 2nf_o]$, $[11f_i \pm 2nf_o]$ and $[13f_i \pm 2nf_o]$ harmonic families, for a range of output voltage ratios from 1.0 to 0.1, and for a range of load displacement angles, which, [by virtue of the identities of expressions (12.38) through (12.40)], embrace the complete spectrum from 0° to 360°, are shown in Tables 12.1 through 12.4.

These "amplitude tables," when used in conjunction with the "frequency charts" of Figs. 12.14 and 12.15, provide a comprehensive set of quantitative data applicable to the input current waveforms of 3-, 6-, and 12-pulse cycloconverters, with both single- and balanced 3-phase outputs.

THE TOTAL RMS VALUE OF THE INPUT CURRENT WAVEFORM

The total rms value of the input current waveform is given by the following general expression:

$$I_{\text{rms}} = [I_P^2 + I_Q^2 + \sum I_h^2]^{1/2} \qquad (12.41)$$

$\uparrow$ In-phase component $\quad\uparrow$ Quadrature component $\quad\uparrow$ Harmonic components

For a cycloconverter of a given pulse number with a single-phase output, the relationship between the rms input and output currents is inherently fixed, regardless of the operating conditions, or, indeed, of whether the converter produces a d-c or an a-c output. In other words, the right-hand side of (12.41) is invariably constant, even though the amplitudes of the constituent terms change with changing operating conditions.

As has been seen, for a cycloconverter with a balanced 3-phase output, cancellation of harmonic components of input current drawn by the individual converters occurs in the waveform of the net input current due to all three

Tables 12.1 *Showing the Amplitudes of the In-Phase and Quadrature Components, and the Characteristic Harmonic Components, Having Frequencies of $f_i \pm 2nf_o$, in the Input Current of the Cycloconverter, for Various Output Voltage Ratios and Output Load Displacement Angles*

$f_i \pm 2nf_o$

	$\phi_o = 0°, 180°$							$\phi_o = \pm 30°, \pm 150°$							$\phi_o = \pm 60°, \pm 120°$							$\phi_o = \pm 90°$						
$2n \rightarrow$	0	2	4	6	8	10	12	0	2	4	6	8	10	12	0	2	4	6	8	10	12	0	2	4	6	8	10	12
$r \downarrow$	I_P	I_Q						I_P	I_Q						I_P	I_Q						I_P	I_Q					
1.0	.84	.54	.42	.18	.00	.04	.00	.02	.77	.64	.72	.17	.12	.06	.01	.01	.49	.87	.95	.03	.03	.02	.02	.01	1.0	.00	1.0	.00	.00	.00	.00	.00
0.9	.73	.68	.39	.12	.03	.02	.01	.01	.66	.75	.61	.19	.09	.03	.01	.01	.41	.91	.80	.06	.04	.02	.01	.01	1.0	.00	.86	.02	.01	.01	.00	.00
0.8	.64	.77	.37	.10	.03	.02	.01	.01	.57	.82	.56	.20	.08	.03	.02	.01	.35	.94	.71	.09	.04	.02	.01	.01	1.0	.00	.77	.05	.03	.02	.01	.00
0.7	.56	.83	.36	.09	.03	.02	.01	.01	.49	.87	.52	.22	.08	.03	.02	.01	.30	.96	.64	.12	.05	.02	.02	.01	1.0	.00	.69	.09	.04	.02	.01	.01
0.6	.48	.88	.35	.08	.03	.02	.01	.01	.42	.91	.48	.23	.07	.03	.02	.01	.25	.97	.58	.15	.06	.03	.02	.01	1.0	.00	.62	.12	.05	.02	.01	.01
0.5	.40	.92	.34	.08	.03	.02	.01	.01	.35	.94	.45	.25	.07	.03	.02	.01	.20	.98	.53	.18	.06	.03	.02	.01	1.0	.00	.56	.15	.05	.02	.01	.01
0.4	.32	.95	.34	.07	.03	.02	.01	.01	.28	.96	.42	.26	.07	.02	.02	.01	.16	.99	.49	.21	.06	.03	.02	.01	1.0	.00	.51	.19	.06	.03	.01	.01
0.3	.24	.97	.34	.07	.03	.02	.01	.01	.21	.98	.40	.28	.07	.03	.02	.01	.12	.99	.44	.24	.06	.03	.02	.01	1.0	.00	.46	.22	.06	.03	.01	.01
0.2	.16	.99	.34	.07	.03	.02	.01	.01	.14	.99	.38	.30	.07	.03	.02	.01	.08	1.0	.41	.27	.07	.03	.02	.01	1.0	.00	.42	.26	.06	.03	.01	.01
0.1	.08	1.0	.33	.07	.03	.02	.01	.01	.07	1.0	.35	.31	.07	.03	.02	.01	.04	1.0	.37	.30	.07	.03	.02	.01	1.0	.00	.37	.30	.07	.03	.01	.01

Notes: 1. Tables show amplitude of component, as per unit value of the total fundamental component.
2. These components are present in the input current of all cycloconverter circuits, regardless of pulse number. For a balanced 3-phase output, only those components whose numbers are boxed thus ☐ are present.
3. The values of $2n$ except $2n = 2$: The values given apply to both the $f_i + 2nf_o$ and $f_i - 2nf_o$ harmonic components.
   For $\phi_o = 0°, 180°,$ and $2n = 2$: The values given apply to both the $f_i + 2f_o$ and $f_i - 2f_o$ harmonic components.
   For other values of ϕ_o, and $2n = 2$: The values given in the "black" column apply to the $f_i + 2f_o$ component for positive ϕ_o, and to the $f_i - 2f_o$ component for negative ϕ_o.
   The values given in the "white" column apply to the $f_i + 2f_o$ component for negative ϕ_o, and to the $f_i - 2f_o$ component for positive ϕ_o.

Tables 12.2 Showing the Amplitudes of the Harmonic Components Having Frequencies of $2f_i \pm (2n+1)f_o$, and $4f_i \pm (2n+1)f_o$, in the Input Current of the Cycloconverter, for Various Output Voltage Ratios and Output Load Displacement Angles

$2f_i \pm (2n+1)f_o$

$(2n+1) \rightarrow$	$\phi_o = 0°, 180°$					$\phi_o = \pm 30°, \pm 150°$					$\phi_o = \pm 60°, \pm 120°$					$\phi_o = \pm 90°$				
$r \downarrow$	1	3	5	7	9	11	1	3	5	7	9	11	1	3	5	7	9	11	1	3	5	7	9	11
1.0	.35	.23	.09	.02	.02	.01	.48	.16	.33	.13	.08	.03	.01	.51	.03	.46	.03	.02	.02	.02	.01	.50	.00	.50	.00	.00	.00
0.9	.37	.17	.04	.00	.00	.00	.49	.21	.24	.11	.04	.01	.01	.53	.07	.35	.04	.02	.01	.01	.01	.53	.03	.40	.02	.01	.01	.00
0.8	.38	.13	.03	.00	.00	.00	.49	.23	.19	.09	.03	.01	.00	.54	.11	.28	.05	.02	.01	.01	.01	.54	.07	.32	.03	.02	.01	.00
0.7	.38	.11	.02	.00	.00	.00	.48	.26	.15	.08	.03	.01	.00	.53	.15	.22	.06	.02	.01	.01	.01	.54	.11	.26	.04	.02	.01	.00
0.6	.39	.09	.02	.00	.00	.00	.48	.28	.12	.07	.02	.01	.00	.52	.19	.18	.06	.02	.01	.01	.01	.54	.16	.20	.05	.02	.01	.00
0.5	.39	.07	.01	.00	.00	.00	.46	.30	.10	.06	.02	.01	.00	.51	.23	.14	.06	.02	.01	.01	.00	.52	.20	.15	.05	.02	.01	.00
0.4	.39	.06	.01	.00	.00	.00	.45	.32	.07	.05	.01	.00	.00	.49	.26	.10	.05	.02	.01	.01	.00	.50	.24	.11	.05	.02	.01	.00
0.3	.39	.04	.01	.00	.00	.00	.44	.34	.05	.04	.01	.00	.00	.47	.30	.07	.04	.01	.01	.00	.00	.48	.28	.08	.04	.01	.00	.00
0.2	.39	.03	.00	.00	.00	.00	.43	.36	.03	.03	.01	.00	.00	.45	.33	.04	.03	.01	.01	.00	.00	.45	.32	.05	.03	.01	.00	.00
0.1	.39	.01	.00	.00	.00	.00	.41	.38	.02	.02	.00	.00	.00	.42	.36	.02	.02	.00	.00	.00	.00	.43	.36	.02	.02	.00	.00	.00

$4f_i \pm (2n+1)f_o$

$(2n+1) \rightarrow$	$\phi_o = 0°, 180°$					$\phi_o = \pm 30°, \pm 150°$					$\phi_o = \pm 60°, \pm 120°$					$\phi_o = \pm 90°$				
$r \downarrow$	1	3	5	7	9	11	1	3	5	7	9	11	1	3	5	7	9	11	1	3	5	7	9	11
1.0	.09	.14	.12	.05	.01	.01	.08	.08	.20	.18	.06	.04	.02	.01	.02	.02	.25	.02	.23	.02	.02	.01	.00	.00	.25	.00	.00	.00
0.9	.14	.13	.07	.02	.00	.00	.16	.10	.20	.06	.11	.04	.02	.01	.01	.09	.05	.27	.03	.14	.01	.01	.00	.04	.02	.29	.01	.16	.01	.01	.00
0.8	.16	.11	.05	.01	.00	.00	.19	.10	.18	.05	.07	.03	.01	.00	.14	.05	.25	.02	.09	.01	.01	.00	.09	.04	.28	.02	.10	.01	.01	.00
0.7	.17	.10	.03	.00	.00	.00	.21	.09	.15	.05	.05	.02	.01	.00	.18	.05	.23	.02	.06	.01	.01	.00	.15	.04	.26	.01	.05	.00	.00	.00
0.6	.18	.08	.02	.01	.00	.00	.23	.10	.13	.05	.04	.02	.01	.00	.21	.04	.19	.02	.04	.01	.00	.00	.20	.02	.22	.00	.02	.00	.00	.00
0.5	.18	.07	.02	.00	.00	.00	.23	.11	.10	.04	.02	.01	.00	.00	.24	.04	.15	.02	.02	.01	.00	.00	.23	.00	.18	.01	.01	.00	.00	.00
0.4	.19	.05	.01	.00	.00	.00	.24	.12	.08	.04	.02	.01	.00	.00	.25	.06	.12	.03	.01	.01	.00	.00	.25	.04	.13	.02	.01	.00	.00	.00
0.3	.19	.04	.01	.00	.00	.00	.23	.14	.06	.03	.01	.01	.00	.00	.25	.10	.08	.03	.01	.01	.00	.00	.25	.08	.09	.03	.01	.00	.00	.00
0.2	.19	.03	.00	.00	.00	.00	.22	.16	.04	.03	.01	.01	.00	.00	.24	.13	.05	.02	.01	.01	.00	.00	.25	.12	.06	.02	.02	.00	.00	.00
0.1	.20	.01	.00	.00	.00	.00	.21	.18	.02	.01	.00	.00	.00	.00	.22	.17	.02	.02	.00	.00	.00	.00	.23	.16	.02	.02	.00	.00	.00	.00

NOTES:
1. Tables show amplitude of harmonic, as per unit value of the total fundamental component.
2. These harmonics are present in the input current of the 3-pulse cycloconverter. For a balanced 3-phase output, only those components whose numbers are boxed thus ☐ are present.
3. Where there is only one column for a given value of $(2n+1)$, the values given apply to both the 'plus' and 'minus' $(2n+1)f_o$ components.
   Where there are two columns for a given value of $(2n+1)$: the values given in the 'black' column apply to the 'plus' $(2n+1)f_o$ components for positive ϕ_o, and to the 'minus' $(2n+1)f_o$ components for negative ϕ_o.
   The values given in the "white" column apply to the "plus" $(2n+1)f_o$ components for negative ϕ_o, and to the "minus" $(2n+1)f_o$ components

Tables 12.3 Showing the Amplitudes of the Harmonic Components Having Frequencies of $5f_i \pm 2nf_o$, and $7f_i \pm 2nf_o$, in the Input Current of the Cycloconverter, for Various Output Voltage Ratios and Output Load Displacement Angles

$5f_i \pm 2nf_o$

$\phi_o = 0°, 180°$

$2n \rightarrow$ / $r \downarrow$	⬜0	2	4	⬜6	8	10
1.0	.04	.07	.11	.09	.05	.01
0.9	.09	.11	.09	.05	.02	.00
0.8	.11	.11	.07	.03	.01	.00
0.7	.14	.11	.06	.02	.01	.00
0.6	.15	.10	.05	.02	.01	.00
0.5	.17	.10	.04	.01	.01	.00
0.4	.18	.09	.03	.01	.01	.00
0.3	.19	.08	.02	.01	.00	.00
0.2	.20	.07	.02	.01	.00	.00
0.1	.20	.07	.01	.01	.01	.00

$\phi_o = \pm 30°, \pm 150°$

$2n \rightarrow$ / $r \downarrow$	⬜0	2(■)	2(□)	4(■)	4(□)	⬜6	8	10
1.0	.02	.05	.05	.16	.06	.14	.05	.04
0.9	.05	.09	.06	.15	.05	.08	.03	.01
0.8	.09	.15	.07	.13	.04	.05	.02	.01
0.7	.10	.16	.06	.10	.03	.03	.01	.00
0.6	.12	.16	.06	.08	.02	.02	.01	.00
0.5	.14	.16	.06	.06	.02	.01	.01	.01
0.4	.16	.15	.05	.04	.01	.01	.01	.01
0.3	.18	.13	.04	.03	.01	.01	.01	.01
0.2	.19	.11	.04	.02	.01	.01	.01	.01
0.1	.20	.09	.05	.01	.01	.01	.01	.01

$\phi_o = \pm 60°, \pm 120°$

$2n \rightarrow$ / $r \downarrow$	⬜0	2(■)	2(□)	4(■)	4(□)	⬜6	8	10
1.0	.02	.02	.02	.20	.02	.18	.02	.01
0.9	.05	.09	.03	.21	.01	.10	.01	.00
0.8	.05	.14	.02	.19	.01	.06	.00	.00
0.7	.04	.17	.02	.15	.01	.04	.01	.00
0.6	.05	.19	.03	.12	.01	.02	.01	.00
0.5	.08	.20	.03	.08	.01	.01	.01	.00
0.4	.11	.19	.03	.05	.01	.01	.01	.00
0.3	.13	.17	.02	.02	.00	.00	.01	.00
0.2	.17	.14	.02	.01	.00	.00	.01	.00
0.1	.19	.10	.04	.01	.01	.01	.03	.00

$\phi_o = \pm 90°$

$2n \rightarrow$ / $r \downarrow$	⬜0	2(■)	2(□)	4(■)	4(□)	⬜6	8	10
1.0	.00	.00	.00	.20	.00	.20	.00	.00
0.9	.02	.04	.01	.23	.01	.12	.00	.00
0.8	.02	.11	.00	.21	.00	.06	.00	.00
0.7	.00	.16	.00	.18	.00	.03	.01	.00
0.6	.02	.19	.01	.14	.01	.01	.01	.00
0.5	.05	.21	.02	.09	.01	.01	.01	.00
0.4	.09	.20	.03	.05	.01	.00	.00	.00
0.3	.13	.18	.02	.02	.00	.00	.00	.00
0.2	.17	.15	.00	.00	.00	.00	.00	.00
0.1	.19	.11	.03	.00	.00	.00	.00	.00

$7f_i \pm 2nf_o$

$\phi_o = 0°, 180°$

$2n \rightarrow$ / $r \downarrow$	⬜0	2	4	⬜6	8	10
1.0	.03	.02	.05	.07	.07	.02
0.9	.05	.06	.07	.06	.03	.01
0.8	.06	.07	.07	.04	.02	.01
0.7	.07	.08	.06	.03	.01	.01
0.6	.09	.08	.05	.02	.01	.00
0.5	.10	.08	.04	.01	.01	.00
0.4	.12	.07	.03	.01	.01	.00
0.3	.13	.06	.02	.01	.00	.00
0.2	.14	.06	.02	.01	.01	.00
0.1	.14	.05	.01	.00	.00	.00

$\phi_o = \pm 30°, \pm 150°$

$2n \rightarrow$ / $r \downarrow$	⬜0	2(■)	2(□)	4(■)	4(□)	⬜6	8	10
1.0	.01	.03	.03	.04	.04	.11	.04	.03
0.9	.02	.05	.05	.09	.04	.10	.03	.02
0.8	.03	.07	.05	.10	.03	.07	.03	.01
0.7	.04	.09	.04	.10	.03	.05	.01	.01
0.6	.06	.10	.05	.09	.02	.03	.01	.01
0.5	.07	.10	.05	.07	.02	.01	.01	.00
0.4	.08	.11	.04	.05	.01	.01	.00	.00
0.3	.09	.11	.04	.03	.01	.00	.00	.00
0.2	.11	.10	.03	.02	.01	.00	.00	.00
0.1	.13	.09	.03	.01	.00	.00	.00	.00

$\phi_o = \pm 60°, \pm 120°$

$2n \rightarrow$ / $r \downarrow$	⬜0	2(■)	2(□)	4(■)	4(□)	⬜6	8	10
1.0	.01	.01	.01	.01	.01	.14	.13	.01
0.9	.01	.02	.04	.02	.08	.14	.06	.01
0.8	.00	.03	.04	.02	.12	.11	.03	.00
0.7	.01	.04	.02	.13	.01	.08	.01	.01
0.6	.00	.04	.06	.02	.12	.05	.01	.01
0.5	.03	.03	.09	.04	.12	.02	.00	.00
0.4	.02	.04	.11	.02	.07	.01	.00	.00
0.3	.04	.07	.13	.02	.04	.01	.00	.00
0.2	.07	.11	.11	.04	.01	.01	.00	.00
0.1	.13	.08	.13	.02	.01	.00	.00	.00

$\phi_o = \pm 90°$

$2n \rightarrow$ / $r \downarrow$	⬜0	2(■)	2(□)	4(■)	4(□)	⬜6	8	10
1.0	.00	.00	.00	.00	.00	.14	.14	.00
0.9	.01	.02	.01	.05	.00	.17	.06	.00
0.8	.01	.02	.00	.12	.00	.14	.00	.00
0.7	.00	.01	.02	.15	.00	.09	.01	.00
0.6	.02	.03	.07	.15	.01	.05	.00	.00
0.5	.02	.11	.03	.12	.00	.02	.01	.00
0.4	.06	.14	.02	.08	.00	.02	.01	.00
0.3	.06	.14	.02	.04	.00	.01	.01	.00
0.2	.10	.12	.01	.01	.01	.00	.00	.00
0.1	.13	.09	.03	.01	.00	.00	.00	.00

NOTES:
1. Tables show amplitude of harmonic, as per unit value of the total fundamental component.
2. These harmonics are present in the input current of 3- and 6-pulse cycloconverters. For a balanced 3-phase output, only those components whose numbers are boxed thus ⬜ are present.
3. Where there is only one column for a given value of $2n$ the values given apply to both the "plus" and "minus" $2nf_o$ components. Where there are two columns for a given value of $2n$; The values given in the "black" column apply to the "plus" $2nf_o$ components for positive ϕ_o, and to the "minus" $2nf_o$ components for negative ϕ_o. The values given in the "white" column apply to the "plus" $2nf_o$ components for negative ϕ_o, and to the "minus" $2nf_o$ components for positive ϕ_o.

Tables 12.4a Showing the Amplitudes of the Harmonic Components Having Frequencies of $11f_i \pm 2nf_o$, in the Input Current of the Cycloconverter, for Various Output Voltage Ratios and Output Load Displacement Angles

$11f_i \pm 2nf_o$

$\phi_o = 0°, 180°$

$2n \rightarrow$ $r \downarrow$	0	2	4	6	8	10	12	14	16
1.0	.01	.01	.01	.01	.03	.05	.04	.02	.01
0.9	.02	.03	.03	.04	.04	.03	.01	.00	.00
0.8	.03	.03	.04	.04	.03	.02	.01	.00	.00
0.7	.04	.04	.04	.03	.02	.01	.00	.00	.00
0.6	.04	.04	.04	.03	.01	.01	.00	.00	.00
0.5	.05	.05	.04	.02	.01	.01	.00	.00	.00
0.4	.06	.05	.03	.01	.01	.00	.00	.00	.00
0.3	.07	.05	.02	.01	.00	.00	.00	.00	.00
0.2	.08	.04	.01	.00	.00	.00	.00	.00	.00
0.1	.09	.03	.01	.00	.00	.00	.00	.00	.00

$\phi_o = \pm 30°, \pm 150°$

$2n \rightarrow$ $r \downarrow$	0	2	4	6	8	10	12	14	16
1.0	.01	.01	.01	.01	.02	.02	.02	.02	.01
0.9	.02	.02	.02	.02	.06	.05	.07	.01	.00
0.8	.03	.03	.03	.03	.06	.03	.01	.00	.00
0.7	.04	.04	.04	.05	.04	.01	.01	.00	.00
0.6	.04	.04	.06	.06	.03	.02	.00	.00	.00
0.5	.05	.05	.06	.05	.01	.01	.00	.00	.00
0.4	.06	.06	.05	.04	.01	.00	.00	.00	.00
0.3	.07	.07	.04	.02	.00	.00	.00	.00	.00
0.2	.07	.07	.02	.01	.01	.00	.00	.00	.00
0.1	.05	.05	.01	.00	.00	.00	.00	.00	.00

$11f_i \pm 2nf_o$

$\phi_o = \pm 60°, \pm 120°$

$2n \rightarrow$ $r \downarrow$	0	2	4	6	8	10	12	14	16
1.0	.01	.01	.01	.01	.01	.09	.09	.01	.01
0.9	.01	.02	.02	.02	.01	.08	.03	.00	.00
0.8	.01	.02	.03	.04	.07	.05	.01	.00	.00
0.7	.02	.03	.03	.01	.08	.02	.00	.00	.00
0.6	.02	.03	.05	.07	.06	.01	.00	.00	.00
0.5	.02	.03	.01	.08	.04	.00	.00	.00	.00
0.4	.03	.05	.08	.06	.02	.00	.00	.00	.00
0.3	.03	.08	.08	.03	.00	.00	.00	.00	.00
0.2	.02	.08	.06	.01	.01	.00	.00	.00	.00
0.2	.05	.09	.03	.00	.00	.00	.00	.00	.00
0.1	.08	.06	.01	.01	.00	.00	.00	.00	.00

$\phi_o = \pm 90°$

$2n \rightarrow$ $r \downarrow$	0	2	4	6	8	10	12	14	16
1.0	.00	.00	.00	.00	.00	.09	.00	.00	.00
0.9	.00	.01	.00	.02	.06	.10	.09	.00	.00
0.8	.00	.01	.00	.02	.10	.05	.03	.00	.00
0.7	.01	.01	.00	.08	.08	.02	.01	.00	.00
0.6	.01	.02	.00	.09	.04	.01	.00	.00	.00
0.5	.01	.01	.00	.07	.01	.00	.00	.00	.00
0.4	.02	.04	.01	.03	.00	.00	.00	.00	.00
0.3	.02	.08	.01	.01	.00	.00	.00	.00	.00
0.2	.03	.09	.01	.00	.00	.00	.00	.00	.00
0.1	.07	.07	.00	.00	.00	.00	.00	.00	.00

NOTES:
1. Tables show amplitude of harmonic, as per unit value of the total fundamental component
2. These harmonics are present in the input current of 3-, 6-, and 12-pulse cycloconverters. For a balanced 3-phase output, only those components whose numbers are boxed thus □ are present
3. Where there is only one column for a given value of $2n$, the values given apply to both the "plus" and "minus" $2nf_o$ components. Where there are two columns for a given value of $2n$: The values given in the "black" column apply to the "plus" $2nf_o$ components for positive ϕ_o, and to the "minus" $2nf_o$ components for negative ϕ_o. The values given in the "white" column apply to the "plus" $2nf_o$ components for negative ϕ_o, and to the "minus" $2nf_o$ components for positive ϕ_o.

Tables 12.4b Showing the Amplitudes of the Harmonic Components Having Frequencies of $13f_i \pm 2nf_o$ in the Input Current of the Cycloconverter, for Various Output Voltage Ratios and Output Load Displacement angles

$13f_i \pm 2nf_o$

$\phi_o = 0°, 180°$

$2n \to$ $r \downarrow$	0	2	4	6	8	10	12	14	16
1.0	.01	.01	.01	.01	.01	.02	.04	.04	.02
0.9	.02	.02	.02	.02	.03	.03	.02	.01	.00
0.8	.02	.02	.02	.03	.03	.02	.01	.00	.00
0.7	.03	.03	.03	.03	.02	.01	.01	.00	.00
0.6	.03	.03	.03	.03	.01	.01	.00	.00	.00
0.5	.03	.04	.03	.02	.01	.01	.00	.00	.00
0.4	.04	.04	.03	.01	.01	.00	.00	.00	.00
0.3	.05	.04	.02	.01	.00	.00	.00	.00	.00
0.2	.07	.04	.01	.00	.00	.00	.00	.00	.00
0.1	.07	.03	.01	.00	.00	.00	.00	.00	.00

$\phi_o = \pm 30°, \pm 150°$

$2n \to$ $r \downarrow$	0	2	4	6	8	10	12	14	16
1.0	.01	.01	.01	.01	.01	.02	.06	.06	.02
0.9	.02	.02	.02	.02	.02	.05	.04	.02	.00
0.8	.02	.02	.02	.03	.03	.04	.02	.01	.00
0.7	.02	.03	.03	.03	.05	.04	.01	.00	.00
0.6	.03	.03	.04	.04	.04	.03	.01	.00	.00
0.5	.02	.04	.04	.05	.03	.00	.00	.00	.00
0.4	.02	.05	.05	.04	.02	.00	.00	.00	.00
0.3	.01	.05	.04	.03	.01	.00	.00	.00	.00
0.2	.01	.06	.04	.01	.01	.00	.00	.00	.00
0.1	.02	.06	.02	.01	.00	.00	.00	.00	.00
	.04	.05	.02	.01	.00	.00	.00	.00	.00

$13f_i \pm 2nf_o$

$\phi_o = \pm 60°, \pm 120°$

$2n \to$ $r \downarrow$	0	2	4	6	8	10	12	14	16
1.0	.01	.01	.01	.01	.01	.01	.07	.07	.01
0.9	.01	.01	.01	.01	.01	.06	.06	.02	.01
0.8	.01	.02	.01	.02	.05	.06	.03	.00	.00
0.7	.01	.02	.01	.03	.06	.04	.01	.00	.00
0.6	.02	.02	.01	.05	.05	.02	.00	.00	.00
0.5	.01	.03	.01	.06	.03	.00	.00	.00	.00
0.4	.02	.03	.00	.06	.03	.00	.00	.00	.00
0.3	.02	.06	.01	.04	.02	.00	.00	.00	.00
0.2	.03	.07	.01	.02	.00	.00	.00	.00	.00
0.1	.06	.06	.01	.00	.00	.00	.00	.00	.00

$\phi_o = \pm 90°$

$2n \to$ $r \downarrow$	0	2	4	6	8	10	12	14	16
1.0	.00	.00	.00	.00	.00	.00	.08	.08	.00
0.9	.00	.00	.00	.00	.01	.07	.07	.02	.00
0.8	.00	.00	.01	.02	.04	.08	.03	.00	.00
0.7	.00	.00	.02	.01	.08	.05	.01	.00	.00
0.6	.01	.01	.01	.06	.06	.02	.00	.00	.00
0.5	.01	.02	.04	.07	.03	.00	.00	.00	.00
0.4	.00	.00	.04	.05	.03	.00	.00	.00	.00
0.3	.01	.05	.07	.02	.01	.00	.00	.00	.00
0.2	.02	.08	.07	.00	.00	.00	.00	.00	.00
0.1	.06	.06	.04	.01	.00	.00	.00	.00	.00

NOTES: 1. Tables show amplitude of harmonic, as per unit value of the total fundamental component
2. These harmonics are present in the input current of 3-, 6-, and 12-pulse cycloconverters. For a balanced 3-phase output, only those components whose numbers are boxed Thus ☐ are present
3. Where there is only one column for a given value of $2n$, the values given apply to both the "plus" and "minus" $2nf_o$ components Where there are two columns for a given value of $2n$: The values given in the "black" column apply to the "plus" $2nf_o$ components for positive ϕ_o, and to the "minus" $2nf_o$ components for negative ϕ_o
The values given in the "white" column apply to the "plus" $2nf_o$ components for negative ϕ_o, and to the "minus" $2nf_o$ components for positive ϕ_o

374   CYCLOCONVERTER INPUT CURRENT ANALYSIS

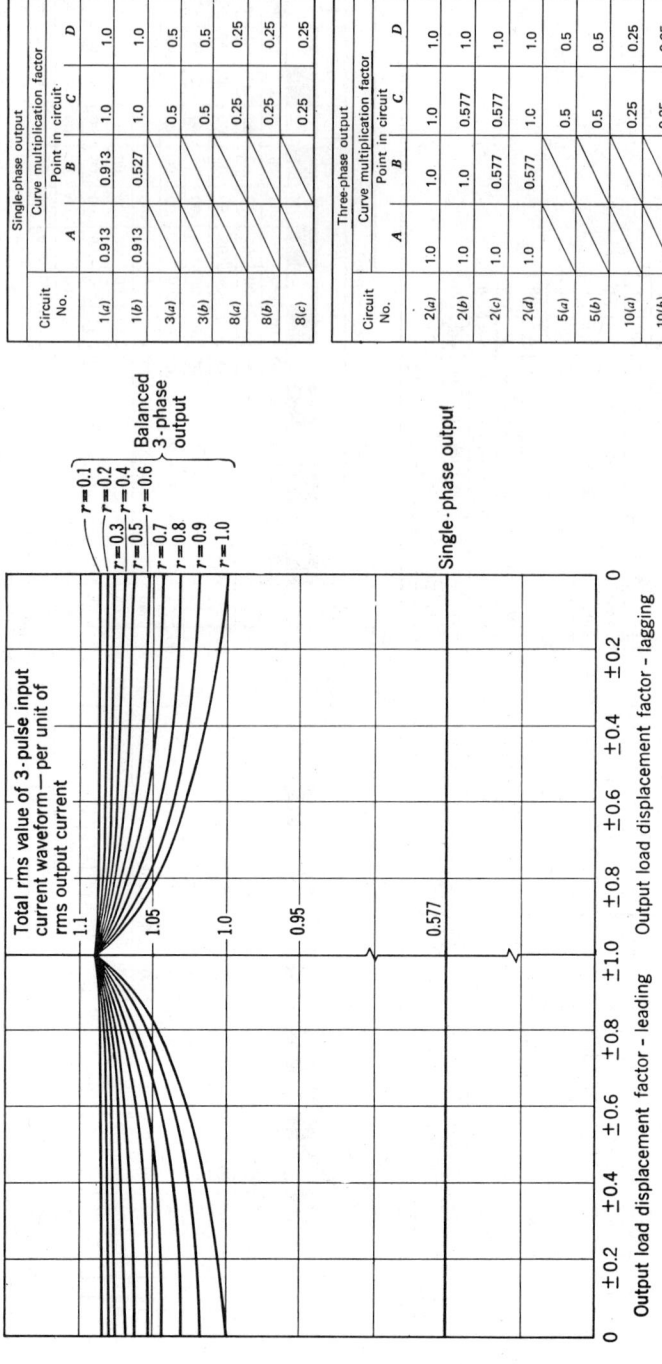

Figure 12.16. Curves showing the relationships between the per unit total rms value of a 3-pulse current waveform at the input of the cycloconverter, the output load displacement factor, and the output voltage ratio.

TOTAL RMS VALUE OF THE INPUT CURRENT

converters. Thus, the net input current is relatively free of harmonics, and, as a result, its rms value is invariably less than $3 \times$ the rms value of the input current waveforms drawn by the individual converters.

Further, since the amplitudes of the "uncancelled" harmonic components depend upon the output voltage ratio, and the load displacement factor, the rms value of the input current of the cycloconverter with a 3-phase output does not bear a fixed relationship to the output current, but it is a function both of the output voltage ratio and the load displacement factor.

The relationships between the total rms value of the input current waveform (as given by expression 12.41), the load displacement factor, and the output voltage ratio, are shown graphically in Figs. 12.16 through 12.18, for both 1- and balanced 3-phase outputs, for 3-, 6-, and 12-pulse waveforms respectively.

THE DISTORTION FACTOR OF THE INPUT CURRENT WAVEFORM

The distortion factor of the input current waveform is given by the following general expression:

$$\mu = \frac{I_1}{I_{\text{rms}}}$$

where I_1 is given by (12.28) and I_{rms} is given by (12.41).

The relationships between the distortion factor of the input current waveform, the load displacement factor, and the output voltage ratio, are shown graphically in Figs. 12.19 through 12.21, for both 1- and balanced 3-phase outputs, for 3-, 6-, and 12-pulse waveforms respectively.

GENERAL CONCLUSIONS ON THE DISTORTION OF THE INPUT CURRENT WAVEFORM

From the quantitative analytical data presented in the foregoing sections, the following general conclusions can be made concerning the harmonic content of the input current of the phase-controlled cycloconverter.

Cycloconverter with One-Phase Output

For a cycloconverter with a single-phase output, the characteristic $(f_i + 2f_o)$ and $(f_i - 2f_o)$ harmonic components are invariably present in the input line current, regardless of the pulse number of the converter. Since the amplitudes

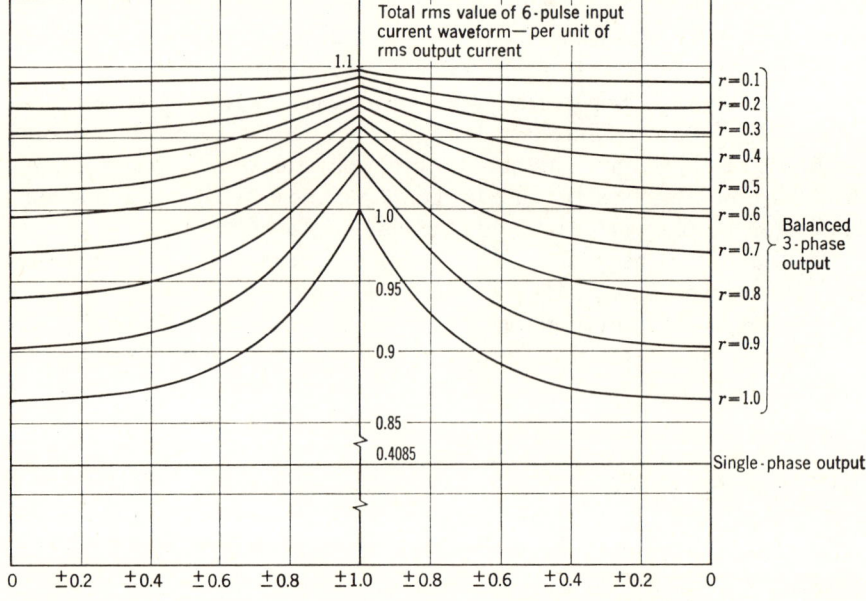

Figure 12.17. Curves showing the relationships between the per unit total rms value of a 6-pulse current waveform at the input of the cycloconverter, the output load displacement factor, and the output voltage ratio.

of these particular components are quite substantial in relation to the amplitude of the fundamental component, the distortion factor of the input current of the cycloconverter with a single-phase output is invariably relatively low, and has a limiting maximum value (i.e., for a cycloconverter with an "infinite" pulse number) of approx. 0.83 (for $r = 1.0$). Thus, for a cycloconverter with a single-phase output, unlike the phase-controlled converter with a d-c output, it is not possible, by the expedient of using a converter with a sufficiently high pulse number, to produce a "pure" current waveform at the input.

Apart from giving rise to an input current waveform which inherently has a quite substantial harmonic content, the presence of the $(f_i - 2f_o)$ characteristic harmonic frequency is also potentially objectionable, from the viewpoint that its frequency is less than that of the input supply. This subharmonic component of current has to flow through the input transformer—(unless of course, a suitable by-pass filter circuit is connected across the converter input terminals). With output-to-input frequency ratios of, or in the vicinity of, 1:2, the $(f_i - 2f_o)$ harmonic component has zero, or very low, frequency: since the amplitude of this component is typically in the order of 50% (or more) of the fundamental component, it can be appreciated

TOTAL RMS VALUE OF THE INPUT CURRENT 377

Circuit No.	Single-phase output Curve multiplication factor Point in circuit				Circuit No.	Three-phase output Curve multiplication factor Point in circuit			
	A	B	C	D		A	B	C	D
3(a)	1.0	1.0			5(a)	1.0	1.0		
3(b)	1.0	0.577			5(b)	1.0	0.577		
4(a)	2.0	2.0	2.0	2.0	6(a)	2.0	2.0	2.0	2.0
4(b)	2.0	2.0	1.154	2.0	6(b)	2.0	2.0	1.154	2.0
4(c)	2.0	1.154	1.154	2.0	6(c)	2.0	1.154	1.154	2.0
4(d)	2.0	1.154	2.0	2.0	6(d)	2.0	1.154	2.0	2.0
7(a)			2.0	2.0	7(a)	2.0	2.0		
7(b)			1.154	2.0	7(b)	2.0	2.0		
7(c)			1.154	2.0	7(c)	2.0	1.154		
7(d)			2.0	2.0	7(d)	2.0	1.154		
8(a)			① 0.5 ② 0.288		10(a)			① 0.5 ② 0.288	
9(a)			① 2.0 ② 1.154	2.0					
9(b)			① 2.0 ② 1.154	2.0					
9(c)			2.0	2.0					
9(d)			2.0	2.0					
11(a)			① 2.0 ② 1.154	2.0					
11(b)			① 2.0 ② 1.154	2.0					
11(c)			2.0	2.0					
11(d)			2.0	2.0					

Figure 12.17 (*Continued*)

that, in the vicinity of this critical frequency ratio, the cycloconverter with a single-phase output presents a rather objectionable load to the input system. Thus, the presence of the $(f_i - 2f_o)$ harmonic component can be visualized as constituting a natural (but not unsurpassable) obstacle to a continuous control of the output-to-input frequency ratio through a critical region in the vicinity of 2:1. Of course, for increasing frequency ratios beyond 2:1, the frequency of this harmonic component of input current once again increases.

Cycloconverter with Balanced Three-Phase Output

For a cycloconverter with a balanced 3-phase output, the first term to appear in the characteristic family of harmonic frequencies is the $(f_i \pm 6f_o)$ component. From Table 12.1, it is seen that the amplitude of this component

378 CYCLOCONVERTER INPUT CURRENT ANALYSIS

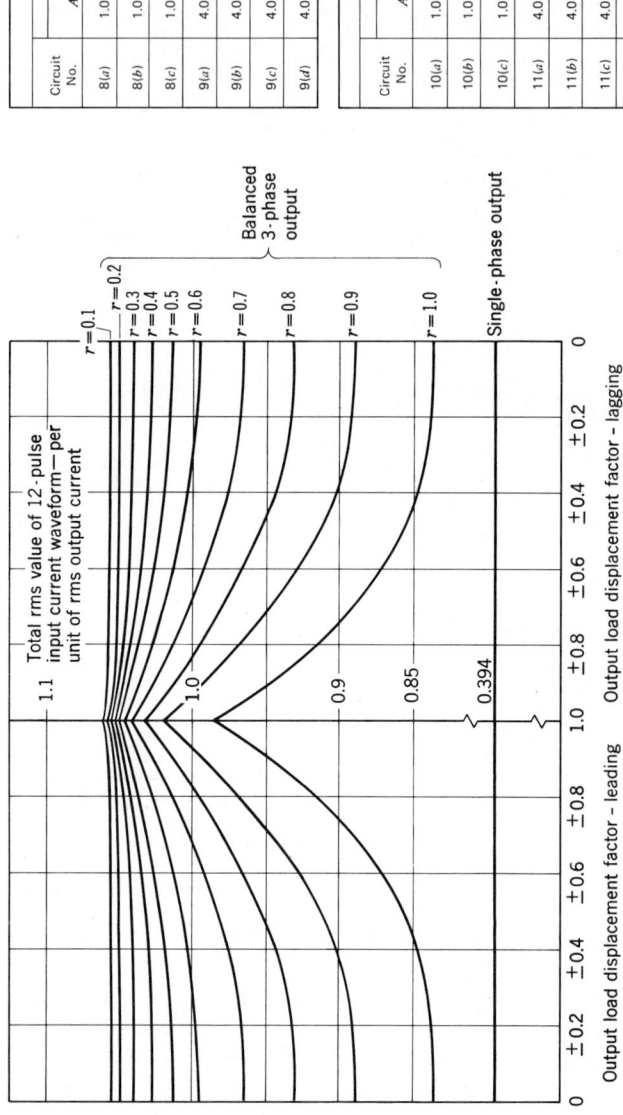

Figure 12.18. Curves showing the relationships between the per unit total rms value of a 12-pulse current waveform at the input of the cycloconverter, the output load displacement factor, and the output voltage ratio.

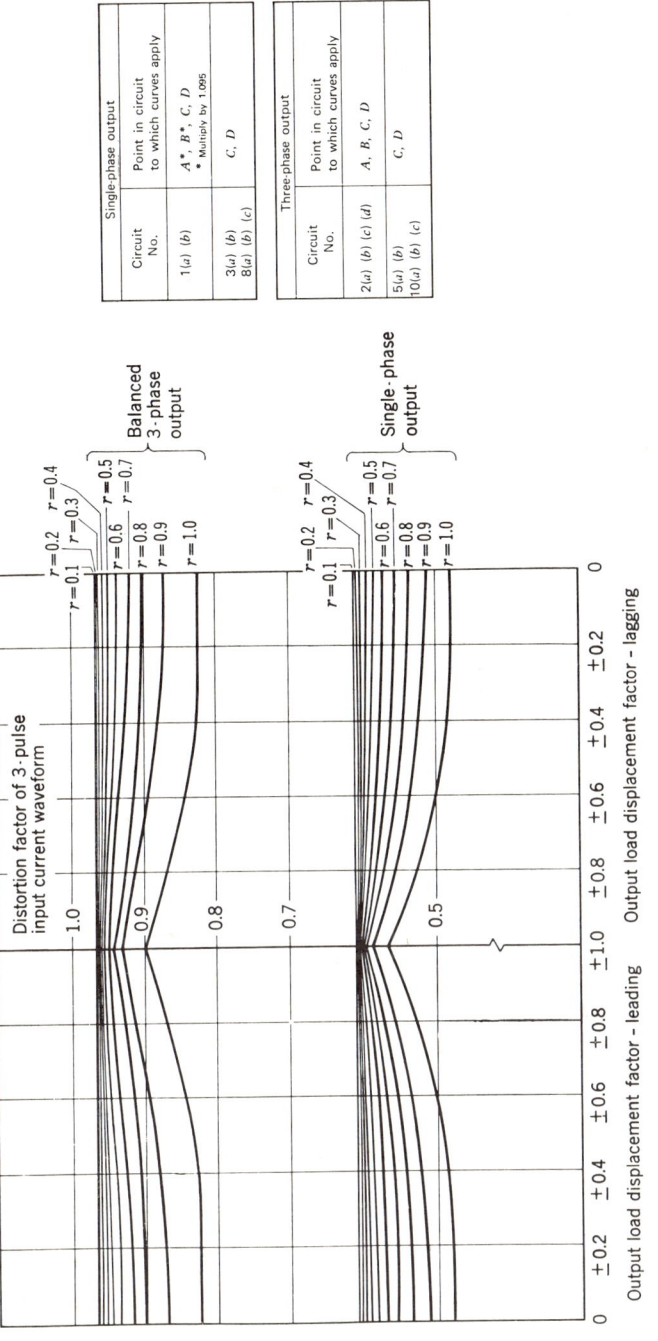

Figure 12.19. Curves showing the relationships between the distortion factor of a 3-pulse current waveform at the input of the cycloconverter, the output load displacement factor, and the output voltage ratio.

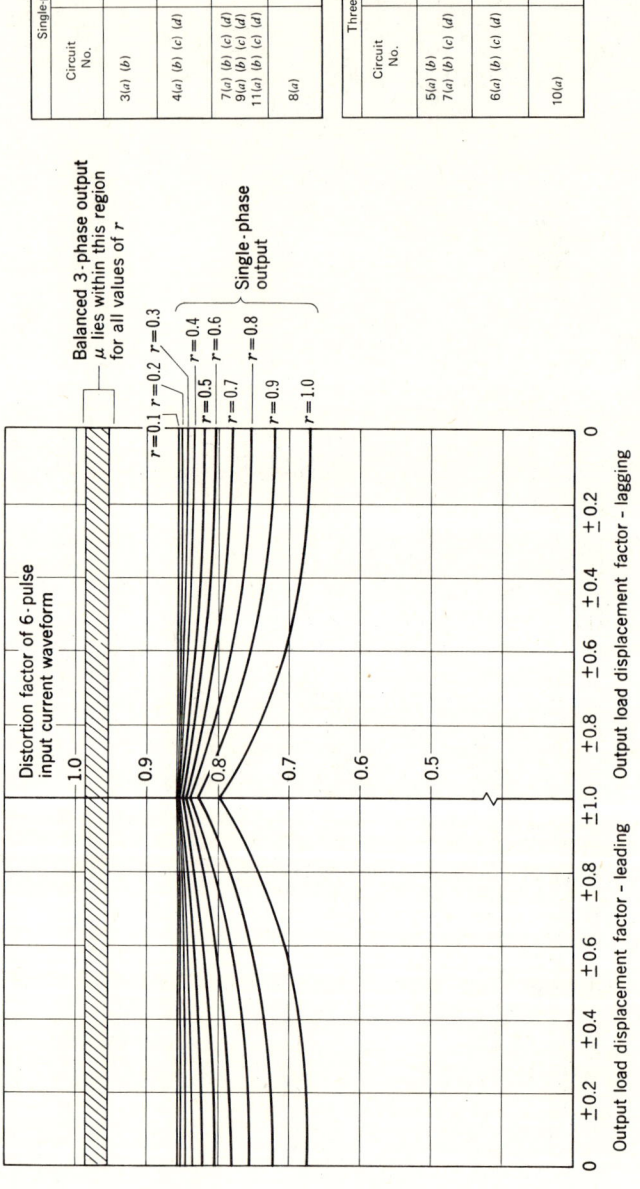

Figure 12.20. Curves showing the relationships between the distortion factor of a 6-pulse current waveform at the input of the cycloconverter, the output load displacement factor, and the output voltage ratio.

DISTORTION FACTOR OF THE INPUT CURRENT 381

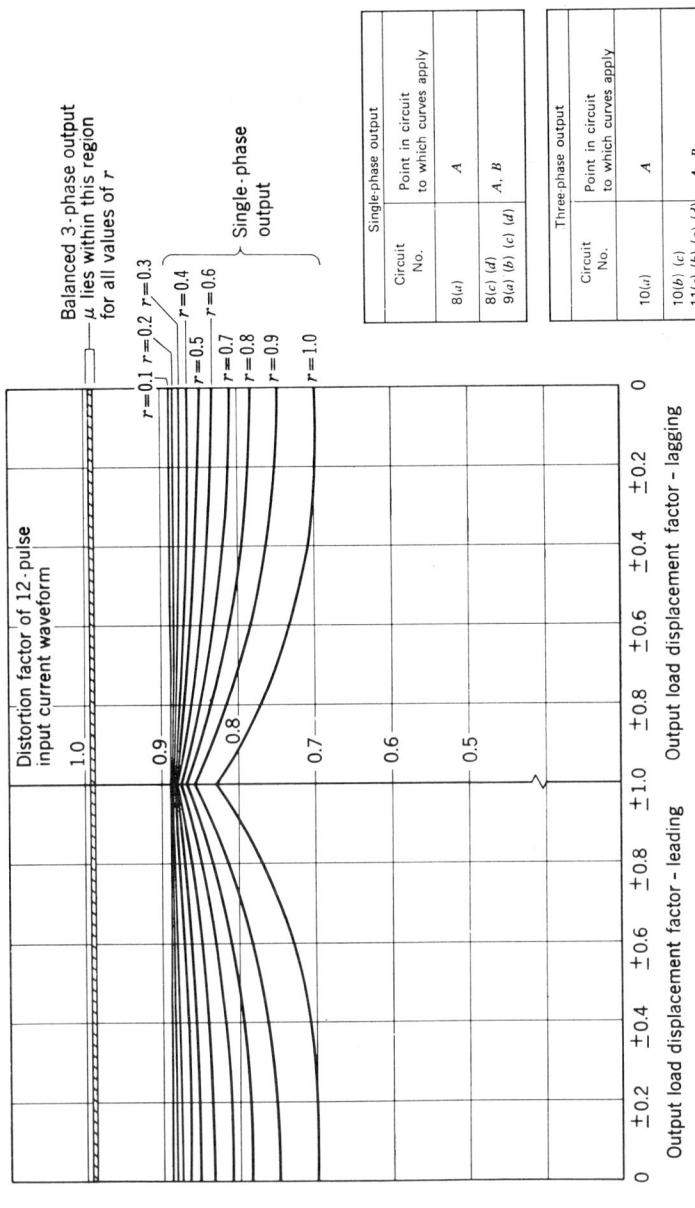

Figure 12.21. Curves showing the relationships between the distortion factor of a 12-pulse current waveform at the input of the cycloconverter, the output load displacement factor, and the output voltage ratio.

is generally a relatively small proportion—typically 3%—of the fundamental component.

Thus, the presence of the characteristic family of harmonic frequencies no longer constitutes any significant practical limitation, since it makes no substantial contribution to the distortion of the input current waveform, nor does it present any appreciable subharmonic load to the input system.

Thus, for a cycloconverter with a balanced 3-phase output, the distortion of the input current waveform is determined substantially by the circuit-dependent harmonic components. In this case, therefore, with circuits of sufficiently high pulse number, it is possible to obtain input current waveforms with very low distortion. Indeed, the distortion of the input current waveform of the cycloconverter with a balanced 3-phase output is generally less than that of the input current waveform of a converter, with the same pulse number, operating with a steady d-c output.

INPUT CURRENT UNBALANCE AT DISCRETE FREQUENCY RATIOS

A point which has been ignored so far is that, at certain discrete output-to-input frequency ratios, an unbalance can exist between the rms currents in the input lines connected to the cycloconverter. The reason for this is that at the discrete frequency ratios in question, the input and output waves are synchronous with one another, and hence any assymetry which inherently exists between the waveshapes of the currents in the 3 input lines, due to the different conduction periods of the associated thyristors, relative to the load current waveforms, is permitted to exist permanently, rather than being, in effect, continuously "rotated" between the input lines, as is normally the case for non-synchronous output and input waves.

This phenomenon is, in fact, illustrated by the harmonic frequency charts of Figs. 12.14 and 12.15, inasmuch as these charts show that at certain discrete output-to-input frequency ratios, several of the harmonic frequencies become equal to one another; moreover, certain harmonic frequencies may actually become equal to the input frequency. (The frequency ratios at which this occurs correspond with those at which "mating" of certain output voltage harmonics with the wanted output voltage component also takes place).

In mathematical terms, the result of this "harmonic mating" is to create "composite" components, whose amplitudes are different for the different input lines, because of the different phase relationships of the mating components, relative to one another. This unbalance between the "mated components" shows up directly as an unbalance between the rms currents in the different input lines.

UNBALANCE AT DISCRETE FREQUENCY RATIOS

The amount of unbalance depends upon the circuit pulse number, the number of output phases, the displacement factor of the load, and the particular output-to-input frequency ratio at which the "harmonic mating" takes place.

Generally, the unbalance is relatively insignificant in the input current waveform of a 6- (or higher) pulse cycloconverter with a balanced 3-phase output, but it may, in theory, be quite significant, at certain specific frequency ratios, for a single-phase output, particularly for a 3-pulse cycloconverter.

The curves of Figs. 12.22 and 12.23 show the quantitative result of the "harmonic mating" for 3- and 6-pulse current waveforms, for the particular output-to-input frequency ratios of 1/3 and 1/2 respectively. Consider, for example, the 3-pulse input current waveform, at a frequency ratio of 1/3. According to the curves of Fig. 12.22, the rms current in a given input line

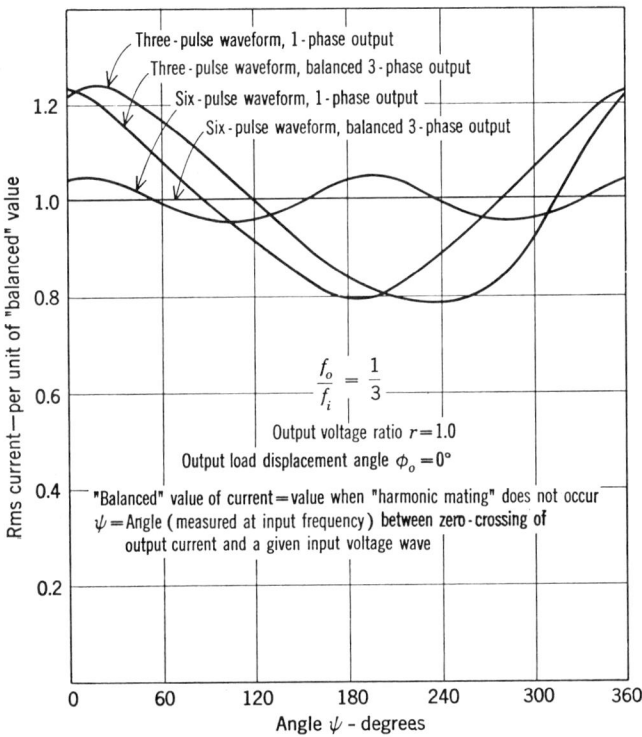

Figure 12.22. Curves showing the approximate variation of the per unit rms value of the input current of the cycloconverter, with the phase displacement between the input and the output waves, for $f_o/f_i = \frac{1}{3}$.

may have a value ranging anywhere between approximately 1.25 and 0.78 × the "balanced" value, (for both 1- and 3-phase outputs), depending upon the exact phase-position of the output current wave, relative to the input voltage wave.

The curves for each of the 3 input lines are similar, but mutually displaced by 120°. Thus, even if the phase relationships between the input and output waves are such that the rms value of the current in one of the input lines is equal to the "balanced" value, under this condition the currents in the other two lines necessarily must be "unbalanced."

In practice, unless a deliberate synchronization between the output and input frequencies is made, some degree of "phase drift" between the input and output waves almost inevitably occurs, (in other words, in practice, the angle ψ almost inevitably drifts), so that the unbalanced loading of the 3 input lines is, in the course of time, "rotated" between the input phases. Thus, for most practical purposes, this phenomenon does not have any substantial

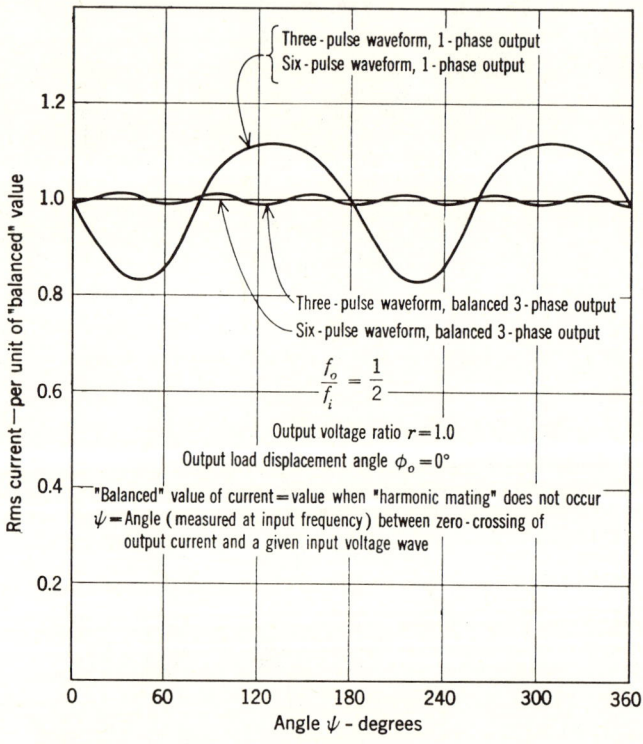

Figure 12.23. Curves showing the approximate variation of the per unit rms value of the input current of the cycloconverter, with the phase displacement between the input and output waves, for $f_o/f_i = \frac{1}{2}$.

CONTROL OF THE QUADRATURE INPUT CURRENT 385

bearing upon the required ratings of the input transformer windings and input lines.*

REDUCTION OF THE QUADRATURE INPUT CURRENT BY MEANS OF SPECIAL CONTROL TECHNIQUES

In Chapter 4 it was seen that for a 2-quadrant phase-controlled converter providing a controllable d-c output, a substantial reduction in the amplitude of the quadrature component of input current at reduced output voltage can be obtained, by means of a process of "consecutive" firing angle control. The basic principle, applicable to two or more converter "stages" connected in series, is to control the voltages of the individual stages "consecutively," so that only one at a time operates at a reduced voltage ratio, with a corresponding reduced input displacement factor, while the remainder operate at maximum voltage ratio.

It has been seen that for the cycloconverter with a normal "concurrent" voltage control, the relative amplitude of the fundamental component of input current actually increases somewhat as the output voltage ratio is reduced (Fig. 12.12). Thus, for certain applications, it may be particularly desirable to employ some special control method to reduce the quadrature component of input current at reduced output voltage, thereby avoiding the need to rate the supply and input transformer for a current in excess of that required at maximum output voltage.

For simplicity, the present discussion will be restricted to the case of two cycloconverter stages connected in series with one another; however, the same general principles are applicable to any number of converter "stages" connected together.

The logical method of applying the consecutive firing angle control principle to the control of a cycloconverter, would be to regulate the combined output voltage of the two stages by means of operating one with unity output voltage ratio, whilst the voltage ratio of the other is controlled from maximum to zero, and then back to maximum, with reversed phase. This principle is illustrated by Fig. 12.24a (Control Method 1).

It is seen from the curves of Fig. 12.10 that for a given load displacement factor, the quadrature component of current at the input of the cycloconverter, relative to the output current, steadily increases with decreasing voltage ratio. Thus, with this control method, the normalized quadrature current at

* It is to be noted, however, that this effect also applies to the thyristor currents. Due to the relatively short thermal time constant of the thyristor, it may be necessary to pay some attention to the unbalanced load in this case, either in terms of rating the thyristors accordingly, or in terms of incorporating an automatic control for producing a controlled "phase drift," and hence an even distribution of load between the thyristors, on a sufficiently short-time basis.

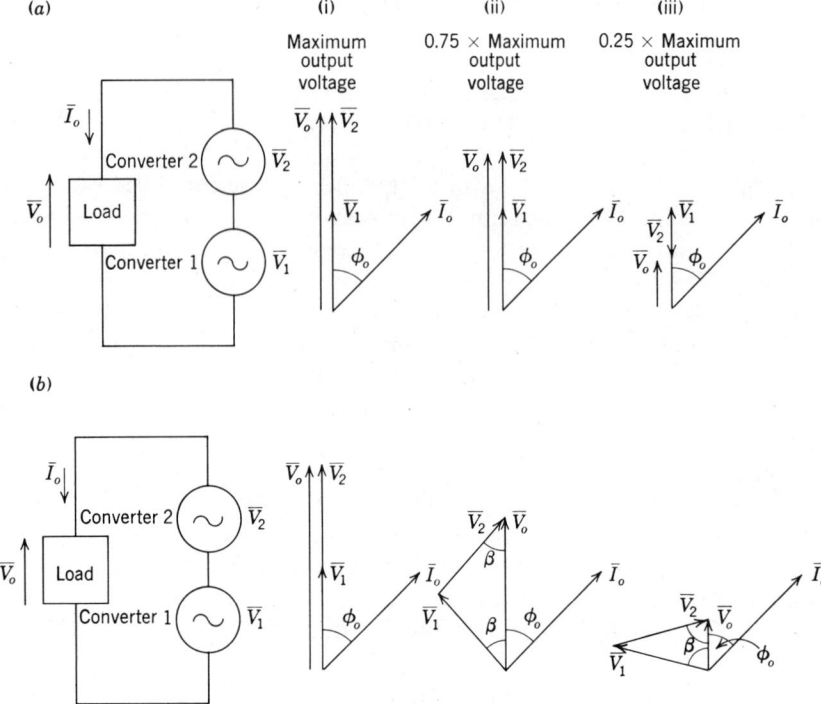

Figure 12.24. Two alternative control methods for two cycloconverters connected in series, for reducing the quadrature component of current at the input at reduced output voltage ratio. (a) Control method 1. (b) Control method 2.

the input has its maximum value at half maximum output voltage (i.e., $V_2 = 0$), and has the same minimum value at each extremity of the voltage control range.

An alternative control method is illustrated in Fig. 12.24b (Control Method 2). The principle here is to keep the voltage ratios of both stages at their maximum value, and to regulate the combined output voltage by means of phase shifting the two voltages with respect to one another. The load displacement angles viewed by the two stages are as follows:

$$\phi_{01} = -\beta + \phi_o$$
$$\phi_{02} = \beta + \phi_o$$

where 2β = angular displacement between the voltages of the 2 stages
  ϕ_o = load displacement angle, (negative for lagging load).

Consider the particular case of $\phi_o = 0°$. In this case, the displacement angles viewed by the two stages are equal and opposite to one another, and these change from $0°$ to $\pm 90°$ as the combined output voltage is controlled for maximum to zero. Thus, for the particular theoretical condition of zero output voltage, the quadrature component of current at the input is appropriate to two stages, each with maximum output voltage and zero load displacement factor. From the curves of Fig. 12.10, it is seen that this quadrature current is less than that which would be obtained with a normal "concurrent" control method (i.e., two stages with $r = 0$, $\phi_o = 0°$); however, under this particular condition, the quadrature current at the input is greater than that obtained with control method 1 (i.e., one stage with $r = 1$, $\phi_o = 0°$, and the other with $r = 1$, $\phi_o = 180°$).

For the more general case when the load displacement factor is less than unity, control method 2 shows a greater advantage in terms of the reduction of quadrature current at the input at reduced voltage ratio. This is because at low output voltage ratio, the displacement factor viewed by each of the individual converters increases as the displacement factor of the load itself decreases. In fact, in the extreme theoretical case of zero load displacement factor and zero output voltage ratio, the two converter stages view load displacement angles of $0°$ and $180°$, and under this condition, the quadrature

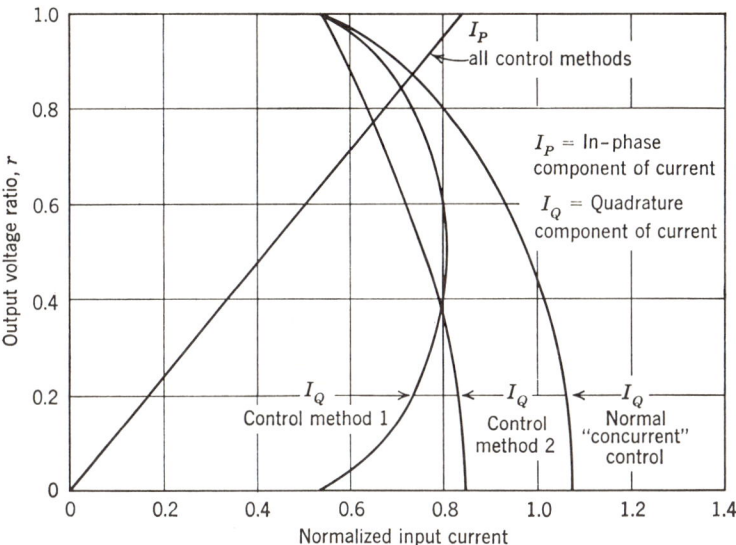

Figure 12.25. Relationships between the output voltage ratio, and the in-phase and quadrature components of input current, for a cycloconverter with two stages connected in series, for various control methods. Output load displacement factor = ± 1.0. Normalized input current =
$$\frac{\text{amplitude of input current component with } I_o \text{ at output}}{\text{amplitude of fundamental input current with } r = 1, I_o \text{ at output}}.$$

component of current at the input is simply the characteristic "phase control" component for $r = 1$, $\phi_o = 0°$.

The relationships between the output voltage ratio and the in-phase and quadrature components of current at the input, for control methods 1 and 2, as well as for the normal "concurrent" control, for output load displacement factors of ± 1.0, and ± 0.6, are shown graphically in Figs. 12.25 and 12.26 respectively.

For each control method, the same linear relationship is obtained between the in-phase component of current at the input, and the output voltage ratio. This is necessarily the case, and it is inherent in the fact that the average input and output powers are equal to one another. It is seen that both control methods 1 and 2 show an advantage over the normal concurrent control, in terms of a reduced quadrature current at the input. For load displacement factors close to unity, control method 1 may have some advantage over control method 2; for reduced load displacement factors, however, control method 2 produces a smaller quadrature current at the input than control method 1.

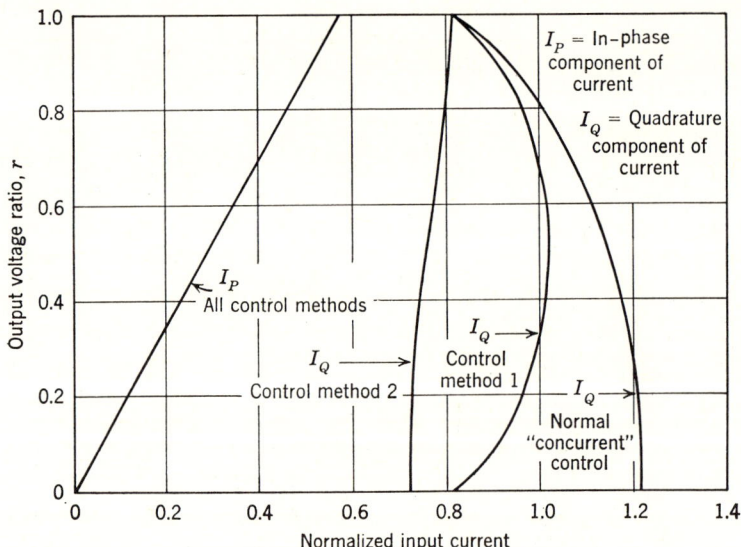

Figure 12.26. Relationships between the output voltage ratio, and the in-phase and quadrature components of input current, for a cycloconverter with two stages connected in series, for various control methods. Output load displacement factor = ± 0.6, leading or lagging. Normalized input current = $\dfrac{\text{amplitude of current component with } I_o \text{ at output}}{\text{amplitude of fundamental input current with } r = 1, I_o \text{ at output}}$.

Chapter Thirteen

The Effect of Input Source Impedance

So far, throughout most of this book, it has been assumed generally that the a-c source voltage which feeds the phase-controlled converter, or cycloconverter, has zero internal impedance. In practice, of course, the a-c source always has some finite amount of internal impedance, and this is usually predominantly inductive. The source inductance may be sufficiently large that it results in an appreciable modification of the process of commutating the current from one thyristor to the next, as compared to the theoretically instantaneous commutation obtained with zero source inductance. Thus, in practice, the commutation process may occupy a quite significant period of time, during which both the "incoming" and "outgoing" thyristors are simultaneously in conduction.

During this "commutation overlap" period, the waveforms of the voltage at the output terminals of the converter, as well as the current and voltage at the input terminals, depart from those obtained with zero source inductance. This has a modifying effect upon the external performance characteristics of the converter. At the output terminals, the effect of the input source inductance is to cause a loss of mean voltage, as well as a modification to the harmonic distortion terms; whilst, at the input terminals, a slight reduction of displacement factor, with respect to the a-c source voltage, as well as a modification to the distortion terms in the current waveform, takes place.

In this chapter, the practical commutation process itself is analyzed; then, the effects of source inductance upon the external performance characteristics of the phase-controlled converter, and of the cycloconverter, are briefly examined.

ASSUMPTIONS AND SCOPE OF DISCUSSION

The first assumption to be made is that the internal impedance of the a-c source is purely inductive. For all practical purposes this is valid, since the internal source resistance is invariably much smaller than the reactance.

As will be seen, the basic effect of the source inductance is to "slow down" the process of commutating the current from one thyristor to the next. The commutation process does not take place instantaneously, but it occupies a finite period of time, the duration of which depends upon the magnitude of the source inductance, the current to be commutated, and the converter firing angle. For most practical applications, within the normal load range, the time period occupied by the commutation process is shorter than the time period between successive converter commutations. Under these conditions, each commutation by itself is a discrete event, and the converter operates with "discrete commutations."

Under abnormal overload conditions, however, (as well as, possibly, under certain "normal" operating conditions, with converters having high enough pulse number), it is possible for successive commutation processes to "merge" into one another, and the converter now operates with "simultaneous commutations."

For the sake of conciseness, the basic intent in this chapter is to consider only those immediate first-order effects of the input source impedance, which have a bearing upon the "normal" practical range of converter operation. For this reason, the discussion and analysis throughout this chapter is *limited specifically* to poeration with *discrete commutations*. Thus, the quantitative data, and equivalent output circuits derived, are valid only for operation with discrete commutations.

Again, for the sake of conciseness, the discussion and analysis for converters with steady d-c output is limited to 2-quadrant circuits. And, for the cycloconverter, the treatment is limited to a mainly qualitative examination of the effect of the source inductance upon the output voltage waveform.

THE COMMUTATION PROCESS

The process of commutating the current from one thyristor to the next, in which the practical effect of the source inductance is taken into account, will be explained by reference to the 3-pulse commutating group, shown in Fig. 13.1. In this circuit, an inductance L is connected in each input line; this represents the total inductance connected between the a-c source voltage and the converter input terminal. In practice, this inductance may be partly a result of the internal impedance of the supply, and partly due to the leakage inductance of the transformer (if any) feeding the converter.

The waveforms of Fig. 13.2 illustrate the operation of the circuit during a period in which the current is commutated from thyristor 1 to thyristor 2 (other commutations are, of course, accomplished in a similar manner). It is assumed here that the converter operates at a steady firing angle α, and that the current I_d at the output terminal of the group is perfectly smooth.

THE COMMUTATION PROCESS 391

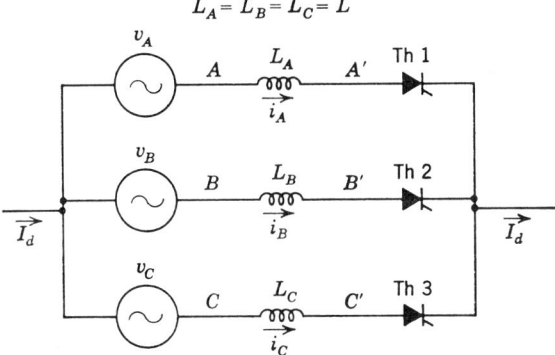

Figure 13.1. Three-pulse commutating group with a-c source inductance.

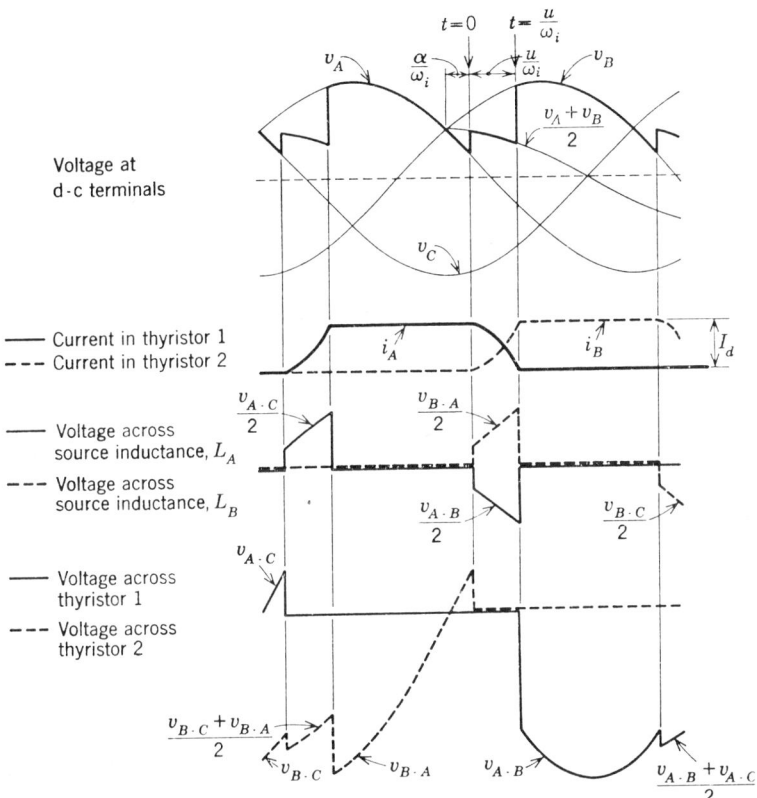

Figure 13.2. Waveforms illustrating the operation of the 3-pulse 2-quadrant phase-controlled converter, with a-c source inductance. Rectifier operation.

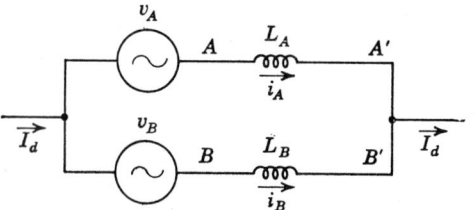

Figure 13.3. Equivalent circuit for 3-pulse group during commutation of current from thyristor 1 to thyristor 2.

Immediately before time $t = 0$, thyristor 1 is on, and thyristors 2 and 3 are off; the output current flows through the A source voltage, inductor L_A, and thyristor 1. The voltage at the output terminal is the A source voltage, v_A, since, at this time, there is no voltage developed across L_A, because there is zero rate of change of current.

At time $t = 0$, thyristor 2 is fired. The B source voltage v_B is more positive than v_A, and therefore there is a tendency for the current in thyristor 1 to decrease, and for the current in thyristor 2 to increase. However, due to the presence of the source inductance, it is not possible for the current in either of these thyristors to change instantaneously, and thus, for the time being, both these thyristors together remain in conduction. So long as this situation exists, the potential at points A' and B' must be virtually the same as one another, and hence, in effect, these points are connected directly together. The equivalent circuit during this "commutation overlap" period is shown in Fig. 13.3.

The distribution of voltage in the circuit can be deduced as follows:
The combined voltage across L_A and L_B is given by

$$L\frac{di_B}{dt} - L\frac{di_A}{dt} = v_{BA} \tag{13.1}$$

But

$$i_A + i_B = I_d \tag{13.2}$$

$$\therefore \quad i_B = I_d - i_A$$

Substituting for i_B into (13.1) gives

$$L\frac{d}{dt}(I_d - i_A) - L\frac{di_A}{dt} = v_{BA}$$

Whence

$$L\frac{di_A}{dt} = \text{voltage across } L_A = -\frac{v_{BA}}{2} \tag{13.3}$$

And, substituting for $L(di_A/dt)$ into (13.1) gives

$$L\frac{di_B}{dt} = \text{voltage across } L_B = \frac{v_{BA}}{2} \qquad (13.4)$$

Thus the line-to-line voltage, v_{BA}, is divided equally across the inductors L_B and L_A, and the voltage at the output terminal of the converter follows a waveform which is the mean of the A and B source voltages, as illustrated in Fig. 13.2.

The expression for the instantaneous current i_A during the commutation overlap period can be found from (13.3):

$$L\frac{di_A}{dt} = -\frac{v_{BA}}{2}$$

$$\therefore i_A = -\frac{\sqrt{3}\,\hat{V}_N}{2L}\int \sin(\omega_i t + \alpha)\,dt$$

$$= \frac{\sqrt{3}\,\hat{V}_N}{2\omega_i L}\cdot\cos(\omega_i t + \alpha) + C \qquad (13.5)$$

when $t = 0$, $i_A = I_d$. Hence

$$I_d = \frac{\sqrt{3}\,\hat{V}_N}{2\omega_i L}\cdot\cos\alpha + C$$

$$\therefore C = I_d - \frac{\sqrt{3}\,\hat{V}_N}{2\omega_i L}\cos\alpha$$

Substituting for C into (13.5) gives

$$i_A = I_d - \frac{\sqrt{3}\,\hat{V}_N}{2\omega_i L}\{\cos\alpha - \cos(\omega_i t + \alpha)\} \qquad (13.6)$$

And, substituting for i_A into (13.2) gives

$$i_B = \frac{\sqrt{3}\,\hat{V}_N}{2\omega_i L}\{\cos\alpha - \cos(\omega_i t + \alpha)\} \qquad (13.7)$$

Equations (13.6) and (13.7) describe the instantaneous currents i_A and i_B, up to the point at which the output current has been completely transferred from the outgoing to the incoming thyristor. This is accomplished at time $t = u/\omega_i$, at which point the current in thyristor 1 becomes zero, and, simultaneously, the current in thyristor 2 becomes equal to I_d. The current in thyristor 2 now remains at I_d, whereas thyristor 1 ceases to conduct, and becomes reverse biased. The voltage at the output terminal now follows the waveform of the B source voltage, up to the point at which the next commutation is initiated.

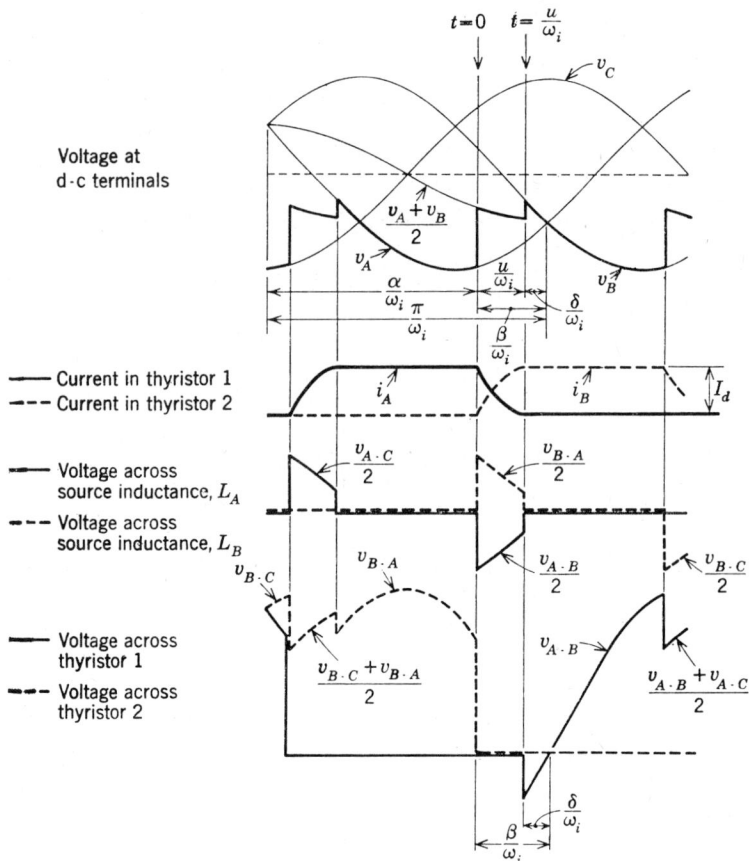

Figure 13.4. Waveforms illustrating the operation of the 3-pulse 2-quadrant phase-controlled converter, with a-c source inductance. Inverter operation.

The relationship between the overlap angle, the firing angle, the direct current, the source voltage, the source inductance, and the supply frequency can be found as follows:

$$i_B = I_d \quad \text{when} \quad \omega_i t = u$$

Substituting these values into (13.7) gives

$$\therefore \quad \cos \alpha - \cos(u + \alpha) = \frac{2\omega_i L I_d}{\sqrt{3}\,\hat{V}_N} \qquad (13.8)$$

In the particular example illustrated by the waveforms of Fig. 13.2, the firing angle α is less than 90°, and the converter operates in its rectifying mode. In point of fact, the mechanism of the commutation process is exactly the same for the inverting mode of operation, and thus the preceding analysis is applicable to the whole range of operation of the converter, from "full rectification" to "full inversion." This is shown by the waveforms of Fig. 13.4, which illustrate the commutation process for a firing angle greater than 90°.

As a matter of convenience, it is usual to define two angles β and δ, for inverter operation, as follows:

$$\beta = \pi - \alpha \qquad (13.9)$$

$$\delta = \pi - (\alpha + u) \qquad (13.10)$$

As illustrated by the waveforms of Fig. 13.4, β is the angle in advance of the zero-crossing of the line-to-line commutating voltage at which the firing pulse is applied, whereas δ is the angle in advance of the voltage zero-crossing at which the commutation of current is completed. Thus δ is the "recovery angle," since it is the angle over which reverse voltage is developed across the thyristor, before the anode voltage swings into the forward direction.

By substitution of expressions (13.9) and (13.10) into expression (13.8), the following relationship is obtained between β, δ, the direct current, the source voltage, the source inductance, and the supply frequency:

$$\cos \delta - \cos \beta = \frac{2\omega_i L I_d}{\sqrt{3}\, \hat{V}_N} \qquad (13.11)$$

A slight rearrangement of expression (13.11) yields the relationship of expression (10.1), the physical significance of which is discussed in Chapter 10.

EFFECT OF SOURCE INDUCTANCE ON THE EXTERNAL PERFORMANCE CHARACTERISTICS OF THE PHASE-CONTROLLED CONVERTER

Some General Observations

From an inspection of the waveforms of Figs. 13.2 and 13.4 it is possible to perceive, in qualitative terms, the effect of the inductance of the a-c source, upon the external performance characteristics of the phase-controlled converter.

First, it is seen that the appearance of the d-c terminal voltage waveform is modified, as compared to the theoretical waveform obtained with no commutation overlap. Thus, at the instant of commutation, the voltage at the d-c terminals jumps to a waveform which is the mean of the voltages of the outgoing and incoming lines, instead of jumping immediately to the incoming voltage wave, as it does with no source inductance; and this "mean voltage wave" is retained at the d-c terminals until the commutation

396 THE EFFECT OF INPUT SOURCE IMPEDANCE

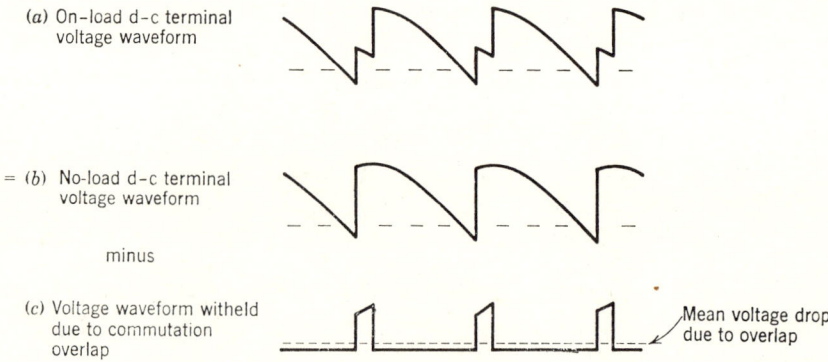

(a) On-load d-c terminal voltage waveform

= (b) No-load d-c terminal voltage waveform

minus

(c) Voltage waveform withheld due to commutation overlap

Mean voltage drop due to overlap

Figure 13.5. Waveforms illustrating that the on-load d-c terminal voltage waveform is composed of the difference between the no-load voltage waveform, and a waveform representing the voltage withheld from the d-c terminals due to commutation overlap. $(a) = (b) - (c)$. (Three-pulse waveforms.)

of the current has been completed. Thereafter, up to the next firing instant, the voltage wave at the d-c terminals is the appropriate source voltage. (This, of course, assumes that the direct current at the output is smooth, and therefore that there is no voltage developed across the source inductance during this time.)

Thus, during the commutation overlap period, a segment of positive voltage, which, in the absence of the source inductance, would otherwise appear, is "withheld" from the converter output terminals. The result is that a loss of average voltage occurs at the output terminals. This is further illustrated by the waveforms of Fig. 13.5.

Since the commutation overlap modifies the shape of the output voltage waveform, it can be deduced that an additional effect of the source inductance is to cause a modification of the harmonic distortion of the output voltage, as compared to that obtained with no commutation overlap. Since the fundamental ripple frequency is the same as with no commutation overlap, the spectrum of harmonic frequencies is also the same; thus the amplitudes of the harmonic components must be altered.

At the input side of the converter, the effect of the source inductance is to cause a "rounding off" of the edges of the waveforms of the line current. This implies that the amplitudes of the higher order harmonic terms in the current waveform are progressively reduced, as compared to the theoretical amplitudes of these components with no source inductance. In addition, the duration of each segment of the waveform of the input current is "stretched," by the overlap angle. The waveforms of Fig. 13.6 illustrate that this "stretching" effect causes a slight lagging phase shift of the fundamental component

EFFECT ON CONVERTER PERFORMANCE

of current, with respect to the input source voltage, as compared to the case with no commutation overlap. This lagging phase shift is, of course, consistent with the presence of inductance in the a-c line; it is also consistent with the reduction of mean voltage which occurs at the output terminals, since this would be expected to be accompanied by a corresponding change of displacement factor at the input.

A further effect of the source inductance is to produce "commutation notches" in the waveforms of the voltages appearing at the converter input terminals. If the proper precautions are not taken, the presence of these notches can lead to interference difficulties in the firing pulse timing circuits, (as well as in other electronic circuits connected directly to the same source terminals). In addition, in multipulse converter circuits, the commutation notches are a potential source of excessive rate of rise of forward voltage across the thyristors, and thus it is necessary to take the proper precautions against spurious anode breakover on account of this effect.

A further important effect of the source reactance is to influence the firing angle "end stop" limit in the inverting region of operation. The reason for this is that it is necessary to initiate the commutation sufficiently early to ensure that it is completed before the time at which the anode voltage of the outgoing thyristor swings positive. If this is not the case, then a commutation failure occurs. This particular aspect of the effect of the input source inductance is discussed fully in Chapter 10.

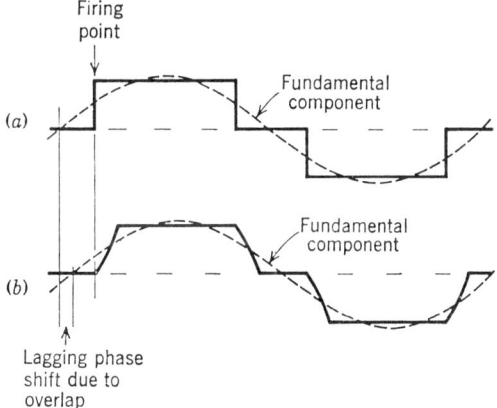

Figure 13.6. Waveforms illustrating that the commutation overlap results in a slight lagging phase shift of the fundamental component of the input current waveform, as compared to the case with no overlap. (*a*) Input current waveform with no commutation overlap; and (*b*) input current waveform with commutation overlap. Same firing angle as at (*a*). (Six-pulse waveforms.)

The Loss of Direct Voltage

It has been seen, in qualitative terms, that the effect of the commutation overlap is to result in a reduction of the mean voltage appearing at the d-c terminals of the converter, as compared to the mean voltage obtained with no overlap (i.e., the no-load voltage).

It is convenient to regard the voltage waveform appearing at the output terminals of the converter as being comprised of the voltage wave obtained with no overlap—that is to say, the no-load voltage wave—minus a waveform representing the segments of voltage withheld from the output terminals, as a result of the commutation overlap. This idea is illustrated by the waveforms of Fig. 13.5. The mean value of the "missing" voltage waveform, shown at c, represents the loss of direct voltage at the output terminals.

The mean value of this "missing" voltage wave can be calculated, by integrating one of the segments, over the appropriate limits of $t = (u + \alpha)/\omega_i$ and $t = \alpha/\omega_i$; the result is then divided by the time between consecutive firing instants, so as to obtain the mean voltage loss, in terms of u and α.

What is of prime interest, however, is the relationship between the loss of mean voltage, the direct current, and the source inductance. This relationship actually is independent of both the firing angle α and the overlap angle u, and in order to obtain the required result, the mathematical approach referred to above, is unnecessary.

Consider the waveforms of Fig. 13.5. Each segment of voltage which is withheld from the converter output terminals, is applied, instead, across the appropriate source inductance. It is the application of this voltage across the source inductance which drives the current in the input line from zero, up to the value of the direct current, I_d.

The volt-second integral required to increase the current from zero to I_d in an inductance L is given by

$$Vt = LI_d \qquad (13.12)$$

This is the volt-second integral of one of the segments of the missing voltage wave of Fig. 13.5c. The average voltage withheld from the converter output terminals is given by

$$V_{av} = \frac{Vt}{\underbrace{\frac{1}{3f_i}}_{\substack{\text{Period between consecutive}\\ \text{firing instants of a 3-pulse converter.}}}}$$

$$= \frac{3\omega_i L}{2\pi} \cdot I_d \qquad (13.13)$$

EFFECT ON CONVERTER PERFORMANCE

And, the absolute value of the mean d-c terminal voltage of the 3-pulse converter is given by

$$V_d = V_{do_\alpha} - \frac{3\omega_i L}{2\pi} \cdot I_d \qquad (13.14)$$

↑ ↑
No load voltage at firing angle α   Equivalent output resistance, due to commutation overlap

THE EQUIVALENT OUTPUT CIRCUIT. Expression (13.14) describes the output voltage/current relationship of a direct voltage source, V_{do_α}, having an equivalent internal resistance R_e, and this resistance is independent of the converter firing angle. It is given by the following expression:

$$R_e = \frac{3\omega_i L}{2\pi} \qquad (13.15)$$

Thus, the equivalent d-c circuit of the 3-pulse phase-controlled converter, with input source inductance, can be represented as in Fig. 13.7.

It should be appreciated that the equivalent output resistance does not exist physically, and there is no power dissipation associated with it; it is simply a convenient means of representing quantitatively the loss of direct voltage which occurs at the output terminals of the converter, due to the commutation overlap.

The equivalent circuit of Fig. 13.7 is applicable to the complete range of operation of the converter, both in the rectifying and inverting regions of operation. For inverter operation, however, it is usually more convenient to represent the "internal counter voltage" of the converter either in terms of the angle β, or the angle δ. (These angles are defined by expressions (13.9) and (13.10) respectively.) The resulting equivalent circuits are shown in Figs. 13.8a and b respectively.

The circuit at a illustrates, for example, that if the converter is operated at a fixed angle β, then the equivalent resistance in the d-c circuit is positive; thus, the d-c terminal voltage increases as the current increases.

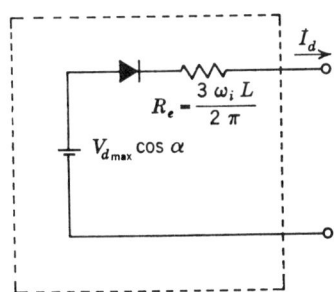

Figure 13.7. Equivalent d-c circuit for the 3-pulse 2-quadrant phase-controlled converter with a-c source inductance, L.

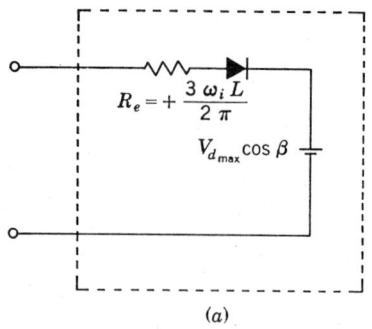

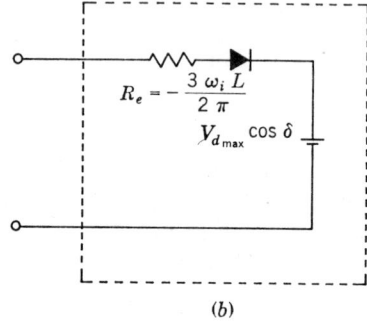

Figure 13.8. Alternative equivalent d-c circuits for the 3-pulse phase-controlled converter in inverter operation.

Conversely, the circuit at *b* illustrates that if the converter is operated at a fixed recovery angle δ, then the equivalent resistance in the d-c circuit is negative; in this event, the d-c terminal voltage decreases as the current increases.

The equivalent output resistance for other converter circuits can be derived in a similar manner as for the 3-pulse commutating group. The equivalent circuits for various configurations of phase-controlled converter, with input source inductance, are illustrated in Figs. 13.9 through 13.11.

THE UNIVERSAL D-C TERMINAL VOLTAGE RATIO/FIRING ANGLE CHARACTERISTICS. The mean d-c terminal voltage of any 2-quadrant phase-controlled converter with equivalent output resistance R_e is given by

$$V_d = V_{d_{\max}} \cos \alpha - I_d R_e \qquad (13.16)$$

↖ Voltage drop due to source inductance

EFFECT ON CONVERTER PERFORMANCE

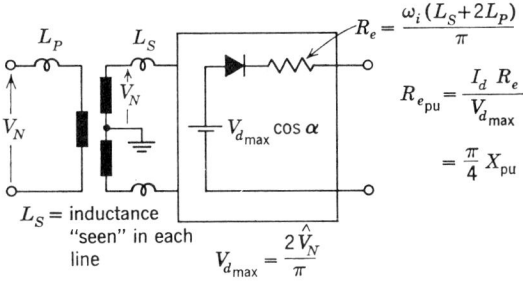

Two-pulse midpoint circuit

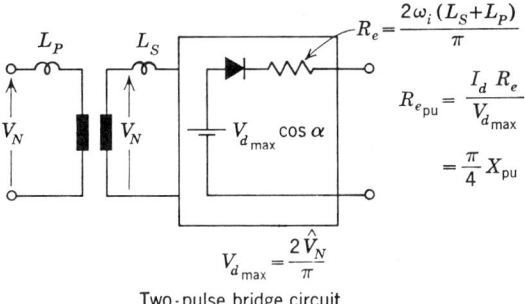

Two-pulse bridge circuit

Figure 13.9. Schematic equivalent circuits of 2-pulse 2-quadrant phase-controlled-converters, showing the relationships between the equivalent output resistance and the input source inductance. X_{pu} = Per unit reactance corresponding to a sinusoidal load current, with amplitude equal to that of the full-load fundamental component consumed by the converter.

Dividing expression (13.16) by $V_{d_{max}}$ gives

$$\frac{V_d}{V_{d_{max}}} = \cos \alpha - \frac{I_d R_e}{V_{d_{max}}} \qquad (13.17)$$

↑ Ratio of d-c terminal voltage (at firing angle α, with commutation overlap), to the maximum d-c terminal voltage (at $\alpha = 0°$), with no overlap.

↑ D-c terminal voltage ratio with no overlap

↑ Ratio of voltage drop caused by overlap, to the maximum d-c terminal voltage (at $\alpha = 0°$) with no overlap

The ratio $I_d R_e / V_{d_{max}}$ may be defined as the equivalent per-unit output resistance, denoted by $R_{e_{pu}}$:

$$R_{e_{pu}} = \frac{I_d R_e}{V_{d_{max}}} \qquad (13.18)$$

(It is ot be noted, however, that $R_{e_{pu}}$ does not define the short-circuit current available at the d-c terminals, for the reason that the equivalent

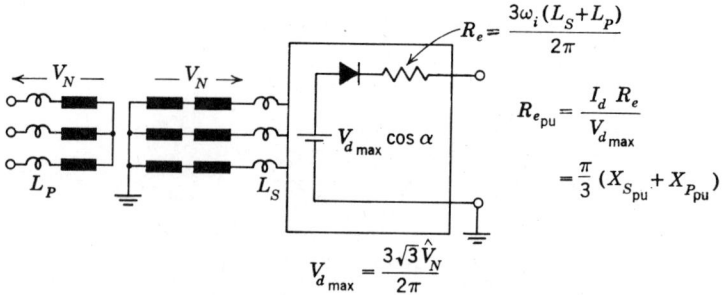

Three-pulse midpoint circuit

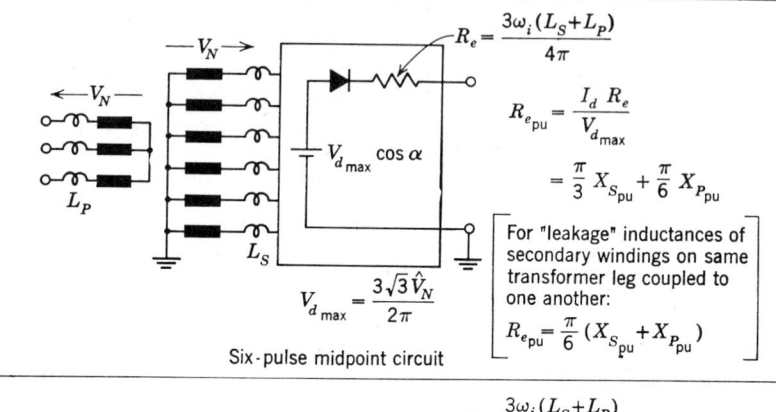

Six-pulse midpoint circuit

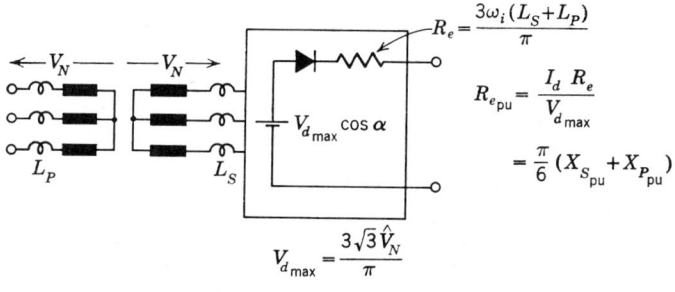

Six-pulse bridge circuit

Figure 13.10. Schematic equivalent circuits of 3- and 6-pulse 2-quadrant phase-controlled converters, showing the relationships between the equivalent output resistance and the input source inductance. $L_{S,P}$ = Inductance "seen" in each line during converter commutation. X_{pu} = Per unit reactance corresponding to a balanced sinusoidal load current, with amplitude equal to that of the full-load fundamental component consumed by the converter.

EFFECT ON CONVERTER PERFORMANCE 403

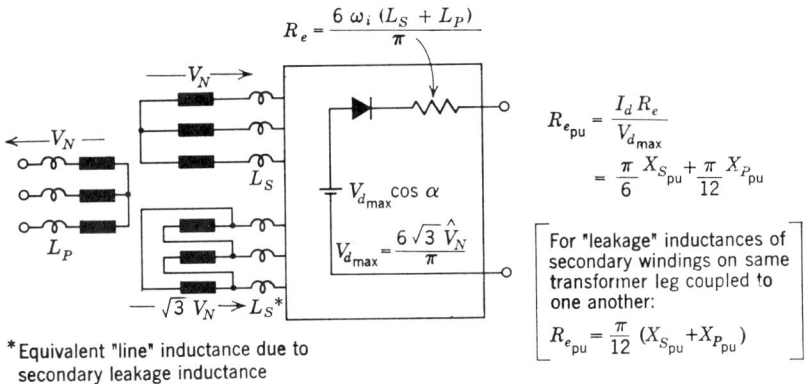

Twelve-pulse bridge circuit (series connected bridges)

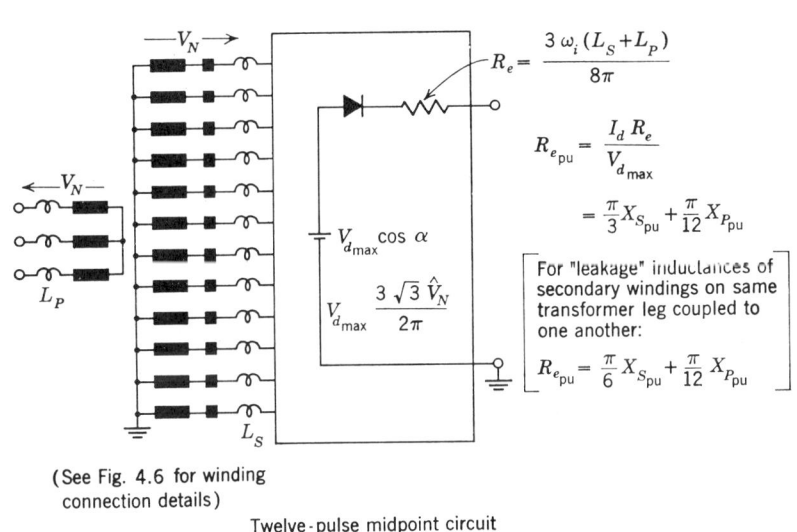

(See Fig. 4.6 for winding connection details)

Twelve-pulse midpoint circuit

Figure 13.11. Schematic equivalent circuits of 12-pulse 2-quadrant phase-controlled converters, showing the relationships between the equivalent output resistance and the input source inductance. $L_{S,P}$ = Inductance "seen" in each line during converter commutation. X_{pu} = Per unit reactance corresponding to a balanced sinusoidal load current, with amplitude equal to that of the full-load fundamental component consumed by the converter.

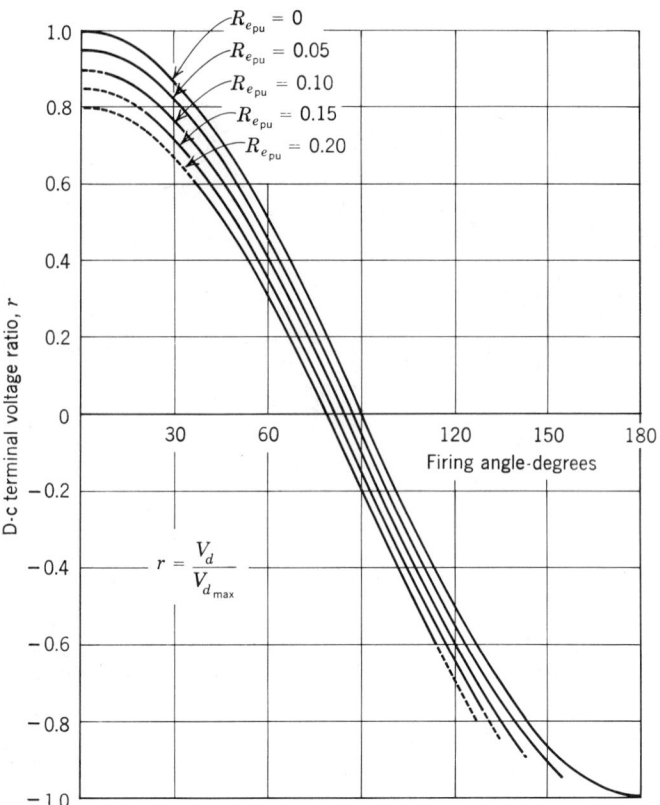

Figure 13.12. Universal relationships between d-c terminal voltage ratio and firing angle, for various values of per unit equivalent output resistance, for 2-quadrant phase-controlled converters. *Note:* "Full" portions of curves are applicable for converter pulse numbers up to twelve. Dashed portions are not applicable to 12-pulse converters, with source inductance common to all commutating groups.

resistance R_e—as defined in Figs. 13.9 through 13.11—is valid only for operation with discrete commutations.)

The curves of Fig. 13.12 show the universal relationships between the d-c terminal voltage ratio and the firing angle, for 2-quadrant phase-controlled converters, for various values of $R_{e_{pu}}$.

It is seen, for any given value of $R_{e_{pu}}$, that the theoretical effect of the commutation overlap is to reduce the d-c terminal voltage ratio by exactly the same amount at each end of the control range; thus, as with no commutation overlap, the d-c terminal voltage ratio at "full rectification" is theoretically equal to the d-c terminal voltage ratio at "full inversion." In practice, of course, in order to prevent commutation failures, a finite recovery angle δ must remain at the extremity of the inversion control range; thus the maximum negative d-c terminal voltage attainable in practice is slightly less than the maximum positive voltage.

It can also be seen that whereas the firing angle for maximum positive voltage is invariably $0°$, the firing angle for maximum negative voltage decreases as $R_{e_{pu}}$ increases. Thus the firing angle/d-c terminal voltage characteristic for finite values of $R_{e_{pu}}$ is not perfectly symmetrical, and the transition from rectification to inversion does not occur exactly at $\alpha = 90°$, but at some angle slightly in advance of this point.

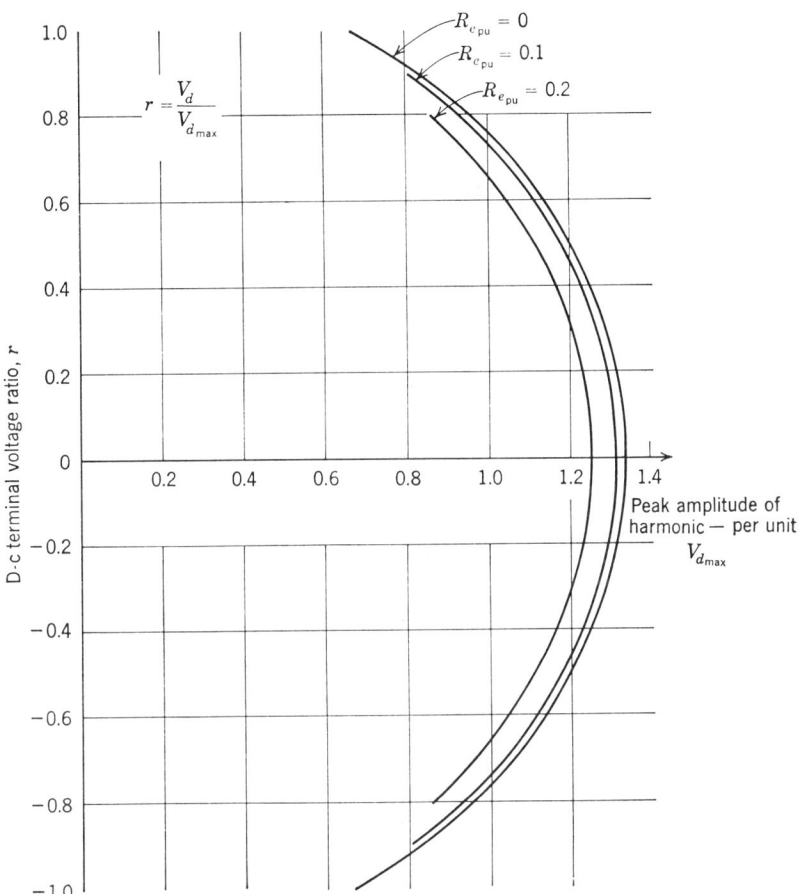

Figure 13.13. Curves showing the relationship between the amplitude of the $2 \times f_i$ harmonic distortion component, and the d-c terminal voltage ratio of the phase-controlled converter, for various values of per unit equivalent output resistance. This harmonic distortion component is present in the d-c terminal voltage of the 2-pulse converter.

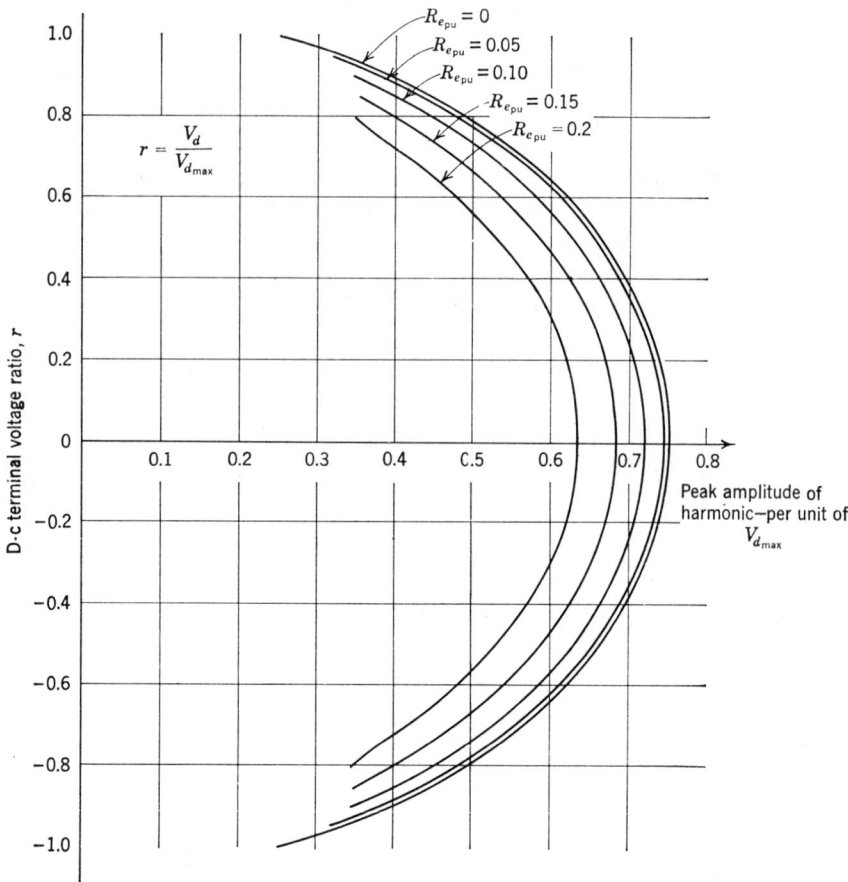

Figure 13.14. Curves showing the relationship between the amplitude of the $3 \times f_i$ harmonic distortion component, and the d-c terminal voltage ratio of the phase-controlled converter, for various values of per unit equivalent output resistance. This harmonic distortion component is present in the d-c terminal voltage of the 3-pulse converter.

It is mentioned that the curves of Fig. 13.12 are applicable only so long as operation with discrete commutations is preserved. Thus the dotted portions of the curves are not applicable to 12- (and higher) pulse number converters with "common" source inductance.

Harmonic Distortion of the d-c Terminal Voltage

The harmonic frequencies present in the d-c terminal voltage with overlap are the same as those present with no overlap. It can be shown that the peak amplitude of the harmonic of order n, relative to the maximum mean d-c

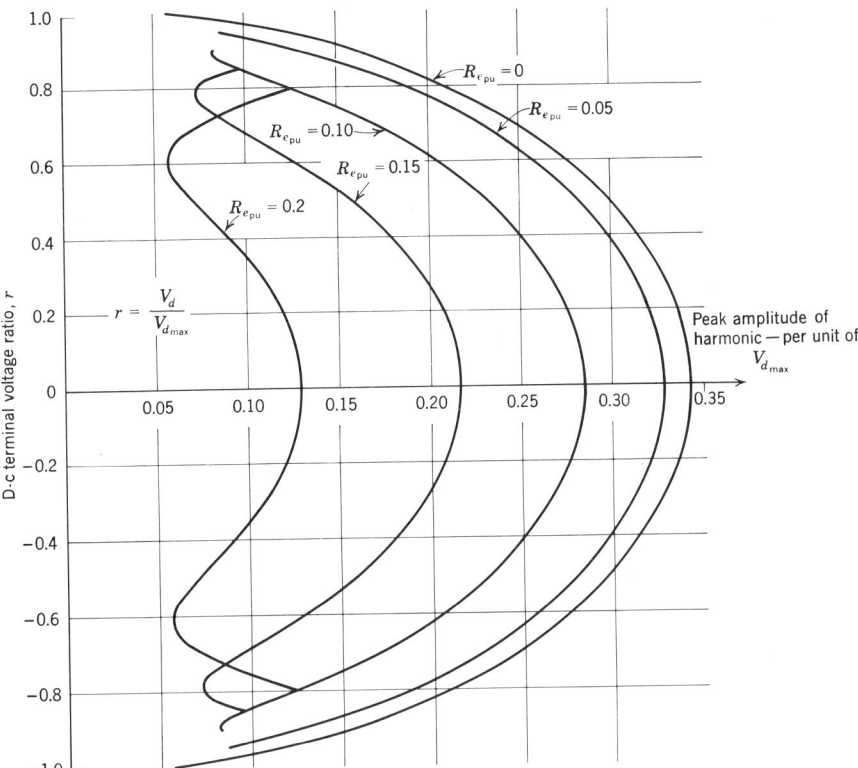

Figure 13.15. Curves showing the relationship between the amplitude of the $6 \times f_i$ harmonic distortion component, and the d-c terminal voltage ratio of the phase-controlled converter, for various values of per unit equivalent output resistance. This harmonic distortion component is present in the d-c terminal voltage of 3- and 6-pulse converters.

terminal voltage with no overlap, is given by the following expression:

Relative amplitude of harmonic of order $n =$

$$\left[\frac{1 + \cos(n+1)u}{2(n+1)^2} + \frac{1 + \cos(n-1)u}{2(n-1)^2} \right.$$
$$\left. - \frac{\cos(2u + 2\alpha) + \cos 2\alpha + \cos(2\alpha + [n+1]u) + \cos(2\alpha - [n-1]u)}{2(n-1)(n+1)} \right]^{1/2}$$

(13.19)

The curves of Figs. 13.13 through 13.16 illustrate the variation, with d-c voltage ratio, of the relative amplitudes of the $2f_i$, $3f_i$, $6f_i$, and $12f_i$ harmonic

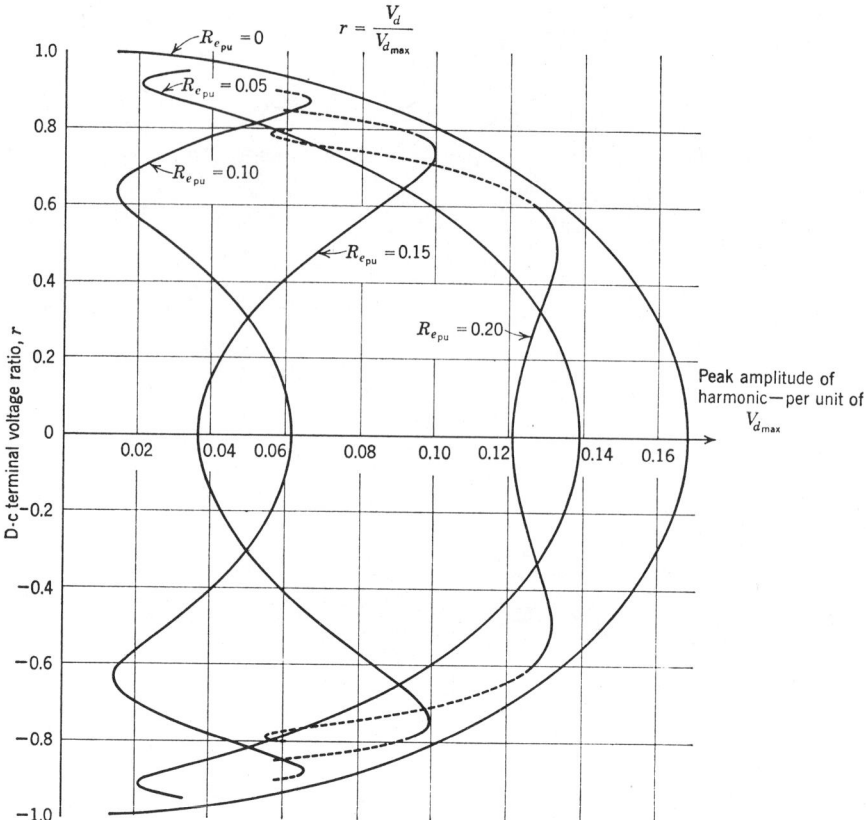

Figure 13.16. Curves showing the relationship between the amplitude of the $12 \times f_i$ harmonic distortion component, and the d-c terminal voltage ratio of the phase-controlled converter, for various values of per unit equivalent output resistance. This harmonic distortion component is present in the d-c terminal voltage of 3-, 6-, and 12-pulse converters. Dashed portions are not applicable to 12-pulse converter with source inductance common to all commutating groups.

components respectively, for various values of the per-unit equivalent output resistance, $R_{e_{pu}}$ (defined by expression (13.18)).

It is of interest to note from these curves that for any given d-c terminal voltage ratio (but not firing angle), the amplitude of the harmonic distortion, with commutation overlap, is less than it is with no overlap. In other words, for a given d-c terminal voltage ratio, the on-load distortion is invariably less than the no-load distortion. It is also of interest to note, (as would be evident from an inspection of the voltage waveform), that each curve is symmetrical about the point of zero d-c terminal voltage.

Relationships Between Per-Unit Equivalent Output Resistance and Per-Unit Input Reactance

It is usual to express the voltage developed across the internal impedance of an a-c source, due to the full load current flowing through it, as a ratio of the source voltage. This ratio is termed the per-unit impedance of the source:

$$Z_{pu} = \frac{I_{FL} Z}{V_N} \tag{13.20}$$

where I_{FL} is the rms value of the full load current
Z is the impedance of the source

If the harmonic components of the input current are neglected, then the full load per-unit reactance of the a-c source which feeds the phase-controlled converter can be defined as follows:

$$X_{pu} = \frac{I_{1_{FL}} X_L}{V_N} \tag{13.21}$$

where $I_{1_{FL}}$ is the rms value of the full load fundamental component of current
X_L is the reactive impedance of the a-c source.

It has been seen that the equivalent resistance at the output of the converter is directly related to the reactance of the a-c source. Moreover, the maximum mean direct output voltage is directly proportional to the a-c source voltage, and the direct current at the output terminals is directly proportional to the fundamental component of the current at the input. It follows, then, that the per-unit equivalent output resistance, as given by (13.18), is directly related to the per-unit input reactance.

The relationships between $R_{e_{pu}}$ and X_{pu}, for various phase-controlled converter circuits, are indicated in the schematic diagrams of Figs. 13.9 through 13.11.

EFFECT OF SOURCE INDUCTANCE ON THE EXTERNAL PERFORMANCE CHARACTERISTICS OF THE CYCLOCONVERTER

General Comments

In general terms, the commutation overlap has a similar modifying effect upon the external performance characteristics of the cycloconverter, as it does on those of the phase-controlled converter with a steady d-c output.

Thus, at the output terminals, a loss of voltage, as well as a modification of the harmonic distortion, takes place. Interestingly, however, the modification to the harmonic distortion takes the form not only of an alteration of the amplitudes of the "beat frequency" components, but, in addition, small amounts of "direct multiple" odd harmonics of the output frequency are introduced.

At the input side of the cycloconverter, the effect of the source inductance is to cause a "rounding off" of the steep edges of the line current waveforms. This results in a reduction in the amplitudes of the higher order harmonic terms, as well as a slight reduction in the input displacement factor, as compared to the case with zero source inductance.

A rigorous quantitative analysis of the modifying effects of the source inductance upon the performance characteristics of the cycloconverter is complicated. This is especially so for a cycloconverter supplying a 3-phase output from a common input source, for which case the mutual effects of the commutations of the 3 converters upon each other are extremely complex. A generalized quantitative analysis of these effects is far beyond the scope of this book. Such an analysis, in any case, would be of relatively minor practical interest, since the phenomena under consideration generally are of a second-order nature.

The following treatment, therefore, is limited mainly to a brief qualitative examination of the effect of the source inductance upon the output voltage wave of the cycloconverter, together with the presentation of an approximate equivalent output circuit, valid for the special case of maximum output voltage ratio. This brief treatment, although far from being comprehensive, should provide an insight into the underlying theoretical reasons for the modifying effect of the source inductance upon the output voltage of the cycloconverter, as well as providing approximate quantitative information related thereto.

Cycloconverter Supplying One-Phase Output

A theoretical waveform of the voltage appearing at the output terminals of a 6-pulse bridge cycloconverter, with input source inductance, is shown in Fig. 13.17. It is assumed here that the firing angle modulation is such as to produce maximum output voltage, and that the output current waveform is a sinusoid which is in-phase with the wanted component of the no-load output voltage wave. (Because of the effect of the internal voltage loss, to be explained, the displacement angle of the external load is not quite 0°, but actually it is slightly leading.)

EFFECT ON CYCLOCONVERTER PERFORMANCE 411

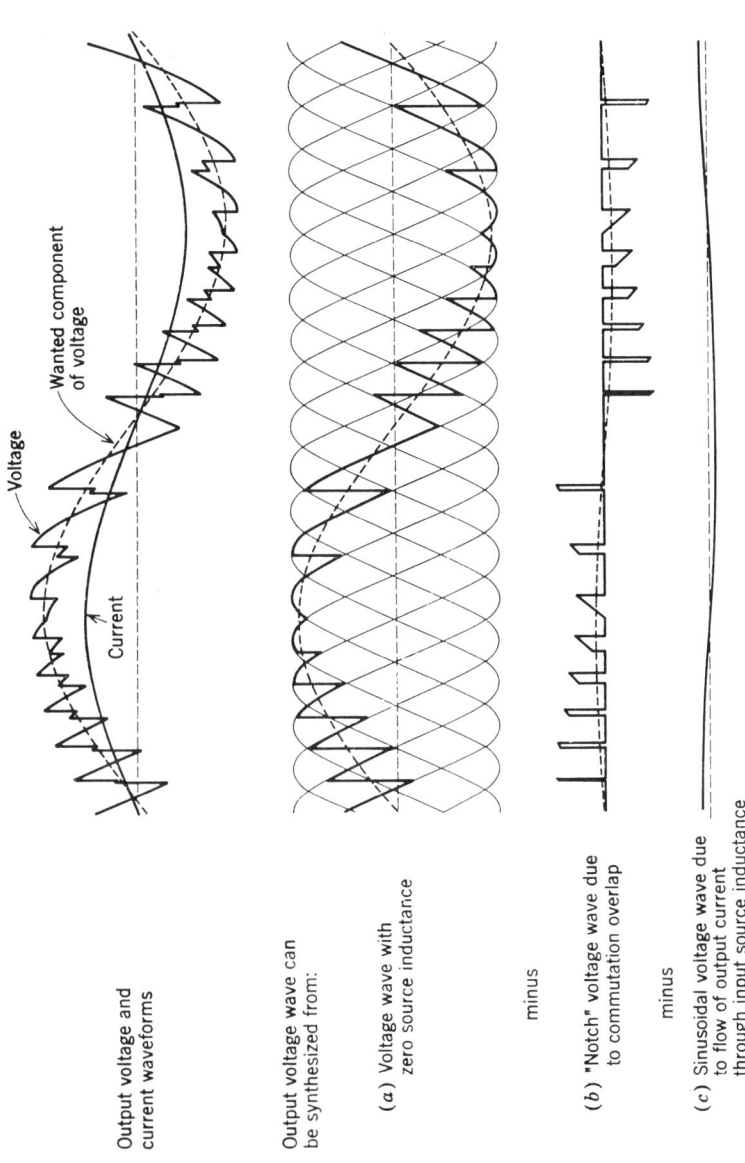

Figure 13.17. Theoretical waveforms for 6-pulse bridge cycloconverter supplying single-phase output, with input source inductance. Load displacement angle near 0°. Waveforms are for maximum theoretical output voltage. Per unit input source reactance ≃ 0.1.

As illustrated, the output voltage waveform can be regarded as being comprised of the difference between the no-load voltage waveform, and two separate "missing" components of voltage.

The first of these "missing" components is a waveform representing the "notches" removed by the commutation overlap. The duration of each of these notches corresponds with the commutation overlap period, and each segment of this waveform follows a voltage which is half of the instantaneous "internal" source voltage existing between the incoming and outgoing lines.

The second "missing" voltage component is simply a sinusoid which results from the flow of the output load current in the input line inductance. (The fact that the output current actually is "rotated" between the input line inductances in sequence is irrelevant—assuming, of course, that all inductances are equal.) Thus, this voltage wave leads the output current by 90°, and its amplitude is equal to the product of the rate of change of the output current, and twice the input line inductance. (For the bridge circuit, the output current flows through two line inductances in series. Strictly speaking, during each commutation overlap period, the output load current is drawn through two line inductances in parallel, and thence back to the source through the third inductance. Thus, during these periods, the amplitude of this component of voltage actually should drop to $\frac{3}{4}$ of the sinusoidal value; this point, however, is of theoretical interest only.)

Consider the component of missing voltage shown at b. During the first quarter cycle of the output current, the level of the output voltage is progressively increasing, and the firing angle is being progressively advanced. Thus, during this period, the frequency of commutations is greater than with a steady firing angle. During the second quarter cycle of the output current, on the other hand, the level of the output voltage is progressively decreasing, and the firing angle is being progressively retarded. Thus, during this period, the frequency of commutations is less than with a steady firing angle. Since, however, for this particular load displacement angle, exactly the same number of commutations occur during the course of one half-cycle of output current, as occur during the same time period with a steady firing angle, it can be deduced, intuitively, that the wanted component of voltage, lost as a result of the commutation overlap, must correspond fairly closely with that given by the equivalent output resistance corresponding to a steady firing angle.

Two distinct effects can be deduced from an inspection of the detailed appearance of the missing "notch" voltage waveform of Fig. 13.17b. First, the asymmetry between the spacing of the notches in the first and second quarter cycles can be presumed to result in the presence of small amounts of direct integer harmonics of the output frequency. This is indeed the case, and in fact it can be shown that these direct integer harmonics constitute a separate harmonic family, distinct from the "beat frequency" harmonic components. (The amplitudes of these harmonic components, however, are generally quite small; typically, the amplitude of the 3rd harmonic does not exceed 1 or 2% of $V_{W_{max}}$.) Second, because of the asymmetry of the waveform, the wanted component of voltage loss is not precisely in-phase with the output current, but leads it by a small angle (in this case, about 12°). Thus, under these particular conditions, the equivalent output impedance which characterizes the missing component of wanted output voltage contained in the "notch" waveform of Fig. 13.17b is not, strictly speaking, a pure resistance, but it has also an inductive component. This inductive component is, however, relatively small (and, in fact, it is invariably considerably smaller than the source inductance itself, which produces the component of voltage loss shown at c in Fig. 13.17).

The asymmetry of the missing "notch" voltage waveform decreases with decreasing output voltage ratio, as well as with decreasing output to input frequency ratio. Thus, it can be deduced that a decrease in either of these parameters results in a corresponding decrease in both the "integer multiple" harmonic distortion, as well as in the reactive component of the equivalent output impedance, which characterizes the asymmetry of the missing notch voltage wave. The resistive component of output impedance, on the other hand, does not change, since the total number of commutations during each half cycle of current remain unaltered.

A further factor which has a bearing both upon the magnitude of the "integer multiple" harmonic distortion, and upon the magnitude and phase of the equivalent output impedance, which characterizes the missing "notch" voltage waveform, is the displacement angle of the external load. This is illustrated by the waveforms of Figs. 13.18 and 13.19.

In Fig. 13.18, theoretical waveforms are shown for an external load whose displacement angle is such that the output current leads the wanted component of the "internal" or "no-load" voltage by exactly 90°. In this case, throughout the complete output cycle, the frequency of commutations is greater than with a steady firing angle; in fact, for the particular theoretical case shown, of maximum output voltage, the firing pulses are produced at the same even rate over the whole span of the output cycle. As a result, the "notch" voltage waveform is substantially symmetrical.

The following deductions can be made. First, because of the symmetrical appearance of the "notch" voltage waveform, the "integer multiple" harmonic content of this wave is substantially zero. By the same token, the component of this waveform having the wanted output frequency is in-phase with the output current; thus, the equivalent output impedance which characterizes the missing component of wanted output voltage contained in the "notch" waveform of Fig. 13.18b, is purely resistive. Second, because of

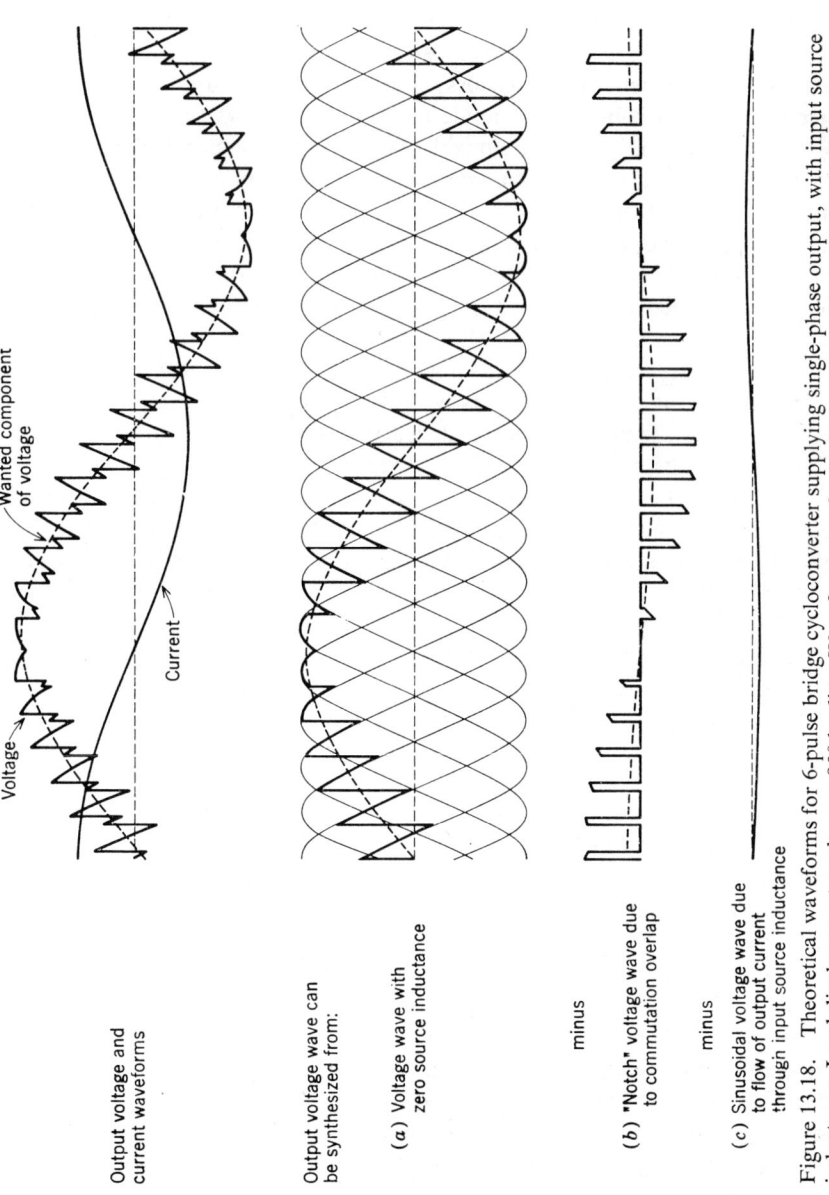

Figure 13.18. Theoretical waveforms for 6-pulse bridge cycloconverter supplying single-phase output, with input source inductance. Load displacement angle near 90° leading. Waveforms are for maximum theoretical output voltage. Per unit input source reactance ≃ 0.1.

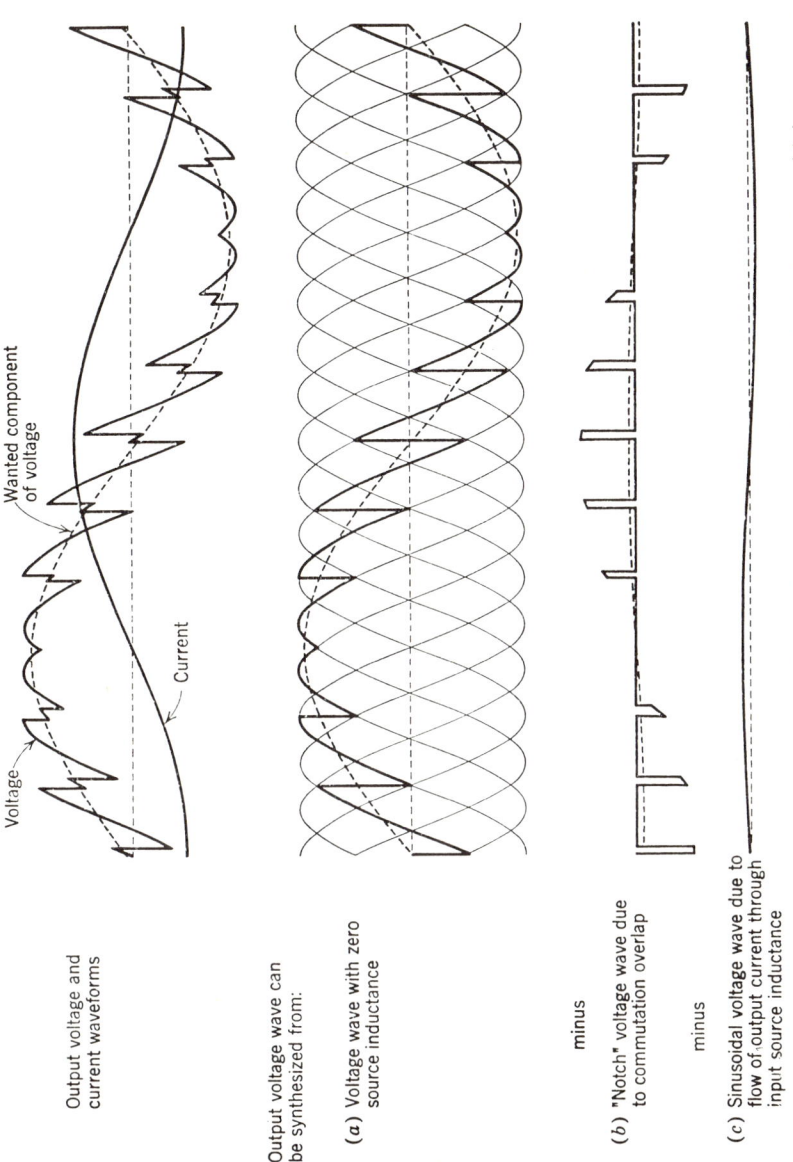

Figure 13.19. Theoretical waveforms for 6-pulse bridge cycloconverter supplying single-phase output, with input source inductance. Load displacement angle near 90° lagging. Waveforms are for maximum theoretical output voltage. Per unit input source reactance ≃ 0.1.

the increased repetition rate of the commutation notches, the amplitude of the equivalent output resistance is greater than that corresponding to a steady firing angle.

Since a decrease in the output voltage ratio, or in the output to input frequency ratio, is accompanied by a corresponding decrease in the frequency of the commutations, it can be deduced that the magnitude of the equivalent output resistance, for this load displacement factor, is a function of both of these parameters. Thus, the value of the equivalent output resistance tends to decrease towards the "steady d-c" value, both with decreasing output voltage ratio, and with decreasing output to input frequency ratio.

In Fig. 13.19, theoretical waveforms are shown for an external load whose displacement angle is such that the output current lags the wanted component of the internal no-load voltage by 90°. Once again, for the theoretical case of maximum output voltage considered, the firing pulses are produced with the same even rate throughout the output cycle; but, in this case, the repetition rate is less than that corresponding to a steady firing angle. And, once again, the "notch" voltage waveform is substantially symmetrical. Thus, it can be deduced that the direct integer harmonic content of this wave is zero, and that the equivalent output impedance is purely resistive. In this case, however, because of the reduced repetition rate of the commutation notches, the amplitude of the equivalent output resistance is less than that obtained with a steady firing angle. Moreover, a decrease in the output voltage ratio, or in the output to input frequency ratio, results in an increase in the magnitude of the equivalent output resistance, towards the "steady d-c" value.

From the preceding discussion, it is clear that the magnitude and phase of the equivalent output impedance which characterizes the missing wanted component of voltage contained within the "notch" voltage waveform, generally is a function of the output to input frequency ratio, the displacement angle of the external load, and the output voltage ratio. For all frequency ratios and load displacement angles, however, the deviation (if any) of the value of the equivalent output impedance, from that obtained with a steady d-c output (i.e., R_e), is greatest at maximum output voltage ratio, and this deviation decreases towards zero as the output voltage ratio decreases towards zero.

THE EQUIVALENT OUTPUT CIRCUIT, FOR MAXIMUM OUTPUT VOLTAGE. By making the simplifying assumption that, within any given subcycle period, the output circuit of the cycloconverter can be regarded as comprising a sinusoidal voltage source, representing the no-load wanted output voltage, connected in series with an equivalent "subcycle output resistance," the value of which is determined by the commutating frequency existing at the time, it is possible to arrive at the "average" equivalent output impedance of

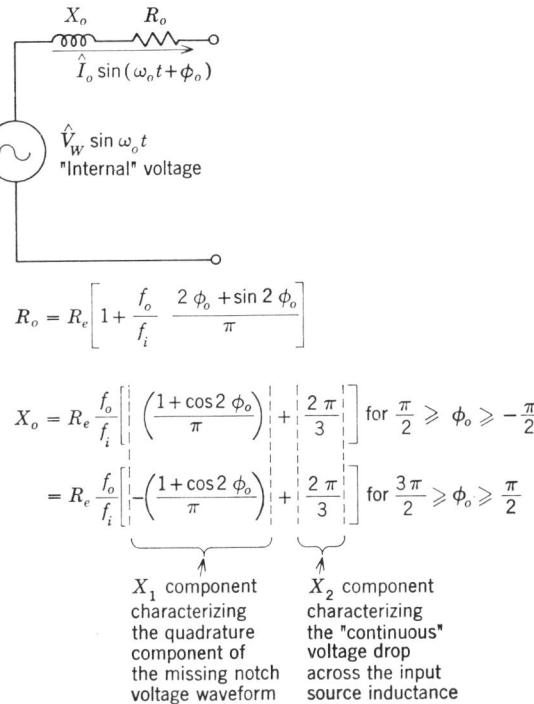

Figure 13.20. Approximate equivalent output circuit of the cycloconverter, with input source inductance. Applicable to maximum theoretical output voltage. R_e for any particular circuit configuration can be obtained from Figs. 13.9 through 13.11. $\phi_o =$ Displacement angle between the "internal" output voltage and the output current.

the cycloconverter, which characterizes the missing wanted component of the "notch" voltage waveform. For the particular condition of maximum output voltage the computation is relatively simple, since the commutating frequency has only two discrete values during the course of each output cycle: $P \times (f_i + f_o)$ during the "fast switching" period, and $P(f_i - f_o)$ during the "slow switching" period, where P is the converter pulse number.

Such a computation leads to the approximate equivalent output circuit for the cycloconverter, supplying a single phase output, for the particular theoretical condition of maximum output voltage, shown in Fig. 13.20.

As already discussed, for any given output to input frequency ratio, as the output voltage ratio decreases, the resistive component of the output impedance of the cycloconverter tends towards the steady d-c value, R_e. The reactive component, on the other hand, tends towards the value of the "X_2 component."

Cycloconverter Supplying Three-Phase Output

With 3 converters operating from a common input source, it is not permissible to consider each individual converter as a completely separate entity. This is because the commutation notches produced by all three converters appear on the common input terminal voltage waveform. Thus, at any one output phase, there appear not only the notches due to the commutations of that output phase, but also, to a lesser or greater extent, the commutation notches due to the other output phases. As a general rule, a detailed derivation of the exact waveforms is very complicated; and, even once the waveforms are obtained, quantitative analysis is very difficult.

In general terms, the modifying effect on the output voltage of a given phase, of the commutations of the other two phases, is two-fold. Firstly a modification (generally an increase) in the direct integer multiple harmonic distortion takes place; and secondly, the wanted component of voltage lost at the output terminals generally increases.

These effects are illustrated by the waveforms of Figs. 13.21 and 13.22. These waveforms are appropriate to a 6-pulse converter, with a load having (approximately) unity displacement factor, and are for the particular output to input frequency ratio of 1:3.

Figure 13.21 is applicable to maximum theoretical output voltage. At a, the missing "notch" voltage wave due to the commutation overlap, and the waveform of the missing voltage due to the "continuous" flow of the output current in the input source inductance, are shown for the condition when a single converter ("output phase 1") operates "on its own," from the input source to which it is connected. In other words, these waveforms are for a cycloconverter supplying a single-phase output, and actually they are identical with those already seen in Fig. 13.17.

At b, the corresponding "output phase 1" waveforms are shown for the condition when two additional converters simultaneously operate from the same input source (the total fundamental load on the input system thereby being increased threefold), producing balanced output voltages (and currents) at $+120°$ and $-120°$ respectively, with respect to "output phase 1." The resulting missing "notch" voltage wave at "output phase 1" can be regarded as being comprised of the "notch" wave obtained when this converter operates by itself (i.e., single-phase output), with a superimposed "notch" wave (shown black), due to the commutations of the converters of the other two output phases. The waveform of the missing voltage, due to the "continuous" voltage drop across the input source inductance, is due to the total input current drawn by all 3 output phases. This waveform, too, can be regarded as being comprised of the sinusoidal component (shown dashed) obtained

EFFECT ON CYCLOCONVERTER PERFORMANCE 419

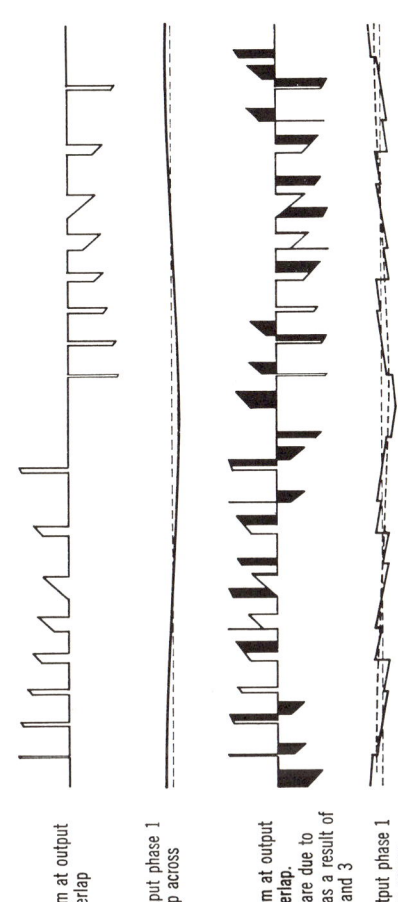

(a)

Missing "notch" voltage waveform at output phase 1 due to commutation overlap

Missing voltage waveform at output phase 1 due to "continuous" voltage drop across input source inductance.

(b)

Missing **notch** voltage waveform at output phase 1 due to commutation overlap. Parts of waveform shown black are due to commutation notches produced as a result of the loading of output phases 2 and 3

Missing voltage waveform at output phase 1 due to "continuous" voltage drop across input source inductance.
Dashed waveform is for output phase 1 only loaded.

Figure 13.21. "Missing" voltage waveforms, due to the input source reactance, at one output phase of a cycloconverter with 3-phase output, for the theoretical condition of maximum possible output voltage. Per unit input source reactance $\simeq 0.3$ (for all three output phases fully loaded). Output load displacement factor $\simeq 1.0$. Six-pulse cycloconverter. $f_o/f_i = \frac{1}{3}$. (a) Output phase 1 only fully loaded—other two output phases unloaded. (b) All three output phases fully loaded.

420   THE EFFECT OF INPUT SOURCE IMPEDANCE

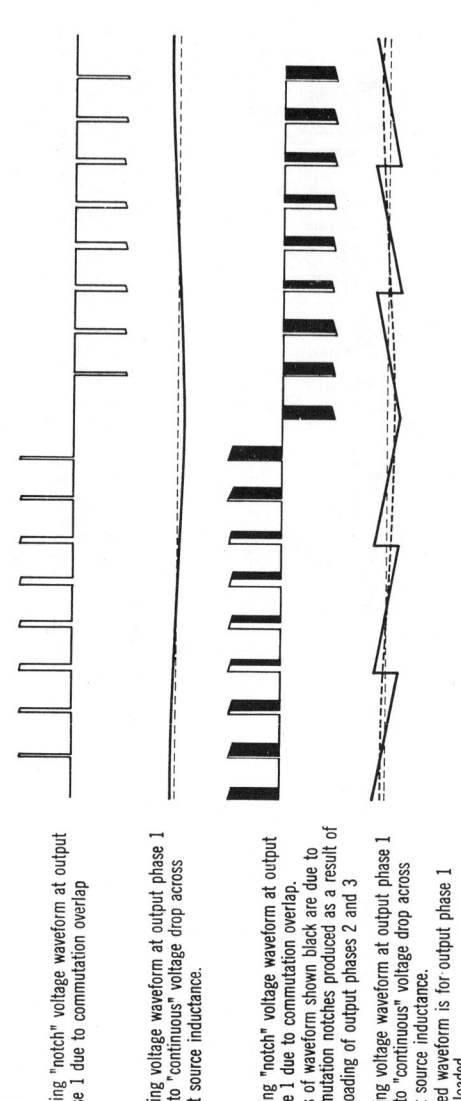

(a)

Missing "notch" voltage waveform at output phase 1 due to commutation overlap

Missing voltage waveform at output phase 1 due to "continuous" voltage drop across input source inductance.

(b)

Missing "notch" voltage waveform at output phase 1 due to commutation overlap. Parts of waveform shown black are due to commutation notches produced as a result of the loading of output phases 2 and 3

Missing voltage waveform at output phase 1 due to "continuous" voltage drop across input source inductance.
Dashed waveform is for output phase 1 only loaded.

Figure 13.22. "Missing" voltage waveforms, due to input source reactance, at one output phase of a cycloconverter with 3-phase output, for the theoretical condition of zero phase modulation of the converter firing angles. Per unit input source reactance $\simeq 0.3$ (for all three output phases fully loaded). Output load displacement factor $\simeq -1.0$. Six-pulse cycloconverter. $f_o/f_i = \frac{1}{3}$. (a) Output phase 1 only fully loaded—other two output phases unloaded. (b) All three output phases fully loaded.

when "output phase 1" operates by itself, with a superimposed component, due to the input currents of the other two output phases.

In Fig. 13.22, corresponding waveforms are shown for the theoretical condition of zero phase-modulation of the converter firing angles (due to the commutation overlap, theoretically this does not quite correspond to the condition of zero output voltage).

An inspection of the waveforms of Figs. 13.21 and 13.22 shows that at maximum output voltage, the wanted component of voltage lost at a given output phase, with a balanced 3-phase output, is not substantially different from the wanted component of voltage lost when the other two phases are unloaded. With zero phase modulation of the firing angles, however, the wanted component of the "notch" waveform lost at a given output phase, due to the commutations of the other two output phases, is theoretically exactly 3 × that obtained when the other two phases are unloaded.

These two specific examples will suffice to illustrate that in general the additional loss of wanted output voltage at any given output phase of the cycloconverter, due to a balanced loading of the other two output phases, is relatively small at high output voltage ratios; but this voltage loss increases (to about 3 × the voltage loss obtained when this is the only output phase which is loaded) as the output voltage ratio decreases towards zero.

Bibliography

This bibliography is not exhaustive. It has been selected as being representative of the recent literature which is most closely complementary to, and compatible with, the subject matter of the present book. A few early works of particular historical interest are also included.

Inasmuch as the theoretical analysis of the operation and performance of the phase-controlled cycloconverter presented in this book is believed to be original, and is published here for the first time, there are no known related works which deal in a similar manner with this particular topic.

ARTICLES AND PAPERS

Eine Unmittelbare Asynchrone Umrichtung fur Niederfrequente Bahnnetze, M. Schenkel, *Electrische Bahnen,* Vol. 8, 1932.

Der Gesteurerte Umrichter, J. Von Issendorff, *Wiss. Veroffente.,* Siemens, Vol. 14, 1935.

The Calculation of Rectifier and Inverter Performance Characteristics, J. C. Read, *J. IEE,* Vol. 92, Part 2, 1945.

The Operation of Several Phase Displaced Inverters on the Same Receiving Network, E. Uhlmann, *Direct Current,* June 1953.

The Influence of the Number of Phases Used in a Converter on Harmonics, F. Busemann, *Direct Current,* December, 1954.

The Current and Voltage Conditions in the Graetz Three-Phase Rectifier Bridge Circuit, W. Dallenbach, *Direct Current,* December, 1958.

Commutation Processes in the Operation of a Three-Phase Bridge Circuit During Transient Phenomena, A. B. Posse, *Direct Current,* June 1958.

Precise Frequency Power Generation from an Unregulated Shaft, K. M. Chirgwin, L. J. Stratton, and J. R. Toth, *AIEE Trans.,* Vol. 79, Part 2, 1960.

Current and Voltage Conditions from No-Load to Short Circuit in Three-Phase Bridge Circuits, F. Hoelters, *Direct Current,* March 1961.

The Universal Characteristic of the Three-Phase Bridge Converter, L. L. Freris, *Direct Current,* October, 1961.

Circle Diagram of an Inverter, Yu. G. Tolstov, *Direct Current,* October, 1961.

The Application of Silicon Controlled Rectifiers to the Control of Electrical Machines, K. G. King, *Proc. IEE,* Vol. 110, No. 1, 1963.

An Analysis of The Three-Phase Bridge Converter, L. L. Freris, *Direct Current,* January, 1963.

The Static Converter as a High Speed Amplifier, N. A. Bjaresten, *Direct Current,* June 1963.

A Chart Showing The Relations Between Electrical Quantities on the AC and DC Sides of a Converter, E. W. Kimbark, *Direct Current,* June 1963.

A Note on the Reverse-Parallel Operation of Rectifier Bridges, K. D. Srivastava and R. M. Davis, *Direct Current,* July 1963.

A Static Frequency Changer Fed Squirrel Cage Motor Drive for Variable Speed and Reversing, R. Heck and M. Meyer, *Siemens Rev.,* November, 1963.

Commutation Reactance of the Transformer in a Static Power Converter, R. Feinberg and W. Y. Chen, *Proc. IEE,* Vol. 111, No. 1, 1964.

Commutation Phenomena in a Static Power Converter, R. Feinberg and W. Y. Chen, *Proc. IEE,* Vol. 111, No. 1, 1964.

Static Frequency Changers for 16 2/3 c/s Railway Networks, W. Faust, *Brown Boveri Rev.,* August 1964.

Varying The Speed of Three-Phase Motors by Means of Static Frequency Changers, A. Schonung, *Brown Boveri Rev.,* August, 1964.

Semiconductor Rectifiers in Circulating Current-free Anti-parallel Connection for Reversing Drives, R. Fischer and H. Geissing, *Siemens Rev.,* August, 1964.

Reversing Thyristor Armature Dual-Converter with Logic Crossover, D. L. Duff, A. Ludbrook, *IEEE Trans.,* Vol. IGA−1, No. 3, 1965.

A Simplified Technique for Analyzing the Three-Phase Bridge Rectifier Circuit, A. Ludbrook and R. M. Murray, *IEEE Trans.,* Vol. IGA−1, No. 3, 1965.

Controlled Rectifier Circuit with an Uncontrolled By-Pass Valve, N. G. Hingorani, *Direct Current,* May 1965.

Interphase Transformer for Multiple Connected Power Rectifiers, O. N. Acosta, *IEEE Trans.,* Vol. IGA−1, No. 6, 1965.

Thyristor Converters for DC Reversing Drives, H. Geissing and G. Moltgen, *Siemens Rev.,* October, 1965.

Optimizing Control Systems for Land Vehicles, W. Slabiak and L. J. Lawson, IEEE Industrial Static Power Conversion Conference Record, November, 1965.

Variable Speed with Controlled Slip Induction Motor, C. J. Amato, IEEE Industrial Static Power Conversion Conference Record, November, 1965.

The Application of Thyristors to the Excitation Circuits of Synchronous Motors, J. D. Edwards, A. J. Gilbert, and E. H. Harrison, IEE Conference Publication No. 17, Part 1, 1965.

Thyristor Converters for Motor Drives—Some Experience in Design and Operation, G. Mellgren, IEE Conference Publication No. 17, Part 1, 1965.

The Application of Thyristors to the Control of DC Machines, J. R. G. Schofield, G. A. Smith, and M. G. Whitmore, IEE Conference Publication No. 17, Part 1, 1965.

A Modern Rod Mill Main Drive Using Thyristors, J. I. Jones, IEE Conference Publication No. 17, Part 1, 1965.

Thyristor Equipments in the Steel Industry, R. W. Sturland and M. Allison-Beer, IEE Conference Publication No. 17, Part 1, 1965.

The Application of a Cycloconverter to the Control of Induction Motors, P. Bowler, IEE Conference Publication No. 17, Part 1, 1965.

Thyristor Amplifiers for Machine Field Excitation, R. D. Pettit, IEE Conference Publication No. 17, Part 1, 1965.

Thyristor Converters with Natural Commutation, F. Wesselak, Siemens Rev., December, 1965.

Use of Capacitors for Reduction of Commutation Angle in Static High Power Converters, N. G. Hingorani and J. K. Hall, *Proc. IEE,* Vol. 112, No. 12, 1965.

Analysis of a Hybrid Bridge Rectifier, L. L. Freris, *Direct Current,* February, 1966.

Voltage Control by Means of Power Thyristors, D. W. Borst, E. J. Diebold, and F. W. Parrish, *IEEE Trans.,* Vol. IGA–2, No. 2, 1966.

Variable Speed-Constant Frequency Aricraft Generating System Performance, M. B. Wall, IEEE Aerospace System,Conference, July 1966.

Precise Control of a Three-Phase Squirrel Cage Induction Motor Using a Practical Cycloconverter, W. Slabiak and L. J. Lawson, *IEEE Trans.,* Vol. IGA–2, No. 4, 1966.

AC Motor Supply with Thyristor Converters, L. A. Graham, J. Forster, and G. Schliephake, *IEEE Trans.,* Vol. IGA–2, No. 5, 1966.

Thyristorized Drive of a Mine Winder, J. Francoli and G. J. Heim, *Brown Boveri Rev.,* October, 1966.

Thyristor Charging Equipment for Traction Batteries, C. Lentz and W. Lossel, *Siemens Rev.,* December, 1966.

Thyristor Adjustable Frequency Power Supplies for Hot Strip Mill Run-Out Tables, R. A. Hamilton and G. R. Lezan, *IEEE Trans.,* Vol. IGA–3, No. 2, 1967.

Thyristor-Controlled Drives for Rotary Printing Presses, F. Spiegelberg, *Brown Boveri Rev.,* May/June 1967.

Sectional Drives Fed by Semiconductor Units for High-Speed Newsprint Machines, E. Wanner, *Brown Boveri Rev.,* May/June 1967.

Veritron Thyristorized Supply Units for Variable-Speed Drives, E. Altzinger and T. Klasen, *Brown Boveri Rev.,* May/June 1967.

Rectifier-Fed Drives for Printing Presses and Paper-Processing Machines, F. Spiegelberg, *Brown Boveri Rev.,* May/June 1967.

The Use of Static Converters to Vary the Speed of Three-Phase Drives Without Losses, H. Stemmler, *Brown Boveri Rev.,* May/June 1967.

The Phase-Locked Oscillator – A New Control System for Controlled Static Converters, J. D. Ainsworth, IEEE Paper No. 31TP67–499, July 1967.

Harmonic Instability Between Controlled Static Converters and AC Networks, J. D. Ainsworth, *Proc. IEE,* Vol. 114, No. 7, 1967.

Effects of Interaction Among Groups in a Multigroup AC–DC Converter, L. L. Freris, *Proc. IEE,* Vol. 114, No. 7, 1967.

A True No-Break, Off-Line Uninterrupted Power Supply, L. J. Lawson, IEEE/IGA Conference Record, 1967.

Semiconductor Rectifiers Go High Power, D. L. Duff and A. Ludbrook, *IEEE Trans.,* Vol. IGA–4, No. 2, 1968.

Electronic Controls for D–C Propeller Motor Drives on Ships, N. Ettner and J. Stiglitz, *Siemens Rev.,* March 1968.

Application Considerations for SCR D–C Drives and Associated Power Systems, A. P. Jacobs and G. W. Walsh, *IEEE Trans.,* Vol. IGA–4, No. 4, 1968.

Redesign of D–C Motors for Application with Thyristor Power Supplies, C. E. Robinson, *IEEE Trans.,* Vol. IGA–4, No. 5, 1968.

Phase Shifting of Harmonics in AC Circuits of Rectifiers, I. K. Dortort, *IEEE Trans.*, Vol. IGA—4, No. 6, 1968.

Interaction Between SCR Drives, B. P. Stahl, *IEEE Trans.*, Vol. IGA—4, No. 6, 1968.

Commutation dv/dt Effects in Thyristor Three-Phase Bridge Converters, J. B. Rice and L. E. Nickels, *IEEE Trans.*, Vol. IGA—4, No. 6, 1968.

VSCF for High Quality Electrical Power and Reliability, D. A. Fisk., ASME Aviation and Space Division Conference, June 1968.

A Subsynchronous Static Converter Cascade for Variable-Speed Boiler Feed Pump Drives, H. Elger and M. Weib, *Siemens Rev.*, October, 1968.

Thyristor-Supplied Tandem Cold Mill, B. W. Draper and R. J. Goodridge, *Proc. IEE*, Vol. 115, No. 10, 1968.

Simoreg Drives for Single-Quadrant Operation, G. Kliesch and H. Ludwig, *Siemens Rev.*, December 1968.

Concepts of Gearless Ball-Mill Drives, E. A. E. Rich, *IEEE Trans.*, Vol. IGA—5, No. 1, 1969.

Bridge Rectifiers with Double and Multiple Supply, A. Yair, W. Alpert, and J. Ben Uri, *Proc. IEE*, Vol. 116, No. 5, 1969.

Direct Digital Control of Thyristor Amplifiers, F. Fallside and R. D. Johnson, *Proc. IEE*, Vol. 116, No. 5, 1969.

Accelerated Recovery from Commutation Faults in Bridge Connected A—C—D—C Converters, J. Reeve and G. E. Burdett, IEE Conference Publication No. 53, Part 1, 1969.

Thyristor Converters for DC Motor Drives, J. A. Davies, A. C. Kidd, R. E. Beadle, and G. Tilstone, IEE Conference Publication No. 53, Part 1, 1969.

Thyristor Excitation of Alternators, W. Fairney, I. Lodge, and J. E. Toms, IEE Conference Publication No. 53, Part 1, 1969.

Speed Control of Large Induction Motors by Thyristor Converters, D. A. Paice, *IEEE Trans.*, Vol. IGA—5, No. 5, 1969.

Fault-Development Control in A—C—D—C Converters, J. Arrillaga and G. Galanos, *Proc. IEE*, Vol. 116, No. 7, July, 1969.

Recent Developments in the Field of Variable-Speed Drives, M. Meyer, *Siemens Rev.*, December 1969.

Thyristor-Fed Main Drives in the Blooming Mill at the Hunedoara Iron and Steel Combine, Romania, J. Wokusch, *Siemens Rev.*, January 1970

Static Frequency Changer Supply System for Synchronous Motors Driving Tube Mills, J. Langer, *Brown Boveri Rev.*, March 1970.

Drive System and Electronic Control Equipment of the Gearless Tube Mill, H. Stemmler, *Brown Boveri Rev.*, March 1970.

Fault Detection Scheme for Direct Digital Control of A—C—D—C Interconnections, J. Arrillaga and G. Galanos, *Proc. IEE*, Vol. 117, No. 4, April, 1970.

Capacitive Compensation of Thyristor-Controlled Slip-Energy-Recovery Systems, W. Shepard, and A. Q. Khalil, *Proc. IEE*, Vol. 117, No. 5, May 1970.

Generalized Theory of Static Power Frequency Changers, L. Gyugyi, Ph.D. Thesis, University of Salford, October 1970.

BOOKS

The Fundamental Theory of Arc Converters, H. Rissik, Chapman and Hall, 1939.

High Voltage Direct Current Power Transmission, C. Adamson and N. Hingorani, Garraway, Ltd., 1960.

Principles of Inverter Circuits, B. D. Bedford and R. G. Hoft, Wiley, 1964.

Rectifier Circuits, J. Schaefer, Wiley, 1965.

High Voltage Direct Current Converters and Systems, B. J. Cory, Ed., MacDonald, 1965.

Index

Amplifier, operation of cycloconverter as, 23, 186, 231, 240, 305–306, 310
  operation of dual-converter as, 117, 127
  operation of phase-controlled converter as, 12, 117, 229

Bank selection, for cycloconverter, by firing pulse overlap method, 190–198
  by first current-zero method, 199–201
  by fundamental current-zero method, 201 203
  by voltage sensing method, 203–206
  problems connected with, 181–182
  for dual-converter, 114–126
Blocking oscillator firing pulse stage, 274–275

Carrier frequency pulse isolation techniques, 274–277
Characteristic harmonic frequencies, in input current of cycloconverter, 166–167, 175, 177, 360–361, 364, 365, 369, 375–377, 382
Circulating current-free operation, of cycloconverter, 151, 152–155, 182–187, 199–206
  of dual-converter, 114–126
Circulating current operation of cycloconverter, 151, 156–161, 187–198
  of dual-converter, 114, 126–144
Circulating current reactor, 114, 127, 128, 150, 192, 207, 208
Closed loop control, of output of cycloconverter, 186–187, 203–206, 239, 240–241, 309

  of output of dual-converter, 119, 121–126, 142
  of output of phase-controlled converter, 243–247, 254–259
Commutation failure, 32, 259, 260, 264
Commutation notches, 242, 243–244, 266, 397
Commutation overlap, 261–263, 389, 390–395
Constant frequency power supplies, 25–26
Control schemes, functional, for cycloconverter, 190–192, 203–206
  for dual-converter, 116–118, 121–126, 127–128
Control techniques, for cycloconverter, 186–206
  for dual-converter, 114–118, 121–126, 127–128, 142
  for reducing reactive load consumed, by two-quadrant converter, 109–110
  by cycloconverter, 385–388
Converter, simple six-pulse midpoint, 38
  six-pulse bridge, 44–46
    circuit, 45
    dc voltage ratio, 82, 92, 400–402, 404–405
    harmonic composition of dc terminal voltage, 77, 83, 94–95, 407, 408
    harmonic composition of input current, 88, 94–95
    input displacement factor, 85, 92–93
    input distortion factor, 87
    waveforms, 46, 87
  six-pulse half-controlled, 62–65
    circuit, 63
    dc voltage ratio, 105
    harmonic composition of dc terminal voltage, 100

428 INDEX

harmonic composition of input current, 100
input displacement factor, 105
input power factor, 102
waveforms, 64–65
six-pulse midpoint with interphase reactor, 38–41
   circuit, 39
   dc voltage ratio, 82, 92, 400–402, 404–405
   harmonic composition of dc terminal voltage, 77, 83, 94–95, 407, 408
   harmonic composition of input current, 88, 94–95
   input displacement factor, 85, 92–93
   input distortion factor, 86
   waveforms, 40–86
six-pulse with one freewheel diode, 65–67
   circuit, 66
   dc voltage ratio, 106
   harmonic composition of dc terminal voltage, 101
   harmonic composition of input current, 101
   input displacement factor, 107
   input power factor, 102
   waveforms, 67
six-pulse with two freewheel diodes, 67–68
   circuit, 66
   dc voltage ratio, 106
   harmonic composition of dc terminal voltage, 101
   harmonic composition of input current, 101
   input displacement factor, 107
   input power factor, 102
   waveforms, 67
three-pulse half-controlled bridge, 59–61
   circuit, 59
   dc voltage, 105
   harmonic composition of dc terminal voltage, 98
   harmonic composition of input current, 98
   input displacement factor, 105
   input power factor, 102
   waveforms, 60
three-pulse midpoint, 34–38
   circuit, 34
   dc voltage ratio, 76, 92, 400–402, 404–405
   harmonic composition of dc terminal voltage, 73–76, 77, 94–95, 406–408
   harmonic composition of input current, 77–79, 94–95
   input displacement factor, 80, 92–93
   input distortion factor, 80
   input power factor, 80
   waveforms, 36–37
three-pulse midpoint with freewheel diode, 61–62
   circuit, 62
   dc voltage ratio, 106
   harmonic composition of dc terminal voltage, 99
   harmonic composition of input current, 99
   input displacement factor, 107
   input power factor, 102
   waveforms, 62
twelve-pulse bridge, 47
   circuit, 47
   dc voltage ratio, 82, 92, 400–401, 403–405
   harmonic composition of dc terminal voltage, 77, 84, 94–95, 407–408
   harmonic composition of input current, 91, 94–95
   input displacement factor, 85, 92–93
   input distortion factor, 90
   waveforms, 48–90
twelve-pulse midpoint with interphase reactors, 41–44
   circuit, 42
   dc voltage ratio, 82, 92, 400–401, 403–405
   harmonic composition of dc terminal voltage, 77, 84, 94–95, 407–408
   harmonic composition of input current, 91, 94–95
   input displacement factor, 85, 92–93
   input distortion factor, 89
   waveforms, 43, 89
two-pulse bridge, 32–33
   circuit, 33
   dc voltage ratio, 92, 400–401, 404–405
   harmonic composition of dc terminal voltage, 77, 94–95, 405, 407
   harmonic composition of input current, 94–95
   input displacement factor, 92–93
two-pulse half-controlled, 55–58
   circuit, 55
   dc voltage ratio, 105
   harmonic composition of dc terminal voltage, 97

INDEX 429

harmonic composition of input current, 97
input displacement factor, 105
input power factor, 102
waveforms, 57
two-pulse midpoint, 27–32
  circuit, 28
  dc voltage ratio, 91, 400–401, 404–405
  harmonic composition of dc terminal voltage, 77, 94–95, 405, 407
  harmonic composition of input current, 94–95
  input displacement factor, 92–93
  waveforms, 30–31
two pulse midpoint with freewheel diode, 58–59
  circuit, 59
  dc voltage ratio, 106
  harmonic composition of dc terminal voltage, 97
  harmonic composition of input current, 97
  input displacement factor, 107
  input power factor, 102
Cosine wave crossing control, basic principle, 229–231
  for cycloconverter, 232–241, 289–291, 305–306
  schemes using, 249–254
  self-regulating property, 231, 306
  unique property for cycloconverter, 235–238, 327
Crossover distortion, of output of cycloconverter, 182–206
  of output of dual-converter, 115, 119–121, 125
Cycloconverter, bank selection, 181–182, 190–206
  basic operating characteristics, 17–18
  basic principle of operation, 16–17, 146–150
  composition of input current, 166–180, 328–388
  control difficulties, 181–186
  crossover distortion of output, 182–206
  firing pulse timing control, 232–241
  functional control schemes, 190–192, 203–209
  harmonic distortion of output voltage, 161–166, 278–327, 410–416, 418

operation with circulating current, 151, 156–161, 187–198
operation with no circulating current, 151, 152–155
output frequency control, 145, 149–150, 232–234
output voltage control, 145, 149–150, 151, 232–234
six-pulse bridge, 210, 213–214
  circuit (isolated loads), 213
  circuit (non-isolated loads), 214
  fundamental component of input current, 349–357
  harmonic composition of input current, 341–345, 359–365, 368, 369, 371–373
  harmonic composition of output voltage, 302–303, 308, 310–311, 312, 319–322
  input displacement factor, 358–359
  input distortion factor, 375, 380
  input transformer connections, 341, 344
  rms value of input current, 368, 375, 376, 377
  wanted component of output voltage, 305–306
six-pulse bridge open-delta, 219
  circuit, 221
  input displacement factor, 228
  waveforms, 218
six-pulse bridge ring connected, 228
  circuit, 227
  input displacement factor, 228
  waveforms, 224–225
six-pulse midpoint, 210
  circuit, 211
  fundamental component of input current, 349–357
  harmonic composition of input current, 341–345, 359–365, 368, 369, 371–373
  harmonic composition of output voltage, 302–303, 308, 310–311, 312, 319–322
  input displacement factor, 358–359
  input distortion factor, 375, 380
  input transformer connections, 341, 344
  rms value of input current, 368, 375, 376, 377
  wanted component of output voltage, 305–306
three-pulse midpoint, 208–209
  circuit, 209
  fundamental component of input current, 349–357

430 INDEX

harmonic composition of input current, 334–340, 359–365, 368, 369, 373
harmonic composition of output voltage, 293, 300–301, 307–310, 311, 312, 318–322
input displacement factor, 358–359
input distortion factor, 375, 379
input transformer connections, 337, 340
rms value of input current, 368, 374, 375
wanted component of output voltage, 305–306
three-pulse open-delta, 219
  circuit, 220
  input displacement factor, 228
three-pulse ring connected, 227
  circuit, 226
  input displacement factor, 228
twelve-pulse bridge, 216
  circuit, 215
  fundamental component of input current, 349–357
  harmonic composition of input current, 345–349, 359–361, 363–365, 368, 369, 372–373
  harmonic composition of output voltage, 303–304, 308, 311, 312, 313, 320–322
  input displacement factor, 358–359
  input distortion factor, 375, 381
  input transformer connections, 346, 349
  rms value of input current, 368, 375, 378
  wanted component of output voltage, 305–306
twelve-pulse midpoint, 210
  circuit, 212
  fundamental component of input current, 349–357
  harmonic composition of input current, 345–349, 359–361, 363–365, 368, 369, 372–373
  harmonic composition of output voltage, 303–304, 308, 311, 312, 313, 320–322
  input displacement factor, 358–359
  input distortion factor, 375, 381
  input transformer connections, 346, 349
  rms value of input current, 368, 375, 378
  wanted component of output voltage, 305–306
Cycloconverter circuits, 207–228

open delta, 216–220, 221
ring connected, 220–228
"symmetrical," 208–216
Cycloconverter constant frequency power supplies, 25–26

D–C voltage loss, due to commutation overlap, 396, 398–406
D–C voltage ratio, definition of, 69
effect of source inductance upon, 396, 398–406
for specific circuit, see specific converter circuit title
universal relationship to firing angle and displacement factor, for all half-controlled converters, 96, 103–105
universal relationship to firing angle and displacement factor, for all two-quadrant converters, 92–93
Discontinuous current, cycloconverter control difficulties arising from, 181–187
dual-converter control difficulties arising from, 118–121
firing pulse requirements for, 275
operation of two-quadrant converter with, 49–54
Discrete commutations, definition of, 390
Displacement angle, input, definition of, 70
of cycloconverter, 167–173
of open-delta cycloconverter, 217–219
of ring connected cycloconverter, 223–225
see also Displacement factor
Displacement factor, input, definition of, 70
of "ideal" cycloconverter, 179
of open-delta cycloconverters, 217–219, 228
relationship to ac power, 71
relationship to power factor and distortion factor, 71–72
of ring connected cycloconverters, 223–225, 228
for specific circuit, see specific converter or cycloconverter circuit title
of all symmetrical cycloconverters, 358–359
universal relationship to dc voltage ratio and firing angle, for all half-controlled converters, 96, 103–105

INDEX 431

for all two-quadrant converters, 92–93
Distortion, harmonic, *see* Harmonic distortion
Distortion factor, input, definition of, 70
   of cycloconverter input current, 375–376, 382
   relationship to power factor and displacement factor, 71
   for specific circuit, *see* specific converter or cycloconverter
Dual-converter, 111–144
   bank selection, 114–126
   basic arrangement, 11
   basic principles of operation, 111, 113–114
   basic schemes for dc motor speed control, 14–16
   complementary firing pulse generators for, 271–272
   control difficulties, 118–121
   functional control schemes for, 116–118, 121–126, 127–128
   operation with circulating current, 114, 126–144
   operation with no circulating current, 114–126
   typical circuits, 112

End-stop control, 259–271
Equivalent circuit, for cycloconverter, 146–150, 158–160, 173–180
   with commutation overlap, 416–417
   for dual-converter, 111, 113
   for half-controlled converter, 103–104
   for open-delta cycloconverter, 216–217
   for ring connected cycloconverter, 221–223
   for two-quadrant converter, 94
   with commutation overlap, 399–403
Equivalent impedance, in cycloconverter output due to commutation overlap, 416–417
Equivalent resistance, in converter output, due to commutation overlap, 399–405
   per unit value of, 401
   relationship of per unit value to per unit source impedance, 401–403, 409
"Extended" firing pulses, 248
   production of, 274–277

Feedback control, *see* Closed loop control
Firing angle, cosine wave crossing control of, 229–241, 249–254
   definition of, 33
   end-stop control, 259–271
   "integral" control of, 242–245, 254–255, 256–257
   phase-locked oscillator control of, 245–247, 255, 258–259
   universal relationship to dc voltage ratio and displacement factor, for all half-controlled converters, 96, 103–105
   universal relationship to dc voltage ratio and displacement factor, for all two-quadrant converters, 92–93
Firing pulse isolating output stages, 272–277
Firing pulse timing control principles, 229–247
Firing pulse timing control schemes, 249–259
   for dual-converter, 271–272
   end-stop control, 263–271
   using cosine wave crossing control, 249–254
   using "integral" control, 254–255, 256–257
   using phase-locked oscillator, 255, 258–259
Firing pulse timing errors, effect on cycloconverter performance, 239, 240
Fourier analysis, limitations of for cycloconverter, 162–280
Four-quadrant converter, basic description, 10–12, 111, 113–114
Freewheel diode, 54, 58–59, 61–62, 65–68
Frequency control, of cycloconverter output, 145, 150, 232–233, 234
Frequency ratio, maximum, of cycloconverter, 323–327
Fundamental component of input current, 70–72
   of converters with freewheel diodes, 108–109
   of cycloconverter, 334, 335, 336, 337, 342, 345, 349–359
   of half-controlled converter, 103–104, 106, 107
   of two-quadrant converter, 28, 29, 30–31, 35, 36–37, 40, 43, 46, 48, 77, 79, 85, 88, 91–94, 106, 108, 109–110
   effect of source inductance upon, 396–397

Half-controlled converters, 54, 55–58, 59–61, 62–65
   for specific circuit, *see* specific converter circuit title
   universal relationships between firing angle, dc voltage ratio, input displacement factor, 96, 103–105

Harmonic distortion, cancellation of, by increasing circuit pulse number, 33–34, 82, 85, 313, 367
   in input current of cycloconverter with three-phase output, 167, 168–173, 179–180, 333, 338–339, 360, 361, 362, 363, 364–365, 368, 375, 377, 382
     derivation of, for input current of cycloconverter, 329–343, 345, 347–349
       for input current of phase-controlled converter, 72, 77–78
       for output voltage of cycloconverter, 279–304
       for output voltage of phase-controlled converter, 72–76
       for specific circuit, *see* specific converter or cycloconverter circuit title
   of converter output voltage, effect of source inductance upon, 396, 405, 406–408
   of cycloconverter input current, 166–167, 168–173, 175, 177, 180, 328–385
   of cycloconverter output voltage, 145–146, 161–166, 182–190, 197–198, 278–327
     effect of source inductance upon, 410–421
   of open-delta cycloconverter output voltage, 218, 219
   of output voltage of dual-converter in circulating current operation, 142–144
   universal relationships between output voltage and input current, for cycloconverters, 365–366
     for two-quadrant converters, 94–95
Harmonic frequencies, in input current, of cycloconverter, 166–167, 177, 180, 359–366, 375–377, 382
   of two-quadrant converter, 94–95
   in output voltage, of cycloconverter, 161–165, 239, 306–313, 365–366, 410, 412–413
   of two-quadrant converter, 77, 94–95
Harmonic frequency charts, for cycloconverter input current, 363–365
   for cycloconverter output voltage, 307–313

Induction motor speed control, principles of, 19A23
In-phase component of input current, of converters with freewheel diodes, 108–109
   of cycloconverter, 167, 173–180, 349–350, 352
   of half-controlled converter, 103–104, 106, 107
   of two-quadrant converter, 94, 106, 109–110
Integer multiple harmonic distortion, in cycloconverter input current, 360
   in cycloconverter output voltage, 165, 306, 310, 327, 410, 412–413, 416, 418
"Integral control" of firing angle, basic principle, 242–245
   scheme using, 254–255, 256–257
Interference, in converter pulse timing control, 242, 243–244, 397
Interphase reactor, 38–39, 41, 208
Inversion end-stop, 261–267, 268, 271, 397
"Inverter operation" of two-quadrant converter, 29, 31–32, 37–38, 93, 147, 148, 149, 151, 394–395, 399–400
Isolation of firing pulses, 272–277

"Multipulse" converters, 47–49
   cancellation of harmonics, in dc terminal voltage of, 82
     in input current of, 85
   dc terminal voltage of, 81–82
   derivation of performance characteristics of, 81–85, 88, 91–95

Natural commutation, 10, 29, 146, 325, 390–395

"Objectionable" distoriton terms, in cycloconverter input current, 363–364, 376–377
   in cycloconverter output voltage, 240–241, 279, 309–310
One-quadrant converter, basic description, 10, 11, 54–55
   for specific curcuit, *see* specific converter circuit title
Open-delta cycloconverter circuits, 216–220, 221
   input displacement factor fo, 217–219, 228
Open-loop pulse timing, limitation of, for cycloconverter, 238–240, 279

Phase-controlled converter, basis operating characteristics, 12
   basic principle of operation, 9–10, 27–32, 34–38
   for specific circuit, *see* specific converter circuit title

INDEX

Phase-locked oscillator control of firing angle, basic principle, 245–247
  scheme using, 255, 258–259
Phase modulation, of cycloconverter firing angle, 145, 147, 149–150, 232–233, 234, 283–284, 290–291
Power factor, input, definition of, 70
  of cycloconverter, compared to converter with dc output, 176
  relationship to distortion factor and displacement factor, 71
  for specific circuit, *see* specific converter circuit title
Power flow, through cycloconverter, 17, 145, 147–150, 151, 152–155, 166–180, 350, 360
  through two-quadrant converter, 10, 32, 80–81, 92–93
Pulse number, definition of, 34

Quadrature component of input current, of converters with freewheel diodes, 108–109
  of cycloconverter, 167–168, 173–180, 349, 351–355, 356
  of half-controlled converter, 103–104, 106, 107
  of two-quadrant converter, 94, 106, 109–110
  reduction of, for cycloconverters, 385–388
  for two-quadrant converters, 109–110

Reactive loading of supply, *see* Quadrature component
Recovery angle, 261–265
"Rectifier operation," of two-quadrant converter, 28–29, 32, 37–38, 93, 147, 148, 151
"Regenerative" loads, on cycloconverter, 23, 25, 172–173, 358–359
  on dual-converter, 118, 126
  on two-quadrant converter, 15
Ring connected cycloconverter circuits, 220–228
  input displacement factor of, 224–225, 228
RMS distortion of cycloconverter output voltage, 235–238, 290
RMS value of cycloconverter input current, 368, 374, 375, 376, 377, 378, 382–385

Self-induced circulating current in cycloconverter, 156–161, 190, 192–195
  in dual-converter, 135, 136–139, 140–141
Silicon Controlled Rectifier, *see* Thyristor
Source impedance, 389
  effect of, on average converter output voltage, 389, 395–396, 398–405
  on commutation process, 390–395
  on converter input current, 389, 396–397
  on converter input voltage, 397
  on converter output voltage distortion, 389, 396, 405, 406–408
  on cycloconverter input current, 410
  on cycloconverter output voltage, 410–421
  on inversion end-stop, 261–263, 397
  per unit value of, 409
  relationship of per unit value of, to per unit equivalent output resistance of converter, 401, 402–403, 409
Speed control, of ac machines with cycloconverters, 18–25
  of dc machines, with dual-converter, 117–118, 120–126
  with various types of converter, 13–16
Subharmonic distortion, in input current of cycloconverter, 363–365, 375–377, 382
  in output voltage of cycloconverter, 163, 239, 240, 278, 279, 308, 309, 312, 324–327
Switching function, 72, 73–75, 280, 282, 283, 286, 287, 288, 290–291, 330, 332, 333

Thyristor, static anode/cathode characteristics, 2–4
  turn-off characteristics, 6–8
  turn-on characteristics, 4–6
  voltage and current ratings, 1
Timing wave, 230, 231, 232–233, 242, 249–254, 291
Transformer connections, converter input, 28, 34, 38, 39, 42, 47, 62, 63, 66, 86–87, 89–90
  cycloconverter input, 208–216, 219–221, 226–227, 337, 340, 341, 344, 346, 349
Transformer ratings, converter input, 86–87, 89–90
  cycloconverter input, 209, 211–215, 220–221, 226–227, 328

data related to, 352–358, 364–365, 369–374, 376–381
Two-quadrant converter, basic description, 10, 11
  for specific circuit, see specific converter circuit title
  universal relationships, between firing angle, dc voltage ratio, input displacement factor, 92–94
  between input and output harmonics, 94–95

Unbalanced input current, of cycloconverter at discrete frequencies, 382–385

Voltage control, basic principle, of converter output, 9–10, 28–32, 35–38, 92–93
  of cycloconverter output, 149–150, 232–233, 234, 305–306
  of dual-converter output, 113–114

Wanted component of output voltage of cycloconverter, 145, 146–149, 151, 152–155, 156, 157, 236–238, 305–306
Waveforms, for cycloconverter, with commutation overlap, 411, 414–415
  with continuous circulating current, 157, 189
  with no circulating current, 152–155, 164, 169–172, 189, 200, 201
  with "partial" circulating current, 196, 197
  for open-delta cycloconverter, 218
  for ring connected cycloconverter, 224–225
  for specific circuit, see specific converter circuit title